Probabilities, Laws, and Structures

[The Philosophy of Science in a European Perspective, Vol. 3]

Proceedings of the ESF Research Networking Programme

THE PHILOSOPHY OF SCIENCE IN A EUROPEAN PERSPECTIVE

Volume 3

EUROPEAN SCIENCE FOUNDATION

www.pse-esf.org

Dennis Dieks • Wenceslao J. González
Stephan Hartmann • Michael Stöltzner
Marcel Weber
Editors

Probabilities, Laws, and Structures

Springer

Editors
Dennis Dieks
Institute for History and Foundations
of Science Utrecht
University Budapestlaan 6
3584 CD, Utrecht
The Netherlands

Wenceslao J. González
Faculty of Humanities
University of A Coruña,
Campus de Esteiro, s/n
C.P. 15403, Ferrol
Spain

Stephan Hartmann
Center for Logic and Philosophy
of Science Tilburg
University PO Box 90153 5000 LE,
Tilburg The Netherlands

Michael Stöltzner
Department of Philosophy
University of South Carolina
James F. Byrnes Building
SC 29208, Columbia
USA

Marcel Weber
Département de Philosophie
Université de Genève
2, rue de Candolle
1211, Genève
Switzerland

ISBN 978-94-007-3029-8 e-ISBN 978-94-007-3030-4
DOI 10.1007/978-94-007-3030-4
Springer Dordrecht Heidelberg London New York

Library of Congress Control Number: 2012931558

Printed on acid-free paper

Springer is part of Springer Science+Business Media (www.springer.com)

Table of Contents

Team E: History of the Philosophy of Science

Preface

It is my pleasure and privilege to present the third volume of the series *Philosophy of Science in a European Perspective* produced by the European Science Foundation (ESF) Research Networking Programme that runs under the same name. Like the first two volumes, *The Present Situation in the Philosophy of Science* (2010) and *Explanation, Prediction, and Confirmation* (2011), also published by Springer, it collects selected papers given at a series of workshops organized by the five teams of the programme from one year, in this case 2010. For the present volume, these workshops included the following events, all funded by the ESF with some further support from the host institutions:

Team A, Formal Methods: *Pluralism in the Foundations of Statistics* (University of Kent, organized by Stephan Hartmann and David Corfield, September 9-10, 2010)

Team B, Philosophy of the Natural and Life Sciences and **Team D**, Philosophy of the Physical Sciences, joint workshop: *Points of Contact Between the Philosophy of Physics and the Philosophy of Biology* (London School of Economics, organized by Miklos Redei, Dennis Dieks, Hanne Andersen and Marcel Weber, 13-15 December, 2010)

Team C, Philosophy of the Cultural and Social Sciences: *The Debate on Mathematical Modeling in the Social Sciences* (University of A Coruña, Ferrol Campus, organized by Wenceslao J. Gonzalez, 23-24 September, 2010)

Team E, History of the Philosophy of Science: *Historical Debates about Logic, Probability and Statistics* (University of Paderborn, organized by Michael Stöltzner, Thomas E. Uebel, Volker Peckhaus, Katharina Gefele, and Anna-Sophie Heinemann, 9-10 July, 2010)

As in the previous years of the ESF programme, these workshops brought together scholars from all across Europe, including a substantial proportion of junior researchers as well as graduate students. The workshops generated considerable interest from local students and faculty at the respective workshop venues. While the programme's core topic for the year 2010 was probability and statistics, most of the five teams embraced the opportunity of building bridges to more or less closely connected issues in general philosophy of science, philosophy of physics and philosophy of the special sciences. However, papers that use or analyze the concept of probability for various philosophical purposes are clearly a major theme in this volume, as it was in the previous volume. This reflects the impressive productivity of probabilistic approaches in the philosophy of science, which form an important part of what has become known as formal epistemology (although, of course, there are non-probabilistic approaches in formal epistemology as well). It is probably fair to say that Europe has been particularly strong in this area of philosophy in recent years.

The papers from **Team A** focus on the foundations of statistics. While the importance of statistical methods in many areas of science is undisputed, debate on the proper foundations and the scope of these methods continues among both practitioners and philosophers. Is statistics something like a logic of inductive inference, as it was envisioned by some members of the Vienna Circle, or is it more properly viewed as a decision theory for choosing among alternative courses of action? Can null hypotheses be supported by statistical data? Does subjective Bayesianism provide a general framework for statistical testing, or should statisticians strive for objective probabilities such as Neyman-Pearson frequentist error probabilities? Should we be pluralists about the foundations of statistics? These are some of the questions discussed in the first section of this volume.

Teams B and D decided to join forces to discuss points of common interest in the philosophy of physics and philosophy of biology. When organizing the corresponding workshop, it quickly became clear that there are much more points of contact that one might have thought, given that these two areas of philosophy of science have developed largely independently of each other in recent years. Of course, the philosophy of biology has had to struggle hard to free itself from a philosophy of science that was strongly physics-centered, but it is now time to put these quarrels behind us and to take a fresh look at some problems that concern both areas of science. Probability and statistical methods are, of course, one such topic, but we decided to also take the opportunity of addressing other themes that are vital both in physics and biology, including the perennial topics of laws and natural kinds. As it became clear at the workshop, the concept of structure (as in mathematical structure) has become increasingly important in the philosophy of both areas and is at the center of exciting new developments.

Team C focused on mathematical modeling in the social sciences, construed to include economics, political science, cognitive science, and the law. With the exception of economics, these disciplines have to – my knowledge – hardly been investigated by philosophers of science with such a focus, which makes these papers particularly welcome. They reveal impressively how diverse and yet closely connected the sciences are today, at least with respect to the role of mathematical models (including the use of "techno-mathematical" models in social sciences). One of the most difficult problems for mathematically formalized theories and models has to do with the question of how the magnitudes that feature in them are connected to the real world. Many of the considerations in the contributions may be seen as seeking answers to this question. Furthermore, this section contains papers on such topics as the use of experiments in political science or of probabilistic thinking in the courtroom.

The contributions from **Team E** take a new look at the formative years of modern philosophy of science, which, of course, are situated in the late 19th and early 20th Century. As these papers make clear, much of the current debates not only with respect to the foundations of statistics and probability, but also on induction, indeterminism vs. determinism, laws of nature, and the role of mathematics and formal methods in science as well as in epistemology have their historical roots in these years. Of course, members of the Vienna Circle such as Otto Neurath or Rudolf Carnap played a major role in shaping these debates, but also physicists such as Erwin Schrödinger, mathematicians such as John von Neumann, Richard von Mises and Ernst Schröder, physiologists such as Johannes von Kries or social scientists such as Adolphe Quetelet. It is fascinating to see how much of the current debates were already anticipated by these thinkers – which, of course, is not to deny that there has also been progress, which the papers of this volume jointly document.

I hope that readers will be as impressed as I am about the diversity as well as the quality and depth of current research in philosophy of science in Europe.

On behalf of all the editors, I wish to close by thanking Maria Carla Galavotti, Cristina Paoletti and Beatrice Collina for their patience and sometimes insistence in running this ESF networking programme, which is more complex than one might think and always tends toward a state of higher system entropy. Furthermore, I wish to thank Robert Kaller for producing the manuscript and the European Science Foundation and the Universities involved in the various workshops for their financial support.

Konstanz, June 2011 *Marcel Weber*

Team A
Formal Methods

CHAPTER 1

SEAMUS BRADLEY

DUTCH BOOK ARGUMENTS AND IMPRECISE PROBABILITIES

1.1 FOR AND AGAINST IMPRECISE PROBABILITIES

I have an urn that contains 100 marbles. 30 of those marbles are red. The remainder are yellow. What sort of bets would you be willing to make on the outcome of the next marble drawn from the urn? What odds would you accept on the event "the next marble will be yellow"? A reasonable punter should be willing to accept any betting quotient up to 0.7. I define "betting quotient" as the ratio of the stake to the total winnings. That is the punter should accept a bet that, for an outlay of 70 cents, guarantees a return of 1 euro if the next marble is yellow. And the punter should obviously accept bets that cost less for the same return, but what we are really interested in is the *most* the punter would pay for a bet on an event.

I am making some standard simplifying assumptions here: agents are risk neutral and have utility linear with money; the world of propositions contemplated is finite. The first assumption means that expected monetary gain is a good proxy for expected utility gain and that maximising monetary gain is the agents' sole purpose. The second assumption is made for mathematical convenience.

Now consider a similar case. This case is due originally to Daniel Ellsberg (Ellsberg 1961), this is a slightly modified version of it due to Halpern (2003). My urn still contains 100 marbles, 30 of them red. But now the remainder are either yellow or blue, in some unknown proportion. Is it rational to accept bets on Yellow at 0.7? Presumably not, but what is the highest betting quotient the punter should find acceptable? Well, you might say, there are 70 marbles that could be yellow or blue; his evidence is symmetric so he should split the difference:[1] a reasonable punter's limiting betting quotient should be 0.35. Likewise for Blue. His limiting betting quotient for Red should be 0.3.

What this suggests is that this punter considers Yellow more likely than Red, since he's willing to pay more for a bet on it. So, as a corollary, he should prefer a bet on Yellow to a bet on Red. And thus, if offered the chance to bet on Red or to bet on Yellow, for the same stakes, he should prefer the bet on Yellow.

1 I am studiously avoiding mentioning the "principle of indifference" since I use "indifference" to mean something else in the main text.

D. Dieks et al. (eds.), *Probabilities, Laws, and Structures*, The Philosophy of Science in a European Perspective 3, DOI 10.1007/978-94-007-3030-4_1,

Empirical studies show that many people prefer the bet on Red to the bet on Yellow, but are indeed indifferent between Yellow and Blue (Camerer and Weber 1992). This behaviour seems to contradict the good classical Bayesian story I have been telling above. And it seems that preferring to bet on Red has some intuitive appeal: you know more about the red marbles; you are more certain of their number.

In the first example, there was uncertainty[2] about which marble would be drawn. In the second example, as well as that uncertainty, there was *ambiguity* about what the chance set-up was. This is uncertainty of a different kind. It is accommodating this second kind of uncertainty that motivates studies of "imprecise probabilities". Instead of the standard Bayesian approach of representing uncertainty by a probability measure, the advocate of imprecise probabilities represents uncertainty by a *set* of probability measures. This sort of approach has been explored by, among others,[3] Isaac Levi (Levi 1974, 1986), Peter Walley (Walley 1991, 2000), and Joseph Halpern (Halpern 2003, 2006).

The precise probabilist has his belief represented by a probability function, for example $\mathbf{pr}(R) = 0.3, \mathbf{pr}(Y) = \mathbf{pr}(B) = 0.35$ for the "split the difference" probabilist. The imprecise punter has a *set* of probability measures $\mathcal{P}$ representing her belief. $\mathcal{P}(Y)$ is the set of values those probability measures give the event Yellow. For example, if the imprecise probabilist considers possible every possible combination of yellow and blue marbles, her credal state might be characterised as follows: $\mathcal{P}(R) = \{0.3\}, \mathcal{P}(Y) = \mathcal{P}(B) = \left\{\frac{0}{100}, \frac{1}{100}, \dots \frac{70}{100}\right\}$.

Some advocates of imprecise probabilities – Levi, for example – insist that the $\mathcal{P}(X)$ should be a convex set. That is, they would demand that $\mathcal{P}(Y) = [0, 0.7]$ the whole interval between the least and the most the probability might be. I don't subscribe to this view. Consider representing my uncertainty in whether a strongly biased coin will land heads: if I don't know which way the bias goes then any convex credal state will include a 0.5 chance. But this is exactly the sort of chance I know I can rule out, since I know the coin is biased.[4]

One might reason that each $\mathbf{pr}$ in $\mathcal{P}$ is equally likely, so using a uniform "second-order probability" I can derive a single probability. This is a more formal version of the "split the difference" intuition. I think it makes an unwarranted assumption about the chance set up when it assumes that each $\mathbf{pr}$ is equally likely.

One criticism that has been levelled at this approach is that the imprecise probabilist is vulnerable to a Dutch book, and is therefore irrational. A Dutch book is a set of bets that always lose you money. A punter is vulnerable to the Dutch book if there is a set of bets that she considers acceptable – which she would take – that

2 Economists are wont to distinguish "risk" and "uncertainty"; the former being where the probabilities are known, the latter where they are unknown. I prefer to use "uncertainty" as a catch-all term for ways one might fail to be certain, reserving "ambiguity" for cases of unknown, or incompletely known probabilities.

3 It would be futile to try and list all those who have contributed to this area, so I list only those whose work informs the current paper.

4 See also §4 of Kyburg and Pittarelli (1992).

is a Dutch book. Accepting a set of bets that always lose you money is clearly an irrational thing to do, so avoiding Dutch books is an indicator of rationality.

The plan is to outline exactly how this Dutch book challenge is supposed to go, and show that it is flawed: that one of the premises it rests on is too strong. First I outline the Dutch book argument, making clear its premises and assumptions. Then I argue that one of the conditions on rational preference is too strong in the presence of ambiguity and that therefore the imprecise probabilist is not vulnerable to a Dutch book. This leads on to a discussion of decision-making with imprecise probabilities, and I defend the imprecise view against related criticisms.

1.2 The Dutch Book Argument

In this section, I set out a fairly detailed characterisation of the Dutch book theorem. Note that I am concerned only with a *synchronic* Dutch book in this paper. All bets are offered and accepted before any marbles are drawn from the urn. Once there is learning, things become much more tricky. Indeed, learning in the imprecise framework brings with it its own problems.[5]

Before we can discuss the argument, we need some formal structure. We need a characterisation of formal theories of degree-of-belief, of betting and of preference among bets.

1.2.1 Formalising Degrees of Belief

We have an algebra of events: $\mathbb{E}$. I take $\mathbb{E}$ to be a set of propositions[6] closed under negation, disjunction and conjunction (formalised $\neg, \vee, \wedge$ respectively). One might also take the algebra of events to be a collection of sets of possible worlds closed under complementation, union and intersection.[7] These are the events the punter is contemplating bets on. Red, Blue and Yellow are the events contemplated in the examples at the beginning. There are two important events: the necessary event and the impossible event. These are formalised as $\top$ and $\bot$ respectively.

We are interested in functions that represent *degree of belief*. As a first approximation of this idea of modelling degree of belief, consider functions that map events to real numbers. The larger the number, the stronger the belief. Let $\mathbf{B}$ be the set of all functions $\mathbf{b}\colon \mathbb{E} \to \mathbb{R}$. The question becomes which subsets of $\mathbf{B}$ are of particular interest? It is typically claimed that the probability functions are the only rational ones.

One class of functions that will be of particular interest are the truth valuations. The function $\omega\colon \mathbb{E} \to \{0,1\}$ is a truth valuation if, for all X, Y:

- $\omega(X \vee Y) = \max\{\omega(X), \omega(Y)\}$

---

5 See: Seidenfeld and Wasserman (1993) and Wheeler (forthcoming).

6 Strictly speaking, we need the Lindenbaum algebra of the propositions: we take equivalence classes of logically equivalent propositions ....

7 The two views are more or less equivalent, see the appendix.

- $\omega(X \wedge Y) = \min\{\omega(X), \omega(Y)\}$
- $\omega(\neg X) = 1 - \omega(X)$

Call the set of functions that satisfy these constraints $\mathbf{V}$. I will sometimes call ω a "world", since specifying truth values of all propositions singles out a world. One particular world is actualised, and this determines which bets pay out. So if a red marble is drawn from the urn, the world that has $\omega(R) = 1, \omega(B) = \omega(Y) = 0$ is actualised. And so, for example, $\omega(R \vee B) = \max\{1, 0\} = 1$, as one would expect.

Another class of functions of particular importance are the probability functions. These also map events to real numbers and satisfy the following restrictions for all X, Y:

- $\mathbf{pr}(\top) > \mathbf{pr}(\bot)$
- $\mathbf{pr}(X)$ is in the closed interval bounded by $\mathbf{pr}(\top)$ and $\mathbf{pr}(\bot)$
- $\mathbf{pr}(X \vee Y) + \mathbf{pr}(X \wedge Y) = \mathbf{pr}(X) + \mathbf{pr}(Y)$

What is nice about this non-standard characterisation due to Joyce (2009) is that it makes clear that setting the probability of the necessary event to 1 is a matter of convention, not mathematical necessity. The important aspects of probability theory as a model of belief are that the functions are bounded and additive: setting $\mathbf{pr}(\top)$ to 1 gives us the standard probability axioms. Let $\mathbf{PR}$ be the collection of all functions satisfying these constraints. It should be clear that $\mathbf{V} \subset \mathbf{PR} \subset \mathbf{B}$.

But *which* probability measure to take as one's degree of belief in the Ellsberg case seems underdetermined. The "split the difference" reasoning used by the precise probabilist seems to go beyond his evidence of the situation. I claim that modelling belief by *sets* of probability functions is often better than using a single function. Instead of resorting to "split the difference" reasoning to home in on one probability function to represent my uncertainty, I think it better to represent that ambiguity by the set of probability functions consistent with the evidence.

But why ought the functions in that set be probability functions, rather than any old functions in $\mathbf{B}$? Because probability functions are still a kind of regulative ideal: the more evidence I accumulate the sharper my imprecise probabilities should become. That is, the more evidence I have, the narrower the range of values my set of probabilities should assign to an event. In the ideal limit, I should like to have a probability function; in the absence of ambiguity I should have a probability function.[8]

8 For another argument in favour of probabilities as epistemically privileged, see Joyce (1998).

1.2.2 Bets and Betting

Now we know how we are characterising degree of belief, let's turn to how to represent betting. This framework is from Halpern (2003) but see also Döring (2000). A bet is, for our purposes, an ordered pair of an event in $\mathbb{E}$ and a "betting quotient". Bets will be ordered pairs of the form (X, α) where $X \in \mathbb{E}$ and $\alpha \in \mathbb{R}$. What is relevant about a bet is the betting quotient and the event in question. The higher the α the punter would accept the more likely she thinks the event in question is. The greater the proportion of the winnings a punter is willing to risk on a bet, the more likely she thinks the event is.

A bet (X, α) pays out 1 euro if X turns out true by the light of truth valuation ω and pays out nothing otherwise. Or more succinctly, (X, α) pays out $\omega(X)$. The bet costs α and you don't get your stake returned when you win. So the net gain of the bet (X, α) is $\omega(X) - \alpha$. The bet $(\neg X, 1 - \alpha)$ is called the complementary bet to (X, α). Think of the complementary bet $(\neg X, 1 - \alpha)$ as "selling" the bet (X, α). Whenever the punter takes a bet (X, α), the bookie is effectively taking on the complementary bet $(\neg X, 1 - \alpha)$. Table 1.1 illustrates the "mirror image" quality that the payoffs of complementary bets have.

Table 1.1: Payoffs for a bet and its complement

	$\omega(X) = 1$	$\omega(X) = 0$
(X, α)	$1 - \alpha$	$-\alpha$
$(\neg X, 1 - \alpha)$	$-(1 - \alpha)$	α

A set of bets $B = \{(X_i, \alpha_i)\}$ costs $\sum \alpha_i$ and pays out $\sum \omega(X_i)$ in world ω. That is, you get 1 for every event that you bet on that ω makes true. So the value of a set of bets B at world ω is $\tau_\omega(B) = \sum (\omega(X_i) - \alpha_i)$. The value of the bet at a world is how much it pays out minus what the bet cost.

For set of bets B let its complement[9] be $B^{\mathrm{C}} = \{(\neg X_i, 1 - \alpha_i)\}$. It is easy to show that $\tau_\omega(B^{\mathrm{C}}) = -\tau_\omega(B)$. The "mirror image" quality of Table 1.1 also holds for sets of bets.

A Dutch book in this context is a set of bets, B such that, for every $\omega \in \mathbf{V}$, we have $\tau_\omega(B) < 0$. That is, the pay out for the bet is negative *however the world turns out.*

1.2.3 Constraints on Rational Betting Preference

We are interested in preference among bets, so define a relation "$A \succeq B$" which is interpreted as meaning "A is at least as good as B", where A and B are bets. We will later put constraints on what preferences are reasonable. As I said above,

9 This is a lazy way of talking, B^{C} is not the complement of B in the sense of set-theoretic complement in the set of bets, but rather the set of bets complementary to those in B.

what is of particular interest is the punter's maximum willingness to bet. For an event X, define α_X by $\sup\{\alpha : (X,\alpha) \succeq (\neg X, 1-\alpha)\}$. These maximum betting quotients are interpreted as characterising the punter's belief state and it will often be useful to talk about the belief function corresponding to these α_Xs. Define $\mathbf{q}(X) := \alpha_X$. You are vulnerable to a Dutch book unless your $\mathbf{q} \in \mathbf{PR}$. That is, unless your (limiting) betting quotients have the structure of a probability function, there is a set of bets – acceptable by the lights of your $\mathbf{q}$ – that guarantees you a loss of money.

Halpern sets out four constraints on what sort of preferences it is rational to have among bets. These are sufficient to force any agent satisfying them to have betting quotients that have the structure of a probability measure. That is, failing to satisfy the axioms of probability makes your betting preferences incompatible with the constraints. In the strict Bayesian picture, $\mathbf{q}$ and degree of belief $\mathbf{pr}$ are used more or less interchangeably. It will be important in what follows that one's willingness to bet and one's degrees of belief are distinct, if related concepts.

Strictly speaking, the preference is among sets of bets, so when discussing single bets I should say "$\{(X,\alpha)\} \succeq \{(Y,\beta)\}$", but the preference relation induces an obvious relation among singletons, so I omit the braces. I don't make much of a distinction in what follows between a bet and a set of bets.

The first of Halpern's requirements says that if one bet B_1 always pays out more money than another B_2, then you should prefer B_1.

$$\text{If, for all } \omega \in \mathbf{V} \text{ we have } \tau_\omega(B_1) \geq \tau_\omega(B_2) \text{ then: } B_1 \succeq B_2 \qquad \text{(DOMINANCE)}$$

Note that this condition does not force the punter to prefer bets with higher *expected* value: only to prefer bets with a higher *guaranteed* value. Preferring bets guaranteed to give you more money seems eminently reasonable.

The second of Halpern's conditions is simply that the preference relation be transitive.

$$\text{If } B_1 \succeq B_2 \text{ and } B_2 \succeq B_3 \text{ then } B_1 \succeq B_3 \qquad \text{(TRANSITIVITY)}$$

Again, this condition seems reasonable.

The third of Halpern's conditions – the one I will take issue with later – is COMPLEMENTARITY.

$$\begin{gathered}\text{For all } X \in \mathbb{E} \text{ and } \alpha \in \mathbb{R} \text{ either,} \\ (X,\alpha) \succeq (\neg X, 1-\alpha) \text{ or } (\neg X, 1-\alpha) \succeq (X,\alpha)\end{gathered} \qquad \text{(COMPLEMENTARITY)}$$

Note that this is weaker than what is often assumed of rational preference: COMPLEMENTARITY does not require that the punter's preference relation be complete or total.[10] It need only be complete with respect to complementary bets, but this is still too much for me. I will discuss why I find this too strong a condition in the next section. Note that this condition is specified in terms of single bets, but in the presence of PACKAGE below, it extends somewhat to sets.

10 A relation R is total or complete when xRy or yRx for all x, y.

The final condition, sometimes known as the "package principle"[11] is, as Halpern[12] puts it, that "preferences are determined pointwise".

$$\text{If } (X_i, \alpha_i) \succeq (Y_i, \beta_i) \text{ for each } 1 \leq i \leq n \text{ then:} \quad \{(X_i, \alpha_i)\} \succeq \{(Y_i, \beta_i)\} \qquad \text{(PACKAGE)}$$

Note this is quite a restricted principle. For example, it does not in general allow that if $A \succeq B$ and $C \succeq D$ then $A \cup C \succeq B \cup D$.

The Dutch book theorem says that if a punter's preference among bets satisfies DOMINANCE, TRANSITIVITY, COMPLEMENTARITY and PACKAGE, then that punter's betting quotients $\mathbf{q}$ will have the structure of a probability function. Or to put it another way, if the punter's betting quotients violate the axioms of probability, then this leads to a preference incompatible with the above conditions.

1.3 AMBIGUITY AND COMPLEMENTARITY

We have seen what is necessary in order to prove the Dutch book theorem (the proof itself is in the appendix). In this section I argue that one particular premise of the theorem – COMPLEMENTARITY – is too strong. It is not warranted in the case of ambiguity.

I use preferring a bet to its complement as a proxy for acceptance of a bet. "This seems unintuitive," one might say, "I prefer lukewarm coffee to cold coffee, but I wouldn't accept either of them by choice." But remember, we are dealing with preference for one bet *over its complement*. So the analogous example would be something like "preferring lukewarm coffee to no lukewarm coffee", and here it seems that *that* preference is tantamount to accepting lukewarm coffee.

So why is COMPLEMENTARITY unreasonable? First let's see why it does seem reasonable in the first example from the introduction. Recall that there we had 100 marbles in an urn, 30 red and the remainder yellow. The punter's maximum betting quotient on red, $\mathbf{q}(R) = 0.3$. That is, 0.3 is the largest value for which a bet on red is preferred to a bet against red. You have confidence that Red will come up about 30% of the time, and since if it's not Red, it's Yellow, $\mathbf{q}(Y) = 0.7$.

Compare this to the Ellsberg case, where instead of the remainder being yellow, the remainder are yellow and blue in some unknown proportion. The probabilist splits the difference[13] and sets his betting quotients as follows: $\mathbf{q}(R) = 0.3, \mathbf{q}(Y) = \mathbf{q}(B) = 0.35$. The imprecise probabilist claims we can act differently. $\mathbf{q}(R) = 0.3$ still seems acceptable, our evidence about Red hasn't changed. But it seems that "ambiguity aversion" about evidence for Yellow and Blue suggests that the punter's maximal betting quotients for each should be lower. Perhaps

11 See e.g. Schick (1986).

12 (Halpern 2003, p. 22).

13 Of course, a subjectivist could set any particular probabilistically coherent value to the events, but what is objectionable to the imprecise probabilist is the suggestion that $\mathbf{q}(B) = 0.7 - \mathbf{q}(Y)$.

even $\mathbf{q}(Y) = \mathbf{q}(B) = 0$, since if you entertain the possibility that the chance of Yellow might be 0, then you should refuse to buy bets on Yellow. I focus on this extreme case. But first, two caveats. $\mathbf{q}$ *is not a representation of belief.* It is a representation of willingness to bet. Part of the conceptual baggage of precise probabilism that the imprecise probabilist needs to get away from is too close a connection between willingness to bet and belief. Obviously they are related, but not as tightly as they are in the precise framework. The belief is represented by the *set* of probabilities $\mathcal{P}$. Also, I am not *endorsing as rational* the extreme view ($\mathbf{q}(Y) = \mathbf{q}(B) = 0$), merely using it for illustrative purposes.

Say $\mathbf{q}(Y) = \mathbf{q}(B) = 0$. Then COMPLEMENTARITY demands that $\mathbf{q}(\neg Y) = 1$. Or, in other words, if $\alpha_Y = 0$ COMPLEMENTARITY demands that $(\neg Y, 1-0.1)$ is preferred to its complement. Similarly for B. This bet should also be acceptable: $(\neg R, 1-0.4)$. Together, these bets form a Dutch book (see Table 1.2). Call this set D.

Table 1.2: Dutch booking an imprecise probabilist

	R	B	Y
$(\neg R, 1-0.4)$	-0.6	$1-0.6$	$1-0.6$
$(\neg B, 1-0.1)$	$1-0.9$	-0.9	$1-0.9$
$(\neg Y, 1-0.1)$	$1-0.9$	$1-0.9$	-0.9
Total	-0.4	-0.4	-0.4

But to demand that the imprecise probabilist conform to COMPLEMENTARITY (and therefore, accept these bets) is to misunderstand the nature of the uncertainty being encoded in $\mathbf{q}(Y) = 0$. If there is ambiguity – uncertainty about the chance set-up itself – low confidence in an event does not translate into high confidence in its negation. There is an important distinction between the balance of evidence and the weight of evidence:[14] how conclusively the evidence tells in favour of a proposition (balance) versus how much evidence there is for the conclusion (weight). COMPLEMENTARITY assumes that the unwillingness to bet on an event is due to the balance of evidence telling against it and if this is the case then it is a reasonable condition. If on the other hand the refusal to accept the bet is due to the lack of *weight* of evidence, then the condition is not reasonable. Because of the ambiguous nature of the chance set-up (the lack of weight of evidence), the punter is simply not willing to bet either way most of the time. So the imprecise probabilist will not want to conform to COMPLEMENTARITY and therefore, will not be subject to a Dutch book in this way. In the appendix, I show that in the absence of COMPLEMENTARITY it is reasonable to have betting quotients satisfying restrictions weaker than those demanded of probabilities.

14 Joyce (2005).

One might still argue that if the punter were forced to choose one side or the other of any given bet (X, α) – that is, if the punter were forced to obey COMPLEMENTARITY – then she would obey the probability axioms or be subject to a Dutch book. This is true, but I don't see how this procedure elicits a fair reflection of the punter's credal state. If I forced the punter to take bets where the αs were all 1s or 0s, the resulting betting quotients would be a valuation function or the punter would be open to a Dutch book: that does not mean that the punter's degrees of belief are truth valuations. The punter's actions only reflect the punter's belief when her actions are not too restricted. So the conditions on betting preference have to be independently reasonable for them to form the basis of a Dutch book argument. COMPLEMENTARITY is not independently reasonable unless the chance set-up is unambiguous.

1.4 DECISION WITH IMPRECISION

To further explore this issue, we need to say something about what decision theory looks like from the imprecise probabilities perspective. The precise probabilist acts in accordance with the rule "maximise expected value with respect to **pr**"; where **pr** is the precise punter's probability. Let's say the expectation of the set of bets B is $E(B)$. Restricted to the case of choosing whether to buy a set of bets B or its complement B^{C}, this amounts to accepting B if $E(B) = \sum_i (\mathbf{pr}(X_i) - \alpha_i) > 0$ and taking B^{C} otherwise. It's easy to show that $E(B^{\mathrm{C}}) = -E(B)$ so one and only one bet has positive value in any pair of complementary bets, unless both bets have 0 expected value. So the precise probabilist always prefers one bet to the other, unless he is indifferent between them. This just follows from his being opinionated.

What about the imprecise probabilist? How is she to decide? I don't intend to suggest a fully worked out decision theory for imprecise probabilities, but simply offer enough of a sketch to explain how the imprecise probabilist avoids Dutch books. So let's turn the discussion in the last section on its head and start from a punter's belief state and derive what decisions that punter would make.

Recall the imprecise punter's credal state $\mathcal{P}$ is a set of probability measures, $\mathcal{P}(Y)$ is the set of values assigned to Y by members of $\mathcal{P}$. This $\mathcal{P}(Y)$ is already a "summary statistic" in a sense: it doesn't make clear that for any $\mathbf{pr} \in \mathcal{P}$ whenever $\mathbf{pr}(Y)$ is high, $\mathbf{pr}(B)$ is low and vice versa. So it is $\mathcal{P}$ that represents the punter's belief state, and $\mathcal{P}(Y)$ and so on are only shorthands, summaries that miss out some information. This is an important point, and one that is not often made.

We can define expectation of set of bets $B = \{(X_i, \alpha_i)\}$ for the imprecise probabilist as follows:

$$\mathcal{E}(B) = \left\{\sum \mathbf{pr}(X_i) - \alpha_i : \mathbf{pr} \in \mathcal{P}\right\}$$

This is the set of expected values of the bet with respect to the set of probabilities in $\mathcal{P}$. So the idea of maximising expected value isn't well defined for the imprecise probabilist.

Another standard summary statistic for imprecise probabilities are the lower and upper envelopes:

- $\underline{\mathcal{P}}(X) = \inf\{\mathbf{pr}(X) : \mathbf{pr} \in \mathcal{P}\}$
- $\overline{\mathcal{P}}(X) = \sup\{\mathbf{pr}(X) : \mathbf{pr} \in \mathcal{P}\}$

Likewise we can define $\underline{\mathcal{E}}$ and $\overline{\mathcal{E}}$ to be the infimum and supremum of $\mathcal{E}$ respectively.

These summary statistics have their own interesting formal properties.[15] For example, $\underline{\mathcal{P}}$ is superadditive. That is, for incompatible X, Y we have $\underline{\mathcal{P}}(X \vee Y) \geq \underline{\mathcal{P}}(X) + \underline{\mathcal{P}}(Y)$.

Again, these are summarising some aspect of the punter's credal state, but they are misrepresentative in other ways: considering the upper and lower envelopes to represent the agent's credal state is a mistake.[16]

So, how should a punter bet? If $\underline{\mathcal{E}}(B) > 0$ then B looks like a good bet: every probability measure in $\mathcal{P}$ thinks that this bet has positive expected value. Likewise, if $\overline{\mathcal{E}}(B) < 0$ then B^{C} looks promising. So any decision rule for imprecise probabilities should take into account these two intuitions. But this still leaves a lot of room for manoeuvre: what should the punter do when $\underline{\mathcal{E}}(B) < 0 < \overline{\mathcal{E}}(B)$?

The more general question of what decision rule the imprecise probabilist should use is left open. She could maximise $\underline{\mathcal{E}}$, she could use Levi's rule (Levi 1986), she could use the Hurwicz criterion (Hurwicz 1951).... I leave this bigger problem unanswered for now. I don't need to offer a fully worked out decision rule to show how the imprecise probabilist can avoid the Dutch book.

As a first approximation, let's imagine an extreme case of the ambiguity averse imprecise probabilist. She refuses to take either side of any bet (X, α) if $\underline{\mathcal{E}}(X, \alpha) < 0 < \overline{\mathcal{E}}(X, \alpha)$. That is, she has no preference between (X, α) and its complement if α is in the interval $\left[\underline{\mathcal{P}}(X), \overline{\mathcal{P}}(X)\right]$. And she will obey the two concerns discussed above: take bets on X if α is low enough, bet against X – bet on $(\neg X, 1 - \alpha)$ – if α is big enough. This punter is disobeying COMPLEMENTARITY, but can act in accordance with the other three conditions. She is disobeying COMPLEMENTARITY because there are values of α for which she has no preference between the complementary bets: she would accept neither.

A punter who obeys the three other conditions but not COMPLEMENTARITY has her limiting betting quotients have the structure of a lower envelope like $\underline{\mathcal{P}}$. I prove this in the appendix. This is not to say that $\underline{\mathcal{P}}$ represents that punter's beliefs, but rather that this particular elicitation procedure (betting) can only give us so much information about the punter's credal state.

However, one might object that this maximally ambiguity-averse punter is still irrational in not accepting a collection of bets that guarantees her a positive value whatever happens: she will not accept what Alan Hájek calls a "Czech book" (Hájek 2008). This is related to the idea in economics and finance of "arbitrage".

15 Halpern (2003) and Paris (1994).

16 Joyce offers nice examples of this in: Joyce (2011).

Consider the set of bets: $C = \{(Y, 0.1), (B, 0.1), (R, 0.3)\}$ in the Ellsberg example. Whatever colour marble is picked out of the urn, C makes you a profit of 0.5, so it would be crazy not to take it! However, our imprecise probabilist punter will not want to accept either of the first two bets, since $\underline{\mathcal{P}}(Y) < 0.1 < \overline{\mathcal{P}}(Y)$ and similarly for Blue. So does this mean she will refuse a set of bets guaranteed to make her money? Isn't this just as irrational as accepting a collection of bets that always lose you money?

Let's say this punter still conforms to DOMINANCE which guarantees that she will prefer bets that always pay more. This condition means that the imprecise punter will still accept the Czech book, even if every stake is between $\underline{\mathcal{P}}$ and $\overline{\mathcal{P}}$. This is because $\tau_\omega(C) = 0.5 > 0$ for any ω, so this set of bets is preferred to its complement by DOMINANCE. So if we take DOMINANCE as a necessary condition on imprecise decision rules, the imprecise punter can accept Czech books. Perhaps the other two conditions that I accept – TRANSITIVITY and PACKAGE – should also be taken into account when thinking about imprecise decision rules.

This brings home the point that the lower and upper envelopes are not all there is to an imprecise probabilist's belief state: every $\mathbf{pr} \in \mathcal{P}$ assigns positive expected value to C (indeed, positive guaranteed value) so whatever the chance set up, this is a good bet. But if the punter were acting just in accordance with her upper and lower probabilities, this fact might get lost in the summarising. So again we see that just focusing on the spread of values misses out important information about the punter's credal state.

The imprecise decision problem has been discussed extensively and many solutions have been proposed. Brian Weatherson reviews some solutions, and argues that many of them fail (Weatherson m.s.). Indeed, he argues that no decision rule which relies solely on $\underline{\mathcal{E}}$ and $\overline{\mathcal{E}}$ can ever be plausible. The whole project of imprecise probabilities has been criticised on the grounds that it cannot offer *any* adequate decision rules. Elga (2010) argues that no decision rule for imprecise probabilites is satisfactory. There is obviously a lot more to be said on the subject of decision making in the imprecise framework.

Avoiding Dutch books is taken to be necessary for being rational. So being vulnerable to Dutch books is sufficient for being irrational. I have shown that the imprecise probabilist isn't as vulnerable to Dutch books as is sometimes suggested. So, I claim, this particular avenue for criticising imprecise models isn't available. I have also suggested that more work needs to be done in exploring how best to make imprecise decisions. I think the increased expressive power of imprecise frameworks, and the fact that we *do* have to deal with several kinds of uncertainty means that imprecise probabilities are a worthwhile area for research.

APPENDIX: PROOF

Recall that α_X is defined as the most that a punter will buy a bet on X for. The Dutch book theorem claims that the function $\mathbf{q}(X) = \alpha_X$ is a probability function.

What I prove here is that $\underline{\mathbf{q}}(X) = \alpha_X$ is a lower probability[17] without using COMPLEMENTARITY. That is, I show that $\underline{\mathbf{q}}$ acts like one of the summary statistics of the "sets of probabilities" approach I favour. Define $\beta_X = \inf\{\beta : (\neg X, 1 - \beta) \succeq (X, \beta)\}$. I also show that $\overline{\mathbf{q}}(X) = \beta_X$ is an upper probability.

If $X \models Y$ then $(Y, \alpha) \succeq (X, \alpha)$ and $(\neg X, 1-\alpha) \succeq (\neg Y, 1-\alpha)$. This follows from DOMINANCE and the fact that if $X \models Y$ then $\neg Y \models \neg X$. Remember, $(X, \alpha_X) \succeq (\neg X, 1-\alpha_X)$ by definition, so by the above result and TRANSITIVITY, if $\gamma \leq \alpha_X$ then $(X, \gamma) \succeq (\neg X, 1-\gamma)$. This is in fact an "if and only if" result, since α_X is *defined* as the largest value for which this preference holds. A similar result holds for β_X. The above results allow us to show that if $X \models Y$ then $\alpha_X \leq \alpha_Y$ and $\beta_X \leq \beta_Y$.

α_X and $\beta_{\neg X}$ are related to each other. $(\neg\neg X, 1 - \beta_{\neg X}) \succeq (\neg X, \beta_{\neg X})$ by definition. Since $\neg\neg X = X$ it follows that $1 - \beta_{\neg X} \leq \alpha_X$. We also have $1 - (1 - \alpha_X) = \alpha_X$, so by definition $(\neg\neg X, 1 - (1 - \alpha_X)) \succeq (\neg X, 1 - \alpha_X)$ so $1 - \alpha_X \geq \beta_{\neg X}$. Thus $1 - \beta_{\neg X} \geq \alpha_X$. These two inequalities together imply that $\alpha_X = 1 - \beta_{\neg X}$.

The set of bets $\{(X, \alpha_X), (\neg X, 1 - \beta_X)\}$ is preferred to its complement by PACKAGE. This set always pays out 1, and costs $\alpha_X + 1 - \beta_X$. So the net gain of this bet is always $\beta_X - \alpha_X$. If $\alpha_X > \beta_X$, the net gain would be negative, so the net gain of the complementary bets would be positive. So the preference for this set over its complement would contradict DOMINANCE. So for any X we know that $\alpha_X \leq \beta_X$.

For logically incompatible propositions, X, Y consider the bet $B = (X \vee Y, \alpha_X + \alpha_Y)$. Now compare this with $C = \{(X, \alpha_X), (Y, \alpha_Y)\}$. These bets always have the same payout. So by DOMINANCE, we know that $B \succeq C$. We also know that $C^C \succeq B^C$ for the same reason. Now, $C \succeq C^C$ by PACKAGE. $B \succeq C \succeq C^C \succeq B^C$ so, by TRANSITIVITY we know that $B \succeq B^C$. By definition, $\alpha_{X \vee Y}$ is the maximum value for which $B \succeq B^C$. So $\alpha_{X \vee Y} \geq \alpha_X + \alpha_Y$. A similar chain of reasoning leads to the conclusion that $\beta_{X \vee Y} \leq \beta_X + \beta_Y$.

This demonstrates that $\underline{\mathbf{q}}(X) = \alpha_X$ is superadditive and $\overline{\mathbf{q}}(X) = \beta_X$ is sub-additive, as is characteristic of lower and upper probabilities. This proof makes no use of the COMPLEMENTARITY condition. However, to make the connection with Sect. 7.1 of Cozman's characterisation of lower probabilities, we also need the assumption that $\alpha_\perp = 0$ and $\alpha_\top = 1$. This isn't needed once COMPLEMENTARITY is in place. Using this condition it is easy to show that $\alpha_X = \beta_X$ for all X and thus that $\underline{\mathbf{q}}(X) = \overline{\mathbf{q}}(X) = \mathbf{q}(X)$ and that this function is additive, non-negative and normalised: a probability measure.

One can, in fact, construct a mass function out of the α_Xs and show that $\underline{\mathbf{q}}$ is a Dempster-Shafer belief function (this again without using COMPLEMENTARITY).

The last part is easier to do in terms of sets of worlds, rather than in terms of propositions, so I sketch the translation between the two paradigms here. Define $[X]$ as the set of valuations that make X true: $[X] = \{\omega \in \mathbf{V} : \omega(X) = 1\}$. $[X] \subseteq [Y]$ if and only if $X \models Y$, so we have an structure preserving bijection

17 In the terminology of Cozman (n.d.).

between propositions and sets of "worlds". I drop the square brackets in what follows.

Define a mass function as follows:

$$m(X) = \alpha_X - \sum_{Y \subsetneq X} m(Y)$$

That is, $m(X)$ picks up all the mass not assigned to subsets of X. This should, strictly speaking be an inductive definition on the size of X, but I take it that it is obvious what is meant here. If we now consider the following equation:

$$\underline{\mathbf{q}}(X) = \sum_{Y \subseteq X} m(Y)$$

This is equivalent to the above characterisation of $\underline{\mathbf{q}}(X) = \alpha_X$. That $\underline{\mathbf{q}}$ has this mass function associated with it means that it is a Dempster-Shafer belief function. This is a particular kind of lower probability: it is an infinite-monotone lower probability.

Acknowledgements: Thanks to Jonny Blamey, Chris Clarke, Erik Curiel and Katie Steele for comments on a previous version of this paper and to Roman Frigg and Richard Bradley for comments on this version. Thanks also to participants at the following conferences where versions of this talk were presented: Philosophy of Probability II conference at LSE 8–9 June 2009; DGL Paris 9–11 June 2010; "Pluralism in Foundations of Statistics" ESF workshop Canterbury 9–10 September 2010.

REFERENCES

Camerer, C. and Weber, M. (1992). Recent developments in modeling preferences: Uncertainty and ambiguity. *Journal of Risk and Uncertainty 5*, 325–370.

Cozman, F. (n.d.). A brief introduction to the theory of sets of probability measures. http://www.poli.usp.br/p/fabio.cozman/Research/CredalSetsTutorial/quasi-bayesian.html.

Döring, F. (2000). Conditional probability and Dutch books. *Philosophy of Science 67*, 391–409.

Elga, A. (2010). Subjective probabilities should be sharp. *Philosophers' Imprint 10*.

Ellsberg, D. (1961). Risk, ambiguity and the Savage axioms. *Quarterly Journal of Economics 75*, 643–696.

Hájek, A. (2008). Arguments for—or against—probabilism? *British Journal for the Philosophy of Science 59*, 793–819.

Halpern, J. Y. (2003). *Reasoning about uncertainty*. MIT press.

Halpern, J. Y. (2006). Using sets of probability measures to represent uncertainty. arXiv:cs/0608028v1.

Hurwicz, L. (1951). Optimality criteria for decision making under ignorance. Cowles Commission Discussion Paper Statistics 370.

Joyce, J. M. (1998). A nonpragmatic vindication of probabilism. *Philosophy of Science 65*, 575–603.

Joyce, J. M. (2005). How probabilities reflect evidence. *Philosophical Perspectives 19*, 153–178.

Joyce, J. M. (2009). Accuracy and coherence: Prospects for an alethic epistemology of partial belief. In F. Huber and C. Schmidt-Petri (Eds.), *Degrees of Belief*, pp. 263–297. Springer.

Joyce, J. M. (2011). A defense of imprecise credence. *Oxford Studies in Epistemology 4*. Forthcoming.

Kyburg, H. and Pittarelli, M. (1992). Set-based Bayesianism. Technical Report UR CSD;TR407, University of Rochester, Computer Science Department. http://hdl.handle.net/1802/765.

Levi, I. (1974). On indeterminate probabilities. *Journal of Philosophy 71*, 391–418.

Levi, I. (1986). *Hard choices: decision making under unresolved conflict*. Cambridge University Press.

Paris, J. (1994). *The uncertain reasoner's companion*. Cambridge University Press.

Schick, F. (1986). Dutch bookies and money pumps. *Journal of Philosophy 83*, 112–119.

Seidenfeld, T. and Wasserman, L. (1993). Dilation for sets of probabilities. *Annals of Statistics 21*, 1139–1154.

Walley, P. (1991). *Statistical Reasoning with Imprecise Probabilities*, Volume 42 of *Monographs on Statistics and Applied Probability*. Chapman and Hall.

Walley, P. (2000). Towards a unified theory of imprecise probabilities. *International Journal of Approximate Reasoning 24*, 125–148.

Weatherson, B. (m.s.). Decision making with imprecise probabilities. Available http://brian.weatherson.org/vdt.pdf.

Wheeler, G. (forthcoming). Dilation demystified. In A. Cullison (Ed.), *The Continuum Companion to Epistemology*. Continuum.

Department of Philosophy, Logic and Scientific Method
London School of Economics
Houghton Street
WC2A 2AE, London
UK
s.c.bradley@lse.ac.uk

CHAPTER 2

TIMOTHY CHILDERS

OBJECTIFYING SUBJECTIVE PROBABILITIES: DUTCH BOOK ARGUMENTS FOR PRINCIPLES OF DIRECT INFERENCE

2.1 INTRODUCTION

A Principle of Direct Inference licenses an inference from the frequency of the occurrence of attributes in a population to the probability of particular occurrence of an attribute in a sample. From a Bayesian point of view, such a Principle requires that if we have knowledge of the relative frequency of an attribute in a population, our degree of belief in the occurrence of that attribute in the population be equal to this frequency (or that this knowledge should somehow constrain our degrees of belief about the occurrence in a sample).[1] This might seem so painfully self-evident as to not need any justification. However, Bayesian justifications for constraining degrees of belief are usually based on Dutch Book arguments, and indeed several such arguments for Principles of Direct inference have been offered. I will discuss three, and find them wanting. Subjective probabilities therefore remain subjective even when conditioned on knowledge of objective probabilities.

2.2 THE FINITE CASE

It would seem that if there is a straightforward justification for a Principle of Direct Inference it can be found in the finite case. Focussing on the finite case allows us to ignore to some degree difficulties with different interpretations of the Dutch Book arguments. For example, we need no longer worry about limitations of a finite agent, and so need not concern ourselves with, say, countable additivity.

I will use as my target an argument from Kyburg (1981), although I doubt he would have seriously endorsed it[2]:

---

1 Given an exact formulation of such Principles turns out to very tricky, witness the large literature on Lewis's version of the Principle of Direct Inference. Luckily we will not need an exact formulation.

2 Kyburg puts this argument forward in a discussion of the Principal Principle, Lewis's oddly named version of a Principle of Direct Inference. Kyburg was not a Bayesian, and was sceptical of Lewis's Principle.

D. Dieks et al. (eds.), *Probabilities, Laws, and Structures*, The Philosophy of Science in a European Perspective 3, DOI 10.1007/978-94-007-3030-4_2,

As above, but you know that the coin was tossed 100 times, and landed heads 86 times. To what degree should you believe the proposition that it landed heads on the first toss?

Answer: 86 per cent.

86 : 14 are the only odds that will protect you from a Dutch book if you are going to treat the tosses all the same and are going to cover an arbitrary set of bets concerning various tosses made this morning. (Kyburg 1981, 774)

Kyburg does not actually provide the Dutch Book argument, leaving it for us to fill in. I shall now try to do so. First, I assume that we are given the proportion of heads.[3] Hence, we should add is that "all that you know is that the coin was tossed 100 times, and landed heads 86 times". We are given a population in which we know the mean, in this case, 0.86. We set odds in advance for each of the tosses made this morning, and a Bookie will choose (at random) among these bets. The claim is that the Bookie can subject us to a sure loss if we set a betting quotient different than 86:14.

As usual, a bet on a the occurrence of an event X (or the truth of a proposition, if you wish) is a contract between a bettor and a Bookie, with the bettor receiving some good a from the Bookie, should the event occur and giving some good b should it not. Bets are said to capture degrees of belief: the stronger the bettor believes, the longer the odds he or she should be willing to offer. This gives the obvious payoff table:

X	Payoff
T	$+a$
F	$-b$

Using S to denote the total amount of money at *stake*, $a + b$, and denoting the *betting quotient* $\frac{b}{b+a}$ as p, we can rewrite the table as usual:

X	Payoff
T	$S(1-p)$
F	$-Sp$

To change the direction of the bet, i.e., to bet against X, change the signs of a and b. A *fair* betting quotient does not guarantee that one side of the bet will win. One way of illustrating this is to say that even if the Bookie and bettor have the same information and capabilities, if the bettor offers an unfair betting quotient, the Bookie can subject him or her to a sure loss by changing the direction of the bet. The Ramsey-de Finetti theorem is often taken as establishing that this unfortunate outcome comes about if fair betting quotients are not probabilities. A converse argument establishes that sticking to fair betting quotients is a sufficient condition for avoiding sure losses at the hands of a direction-changing Bookie. Vulnerability to a Dutch Book is said to be indifference between a sure loss and a sure gain. This

3 But not the order of the outcomes, for then the proper betting quotient on a particular outcome would be either a one or a zero.

indifference is sometimes termed "incoherence". The point is that it is bad to have incoherent degrees of belief.[4]

Now to a Dutch Book argument for a Principle of Direct Inference. Let us take the very simplest case to start with, namely, that the Bookie can require us to bet on each of the 100 outcomes. Assume without loss of generality a stake of 1, denominated in your favourite currency. For each bet, you offer the same betting quotient p. 86 times you will win, 14 times you will lose, i.e., your gain will be $86(1-p)+14(-p)$. Setting this equal to zero, the only solution for p is, of course, $86/100$, and hence $86/100$ is the only fair betting quotient.

More generally, for a finite population of size N consisting of binomial trials, m of which are successes, and given a constant betting quotient of p, the payoff will be $m(1-p)+[(N-m)(-p)]$, which some quick algebra will show is only 0 when $p = m/N$. Hence p is fair only if is m/N since it is the only betting quotient that gives no advantage to either side of the bet. The converse Dutch Book argument, that expected gain is 0 only when $p = m/N$, follows easily.

This argument concerns only the trivial case of betting on each of the outcomes of a finite amount of trials. The result is hardly surprising: if we are betting on a known truth, then the probability of that truth must be 1. And in the case of betting on all outcomes we know the truth about the relative frequency of the attribute. But this is hardly a Principle of Direct Inference, since it doesn't tell us about sampling. Sampling without replacement from a finite population is the first non-trivial case of such a Principle.

A simple urn model can serve as the basis of (what I take to be Kyburg's) Dutch Book argument. The drawing of a ball (and noting its colour) serves as a trial, taking, say, the drawing of a white ball as a success (in Kyburg's example as heads and tails). Draws are random, i.e., any sequence is as likely to be drawn as another. The total population is N with total m white balls, sample size n. Y_n is the random variable that $Y_n = \sum_n X_n$, where X_i the indicator variable of the ith trial.

We can derive the distribution for $Y_n = k$ as follows: there are $\binom{m}{k}$ ways to draw the white balls, and $\binom{N-m}{n-k}$ ways to choose the remaining $n-k$ black balls. The total number of possible sequences of draws of k white balls from a sample of n is then just these two multiplied together. Normalize by $\binom{N}{n}$ to get

$$p(Y_n = k) = \frac{\binom{m}{k}\binom{N-m}{n-k}}{\binom{N}{n}}.$$

This is, of course, a familiar hypergeometric distribution. It is elementary that $p(X_i) = \frac{m}{N}$ and that $E(Y_n) = n\frac{m}{N}$. The expectation of the sample being the population mean, expected loss/gain is only zero when the betting quotient is

4 There are many variants of this argument, some involving gun toting Book-makers accosting you personally and requiring a bet in some utility currency or your life. These variants attempt to get around problems with the implementation of actually, or counterfactually, betting, such as unwillingness to bet or the non-linear utility of money. The issue I am pursuing is internal to the Dutch Book enterprise, so I will ignore the rather obvious, and strong, objections to Dutch Book arguments.

equal to the mean. Utterly trivially, if you're drawing all balls from the urn, the only fair betting quotient is the population relative frequency (although we will return to this triviality).

But this is not, again, a very interesting conclusion: successfully betting on an entire population of outcomes, given a known mean, requires offering a betting quotient equal to that mean. The case of interest is that of sampling. This might seem again trivial: the mean of the sample is the mean of the population (which is why the mean is called an unbiased estimator). It is easy to show that any deviation from the mean by p leads to an expected loss. Where $\bar{x} = \frac{1}{n}\sum_{i=1}^{n} X_i$ and $\mu = E(X_i)$, the sample and population means, $E(\bar{x}) = E\left(\frac{1}{n}\sum_{i=1}^{n} X_i\right) = \frac{1}{n}E\left(\sum_{i=1}^{n} X_i\right) = \frac{1}{n}E\left(\sum_{i=1}^{n} \mu\right) = \mu$. So in a sample of n, expected gain is $\left(n\frac{m}{N}(1-p)\right) - \left[n\frac{N-m}{N}p\right] = n\left(\frac{m}{N} - p\right)$.

This, however, is only the converse Dutch Book argument. It shows that setting probabilities to relative frequencies (in this case) is sufficient to prevent sure loss/gain. But can we go the other way? Jeffrey (1983), following de Finetti, points out that if the probabilities are equal, we can derive them from knowledge of the expectation. For a sample size equal to population size, $n = N$, m successes, $E\left(\sum_i^N X_i\right) = m$. Since the X_i are binomial random variables, $E\left(\sum_i^N X_i\right) = \sum_i^N p_i X_i = \sum_i^N 0 \times p_i + 1 \times p_i = \sum_i^N p_i$. Since the p_i are equal, this is Np_i. So we get that $E\left(\sum_i^N X_i\right) = m = Np_i$ and so $p_i = \frac{m}{N}$.

So it seems that we can go the other way. According to Jeffrey this result "...explains to the Bayesian why knowledge of frequencies can have such powerful effects on our belief states" (Jeffrey 1983, 145). In particular, he also claims that this trivially shows that degrees of belief must be constrained by knowledge of relative frequencies. By contrast, however, de Finetti is quite cautious, claiming only practical usefulness in settling our degrees of belief:

> If, in the case where the frequency is known in advance, our judgement is not so simple, the relation is still very useful to us for evaluating the n probabilities, for by knowing what their arithmetic mean has to be, we have a gross indication of their general order of magnitude, and we need only arrange to augment certain terms and diminish others until the relation between the various probabilities corresponds to our subjective judgement of the inequality of their respective chances. (De Finetti 1964, 114)

De Finetti is right, of course: we might be able to go in the other direction, but we need not. A simple example suffices to show that a Bookie cannot guarantee a sure loss for any bets taken on samples of the population at rates other than the population mean. Consider an urn with 10 balls, 5 of which are white, the rest black. A sample is drawn of five balls. Obviously, the expected number white balls drawn is 2.5. But suppose I offer $p(X_i) = 1$, for $p(S_5 = 5) = 1$. This in no way guarantees a sure loss: I might just draw five white balls. But with no sure loss, there is no indifference between a sure loss and gain, no Dutch Book, and so no basis for casting aspersions on anyone's rationality.

To belabour the point, consider the case in which we are drawing two balls from an urn that contains two black ❶❷ and two white ③ ④ balls. Below is a payoff table for a bet on black for a variety of odds:

Draws	1:1	2:1	3:1	0:10
❶❷	2	4	6	0
❶③	0	1	2	−10
❶④	0	1	2	−10
❷③	0	1	2	−10
❷④	0	1	2	−10
③④	−2	−4	−6	−20

Even at odds of 3:1, sure loss or gain is by no means guaranteed.

If your sample were the full population, and your probabilities were equal, your degrees of belief would, by Dutch-Bookability, be fully constrained. It is also clear that as the sample gets larger room for manoeuvre for fiddling with degrees of belief away from the mean gets smaller. Nonetheless, there is no Dutch Book if you do not take the sample to be an exact little copy of the population. Granted, believing that you can predict deviations from the mean is odd, but the whole point of subjective Bayesianism is that your degrees of belief are wholly yours: it is not the Bayesian's business to lecture you on them.[5]

The chart above does suggest a possible way out: utility is maximized only at odds of 1 : 1. All other odds yield lower expected utility, so why not stick with 1 : 1 odds? The appeal to expected utility does not seem very hopeful. The use of expected utility as a decision rule is in need of justification itself. This justification hardly seems forthcoming, given the poor standing of the independence axioms (as is seen rather clearly with the so-called Allais paradox).

If there is to be no Dutch Book justification for a Principle of Direct Inference, then perhaps we could just append one to Bayesianism. This, however, is unappealing. If the Principle is not justified by Dutch Books, then, from the point of view of subjective Bayesianism, it is not justified as a constraint on degrees of belief. That's simply the point of subjective Bayesianism.

However, doing without a Principle of Direct Probabilities for finite relative frequencies is also unappealing, since they are rather useful. For example, opinion surveys give finite relative frequencies (say, of the approval ratings of politicians), field biologists deal with finite relative frequencies (of, say, species of fish in a lake). Thus we have reason to continue a search for a justification of such a Principle.

5 I have not challenged the assumption of the equiprobability of draws: but I could, and for exactly the same reasons. Then the Dutch Book argument wouldn't even work in the case where the sample is the entire population.

2.3 The Infinite Case

There might be hope for a utility theoretic justification in a milder sense. In the single case, we can ignore a possible loss of utility, but maybe in the long(er) run, it will become a risk impossible to ignore. In other words, we might look to repeated sampling with replacement in place of sampling without replacement. This is leads us to consider the infinite case (the hypergeometric distribution of course approaches the binomial as sample size and population go to infinity). This leads nicely to two Dutch Book arguments for a Principle of Direct Inference, one from Howson and Urbach (1993), and one from Mellor (1971).

I begin with Mellor's argument (from Mellor 1971, 160–164). Consider a series of N trials, each with the same fixed betting quotient p, to capture the notion of a series of identical trials. In this set up, the Bookie chooses the stake, to be equally divided among the trials, so that the payoff on each trial is, of course, the same, i.e., $\frac{1}{N}$. The bookie then picks the direction of the bets after learning the outcomes of the trials.[6]

Given this set up, it is seemingly trivial that a betting quotient can only be fair if it is the same as the relative frequency in the sample. Since the payoff is $-N\left|\frac{S_N}{N} - p\right|$, by the Strong Law of Large Numbers, $p\left(\lim_{n\to\infty} \frac{S_N}{N} = \mu\right) = 1$. So, in a large enough series of trials, betting quotients that differ from the mean lead to a sure loss.

The argument is not as simple as it seems, however. I will concentrate on two problems with Mellor's argument. First, it employs the Strong Law, which holds with probability one. But there are some very odd sequences of trials. For example, Ville famously constructed sequences that unilaterally approach a limit, and so for any finite initial segment, the relative frequency will differ from the limiting relative frequency. In betting terms, this means that one side of the bet would always make a gain for any finite sequence. Such sequences have, of course, measure zero. But that doesn't mean that they don't exist.

Still, it might be countered, if you bet infinitely often, you will lose. But this leads to another problem: the argument becomes bad science fiction. Perhaps I will start to lose at some point, but this could be after the Sun swells into a red dwarf. Or, if I am condemned to an eternal betting hell, why should I care how I bet? Apparently I have live forever, and have unlimited amounts of money. The rest of the story needs to be filled in somehow. But no matter how it is filled in, it can hardly be said to be a pragmatic argument.[7]

---

6 Recall that we use a unit stake. Mellor's requirement that the stakes be divided equally among the trials is non-standard, but it does guarantee the desired result. It does so however at the cost of violating the subjective Bayesian ethos, since the Bookie can choose the direction of the bets *after* learning the outcomes of the trials, and so the Bookie has more information than the Bettor.

7 We could try to argue that the convergence will be speedy. The Law of Iterated Logarithm gives a precise meaning to 'speedy' (i.e., not very). No matter – such an argument will not work in lieu of a solution to the problem of induction.

A more hopeful attempt at a Dutch Book argument for a Principle of Direct Probabilities is made by Howson and Urbach (1993, 345). Their argument stresses the notion of consistency. They concentrate on the conditional probability of an outcome given sampling from a von Mises collective with convergent relative frequency p. If you offer a betting quotient different from p you contradict the condition that you are sampling from a von Mises collective:

Suppose that you are asked to state your degree of belief in a toss of this coin landing heads, conditional upon the information *only* that were the tosses to continue indefinitely, the outcomes would constitute a von Mises collective, with probability of heads equal to p. And suppose you were to answer by naming some number p' not equal to p. Then... you believe that there would be no advantage to either side of a bet on heads at that toss at odds $p' : 1 - p'$. But that toss was specified *only* as a member of the collective characterised by its limit-value p. Hence you have implicitly committed yourself to the assertion that the fair odds on heads occurring at *any* such bet, conditional just on the same information that they are members of a collective with probability parameter p are $p' : 1 - p'$... [B]y assumption the limit of [the collective] is p, and p differs from p', you can infer that the odds you have stated would lead to a loss (or gain) after some finite time, and one which would continue thereafter. Thus you have in effect contradicted your own assumption that the odds $p' : 1 - p'$ are fair. (Howson and Urbach, 1993, 345)

A Mellor-style argument (but one using the Weak Law only) gives a convergence of opinion result, claimed to give empirical content to relative frequency theories.

This argument has several nice features. It need not be dependent on von Mises's account of objective probabilities. It only requires a theory that postulates convergence of frequencies, and is of therefore broad applicability. Even better, it can also support a Principle of Direct Inference for finite populations. If you accept the hypergeometric distribution as your model, then to give a different relative frequency is simply to accept a contradiction: to accept that the mean of a population is p, and then to deny that it is simply a contradiction. If you hold that $p(Y_n = k) = \frac{\binom{m}{k}\binom{N-m}{n-k}}{\binom{N}{n}}$, and that the draws are independent, and yet deny that $p(X_i) = \frac{m}{N}$, you embrace a contradiction – a straightforward *logical* contradiction.

Howson and Urbach's Dutch Book for their Principle of Direct Inference has not, of course, escaped criticism. I will discuss two. The first is that their proof relies on symmetry principles, in the guise of the assumption of the equiprobability of trials. Since symmetry principles are not part of subjective Bayesianism, Howson and Urbach's argument would fail to give a justification of a Principle of Direct Inference. This criticism is made by Strevens (1999). It seems to me, however, to misunderstand the role of symmetry in the argument. Use of symmetry is not ruled out in subjective Bayesianism, it is simply not taken as an additional, obligatory, principle. And it is just in the case of modeling objective probability that you would appeal to the symmetries found in the notion of random trials. It is therefore not an illicit use of the symmetry, but one appropriate to a particular model. In this case, the symmetry comes from the randomness of the von Mises collective, which guarantees that the constant probabilities necessary for the Dutch Book. But randomness is not essential: i.i.d. trials or various urn models could serve the same purpose. So symmetry does not play a central role any more than it does in any result concerning convergence in stochastic processes.

A more troublesome objection is that since any initial segment is compatible with any particular relative frequency there is no inconsistency from ignoring a Principle of Direct Inference. As Albert puts it "...the fact that the sequence has limiting relative frequency $\frac{1}{2}$ implies nothing about its beginning" Albert (2005). According to Albert, this means that there can be no pragmatic inconsistency (in the short run, which as I have just argued is all we get). This objection, however, misses the point of the Howson and Urbach proof. First, they provide a convergence of opinion result to show how it is possible to be sensitive to frequencies. But this should not be taken as implying that they are offering a solution to the problem of induction. It is simply the case that any initial segment is compatible with an infinite sequence of any relative frequency. Howson and Urbach do not claim to offer any solution to this particular problem of induction. However, they do claim, rightly, that if we are by our own lights sampling from a random sequence that converges to p, then that is the only fair betting quotient, by definition.

However, there is another reason to be dissatisfied with the Howson and Urbach argument. The argument is based on Howson's 'logical' interpretation of subjective Bayesianism (explored in Howson 2008). This interpretation takes very seriously the notion of Dutch Book arguments being about logical consistency. So for example, the inconsistency of being indifferent between a sure loss and a sure gain, given an assertion that a betting quotient is fair, is a purely logical inconsistency. Many (but not me) will not find it satisfactory because it is synchronic.

The argument runs as follows: at the time of betting, we offer odds conditional on a model of objective probabilities, say, a collective. This collective specifies the relative frequency of the infinite sequence of outcomes. This frequency is a parameter of the collective, and to give a different parameter leads to contradiction. This means that a book can be made before any outcomes are actually observed, simply on the basis of the contradictory values of the probability. But this means the Dutch Booking takes place at a 'higher' level, at the level of the consistency of probabilities with a model, and not with diachronic accumulation of evidence. (It should be recalled that Howson and Urbach reject diachronic Dutch Book arguments.) The Dutch Book argument, and hence the Principle, is a matter of consistency of beliefs at a particular point, and not at all of the consequences of those beliefs.

This does avoid the odd character of worrying about the consequences of extremely lengthy series of bets. But it does so at the cost of jettisoning all pragmatic considerations. Nor does it rule out any particular model of frequencies. As stated, the Dutch Book is about members of a collective: but if you reject this notion, and adopt a different model (say, you decide you can use clairvoyance to determine the outcomes beforehand) you can only be Dutch Booked à la Howson and Urbach if you violate the constraints of this model (and so Howson and Urbach agree with de

Finetti's take on the problem discussed above). The argument also has nothing to say about those who later change their minds. In particular, it does not imply that we cannot change our minds as to the value of the frequency without being Dutch Bookable. This is an account of subjective probability that is, indeed, *subjective*. There is, of course, a large debate on the undesirability of subjectivism or not. But my purpose has not been to pursue these: it has been to establish that subjective Bayesianism deserves its name.

2.4 Conclusion

There is a Dutch Book argument for obeying the laws of the probability calculus conditional on accepting a certain model of relative frequency in a sychronic setting. This sort of Dutch Book really only establishes that we should keep our logical commitments. But there seems to be no standard Dutch Book justification of a Principle of Direct Probability, at least in the sense of guaranteeing certain loss in a betting situation. In fact, if we simply refuse to accept any model in the finite or infinite case we cannot be Dutch Booked. Whether this counts for or against Bayesianism is another matter.[8]

References

Albert, M.,"Should Bayesians Bet Where Frequentists Fear to Tread?", in: *Philosophy of Science,* 72, 2005, pp. 584–593.

De Finetti, B., "Foresight: Its Logical Laws, its Subjective Sources", in: Henry Kyburg and Howard Smokler (Eds.), *Studies in Subjective Probability*. New York: John Wiley & Sons, Inc. 1964, pp. 97–158.

Howson, C., "De Finetti, Countable Additivity, Consistency and Coherence", in: *The British Journal for the Philosophy of Science*, 59, 2008, pp. 1–23.

Howson, C., and Urbach, P., *Scientific Reasoning: The Bayesian Approach,* 2nd ed. La Salle, IL.: Open Court 1993.

Jeffrey, R., "Bayesianism with a Human Face", in: John Earman (Ed.), *Testing Scientific Theories*. Minneapolis: University of Minnesota Press, 1983, pp. 133–156.

Kyburg, Jr., H. E., "Principle Investigation", in: *The Journal of Philosophy,* 78, 1981, pp. 772–778.

8 I would like to thank Peter Milne, the participants of the ESF workshop Pluralism in the Foundations of Statistics and the anonymous referees for helpful comments. Work on this article was supported by grant P401/10/1504 of the Grant Agency of the Czech Republic.

Lewis, D., "A Subjectivist's Guide to Objective Chance" in: Richard Jeffrey (Ed.) *Studies in Inductive Logic and Probability,* vol II. Berkeley: University of California Press, 1980, pp. 263–293.

Mellor, D. H., *The Matter of Chance*. Cambridge: Cambridge University Press 1971.

Seidenfeld, T., "Calibration, Coherence, and Scoring Rules", in: *Philosophy of Science,* 52, 1985, pp. 274–294.

Strevens, M., "Objective Probability as a Guide to the World.", in: *Philosophical Studies,* 95, 1999, pp. 243–275.

Institute of Philosophy
Academy of Sciences of the Czech Republic
Jilska 1
110 00, Prague
Czech Republic
childers@site.cas.cz

Chapter 3

Ilkka Niiniluoto

The Foundations of Statistics: Inference vs. Decision

In his classical exposition of Bayesian statistics, *The Foundations of Statistics* (1954), L. J. Savage defended the "behavioralistic outlook" against the "verbalistic outlook": statistics deals with problems of deciding what to do rather than what to say. Savage referred to F. P. Ramsey's and Abraham Wald's work on decision theory and to Jerzy Neyman's proposal to replace "inductive inference" with "inductive behavior". All of these approaches were in opposition to R. A. Fisher's formulation of statistical estimation and testing in traditional terms as truth-seeking methods of scientific inference. In spite of the prominence of the decision-theoretic approach, some influential Bayesians (like Dennis Lindley) have preferred to emphasize estimation and testing as procedures of inference. A reconciliation of inference and decision was forcefully proposed by Isaac Levi in his *Gambling With Truth* (1967). Levi argued against "behavioralism" that the tentative acceptance and rejection of scientific hypotheses cannot be reduced to actions that are related to practical objectives. According to Levi's "critical cognitivism", science has its own theoretical objectives, defined by the maximization of expected "epistemic utilities", such as truth, information, and explanatory power. As a development of cognitive decision theory, and in the spirit of critical scientific realism, Ilkka Niiniluoto's *Truthlikess* (1987) suggests that scientific inference is defined by the attempt to maximize expected verisimilitude. This proposal allows us to interpret Bayesian point and interval estimation in terms of decisions relative to loss functions which measure the distances of rival hypotheses from the truth.

3.1 Why I am a Bayesian

Let me start with a personal introduction by telling why I am a Bayesian. In 1973 I defended my Ph.D. thesis on inductive logic by applying Jaakko Hintikka's system to theoretical inferences in science. Inductive logic is a special case of Bayesianism where epistemic probabilities are defined by symmetry assumptions concerning states of affairs expressible in a formal language (for a survey, see Niiniluoto, forthcoming).

My commitment to Bayesianism has a longer history, however. In 1968 I wrote my Master thesis in mathematics "On the Power of Bayes Tests". My supervisor at the University of Helsinki Professor Gustav Elfving (1908–1984) was one of the first mathematical statisticians in Scandinavia who supported the Bayesian approach (see Nordström 1999). His influence can still be seen today in Helsinki in the lively

D. Dieks et al. (eds.), *Probabilities, Laws, and Structures*, The Philosophy of Science in a European Perspective 3, DOI 10.1007/978-94-007-3030-4_3,

interest in Bayesian reasoning and its applications at the Department of Mathematics and Statistics and the Department of Computer Science.

The key text for a young Bayesian was Leonard J. Savage's *The Foundations of Statistics* (1954) which gives an elegant axiomatic treatment of the subjective expected utility model (SEU). In my attempt to reconstruct Savage's proof of the representation of qualitative personal probabilities, I found a minor mistake in his Theorem 3 (*ibid*., p. 37). During the 4th International Congress for Logic, Methodology, and Philosophy of Science in Bucharest in the summer of 1971, Hintikka introduced me to Savage. I was sitting in a park with this admired hero of the Bayesians trying to explain my observations. Savage, who had problems with his sight, was extremely friendly and encouraging. My paper on qualitative probability was published in *Annals of Mathematical Statistics* (see Niiniluoto 1972) after Savage's untimely death.

In moving from mathematics to philosophy of science, I was attracted by the philosophical position of scientific realism. In my doctoral dissertation, the realist interpretation of theoretical terms was defended by means of Hintikka's system which assigns positive probabilities to genuine laws and theories. A natural ingredient of the realist view was a dualist account which accepts both epistemic and physical probabilities (propensities). Levi's and Hintikka's cognitive approach, which includes truth and semantic information as epistemic utilities, suggested a remedy to the instrumentalist tone of decision theory. When I started to work on the Popperian notion of truthlikeness in the mid-seventies, I adapted the Bayesian framework to the method of estimating verisimilitude by calculating expected degrees of truthlikeness. This idea can then be applied as a special case to statistical problems of point and interval estimation by interpreting the loss function as measuring distances from the truth. This provides a fresh perspective to the traditional debate on inferential and decision-theoretic approaches in statistics.

3.2 Behavioralism

Savage's *opum magnum* made a sharp contrast between "the behavioralistic outlook" and "the verbalistic outlook": according to the former, statistics deals with problems of what to do rather than what to say (Savage 1954, pp. 159–161). Verbalism treats statistics as a mode of inference, analogous to deduction, where assertions are consequences of inductive inferences. One might think that this difference is only terminological, since assertions are also acts, and decisions are assertions to the effect that the act is the best available. Savage dismissed this proposal and concluded that verbalism has led to "much confusion in the foundations of statistics".

Savage referred to Neyman's 1938 article as "the first emphasis of the behavioralistic outlook in statistics". Neyman argued that statistics is concerned with "inductive behavior" rather than "inductive inference" (see Neyman 1957). On the basis of the frequentist account of probability, the Neyman-Pearson theory interprets confidence intervals and significance tests in terms of long run frequencies of errors in repeated applications of estimation or test procedures (see Neyman 1977).

As advocates of the behavioralistic outlook Savage referred also to Wald's (1950) "objectivist" decision theory, based on the minimax approach, and the early work of F. P. Ramsey and Bruno de Finetti in the 1920s and 1930s. In terms of John von Neumann's game theory, decision theory was taken to deal with "games against nature", where the relevant losses are the practical consequences of real-life choices of actions. Besides Savage, this framework of Bayesian decision theory, with special emphasis on business decisions, was developed by Robert Schlaifer, Howard Raiffa, and Patrick Suppes, and exposed in textbooks by Blackwell and Girshick (1954), Chernoff and Moses (1959), and Raiffa and Schlaifer (1961).

Game theory and decision theory were important areas of operations research (OR) which was applied to military problems during World War II. Churchman, Ackoff and Arnoff (1957) formulated OR as a general method of a decision-maker or executive for finding optimal solutions to problems relative to the objectives (needs), resources, and available actions. R. L. Ackoff, in his book *Scientific Method: Optimizing Applied Research Decisions* (1962), advertised OR as the method of science.

So Savage was not alone in his defense of the behavioralistic outlook. One can also observe that the emphasis of behavior or action was in harmony with the tradition of American pragmatism. Even though John Dewey's instrumentalism left room for verbalism with his notion of "warranted assertability", he had argued in his *Logic: The Theory of Inquiry* (1938) that logic should be viewed as a theory of problem-solving. Dewey's influence can be seen in the success of the OR approaches in the 1950s, and its echoes can be found in the 1960s and 1970s in Thomas Kuhn's and Larry Laudan's claims that science is a problem-solving rather than a truth-seeking activity (see Niiniluoto 1984).

In lively debates on statistics, Savage admitted that there is a distinction between inference and decision which is meaningful to a Bayesian: inference is "the art of arriving at posterior probabilities" from priors by Bayes's theorem, while decision is concerned directly with action (see Savage et al. 1962, p. 102). Similarly, D. V. Lindley's influential book on Bayesian statistics defined statistical inference as the method of altering degrees of belief by data, and the posterior distribution can then be used in making decisions (Lindley 1965, 1977). However, Lindley also formulated Bayesian tests and interval estimation as special kinds of inferences based on posterior probability.

With influences from Ramsey and Savage, Rudolf Carnap concluded in the late 1950s that there are no inductive inference rules: the task of inductive logic is to assign probabilities to hypotheses, and these probabilities can then be used in rational decision making by calculating expected utilities (see Carnap 1962).

3.3 Fisher's Defense of Statistical Inference

Savage politely acknowledged that R. A. Fisher's *Statistical Methods for Research Workers* (1925) has had far more influence on the development of statistics than any other publication. Yet for him Fisher was the prime example of the "verbalist" approach: statistics is a tool for research workers in empirical science, such as biology and genetics, and its methods are related to inductive inference and the method of hypothesis (see Fisher 1950, p. 8). In *Design of Experiments* (1935), Fisher developed statistical methods in connection with experimentation in agriculture. In *Statistical Methods and Scientific Inference* (1956), he defended the idea that scientific knowledge is generated by "free individual thought" (p. 7).

Fisher was well known for his criticism of the Bayesian tradition. He claimed that the advocates of inverse probability seem forced to regard mathematical probability "as measuring merely psychological tendencies" which are "useless for scientific purposes" (Fisher 1966, pp. 6–7). He cited Boole and Venn against "conservative Bayesians" (Fisher 1956, p. 34). He admitted that sometimes probabilities a priori can be deduced from data, but when they are not available the "fiducial argument" can be used (*ibid.*, p. 17).

Fisher also sharply attacked Neyman's conception of inductive behavior. He argued that in statistical tests "the null hypothesis is never proved or established, but is possibly disproved", so that errors of the second kind (i.e., acceptance of false null hypothesis) have no meaning with respect to simple tests of significance (Fisher 1966, pp. 16–17). A test should not be regarded as "one of a series of similar tests applied to a succession of similar bodies of data", but the scientific worker "gives his mind to each particular case in the light of his evidence and his ideas" (Fisher 1956, p. 42). The "state of opinion derived from a test of significance" is provisional and revisable, while decisions are final (*ibid.*, p. 99). Thus, statistics is not limited to repeated "acceptance procedures" or rules of action, typical of quality control in commerce and technology, but it gives "improved theoretical knowledge" in experimental research (*ibid.*, pp. 76–77).

Fisher further argued against the introduction of loss functions in statistics: "in the field of pure research no assessment of the cost of wrong conclusions … can conceivably be more than a pretense", and "in any case such an assessment would be inadmissible and irrelevant in judging the state of scientific evidence" (Fisher 1966, pp. 25–26). It is important that "the scientific worker introduces no cost functions for faulty decisions": the purposes to which new knowledge is put are not known in advance, as they may involve "a great variety of purposes by a great variety of men", so that the proper conclusions of inference should be "equally convincing to all freely reasoning minds" (Fisher 1956, pp. 102–103).

Lindley partly agreed with Fisher, when he stated that the scientist in the laboratory does not consider the subsequent decisions to be made on the basis of his discoveries (Lindley 1965, p. 67). But Savage made no such compromise: the dualism of economic contexts and pure science, or practical affairs and science, is incorrect (Savage et al. 1962, pp. 15, 101–102). After Savage, the decision-theoretic approach has been quite prominent in Bayesian statistics.

Compromises between inferential views and the Neyman-Pearson theory have been considered by philosophers who try to relate objective error probabilities to the concepts of support or evidence. Ian Hacking (1965), who advocates statistical chance as propensity, construes support in terms of the likelihood function and likelihood comparisons. Such evidential support of a hypothesis is independent of utility. According to Hacking, deciding that something is the case is different from deciding to do something, so that statistics should have room for "belief-guesses" or estimation aiming at the truth regardless of the consequences.

In the context of debates about the likelihood principle, Allan Birnbaum proposed a distinction between "behavioral" and "evidential" interpretations of statistical decisions (see Birnbaum 1977). The former – advocated by Neyman and Pearson, Wald, and Savage, and criticized by Fisher, Cox, and Tukey – takes decision in the literal sense of deciding to act in a certain way (e.g. a lamp manufacturer decides to place a batch of lamps on the market). The latter considers decisions that one of the alternative hypotheses is true or supported by strong evidence. Birnbaum's recommendation to use the NP formalism with evidential interpretation is followed by Deborah Mayo and Aris Spanos (2006) in their "error statistics" approach: error probabilities guarantee that only statistical hypotheses that have passed severe tests are inferred from the data.

3.4 Cognitive Decision Theory and Truthlikeness

Richard Rudner (1953) argued that the scientist *qua* scientist makes value judgements when he accepts or rejects hypotheses. Generalizing from examples of industrial quality control, Rudner claimed that the scientist's decision to regard the evidence as strong enough to warrant the acceptance of a hypothesis "must be made in the light of the seriousness of the mistake". Richard Jeffrey (1956) replied to Rudner and West Churchman that the job of the scientists is to assign probabilities but not to accept and reject hypotheses. Levi (1960) tried to refute both Rudner's and Jeffrey's positions by showing that scientists do accept hypotheses but the seriousness of mistakes need not be taken to be relative to ethical or practical objectives.

Levi gave a systematic exposition of his "critical cognitivism" in *Gambling With Truth* (1967). Against "behavioralism" and some forms of pragmatism Levi argued in detail that the acceptance and rejection of scientific hypotheses cannot be construed as actions relative to practical utilities. This would reduce the role of a scientist (and a statistician) to a practical decision-maker or a guidance councillor of a decision-maker. Science has its own theoretical objectives, defined by "epistemic utilities" like truth, information, explanatory power, and simplicity. On this basis, Levi disagreed with Carnap and Jeffrey on the possibility of inductive acceptance rules. Such rules give conditions for the tentative acceptance of hypotheses into the revisable body of scientific knowledge at a given time. Levi's cognitive decision theory thus applies the Bayesian SEU framework relative to the maximization of expected epistemic utilities.

Levi's basic insight was that the scientist has "to gamble with truth" in order to obtain relief from agnosticism. If truth were the only epistemic utility, then the expected utility of accepting a hypothesis H on the basis of evidence E would be the posterior probability P(H/E) of H given E, which would recommend the conservative strategy of accepting only tautologies and logical consequences of the evidence. Levi's own acceptance rule was based upon a weighted average of the truth value of H and its information content, where the weight β of the information factor is an "index of boldness" of the scientist in risking error. Variations of this definition were proposed in Finland by Jaakko Hintikka, Risto Hilpinen, and Juhani Pietarinen (see Niiniluoto 1987).

In his campaign against inductive probability, Karl Popper (1963) attempted to define a comparative notion of truthlikeness (verisimilitude) for scientific theories. In my first paper on truthlikeness, presented in the LMPS Congress in 1975, I defined degrees of truthlikeness Tr(H,C*) for theories H in a monadic predicate logic L. Here C* is the complete truth expressible in L, and Tr(H,C*) has its maximum value one if and only if H = C*. Definition of Tr is based a distance function d which tells how close the constituents (complete theories) C_i of L are to each other. When the target C* is unknown, I proposed that the degree Tr(H,C*) can be estimated by calculating the expected truthlikeness of H on evidence E, relative to the posterior probabilities $P(C_i/E)$ of constituents C_i in L:

(1) $\mathrm{ver}(H/E) = \sum_i P(C_i/E)\,\mathrm{Tr}(H,C_i)$.

In the spirit of cognitive decision theory, measure ver(H/E) recommends those hypotheses which have the smallest expected distance from the truth. Generalized to full first-order logic, it allows us to conceptualize scientific inference in terms of decisions which maximize expected verisimilitude (see Niiniluoto 1987; Festa 1993). By combining elements from the Bayesian and Popperian traditions, it serves as a foundation of critical scientific realism (Niiniluoto 1999).

A similar proposal has recently been made by Leitgeb and Pettigrew (2010), who require that epistemic agents ought "to approximate the truth" by minimizing expected global and local inaccuracy.

It turned out that my favorite min-sum measure of truthlikeness, found in 1984, reduces to Levi's 1967 definition of epistemic utility, if the underlying distance measure d between complete theories is trivial, i.e., all false complete answers are equally distant from the truth. However, Levi was not ready to accept that utilities depend on distances from the truth, and he has given interesting defenses of the idea that "a miss is as good as a mile" (see Levi 2007; cf. Savage 1954, p. 231).

As special cases, the notion of truthlikeness should be applicable to singular and general quantitative statements. Then distance from the true value of a real-valued parameter could be chosen as the loss function of a decision problem. In Niiniluoto (1982a), I pointed out that the rule of maximizing expected verisimilitude (1) contains as a special case the theory of point estimation of Bayesian statistics. Similarly, if the distance of an interval from the truth is defined, then Bayesian interval estimation can also be treated in decision-theoretic terms (Niiniluoto 1982b). This idea was independently discovered by Roberto Festa. Papers working out this program were published as Niiniluoto (1986) and Festa (1986) (see also Niiniluoto 1987, pp. 426–441).

3.5 Bayesian Estimation

Let $f(x/\theta)$ be the sample distribution of sample x in sample space X given parameter θ in parameter space Θ. Let $g(\theta)$ be the prior probability distribution of θ. Then Bayes's theorem tells that the posterior distribution $g(\theta/x)$ of θ given data x is proportional to $g(\theta)f(x/\theta)$. Let $L(\theta,a)$ be the loss of action a when θ is the true state of nature. Then the Bayes solution of the decision problem minimizes for each x in X the aposteriori loss

(2) $\int_{\Theta} L(\theta,a)\, g(\theta/x)\, d\theta$.

It can be proved that all good solutions of a decision problem are Bayes with respect to some prior probability distribution (see Ferguson 1967).

Bayes tests with two simple alternative hypotheses H_0 and H_1 are likelihood ratio tests where the critical region depends on the prior probabilities of H_0 and H_1 and the losses of erroneous decisions (see Chernoff and Moses 1959).

For point estimation, the loss $L(\theta,\theta')$ of estimate θ' when θ is the true value of parameter can be defined in many ways. For the zero-one function

(3) $L(\theta, \theta') = 0$ if $|\theta - \theta'| < \varepsilon$
$= 1$ if $|\theta - \theta'| \geq \varepsilon$

(with small ε) the aposteriori loss is minimized by the mode (maximum) of the posterior distribution $g(\theta/x)$. For the linear loss function

(4) $L(\theta, \theta') = |\theta - \theta'|$

the Bayes solution is the median of $g(\theta/x)$; for the weighted linear loss

(5) $L(\theta, \theta') = c_1 |\theta - \theta'|$ if $\theta < \theta'$
$= c_2 |\theta - \theta'|$ if $\theta \geq \theta'$

any fractile of $g(\theta/x)$; and for the quadratic loss

(6) $L(\theta, \theta') = (\theta - \theta')^2$

the mean of $g(\theta/x)$ (see Blackwell and Girshick 1954). Box and Tiao (1973), who think that losses represent "a realistic economic penalty", find the choice of loss functions "arbitrary". But the situation is different, if we require that the loss function should have a natural interpretation as distance from the truth. This condition holds at least for (4) and (6).

According to Savage, there is "no important behavioralistic interpretation of interval estimation" (Savage 1954, p. 261). Lindley (1965) defines 100b% Bayesian confidence intervals as intervals which include the true value of parameter θ with posterior probability b. Box and Tiao (1973) define highest posterior density intervals (HPD) by a similar criterion.

To treat interval estimation in decision-theoretic terms, let $L(\theta,I)$ be the loss of a real-valued interval $I = [c,d]$ when θ is true. Formally this allows us to unify theories of point and interval estimation, since intervals are disjunctions of point estimates and point estimates are included as degenerate interval estimates (e.g., $[c,c]$). Let $S(I) = d - c$ be the length of I, and $m(I) = (c + d)/2$ the mid-point of I. Wald's (1950) proposal

(7) $L_1(\theta, I) = 0$, if $\theta \in I$
$= 1$ otherwise

is not satisfactory, since it gives the same value for all true intervals and the same value for all mistaken intervals. Ferguson (1967) mentions only as an exercise (p. 184) the function

(8) $L_2(\theta, I) = \beta\, S(I) - 1$, if $\theta \in I$
$= \beta\, S(I)$ otherwise

(for $\beta > 0$), which is essentially the same as Levi's (1967) epistemic utility. All false point estimates have the same loss by (8). The Bayes rule (2) with L_2 leads to HPD intervals with points θ such that $g(\theta/x) \geq \beta$. The average distance of I from θ would recommend as the best estimate the degenerate interval consisting of the median of $g(\theta/x)$. The same result is obtained for the measure

(9) $L_7(\theta, I) = \beta\, S(I) + \min(\theta, I)$

if $\beta \geq \frac{1}{2}$, but for $\beta < \frac{1}{2}$ L_7 favors interval estimates. A weighted variant of L_7 has been discussed by Aitchison and Dunsmore (1968). The loss function, which was found to be best for cognitive purposes in Niiniluoto (1986), is a variant of the min-sum-measure of truthlikeness:

(10) $L_{10}(\theta, I) = \beta\, \mathrm{sum}(\theta, I) + \min(\theta, I)^2$

where

$$\mathrm{sum}(\theta, I) = \int_\theta |t - \theta|\, dt\,.$$

The weight β in (10) serves as an index of boldness in Levi's sense. If $\beta \geq 1$, then point estimates dominate intervals: the mean of $g(\theta/x)$ is the best estimate. If $0 < \beta < 1$, then the confidence level α of the recommended 100α % interval estimate increases when the penalty of mistakes β decreases.

More recently, there has been a lot of interest among the Bayesians for the decision-theoretic treatment of interval estimation. A good survey with references is given by Rice, Lumley, and Szpiro (2008) – but without citations to Levi and other philosophers. In the same way as Levi searched a balance between truth and information, their account of estimation is "trading bias for precision". The three main proposals considered are L_2, L_7 with $\beta < \frac{1}{2}$, and

(11) $\gamma\, S(I)/2 + 2(\theta - m(I))^2/S(I)$.

Here (11) resembles L_{10}, but the normalization by S(I) gives an infinite loss for all point estimates.

3.6 Discussion

The question, whether statistics is dealing with inference or decision, has internally divided the two main schools in the foundations of statistics: frequentists (Fisher vs. NP) and Bayesians (Lindley vs. Savage). We have seen that cognitive decision theory reconciles these two perspectives: statistical tests and estimation can be viewed as inferences which are conceptualized as decisions relative to epistemic utilities. Statistics can serve as a tool of a research worker, just as Fisher demanded, but this is not in conflict with the decision-theoretic formulation of statistical problems.

One of Savage's memorable slogans was the claim that "the role of subjective probability in statistics is, in a sense, to make statistics less subjective" (Savage et al. 1962, p. 9). Objective Bayesians have sought canonical ways of fixing prior probability distributions. What I have done in this paper is the complementary goal of formulating constraints for the choice of loss functions in situations where the scientist's cognitive goal is to find truth and avoid falsity.

Yet, the difference between cognitive and practical loss functions remains valid. Even though decision theory is a powerful tool for value-laden choices, all decision problems cannot be reduced to practical or economic situations, as the behavioralistic approach seemed to presuppose. But likewise, problems of pragmatic preference cannot always be reduced to theoretical problems, like Popper suggested, since the most truthlike hypothesis need not be the best solution to a problem of action (Niiniluoto 1982a, 1987). Indeed, the practical loss of an incorrect choice need not always be a linear function of the distance from the truth.

The weighted linear loss function (5) is interesting, since it at the same reflects distances from the truth and other considerations. The expected loss of (5) has its minimum when the cumulative distribution function $G(\theta/x)$ of the posterior distribution $g(\theta/x)$ equals $c_2/(c_1+c_2)$. The symmetric choice $c_1 = c_2$, which leads to the median of $g(\theta/x)$ as the most truthlike estimate, is adequate in theoretical contexts. But non-symmetric choice of the weights c_1 and c_2 may be justified in pragmatic problems where underestimation of some quantity is more dangerous than overestimation (cf. Niiniluoto 1982a). Steel (2010) has discussed examples of cases where it is better to overprotect than underprotect against risks of toxic chemicals to human health. Steel calls the Rudner-Churchman attack on value-neutral science "the argument from inductive risk", and defends against some recent critics the distinction between epistemic and nonepistemic values. His thesis is that nonepistemic values can influence scientific inferences without compromising or obstructing the epistemic goal of the attainment of truth. In my view, this may be valid in applied research which aims at giving conditional recommendations of action with explicitly stated value assumptions as antecedents (see Niiniluoto 1993): we know that the weighted loss function gives results that differ from the theoretically best value, but it is legitimate to act on estimates which make the relevant health risks lower.

References

Ackoff, R. L., *Scientific Method: Optimizing Applied Research Decisions*. New York: John Wiley and Sons 1962.

Aitchison, J. and Dunsmore, I. R., "Linear-Loss Interval Estimation of Location and Scale Parameters", in: *Biometrika* 55, 1968, pp. 141–148.

Birnbaum, A., "The Neyman-Pearson Theory as Decision Theory, and as Inference Theory: With a Criticism of the Lindley – Savage Argument for Bayesian Theory", in: *Synthese* 36, 1, 1977, pp. 19–49.

Blackwell, D. and Girshick, M. A., *Theory of Games and Statistical Decisions*. New York: John Wiley and Sons 1954.

Box, G. E. P. and Tiao, G. C., *Bayesian Inference in Statistical Analysis*. Reading, Mass.; Addison-Wesley 1973.

Carnap, R., "The Aim of Inductive Logic", in: Ernest Nagel, Patrick Suppes and Alfred Tarski (Eds.), *Logic, Methodology, and Philosophy of Science: Proceedings of the 1960 International Congress*, Stanford: Stanford University Press 1962, pp. 303–318.

Chernoff, H. and Moses, L. E., *Elementary Decision Theory*. New York: John Wiley and Sons 1959.

Churchman, C. W., Ackoff, R. L. and Arnoff, E. L., *Introduction to Operations Research*. New York: John Wiley and Sons 1957.

Dewey, J., *Logic: The Theory of Inquiry*. New York: Henry Holt and Co. 1939.

Ferguson, T. S., *Mathematical Statistics: A Decision-Theoretic Approach*. New York: Academic Press 1967.

Festa, R., "A Measure for the Distance Between an Interval Hypothesis and the Truth", in: *Synthese* 67, 1986, pp. 273–320.

Festa, R., *Optimum Inductive Methods: A Study in Inductive Probabilities, Bayesian Statistics, and Verisimilitude*. Dordrecht: Kluwer 1993.

Fisher, R. A., *Statistical Methods for Research Workers*. Edinburgh: Oliver and Boyd 1925. (11th ed. 1950.)

Fisher, R. A., *Design of Experiments*. Edinburgh: Oliver and Boyd 1935. (8th ed. 1966.)

Fisher, R. A., *Statistical Methods and Scientific Inference*, Edinburgh: Oliver and Boyd 1956.

Hacking, I., *The Logic of Statistical Inference*, Cambridge: Cambridge University Press 1965.

Jeffrey, R., "Valuation and Acceptance of Scientific Hypotheses", in: *Philosophy of Science* 23, 1956, pp. 237–246.

Leitgeb, H. and Pettigrew, R., "An Objective Justification of Bayesianism I: Measuring Inaccuracy", in: *Philosophy of Science* 77, 2, 2010, pp. 201–235.

Levi, I., "Must the Scientist Make Value Judgments?", in: *Journal of Philosophy* 58, 1960, pp. 345–347.

Levi, I., *Gambling With Truth: An Essay on Induction and the Aims of Science*. New York: Alfred A. Knopf 1967.

Levi, I., "Is a Miss as Good as a Mile?", in: Sami Pihlström, Panu Raatikainen and Matti Sintonen (Eds.), *Approaching Truth: Essays in Honor of Ilkka Niiniluoto*, London: College Publications 2007, pp. 209–223.

Lindley, D. V., *Introduction to Probability and Statistics from a Bayesian Viewpoint: Part 2. Inference*. Cambridge: Cambridge University Press 1965.

Lindley, D. V., "The Distinction Between Inference and Decision", in: *Synthese* 36, 1, 1977, pp. 51–58.

Mayo, D. and Spanos, A., "Severe Testing as a Basic Concept in a Neyman-Pearson Philosophy of Induction", *The British Journal for the Philosophy of Science* 57, 2, 2006, pp. 323–357.

Neyman, J., "'Inductive Behavior' as a Basic Concept of Philosophy of Science", in: *Review of the International Statistical Institute* 25, 1957, pp. 97–131.

Neyman, J., "Frequentist Probability and Frequentist Statistics", in: *Synthese* 36, 1, 1977, pp. 97–131.

Niiniluoto, I., "A Note on Fine and Tight Qualitative Probabilities", in: *Annals of Mathematical Statistics* 43, 1972, pp. 1581–1591.

Niiniluoto, I., "What Shall We Do With Verisimilitude?", in: *Philosophy of Science* 49, 2, 1982, pp. 181–197.(a)

Niiniluoto, I., "Truthlikeness for Quantitative Statements", in: P. D. Asquith and Thomas Nickles (Eds.), *PSA 1982*, vol. 1. East Lansing: Philosophy of Science Association 1982, pp. 208–216.(b)

Niiniluoto, I., "Paradigms and Problem-Solving in Operations Research", in: *Is Science Progressive?*. Dordrecht: D. Reidel 1984, pp. 244–257.

Niiniluoto, I., "Truthlikeness and Bayesian Estimation", in: *Synthese* 67, 1986, pp. 321–346.

Niiniluoto, I., *Truthlikeness*. Dordrecht: D. Reidel 1987.

Niiniluoto, I., "The Aim and Structure of Applied Research", in: *Erkenntnis* 38, 1, 1993, pp. 1–21.

Niiniluoto, I., "The Development of the Hintikka Program", forthcoming in: Dov Gabbay, Stephan Hartmann, and John Woods (Eds.), *Handbook of the History and Philosophy of Logic, vol. 10: Inductive Logic*. Amsterdam: Elsevier.

Nordström, K., "The Life and Work of Gustav Elfving", in: *Statistical Science* 14, 2, 1999, pp. 174–196.

Popper, K., *Conjectures and Refutations*, London: Hutchinson 1963.

Raiffa, H. and Schlaifer, R., *Applied Statistical Decision Theory*. Cambridge, Mass.: The MIT Press 1961.

Rice, K. M., Lumley, T. and Szpiro, A. A., "Trading Bias for Precision: Decision Theory for Intervals and Sets", in: *UW Biostatistics Working Paper Series*, 2008.

Rudner, R., "The Scientist *Qua* Scientist Makes Value Judgments", in: *Philosophy of Science* 20, 1953, pp. 1–6.

Savage, L. J., *The Foundations of Statistics*. New York: J. Wiley and Sons 1954. (2nd ed. New York: Dover 1972.)

Savage, L. J. et al., *The Foundations of Statistical Inference*. London: Methuen 1962.

Steel, D., "Epistemic Values and the Argument from Inductive Risk", in: *Philosophy of Science* 77, 11, 2010, pp. 14–34.

Wald, A., *Statistical Decision Functions*. New York: John Wiley and Sons 1950.

Department of Philosophy, History, Culture and Art Studies
University of Helsinki
Unioninkatu 40 A
00014, Helsinki
Finland
ilkka.niiniluoto@helsinki.fi

Chapter 4

Roberto Festa

On the Verisimilitude of Tendency Hypotheses

Tendency hypotheses – T-hypotheses, for short –, such as "the individuals of the kind *Y* tend to be *X*", are used within several empirical sciences and play an important role in some of them, for instance in social sciences. However, so far T-hypotheses have received little or no attention by philosophers of science and statisticians.[1] An exception is the work made in the seventies of the past century by the statisticians and social scientists David K. Hildebrand, James D. Laing, and Howard Rosenthal who worked out – under the label of prediction logic –, an interesting approach to the analysis of T-hypotheses.[2]

In this paper our main goal is the introduction of appropriate measures for the *verisimilitude* of T-hypotheses.[3] Our verisimilitude measures will be defined in terms of the feature contrast (FC-) measures of similarity proposed by the cognitive scientist Amos Tverski (1977). We shall proceed as follows. In Sect. 4.1, Tverski's FC-measures of similarity for binary features are illustrated and suitably extended to quantitative features. Afterwards, such measures are applied in the definition of appropriate measures for the verisimilitude of universal and statistical hypotheses (Sect. 4.2) and T-hypotheses (Sect. 4.3).

4.1 Feature Contrast (FC-) Measures of Similarity for Binary and Quantitative Features

The FC-measures of similarity proposed by Tverski are intended to provide a quantitative model of the similarity assessments made by human beings w.r.t. a large variety of items, including physical and conceptual objects. In order to

1 As a consequence of this neglect, no standard name for T-hypotheses can be found in the literature.

2 Prediction logic has been developed in a series of papers culminated in a volume on *Prediction Analysis of Cross Classification* (1976). On the conceptual relations between prediction logic and the research on verisimilitude made by philosophers of science, see Festa (2007a, b).

3 The concept of verisimilitude (denoted also by several equivalents terms, such as "truthlikeness", "approximation to the truth", and "closeness to the truth") was introduced by Popper (1963) who claimed that the main cognitive goal of science is the acceptance of highly verisimilar theories. Afterwards, the problems related to the definition of adequate notions of verisimilitude and their application within scientific methodology have been extensively explored by Niiniluoto (1987), Kuipers (1987, 2000), Oddie (1986) and many other authors.

D. Dieks et al. (eds.), *Probabilities, Laws, and Structures*, The Philosophy of Science in a European Perspective 3, DOI 10.1007/978-94-007-3030-4_4,

assess the similarity between two items "we extract and compile from our database a limited list of relevant features on the basis of which we perform the required task" (Tverski 1977, p. 330).

Suppose that the items of interest $a, b, \ldots$ are described by a set $\mathbf{F} \equiv \{F_1, \ldots, F_n\}$ of *binary features*, where any F_i of $\mathbf{F}$ may be present, or absent, in a given item. The set of binary features of an item a is denoted by "A". Given two items a and b, the set $A \cap B$ of the common features of a and b, the set $A\backslash B$ of the distinctive features of a w.r.t. b, and the set $B\backslash A$ of the distinctive features of b w.r.t. a, will be referred to as the *commonality* between a and b, the *excess* of a w.r.t. b, and the *excess* of b w.r.t. a, respectively.

A *salience function* f for $\mathbf{F}$ is a function expressing the relative importance (for our similarity assessments) of each feature of $\mathbf{F}$, where f satisfies the following condition:

(1) For any member F_i and any subset G of $\mathbf{F}$,
(i) $f(F_i) > 0$, and (ii) $f(G) = \sum_{F_i \in G} f(F_i)$.

The similarity $s(a, b)$ between a and b may be defined as a function of the saliences of the commonality between a and b, of the excess of a w.r.t. b, and of the excess of b w.r.t. a – i.e., as a function of $f(A \cap B)$, $f(A\backslash B)$, and $f(B\backslash A)$:

(2) $s(a, b) \equiv f(A \cap B) - \alpha f(A\backslash B) - \beta f(B\backslash A)$ where $\alpha, \beta \geq 0$.[4]

From (2) it is quite easy to see that $s(a, b)$ is a *feature contrast* (*FC*-) measure of similarity, since it may be construed as a measure of the contrast between the common and distinctive features of a and b.

In some cases the items of interest $a, b, \ldots$ are described on the basis of *quantitative features* which may assume, in principle, any real value. However, we will restrict our attention to nonnegative-valued quantitative features. Given a set $\mathbf{F} \equiv \{F_1, \ldots, F_n\}$ of quantitative features and an item a, the value of F_i for a will be denoted by "A_i". A *salience function* f for $\mathbf{F}$ will satisfy the following condition:

(3) For any value Z_i of any feature F_i of $\mathbf{F}$ and any sequence $Z_1, \ldots, Z_n$ of values of $F_1, \ldots, F_n$, (i) $f(Z_i) = \varphi_i Z_i$ where $\varphi_i > 0$; (ii) $f(Z_1, \ldots, Z_n) = \sum_i f(Z_i)$.

The parameters $\varphi_1, \ldots, \varphi_n$ express the relative importance (for our similarity assessments) of the corresponding features $F_1, \ldots, F_n$. Suppose, for instance, that the individuals $a, b, \ldots$ are described by the set $\mathbf{F} \equiv \{F_1, F_2, F_3\}$, where $F_1 \equiv$ age, $F_2 \equiv$ height, and $F_3 \equiv$ weight. In this case, my salience function for $\mathbf{F}$ might be characterized

4 The values of α and β express the relative weight of the left-to-right direction (from a to b) and the right-to-left direction (from b to a) in the assessment of the similarity between a and b. It should be noted that $s(a, b)$ is symmetrical only in the special case where $\alpha = \beta$.

by the parameters $\varphi_1 = 9$, $\varphi_2 = 3$, and $\varphi_3 = 1$, which reveal that, for my assessments of the similarity between individuals, age is three times more important than height and height is three times more important than weight.

In the case of binary features, the commonality between two items *a* and *b* has been identified with the set $A \cap B$ of their common features. Strictly alike intuitions suggest that, in the case where the items of interest are described by a set of quantitative features, the commonality between *a* and *b* should be identified with the sequence

$$(4)\ \ \min(A_1, B_1), \ldots, \min(A_n, B_n),$$

where $\min(A_i, B_i)$ represents the degree at which F_i is *shared* by *a* and *b*.

In the case of binary features, the excess of *a* w.r.t. *b* has been identified with the set $A \backslash B$ of the distinctive features of *a* w.r.t. *b*. Strictly alike intuitions suggest that, in the case of quantitative features, the excess of *a* w.r.t. *b* should be identified with the sequence

$$(5)\ \ \max(A_1 - B_1, 0), \ldots, \max(A_n - B_n, 0),$$

where $\max(A_i - B_i, 0)$ represents the degree at which A_i exceeds B_i.[5]

Of course, the same intuitions underlying (5) suggest that the excess of *b* w.r.t. *a* should be identified with the sequence

$$(6)\ \ \max(B_1 - A_1, 0), \ldots, \max(B_n - A_n, 0).$$

The similarity $s(a, b)$ between two items *a* and *b*, described by a set of binary features, has been defined as a particular function of the saliences of the commonality between *a* and *b,* of the excess of *a* w.r.t. *b*, and of the excess of *b* w.r.t. *a* (see (2)). In the case where *a* and *b* are described by a set of quantitative features, their similarity $s(a, b)$ can be defined, *mutatis mutandis*, in the same way. Indeed, by replacing the three saliences occurring in (2) with the corresponding saliences for quantitative features (as defined in agreement with (3)–(6)) one obtains the following FC-measures of similarity for quantitative features:

$$(7)\ \ s(a, b) = \sum_{i=1}^{i=n} \varphi_i \left(\min(A_i, B_i) - \alpha \max(A_i - B_i, 0) - \beta \max(B_i - A_i, 0)\right)$$

where $\alpha, \beta \geq 0$.[6]

5 In fact, if A_i exceeds B_i, then $\max(A_i - B_i, 0)$ has a positive value given by $A_i - B_i$ while, if A_i does not exceed B_i, then $\max(A_i - B_i, 0)$ is zero.

6 The FC-measures of similarity for quantitative features are essentially identical to the fuzzy feature contrast measures, suggested by Santini and Jain (1999). However, while Santini et al. work out their measures within the rather complicated conceptual framework of fuzzy set theory, our measures are introduced as an intuitively simple extension of Tverski's FC-measures for binary features.

Suppose that one is interested in the assessment of the similarity between a and b w.r.t. a quantitative feature F_i – more precisely, w.r.t. the singleton $\{F_i\}$. It follows from (7) that such similarity, which will be referred to as $s_i(a, b)$, is given by:

(8) $s_i(a, b) = \varphi_i(\min(A_i, B_i) - \alpha \max(A_i - B_i, 0) - \beta \max(B_i - A_i, 0))$.[7]

Note that $s_i(a, b)$ is positive, zero or negative, depending on the values of A_i, B_i, α, and β. In particular, it follows from (8) that:

(9) (i) If $A_i = B_i > 0$, then $s_i(a, b) > 0$, for any value of α and β.
(ii) If $A_i > B_i$, then $s_i(a, b) >/=/< 0$ if and only if $A_i/B_i </=/> (1+\alpha)/\alpha$.

The intuitive content of (9)(ii) can be understood focussing on the clause "if $A_i > B_i$, then $s_i(a, b) > 0$ if and only if $A_i/B_i < (1+\alpha)/\alpha$". According to this clause, if A_i exceeds B_i, then $s_i(a, b)$ is positive if and only if the excess of A_i over B_i is not too big, in the sense that the measure A_i/B_i of such excess is lower than $(1+\alpha)/\alpha$.

One can easily check that the "global" similarity $s(a, b)$ w.r.t. a set $\mathbf{F} \equiv \{F_1, \ldots, F_n\}$ of quantitative features can be decomposed in n "local" similarities $s_i(a, b)$:

(10) $s(a, b) = \sum_{i=1}^{i=n} S_i(a, b)$

where $s_i(a, b)$ is the similarity between a and b w.r.t. a feature F_i of $\mathbf{F}$.

So far we have been dealing with the case where all the F_i-values of the items a and b are *determined*. However, for our purposes, we should consider also the case where some F_i-values are *undetermined* for at least one of the items a and b. Saying that a (b) is F_i-undetermined amounts to saying that the F_i-value of a (b) is absent, undetectable, or unknown. One may ask which value should be attributed to $s_i(a, b)$ in the case where at least one of the items a and b is F_i-undetermined. We suggest that a general answer to this question is given by the following principle:

(11) If at least one of the items a and b is F_i-undetermined then $s_i(a, b) = 0$.

---

7 As pointed out by an anonymous referee, according to definition (8), if A_i and B_i are equal, then the similarity between a and b is proportional to A_i and, in particular, this similarity is zero if $A_i = B_i = 0$. Moreover, the definition is also sensitive to the choice of the scale measurement of quantity A_i. This unpleasant aspect of definition (8) can be removed by applying the well known fuzzification procedures worked out within fuzzy set theory. Indeed such procedures allow to transform a quantitative variable (for example: height) into a fuzzy linguistic term (for example: tall) defined by a membership function whose value is included in the interval [0, 1]. For more details, see Cevolani et al. (2012).

The intuitive motivation for (11) can be expressed as follows. Suppose that one is assessing the similarity between a and b w.r.t. F_i. Then we may say that a is *F_i-similar* to b in the case where $s_i(a, b) > 0$ and that a is *F_i-dissimilar* from b in the case where $s_i(a, b) < 0$. On the other hand, in the case where $s_i(a, b) = 0$, we will say that a is neither F_i-similar to b nor F_i-dissimilar from b. If at least one of the items a and b is F_i-undetermined, then it seems quite plausible to say a is neither F_i-similar to b nor F_i-dissimilar from b, i.e., that $s_i(a, b) = 0$.

If the principle (11) is adopted, then (10) can be considered as a *general* definition of the similarity $s(a, b)$, i.e., as a definition which is applicable also in the case where some F_i-values are undetermined either for a or for b.

4.2 FC-Measures of Verisimilitude for Universal and Statistical Hypotheses

Suppose that the members of a given universe **U** are classified w.r.t. two characters, or families of predicates, $\mathbf{X} \equiv \{X_1, \ldots, X_c\}$ and $\mathbf{Y} \equiv \{Y_1, \ldots, Y_r\}$, where the predicates of each family are mutually exclusive and jointly exhaustive. Then we will say that **U** is *cross classified* w.r.t. **X** and **Y**. The rc members of $\mathbf{Q} \equiv \mathbf{Y} \times \mathbf{X} \equiv \{Y_1X_1, Y_1X_2, \ldots, Y_rX_c\}$ are often called *Q-predicates*. A Q-predicate Y_iX_j will be denoted with "Q_{ij}", so that $\mathbf{Q} = \{Q_{11}, \ldots, Q_{rc}\}$. The subscripts "$c$" in "$X_c$" and "$r$" in "$Y_r$" stay for "column" and "row", w.r.t. the $r \times c$ table which is commonly used to represent the cross classification of **U** w.r.t. **X** and **Y**. For the sake of illustration, let us consider the case where **U** is cross classified w.r.t. the families $\mathbf{X} \equiv \{X_1, X_2, X_3, X_4\}$ and $\mathbf{Y} \equiv \{Y_1, Y_2, Y_3\}$. This case can be represented by the table in Fig. 4.1, where any cell corresponds to one of the Q-predicates of $\{Q_{11}, \ldots, Q_{34}\}$.

	X_1	X_2	X_3	X_4
Y_1	Q_{11}	Q_{12}	Q_{13}	Q_{14}
Y_2	Q_{21}	Q_{22}	Q_{23}	Q_{24}
Y_3	Q_{31}	Q_{32}	Q_{33}	Q_{34}

Q $\{Q_{11}, \ldots, Q_{34}\}$.

Fig. 4.1 $\mathbf{Q} \equiv \{Q_{11}, \ldots, Q_{34}\}$.

Suppose that the universe **U** under investigation is described by a predicate language **L** whose vocabulary includes the families $\mathbf{X} \equiv \{X_1, \ldots, X_c\}$, $\mathbf{Y} \equiv \{Y_1,$

…, Y_r}, and $\mathbf{Q} = \{Q_{11}, \ldots, Q_{rc}\}$. In many kinds of inquiry, the structure of **U** is described in terms of certain features of the Q-predicates of **L**. For instance, with reference to the example above (see Fig. 4.1), one might ask whether the universal hypothesis $h \equiv$ "All Y_1-individuals are X_2" is true. This amounts to asking whether the Q-predicates Q_{11}, Q_{13}, and Q_{14} are empty, i.e., not instantiated, in **U**. More generally, one might ask which Q-predicates are empty in **U**. An answer to questions of this sort can be given by stating appropriate *universal* (U-) *hypotheses*. A U-hypothesis is a conjunction of k, with $0 \leq k \leq rc$, *basic universal* (*BU-*) *hypotheses*, where a BU-hypothesis b_{ij} says that Q_{ij} is empty.[8] Coming back to the above example, the U-hypothesis $h \equiv$ "All Y_1-individuals are X_2" can be restated as follows: $h \equiv b_{11}$ & b_{13} & b_{14}.

An appropriate measure $s(h, h')$ of the similarity between two U-hypotheses h and h' can be defined by applying the similarity measures for binary features introduced in Sect. 4.1. Indeed, a U-hypothesis h of **L** can be characterized in terms of a set $\mathbf{F} \equiv \{F_{11}, \ldots, F_{rc}\}$ of binary features, where F_{ij} is a feature of h in the case where h implies b_{ij}. Now $s(h, h')$ is defined, in agreement with definition (2), as the FC-similarity between h and h' w.r.t. **F**:

$$(12) \quad s(h, h') \equiv f(H \cap H') - \alpha f(H \backslash H') - \beta f(H' \backslash H),$$

where H and H' are the sets of features of h and h'.

The conjunction h_* of all and only the true BU-hypotheses of **L** is the strongest true U-hypothesis of **L**. Hence we might say that h_* is "the truth" about **U** (in **L**). The opinions of the scientists investigating **U** may be expressed by the strongest U-hypothesis h of **L** that they accept. We might say that the main cognitive goal pursued by scientists, i.e., the identification of the actual world, is fully achieved in the case where $h = h_*$. More generally, the degree at which such goal is reached is measured by the *verisimilitude* $Vs(h)$ of h, which may be defined as the similarity $s(h, h_*)$ between h and h_*.

In many empirical inquiries researchers ask what are the (*relative*) *frequencies* of certain Q-predicates among the members of **U**. An answer to questions of this sort can be given within a suitable statistical language $\mathbf{L}^{\mathbf{S}}$, by stating appropriate *statistical* (S-) *hypotheses*. A S-hypothesis is a conjunction of k, with $0 \leq k \leq rc$, *basic statistical* (*BS-*) *hypotheses*, where a BS-hypothesis b_{ij} specifies the value of the frequency $\mathbf{P}_{ij}$ of Q_{ij} in **U**.[9] We will say that the S-hypothesis h is *complete* in the case where $k = rc$, i.e., in the case where h specifies a possible value $P \equiv (P_{11}, \ldots, P_{rc})$ of the frequency vector $\mathbf{P} \equiv (\mathbf{P}_{11}, \ldots, \mathbf{P}_{rc})$.

Given a S-hypothesis h, let P_{ij} be the value of $\mathbf{P}_{ij}$ specified by (the conjunct b_{ij} of) h. Then an appropriate measure $s(h, h')$ of the similarity between two S-hypotheses h and h' can be defined by applying the similarity measures for quantitative

8 Note that a BU-hypothesis is a special kind of U-hypothesis, with $k = 1$.
9 Note that a BS-hypothesis is a special kind of S-hypothesis, with $k = 1$.

features introduced in Sect. 4.1. Indeed, a S-hypothesis h of $\mathbf{L}^S$ can be characterized in terms of a set $\mathbf{F} \equiv \{F_{11}, \ldots, F_{rc}\}$ of quantitative features such that: (1) if h implies b_{ij}, then the F_{ij}-value of h is P_{ij}, and (2) if h does not imply any b_{ij}, then h is F_{ij}-undetermined. Now $s(h, h')$ may be defined, in agreement with definitions and principles (7)–(11), as the FC-similarity between h and h' w.r.t. $\mathbf{F}$.[10]

It is easily seen that there is a unique true complete S-hypothesis, which will be referred to as h_*. We may say that h_* is "the truth" about $\mathbf{U}$ (in $\mathbf{L}^S$). Along the same lines followed in the case of U-hypotheses, the verisimilitude $Vs(h)$ of a S-hypothesis h is defined as the similarity $s(h, h_*)$ between h and h_*.[11]

4.3 FC-Measures of Verisimilitude for T-Hypotheses

As pointed out in the introduction, T-hypotheses play an important role in social sciences. For instance, Hildebrand et al. (1977, p. 28) quote the following three examples of T-hypotheses drawn from sociological literature:

Loss in competition *tends to* arouse anger [Homans 1961, p. 123].

The introduction of universal suffrage led *almost* anywhere (the United States excepted) to the development of Socialist parties [Duverger, 1954, p. 66].

[A] "high" level of education [...] *comes close to* being a [necessary condition for democracy] [Lipset, 1960, p. 57].

According to Hildebrand et al. the italicized qualifications "tends to", "almost", and "comes close to", occurring in the above quotations, reveal the fact that social scientists believe that the "universal counterparts" of the above T-hypotheses are false. More generally, we suppose that social scientists typically believe that *any* universal (non trivial) sociological hypothesis is false, although the "tendency counterpart" of a false universal hypothesis may be (approximately) true. Therefore, we think that the clarification of the intuitive idea that T-hypotheses may be more or less close to the truth is a very important task. Our proposal for dealing with this task is shortly illustrated below.

Let us come back to the example illustrated above, where the universe $\mathbf{U}$ under investigation is cross classified w.r.t. the families $\mathbf{X} \equiv \{X_1, X_2, X_3, X_4\}$ and $\mathbf{Y} \equiv \{Y_1, Y_2, Y_3\}$. In this case researchers may consider several T-hypotheses about $\mathbf{U}$ such as:

(13) $t \equiv Y_1$-individuals tend to be X_2.

10 Earlier accounts of the verisimilitude of statistical distributions (structure descriptions) include Niiniluoto (1987, pp. 302–303 and 321–322).

11 For a comparison between our measures of the verisimilitude of statistical hypotheses and the standard statistical measures of fit, such as the chi-square, see Cevolani et al. (2012).

Recalling that the Q-predicates $Y_1\&X_1$, $Y_1\&X_3$, and $Y_1\&X_4$ may be denoted with "Q_{11}", Q_{13}", and "Q_{14}", respectively, we suggest that t may be construed as follows:

(14) $t \equiv Q_{11}$, Q_{13}, and Q_{14} are rarely instantiated.

More generally, we suggest that a T-hypothesis is a conjunction of k, with $0 \leq k \leq rc$, *basic tendency* (*BT-*) *hypotheses*, where a BT-hypothesis concerning Q_{ij} says that Q_{ij} is rarely instantiated in **U**. Below we will show that T-hypotheses can be precisely stated within the statistical language $\mathbf{L}^S$ illustrated in Sect. 4.2 and that, on the basis of such reformulation, appropriate verisimilitude measures for T-hypotheses can be defined.

First of all, let us define the (*degree of*) *rarity* of a Q-predicate Q_{ij}, where such rarity is denoted with "R_{ij}". The basic intuition underlying our definition is that the rarity of Q_{ij} ($\equiv Y_iX_j$) should depend on the relation between the frequency P_{ij} of Q_{ij} and the frequencies $P(Y_i)$ and $P(X_j)$ of Y_i and X_j.[12] More specifically, we say that Q_{ij} *is not* rare in the case where Y_i and X_j are either probabilistically independent (i.e., $P_{ij} = P(Y_i)P(X_j)$) or positively relevant to each other (i.e., $P_{ij} > P(Y_i)P(X_j)$), while Q_{ij} *is* rare in the case where Y_i and X_j are negatively relevant to each other (i.e., $P_{ij} < P(Y_i)P(X_j)$). Moreover, we assume that, in the case where Q_{ij} is not rare, R_{ij} is put equal to zero while, in the case where Q_{ij} is rare, R_{ij} is put equal to the difference $P(Y_i)P(X_j) - P_{ij}$, which represents the mutual negative relevance between Y_i and X_j.[13]

The above illustrated intuitions lead to the following definition of R_{ij}:

(15) (i) If $P_{ij} \geq P(Y_i)P(X_j)$, then $R_{ij} \equiv 0$;
(ii) If $P_{ij} < P(Y_i)P(X_j)$, then $R_{ij} \equiv P(Y_i)P(X_j) - P_{ij}$.

It follows from (15) that, given the frequencies $P(Y_i)$ and $P(X_j)$, the maximal value of R_{ij} is $P(Y_i)P(X_j)$, which is obtained in the case where $P_{ij} = 0$. Given a frequency vector $P \equiv (P_{11}, \ldots, P_{rc})$, the corresponding *rarity vector* $R \equiv (R_{11}, \ldots, R_{rc})$ can be determined by applying definition (15).

Now we can show how BT- and T-hypotheses can be stated within $\mathbf{L}^S$. Recalling that a BT-hypothesis concerning Q_{ij} tells that Q_{ij} is rarely instantiated in **U**, we will assume that it can be expressed by the following S-hypothesis b_{ij} of $\mathbf{L}^S$:

---

12 Recall that, for any possible value $P \equiv (P_{11}, \ldots, P_{rc})$ of the frequency vector **P** the following equalities hold: $P(Y_i) = \sum_{i=1}^{i=r} \mathrm{P}_{ij}$ and $P(X_j) = \sum_{j=1}^{j=c} \mathrm{P}_{ij}$.

13 A strictly alike measure of mutual positive relevance has been suggested by Carnap (1950/1962, Ch. VI).

(16) $b_{ij} \equiv (P_{ij} = 0)$.[14]

In words, b_{ij} says that the frequency of Q_{ij} in the investigated universe **U** is zero, i.e., that Q_{ij} is maximally rare in **U**. Now a T-hypothesis t can be expressed within $\mathbf{L}^S$ as a conjunction of k, with $0 \leq k \leq rc$, BT-hypotheses of the kind shown in (16).

The similarity $s(t, h)$ between a tendency hypothesis t and a complete S-hypothesis h can be defined w.r.t. a set $\mathbf{F} \equiv \{F_{11}, \ldots, F_{rc}\}$ of quantitative features representing the rarities of the Q-predicates $Q_{11}, \ldots, Q_{rc}$ in the investigated universe. More precisely, given the frequency and rarity vectors specified by h – to be referred to as $P \equiv (P_{11}, \ldots, P_{rc})$ and $R \equiv (R_{11}, \ldots, R_{rc})$ –, the F_{ij}-values of h and t, i.e., H_{ij} and T_{ij}, are defined as follows:

(17) $H_{ij} \equiv R_{ij}$.

(18) (i) If b_{ij} is a conjunct of t, then $T_{ij} \equiv P(Y_i)P(X_j)$.

(ii) If b_{ij} is not a conjunct of t, then T_{ij} is undetermined.

One can see that the F_{ij}-values defined above are given by the rarities that h and t attribute to the Q-predicates $Q_{11}, \ldots, Q_{rc}$. In particular, the intuitive content of (18) (i) can be expressed as follows. Since b_{ij} says that the frequency P_{ij} of Q_{ij} in the investigated universe **U** is zero (see (16)), it implies, due to (15)(ii), that $R_{ij} = P(Y_i) P(X_j)$ – where $P(Y_i)P(X_j)$ is the maximal value of R_{ij} for the frequencies $P(Y_i)$ and $P(X_j)$.

Now $s(t, h)$ can be defined, in agreement with definitions and principles (7)–(11), as the FC-similarity $s(t, h)$ between t and h w.r.t. the set $\mathbf{F} \equiv \{F_{11}, \ldots, F_{rc}\}$ of quantitative features described above:

(19) $s(t, h) = \sum_{ij=11}^{ij=rc} s_{ij}(t, h),$

where $s_{ij}(t, h)$ is the similarity between t and h w.r.t. the feature F_{ij} of **F**. The value of $s_{ij}(t, h)$ can be determined, in agreement with (8) and (11), as follows:

(20) (i) If b_{ij} is a conjunct of t, then $s_{ij}(t, h) = \varphi_{ij}(\min(T_{ij}, H_{ij}) - \alpha \max(T_{ij} - H_{ij}, 0) - \beta \max(H_{ij} - T_{ij}, 0))$.

(ii) If b_{ij} is not a conjunct of t, then $s_{ij}(t, h) = 0$.

It is very important to recall that, due to (18) (i), $T_{ij} = P(Y_i)P(X_j)$, where $P(Y_i)P(X_j)$ is the maximal value of R_{ij} for the frequencies $P(Y_i)$ and $P(X_j)$ (see (15) (ii)). Hence, due to (17), $T_{ij} \geq R_{ij} \equiv H_{ij}$. This implies that $\max(H_{ij} - T_{ij}, 0) = 0$. Therefore (20) (i) can be restated as follows:

14 For a discussion of (16) and of some alternative assumptions, see Cevolani et al. (2012).

(21) If b_{ij} is a conjunct of t, then $s_{ij}(t, h) = \varphi_{ij}$ (min(T_{ij}, H_{ij}) $- \alpha$ max($T_{ij} - H_{ij}$, 0).

One can see from (20) and (21) that $s(t, h)$ depends *only* on the parameter α of the FC-similarity measure s applied in definition (19).

Now we can determine the value $s_{ij}(t, h)$ in the case where b_{ij} is a conjunct of t. Indeed, it follows from (21), together with (17) and (18), that:

(22) If b_{ij} is a conjunct of t, then $s_{ij}(t, h)$ is a function of P_{ij}, $P(Y_i)$, $P(X_j)$, and α. More precisely:

(i) If $P_{ij} \geq P(Y_i)P(X_j)$, then $s_{ij}(t, h) = -\varphi_{ij}\, \alpha\, P(Y_i)P(X_j)$;
(ii) If $P_{ij} < P(Y_i)P(X_j)$, then $s_{ij}(t, h) = \varphi_{ij}\, (P(Y_i)P(X_j) - (1+\alpha)\, P_{ij})$.

It follows from (22) that, if b_{ij} is a conjunct of t and $P_{ij} \geq P(Y_i)P(X_j)$, then $s_{ij}(t, h) < 0$, for any value of α. Moreover, one can prove that:

(23) If b_{ij} is a conjunct of t and $P_{ij} < P(Y_i)P(X_j)$, then $s_{ij}(t, h)$ >/=/< 0 if and only if $P_{ij}/(P(Y_i)P(X_j))$ </=/> $1/(1+\alpha)$.

The intuitive content of (23) can be understood by considering the claim that, if $P_{ij} < P(Y_i)P(X_j)$, then $s_{ij}(t, h) > 0$ if and only if $P_{ij}/(P(Y_i)P(X_j)) < 1/(1 + \alpha)$. According to this claim $s_{ij}(t, h)$ is positive if and only P_{ij} is *sufficiently smaller* than $P(Y_i)P(X_j)$, i.e., if and only if Q_{ij} is *rare enough*.

When $P_{ij}/(P(Y_i)P(X_j))$ is lower than the threshold $1/(1 + \alpha)$ – and, thereby, $s_{ij}(t, h)$ is positive –, we will say that Q_{ij} is *α-rare*. Since $\alpha \geq 0$, the threshold $1/(1+\alpha)$ is included in the interval (0,1). For instance, if $\alpha = 1$, then $1/(1+\alpha) = 0.5$; hence, Q_{ij} will be *α-rare* if and only if $P_{ij}/(P(Y_i)P(X_j)) < 0.5$. On the other hand, if $\alpha = 9$, then $1/(1+\alpha) = 0.1$; hence, Q_{ij} will be *α-rare* if and only if $P_{ij}/(P(Y_i)P(X_j)) < 0.1$. Such examples suggest that a similarity measure $s(t, h)$ characterized by a high value of α imposes, as it were, severe requirements for the admission to the club of the α-rare Q-predicates.

Suppose that a particular FC-measure of similarity s_α, characterized by a specific value of the parameter α, is used. Then one may ask which T-hypothesis t is maximally similar to h, i.e., which T-hypothesis t has the maximal value of $s_\alpha(t, h)$. An answer to this question immediately follows from definition (19) and theorem (23):

(24) Let the T-hypothesis t_α include all and only the conjuncts b_{ij} such that Q_{ij} is α-rare. Then $s_\alpha(t, h)$ is maximal.

Theorem (24) reveals the intuitive meaning of the choice of a particular measure s_α. For instance, the choice of a high value α, and thereby of a low value of

$1/(1 + \alpha)$, reveals that our cognitive goal consists in the identification of the set of the *very rare* Q-predicates, i.e., in the identification, as it were, of the very strong tendencies working in the universe described by a given complete S-hypothesis.

Finally, the verisimilitude $Vs(t)$ of a T-hypothesis t is defined – along the same lines followed in the case of U- and S-hypotheses – as the similarity $s(t, h_*)$ between t and h_*, where h_* is "the truth" about **U** (in $\mathbf{L}^S$).

4.4 Conclusions

Several issues touched in this paper deserve further research. Firstly, the conceptual foundations of our measures of verisimilitude for U-, S- and T-hypotheses, provided in Sect. 4.1, need more systematic investigation. Secondly, our paper has been almost entirely devoted to the *logical* problem of verisimilitude, i.e., to the definition of the degree of verisimilitude of U-, S- and T-hypotheses in the case where it is assumed that 'the truth' is known (Sects. 4.2 and 4.3). In particular, we have dealt with the logical problem of the verisimilitude of T-hypotheses (Sect. 4.3). Our measures for the verisimilitude of T-hypotheses might provide a sound basis for the analysis of the *methodological* problem of verisimilitude, i.e., for the formulation of appropriate procedures for the estimation – on the basis of the available evidence – of the degree of verisimilitude of T-hypotheses in the actual world, i.e., in the case where "the truth" is (typically) unknown. Thirdly, the conceptual relations between the present approach and other approaches to the verisimilitude of T-hypotheses, starting from Hildebrand et al. (1977) and Festa (2007a, b), should be explored. Finally, the methodological role of T-hypotheses should be carefully investigated, by considering examples drawn not only from social sciences, but also from other empirical sciences, such as epidemiology and other biomedical sciences.

Proofs of theorems

Proof of (22)
It follows from equalities $T_{ij} = P(Y_i)P(X_j)$ and $H_{ij} \equiv R_{ij}$ (see (17) and (18) (i)) that $\max(T_{ij} - H_{ij}, 0) = \max(P(Y_i)P(X_j) - R_{ij}, 0)$. Therefore, equality $s_{ij}(t, h) = \varphi_{ij}(\min(T_{ij}, H_{ij}) - \alpha \max(T_{ij} - H_{ij}, 0)$ in (21) implies that $s_{ij}(t, h) = \varphi_{ij}(\min(P(Y_i)P(X_j), R_{ij}) - \alpha \max(P(Y_i)P(X_j) - R_{ij}, 0))$. This equality allows to prove the clauses (i) and (ii). *Clause (i)*. Suppose that $P_{ij} \geq P(Y_i)P(X_j)$. In this case, due to (15) (i) and (17), $H_{ij} \equiv R_{ij} \equiv 0$. This implies that $\min(P(Y_i)P(X_j), R_{ij}) = 0$ and $\max(P(Y_i)P(X_j) - R_{ij}, 0) = P(Y_i)P(X_j)$. Hence, $s_{ij}(t, h) = -\varphi_{ij}\alpha\, P(Y_i)P(X_j)$. *Clause (ii)*. Suppose that $P_{ij} < P(Y_i)P(X_j)$. In this case, due to (15) (ii) and (17), $H_{ij} \equiv R_{ij} \equiv P(Y_i)P(X_j) - P_{ij}$. This implies that $\min(P(Y_i)P(X_j), R_{ij}) = P(Y_i)P(X_j) - P_{ij}$ and

$\max(P(Y_i)P(X_j) - R_{ij}, 0) = P_{ij}$. Hence, $s_{ij}(t, h) = \varphi_{ij}(P(Y_i)P(X_j) - P_{ij} - \alpha P_{ij}) = \varphi_{ij}(P(Y_i)P(X_j) - (1+\alpha)P_{ij})$.

Proof of (23)

First of all, let us prove the claim that $s_{ij}(t, h) > 0$ if and only if $P_{ij}/(P(Y_i)P(X_j)) < 1/(1+\alpha)$. Recalling that $\varphi_i > 0$ (see (3)), it follows from (22) (ii) that inequality $s_{ij}(t, h) > 0$ amounts to inequality $P(Y_i)P(X_j) - (1+\alpha)P_{ij} > 0$ and, thereby, to $P_{ij}/(P(Y_i)P(X_j)) < 1/(1+\alpha)$. In a strictly alike way, one can prove that $s_{ij}(t, h) =/< 0$ if and only if $P_{ij}/(P(Y_i)P(X_j)) =/> 1/(1+\alpha)$.

References

Rudolf Carnap (1950/1962), *The Logical Foundations of Probability*. Chicago: The University of Chicago Press.

Gustavo Cevolani, Vincenzo Crupi, and Roberto Festa (2012), "Features of verisimilitude", in preparation.

Maurice Duverger (1954), *Political Parties*. New York: John Wiley and Sons.

Roberto Festa (2007a), "Verisimilitude, Cross Classification, and Prediction Logic. Approaching the Statistical Truth by Falsified Qualitative Theories", in: *Mind and Society*, 6, pp. 37–62.

Roberto Festa (2007b), "Verisimilitude, Qualitative Theories, and Statistical Inferences", in: Sami Pihlström, Panu Raatikainen and Matti Sintonen (Eds.), *Approaching the Truth. Essays in Honour of Ilkka Niiniluoto*. London: College Publications, pp. 143–178.

David K. Hildebrand, James D. Laing, and Howard Rosenthal (1977), *Prediction Analysis of Cross Classification*. New York: John Wiley and Sons.

George Homans (1961), *Social Behavior: Its Elementary Forms*. New York: Harcourt Brace Jovanovich.

Theo A. F. Kuipers (Ed.) (1987), *What is Closer-to-the-Truth?*. Amsterdam: Rodopi.

Theo A. F. Kuipers (2000). *From Instrumentalism to Constructive Realism*. Dordrecht: Kluwer.

Seymour M. Lipset (1960), *Political Man: The Social Bases of Politics*. Garden City: Doubleday.

Ilkka Niiniluoto (1987), *Truthlikeness*. Dordrecht: Reidel.

Graham Oddie (1986), *Likeness to Truth*. Dordrecht: Reidel.

Karl R. Popper (1963), *Conjectures and Refutations*. London: Routledge and Kegan Paul.

Simone Santini and Ramesh Jain (1999), "Similarity Measures", in: *IEEE Transactions on Pattern Analysis and Machine Intelligence*, 21, pp. 871–883.

Amos Tversky (1977), "Features of Similarity", *Psychological Review*, 84, pp. 327–352. Reprinted in Amos Tversky, *Preference, Belief, and Similarity*, Cambridge, Mass.: The MIT Press, 2004, pp. 7–45.

Department of Philosophy
University of Trieste
Androna Campo Marzio 10
34123, Trieste
Italy
festa@units.it

Chapter 5

Gerhard Schurz

Tweety, or Why Probabilism and even Bayesianism Need Objective and Evidential Probabilities

Abstract

According to probabilism, uncertain conditionals are to be reconstructed as assertions of high conditional probability. In everyday life one often encounters situations of 'exception'. In these situations two uncertain conditionals have contradicting consequents and both of their antecedents are instantiated or true, respectively. The often cited example of this sort is 'Tweety', who happens to be both a bird and a penguin. We believe that if Tweety is a bird then it probably can fly, and if it is a penguin then it probably cannot fly. If one reconstructs these examples by only one probability function, as is required by strong Bayesianism, they come out as probabilistically incoherent (with or without the existence of a specificity relation between the two antecedents). This result is counterintuitive. I argue that if one intends a coherent reconstruction, one has to distinguish between two probability functions, evidential probabilities which are subjective, and objective probabilities which are backed up by statistical probabilities. Drawing on Hawthorne (2005) I give further reasons why probabilism and even Bayesianism needs this distinction. In the end of the paper I present results of an experimental study on examples of 'exception' which confirm that humans operate with these two distinct probability concepts.

5.1 Tweety and Nixon: Conditional Reasoning with Exceptions Within the Framework of Probabilism

Non-monotonic reasoning is focused on conditional reasoning about situations which involve exceptions. Situations of exceptions are given when two conditionals have instantiated (but distinct) antecedents and opposite consequences:

Situations of exceptions:

$A \Rightarrow B$	As are normally Bs
$C \Rightarrow \neg B$	Cs are normally not-Bs
$A \wedge C$	This is A and C

D. Dieks et al. (eds.), *Probabilities, Laws, and Structures*, The Philosophy of Science in a European Perspective 3, DOI 10.1007/978-94-007-3030-4_5,

The given case ("this") must be an exception to at least one of the two conditionals (in the sense that its antecedent is verified but the consequence falsified by 'this' case). Of course the two conditionals must be understood as being uncertain, for otherwise the situation would not even be logically consistent. In what follows, A, B, … are schematic letters of a propositional or first order language (moreover, F, G, … stand for predicate variables, a, b,¼ for individual constants, x, y, … individual variables); $\Rightarrow$ symbolizes the non-strict (uncertain) conditional, while $\rightarrow$ symbolizes the material conditional which is strict in the sense that (a) $P(A\rightarrow B) = 1$ implies $P(B\,|\,A) = 1$ and (b) $Fx\rightarrow Gx$ is interpreted as $\forall x\,(Fx\rightarrow Gx)$.

The most important subcase of exceptions are exceptions involving a relation of specificity. Here one antecedent is (strictly) more specific than the other. The rule of specificity asserts that in such a case the conditional with the more specific antecedent 'fires' its consequent and the conditional with the less specific antecedent is blocked. The canonical example in the NMR (non-monotonic reasoning) community is the example of Tweety (according to Brewka 1991a, p. 2, Tweety is the most famous animal in AI circles):

Exceptions with specificity:

	Generic formulation:	*Singular formulation*:
$A \Rightarrow B$	$\text{Bird}(x) \Rightarrow \text{Canfly}(x)$	$\text{Bird}(\text{Tweety}) \Rightarrow \text{Canfly}(\text{Tweety})$
$C \Rightarrow \neg B$	$\text{Penguin}(x) \Rightarrow \neg\text{Canfly}(x)$	$\text{Penguin}(\text{Tweety}) \Rightarrow \neg\text{Canfly}(\text{Tweety})$
$C \rightarrow A$	$\text{Penguin}(x) \rightarrow \text{Bird}(x)$	$\text{Penguin}(\text{Tweety}) \rightarrow \text{Bird}(\text{Tweety})$
$A \wedge C$	$\text{Bird}(\text{Tweety}) \wedge \text{Penguin}(\text{Tweety})$	

Unambiguous conclusion according to the rule of specificity:

$\neg B$	$\neg\text{Canfly}(\text{Tweety})$

Almost all systems of NMR agree in the rule of specificity (Brewka 1991a; Gabbay et al. 1994). Moreover, this rule has also been demonstrated to fit with humans intuitive reasoning (Schurz 2005, 2007).

In the given case the specificity conditional $C\rightarrow A$ is strict. I call this case "strict specificity", in distinction to weak specificity in which the conditional $C\Rightarrow A$ is uncertain. The two possible formulations, generic versus singular, will become important later on. In most NMR-papers the chosen framework is a propositional language, and therefore, the singular formulation of 'Tweety' is chosen, although it is usually assumed that the singular conditionals are 'somehow' backed up by the generic conditionals as expressed by our generic formulation[1] (cf. Pearl 1988, p. 483; Brewka 1991a, p. 3; 1991b, pp. 194f; Goldszmidt and Pearl 1996, p. 73; Halpern 2003, p. 294; Pfeifer and Kleiter 2008).

1 Pearl (1988, p. 274) emphasizes that this back-up cannot be a universal quantification, but nevertheless he treats the two formulations as roughly equal.

Now let us turn to the probabilistic interpretation. The so-called doctrine of *probabilism* – which also includes what I call 'weak Bayesianism' – makes two assumptions:

Probabilism (including 'weak' or 'dualistic' Bayesianism):

(a) Epistemic states should be represented by rational degrees of belief (over a given space of propositions or statements) which obey the usual (Kolmogorovian) probability axioms (cf. e.g. Joyce 1998; Hájek 2008), and

(b) uncertain conditionals should be understood as assertions of high conditional probabilities[2] (cf. Adams 1997, 1998; Skyrms 1980; MacGee 1989; Bennett 2003; for psychological confirmations of this thesis cf. Evans et al. 2003 and of Oberauer and Wilhelm 2003).

Under the assumption of probabilism the Tweety example is transformed into the following reconstruction (see e.g. Pearl ibid.; Pfeifer and Kleiter ibid.; Halpern ibid.):

Exceptions with specificity, probabilistic reconstruction:

	Generic formulation:	*Singular formulation*:
$P(B\|A) \geq 1-\varepsilon$	With high probability, birds can fly.	With high probability, Tweety can fly.
$P(\neg B\|C) \geq 1-\varepsilon$	With high probability, penguins cannot fly.	With high probability, Tweety cannot fly, given it is a penguin.
$P(A\|C) = 1\ (\geq 1-\varepsilon)$	(Almost) all penguins are birds.	With (almost) certainty, Tweety is a bird, given it is a penguin.
$P(A \wedge C) = 1\ (\geq 1-\varepsilon)$	With (almost) certainty, Tweety is a bird and a penguin.	

Unambiguous conclusion: With high probability, Tweety cannot fly.

According to probabilism, not only the conditional beliefs but also the factual beliefs ($A \wedge C$) get probabilified. In particular, that we are certain of a factual belief means that we attach a probability of 1 to it. We admit, however, also the case where the factual knowledge is uncertain and has merely a high probability but less than 1.

Note that the inference rule of specificity is probabilistically valid because of the following

2 Note that this thesis is weaker than the so-called 'Adams-Stalnaker' thesis that identifies $P(A \Rightarrow B)$ with $P(B|A)$, which has shown to lead to paradoxical consequences by Lewis (1976).

Specificity theorem: $P(B|A \wedge C) \geq 1-(P(\neg B|C)/P(A|C))$.
Proof: $P(\neg B|C) = P(\neg B|C \wedge A) \cdot P(A|C) + P(\neg B|C \wedge \neg A) \cdot P(\neg A|C)$ is a well-known probability theorem. It entails $P(\neg B|C) \geq P(\neg B|C \wedge A) \cdot P(A|C)$. Hence $P(\neg B|C \wedge A) \leq P(\neg B|C)/P(A|C)$ (provided $P(A|C) > 0$). Substituting $1-P(B|A \wedge C)$ for $P(\neg B|C \wedge A)$ and simple transformation yields the claim. Q.E.D.

The specificity theorem implies for strict specificity (i.e., when $P(A|C)=1$) that $P(B|A \wedge C) \geq P(B|C) \geq 1-\varepsilon$, and for weak specificity that $P(B|A \wedge C) \geq 1-(\varepsilon/1-\varepsilon) = (1-2\varepsilon)/(1-\varepsilon)$. Thus, the probability of CanFly given Bird and Penguin is for strict specificity equal to, and for weak specificity almost as high as the probability of CanFly given Penguin. This result licenses the specificity rule if one adds the principle of total evidence, which goes back to Carnap (1950, p. 211) and Reichenbach (1949, §72; he called it 'narrowest reference class'). In a simplified formulation it says the following:

Principle of total evidence (simple formulation): The actual degree of belief in a singular statement Ba is P(Ba|E(a)) iff E(a) entails all evidence about a (or equivalently:[3] all evidence that is probabilistically relevant to Ba).

Thus, the probability that Tweety can fly given the premises should be identified with $P(\text{Canfly}(Tw) \,|\, \text{Bird}(Tw) \wedge \text{Penguin}(Tw))$.

Before we turn to our major problem we illustrate the second subcase of reasoning about exceptions, which is given when two instantiated conditionals with opposite consequences are *not ordered* by a relation of specificity. The most famous case in the area of NMR is the Nixon-example (Brewka 1991a, p. 14). We immediately present its probabilistic reconstruction – the only difference to the Tweety example is that the specificity conditional is missing:

Exceptions with conflict, probabilistic reconstruction:

	Generic formulation:	*Singular formulation*:
$P(B\|A) \geq 1-\varepsilon$	With high probability, quakers are pacifists.	With high probability, Nixon is a pacifist, given he is a quaker.
$P(\neg B\|C) \geq 1-\varepsilon$	With high probability, republicans are not pacifists.	With high probability, Nixon is not a pacifist, given he is a republican.
$P(A \wedge C) = 1\ (\geq 1-\varepsilon)$	With (almost) certainty, Nixon is a quaker and a republican.	

Conflict: Neither B nor ¬B is entailed with high probability.

3 The two formulations are equivalent because if E(a) is the total evidence about a and E*(a) the evidence which is probabilistically relevant to Ba, then (by definition of 'probabilistic relevance') P(Ba|E(a)) = P(Ba|E*(a)). Hence, the actual degree of belief in Ba satisfies the condition in terms of E(a) iff it satisfies the condition in terms of E*(a).

In the result we have a situation of genuine conflict: each possible conclusion is defeated by the opposite one (also probabilistically); and so, no one of the two possible conclusions is preferred. While some authors argue for remaining skeptical in this case, i.e. one should not draw a conclusion (e.g. Pollock 1994), others have argued for "multiple extension" in the sense that one is allowed to draw one of both possible conclusions (Reiter 1980). But this is not our concern here – our real concern is the question of probabilistic coherence, to which we turn in the next section.

5.2 Tweety and Nixon Are Incoherent in the Bayesian Reconstruction

The presented situations of reasoning with exceptions are completely standard in the areas of NMR and probabilistic conditional reasoning. I don't know of any place where someone would have argued that situations of exceptions are probabilistically inconsistent, i.e. incoherent. And yet they are – at least, if one makes the following additional assumption of 'strong' *Bayesianism*, which strengthens probabilism (or 'weak' Bayesianism) as follows:

'Strong' or 'monistic' *Bayesianism* (personalism, subjective Bayesianism): Rational probabilistic reasoning is based on only *one* probability function, namely on one's actual degrees of belief (hypothetical degrees of belief have to be expressed by conditionalization of the actual belief function).

All situations involving conditionals with exceptions, be it with or without specificity, are probabilistically incoherent, if factual and conditional probabilities are of the same kind, i.e. belong to the same probability space, and the probabilities are high enough. This can be seen as follows:

Incoherence proof: By probability theory it holds that

$$P(B) \geq P(A \wedge B) = P(B|A) \cdot P(A) \geq P(B|A) \cdot P(A \wedge C), \text{ and}$$
$$P(\neg B) \geq P(\neg B \wedge C) = P(\neg B|C) \cdot P(C) \geq P(\neg B|C) \cdot P(A \wedge C).$$

Thus, the set of probability inequalities $\{P(B|A) \geq r, P(\neg B|C) \geq r, P(A \wedge C) \geq r\}$ becomes incoherent for

$$P(B) + P(\neg B) \geq r^2 + r^2 > 1.$$

This is the case iff $r^2 > 1/2$, i.e., iff $r > 1/\sqrt{2} \approx 0.71$. Q.E.D.

The information $P(A|C) \geq r$ is not needed to derive this conflict; it is already implicitly contained in $P(A \wedge C) = P(C) \cdot P(A|C) \geq r$.

The result is surprising. Situations involving exceptions are quite common in everyday life, and they do not at all seem inconsistent or incoherent. Although they come out as such given the assumption of strong Bayesianism. If this is right, then strong Bayesianism is wrong. To obtain a coherent reconstruction of reasoning with exceptions we need to distinguish between *two* probability functions. One probability function reflects the subject's actual degrees of belief and depends on the particular evidence which this subject has – I call this the *subjective-evidential probability function* and formalize it from now on as B_C for 'degree of belief' relative to a set of evidence-including background beliefs C about which the subject is certain. The other probability function intends to reflect objective-statistical regularities or propensities in the real world – I call it *objective-statistical* (or if you think "statistical" is too narrow, then call it: *objective-generic*) probability and denote it as P. The difference between the two probability functions is already reflected in the generic formulation of our examples, in which the conditional probabilities are expressed as generic probabilities (in terms of an individual variable instead of an individual constant). However, we have mentioned above that also in the singular formulation most NMR-authors take it for granted that the singular conditional probabilities are determined by generic probabilities, rather than by the actual probabilities of particular evidences (more on this in Sect. 5.3).

The strong Bayesian has a defense. She or he will deny that the singular and the general formulations are treated on par. In the singular case, she argues, in which we deal with only one probability function, the evidence changes our conditional probabilities. I've often heard Bayesians argue as follows: our subjective probability that Tweety can fly, given she is a bird, sinks to low values, if we get to know that Tweety is not an ordinary bird but an exceptional bird, namely a penguin. This is right insofar the following holds for one's actual degree of belief function B_C:

> $B_C(\text{CanFly(Tw)}|\text{Bird(Tw)}) = B_C(\text{CanFly(Tw)}|\text{Bird(Tw)} \wedge \text{Penguin(Tw)})$,
> provided the background beliefs in C contain "Penguin(Tw)"; moreover
> $B_C(\text{CanFly(Tw)}|\text{Bird(Tw)} \wedge \text{Penguin(Tw)}) = B_C(\text{CanFly(Tw)}|\text{Penguin(Tw)})$
> because of strict specificity.

Nevertheless, it seems unintuitive that the conditional probability that a certain individual can fly, given it is a bird, should be affected by additional evidence about this individual – at least if this conditional probability intends to reflect the strength of *nomological connection* between the two properties "Bird" and "Can-Fly" that are instantiated by Tweety. The situation becomes even more problematic in the Nixon-type examples without specificity. For here, the strong Bayesian is forced to change the values of the two opposing conditional probabilities in a way such that they agree, merely on coherency requirements, but without any real reason for doing so. She may either change the low probability conditional into a high probability conditional, or vice versa, the high probability conditional into a low one – or she may even change both of them. But by doing so, the Bayesian hides important information, namely that there are two opposing conditionals relevant to the consequent. That the strong Bayesian reconstruction *hides conflicting probabilistic information* is in my eyes a strong argument against monistic Bayesianism.

5.3 Objective Likelihoods and Bayesian Updating: Further Reasons Why Bayesianism Needs Objective and Subjective-Evidential Probabilities

There are further reasons why both probabilism and Bayesianism need subjective-evidential and objective probability functions. Some years ago Hawthorne (2005) has demonstrated with admirable clarity that the Bayesian approach to the confirmation of hypotheses presupposes these two probability functions, if confirmation should be based on objective or intersubjectively agreed values for likelihoods. We will see that Hawthorne's arguments are closely related to earlier arguments by de Finetti and Carnap.

In scientific contexts the likelihoods represent what hypotheses say about the evidence. Their objectivity is essential to the objectivity of a science – to a common understanding of what the theory says or implies about the world. Moreover the objectivity of likelihoods is necessary for every account of probabilistic confirmation that should satisfy the following two properties:

(a) convergence to intersubjectivity (objectivity) with increasing evidence,
and
(b) resistance against the problem of old evidence.

According to Bayesianism, the probability of hypotheses given evidences is calculated as follows, where we assume a given partition of hypotheses $\mathbf{H} = \{H_1, \ldots, H_m\}$:

(1) $B_C(H_i|E^n) = B_C(E^n|H_i) \cdot B_C(H_i) / B_C(E^n)$
where E^n is a disjunction of conjunctions (or sequences) of n experimental results $E_1 \wedge \ldots \wedge E_n$ (i.e. each E_i is an element of an associated partition $\mathbf{E}^i$ of outcomes of the ith experiment).

The prior probabilities of the hypotheses $B_C(H_i)$ are the well-known subjective factors in Bayesian statistics, the so-called 'priors' – they reflect personal prejudices, so to speak. Almost all Bayesians – even personalists (cf. Hawthorne 2005, p. 286, who quotes Edwards et al. 1963) – stress that with increasing evidence, i.e. with increasing n, the probability value $B_C(H_i|E^n)$ (for arbitrary i) becomes more and more independent from the subjectively chosen priors.[4] The reason for

4 Several versions of such convergence theorems exist in the literature (cf. Earman 1992, p. 58; Howson and Urbach 1993, ch. 14). An especially nice version is given by Hawthorne (2005, pp. 283f).

this is that the likelihoods, i.e. the probability values $B_C(E^n|H_i)$, are assumed to be intersubjective or objective – typically they are given by objective statistical probabilities. Also this assumption is shared by almost all Bayesians. But as Hawthorne points out, objective likelihoods are impossible if they are understood as an actual degree of belief which depends on one's actual evidence (and other actual beliefs). For example, if I have already observed that on a particular occasion a fair coin landed heads, then my actual probability that this coin landed on heads is not 1/2 – which would be the objective likelihood value – but 1. More specifically, the following holds for actual quasi-likelihood probabilities:

(2) B_C(coin lands heads at time t | P(coin lands heads at some time) = 0.5) =

(i)	= 1	if C includes the observation "coin lands heads at time t".
(ii)	= 0.5	if C doesn't include any observation about this coin at time t.
(iii)	= 8/9	if C includes the evidence "either the coin landed heads at time t or it landed tail in three tossings before t".

The actual probability in (i) is similar to the actual probability that this bird can fly, given it is a bird, when we know that this bird is a penguin. The resulting values of evidence-dependent likelihoods may become quite strange, if the evidence is more involved. This is shown in the result of line (iii) of (2) that is due to Hawthorne (2005, p. 291).[5]

Intersubjective confirmation values for hypotheses would hardly be possible if likelihoods were dependent on one's subjective evidence, which varies from scientist to scientist. Related to this problem is the well-known *problem of old evidence* (cf. Earman 1992, ch. 5), that concerns the probability value $B_C(E^n)$, about we have been silent so far. If B_C measures the actual degree of belief, and if we are in an epistemic state C in which we already know the evidence, i.e. if $B(E^n)$ becomes one, then we obtain the following trivialization result:

(3) $B_C(H_i|E^n) = B_C(H_i)$ if $B_C(E^n) = 1$
[because then: $B_C(E^n) = B_C(E^n|H_i)$,
and recall $B_C(H_i|E^n) = B_C(H_i) \cdot B_C(E^n|H_i)/B_C(E^n)$].

So in an epistemic situation C in which we already know the evidence, the amount of confirmation received by the hypothesis given the evidence as explicated by actual degrees of belief would become zero for the standard incremental

5 With H_t for "tossing heads at t" and "TTT_b" for "tossing three times tailes before t", the proof is this: $B(H_t|H_t \vee TTT_b) = B(H_t \wedge (H_t \vee TTT_b))/B(H_t \vee TTT_b) = B(H_t)/[B(H_t) + B(TTT_b) - B(H_t \wedge TTT_b)] = (1/2)/[(1/2)+(1/8)-(1/16)] = 8/9$.

confirmation measure, which explains confirmation in terms of the difference between $B_C(H_i|E^n)$ and $B_C(H_i)$.[6,7]

As Hawthorne argues, both problems can only be avoided if the probability of a hypotheses conditional on an evidence is explicated in a way that is *independent* from the possession of particular evidences. This probability should not be expressed in terms of subjective-evidential probabilities (i.e. actual degrees of belief), but in terms of what Hawthorne calls support functions or *support probabilities*,[8] for which I use here the expression S. Somewhat differently from Hawthorne, I reformulate the definition of support probabilities as follows:

(4) (i) $S(H_i \mid E^n) = P(E^n \mid H_i) \cdot B(H_i) / S(E^n)$, with
(ii) $S(E^n) = \Sigma_{1 \leq j \leq k} P(E^n \mid H_j) \cdot B(H_j)$ [or $= \int_r P(E^n \mid H_r) \cdot dB(H_r)$].[9]
Thereby, B(–) is a subjective *prior* probability function.

While Hawthorne writes univocally S for B as well as for P, my formulation (4) points out that support probabilities are a function of an objective probability function P and a subjective prior probability B over the hypotheses. The prior $B(H_i)$ is understand as a rational degree of belief in H_i prior to all evidences that are possible outcomes of the associated sequence of n experiments (as explained in (1)).

The probability $P(E^n \mid H_i)$ in (4)(i) is an objective probability, but not an ordinary objective-statistical probability because it is applied to singular events. But I assume that $P(E^n|H_i)$ is backed up by an ordinary objective-statistical probability function by way of the statistical version of the principal principle, which goes back to de Finetti and is explicated by me as follows:

(5) *Statistical principal principle*:
(i) *Unconditional version:* $P(E^n(a_1, \ldots, a_n) \mid H_i) = P_{H_i}(E^n(x_1, \ldots, x_n))$, where H_i is a hypothesis for an unconditional statistical probability function. E.g. $H_i : P_{H_i}(Gx) = r$ and $E^n(a_1, \ldots, a_n)$ asserts the frequency of G's in a sample of size n.

---

6 Or it would be 1 when measured as the ratio between the two; or it would be 0 if measured in terms of the difference between $B_C(H_i \mid E^n)$ and $B_C(H_i \mid \neg E^n)$.

7 A referee argued that to solve the old evidence problem one merely has to require that B(E) must be evaluated in the epistemic state C in which the evidence E is not yet known. As Earman (1992, p. 122) I think this is insufficient, because if E confirms H (in C), and we observe E (thereby passing to epistemic state C_E), then we want still say that E confirms H (in C_E) – but we can't, because of the problem of old evidence.

8 Hawthorne uses the name "support functions", but I prefer the name "support probabilities", because support functions are sometimes associated with non-probabilistic belief functions.

9 The integral formulation is needed if **H** is an uncountable partition of hypotheses specifying statistical probability functions by real numbers; i.e. **H** = $\{H_r: r \in \text{Reals}\}$, and each H_r asserts that $P(F_1x_1 \wedge \ldots \wedge F_nx_n) = r$ for suitably chosen predicates F_i.

(ii) *Conditional version*:
$P(E^n(a_1, \ldots, a_n) \mid H_i \wedge C^n(a_1, \ldots, a_n)) = P_{H_i}(E^n(x_1, \ldots, x_n) \mid C^n(x_1, \ldots, x_n))$
where H_i is a hypothesis for a conditional statistical probability function. E.g. $H_i : P_{H_i}(Gx \mid Fx) = r$; $E^n(a_1, \ldots, a_n)$ asserts the frequency of G's in a sample of size n and $C^n(a_1, \ldots, a_n)$ asserts that all members of this sample are Fs.

Note that, depending on the nature of evidence partition, the hypotheses space can be chosen as fine as is needed to determine objective likelihoods $P(E^n \mid H_i)$. For example, the likelihood $P(Fa \wedge Ga \mid P(Fx) = r)$ is not objectively determined, but the likelihood $P(Fa \wedge Ga \mid P(Fx \wedge Gx\ ) = q)$ is objectively determined by the principal principle.

Let me emphasize that the standard Bayesian identification of the prior probability of E with the sum-term in line (4)(ii) is *only possible* if one does *not* use actual belief functions but support probabilities. For actual belief functions (4)(ii) is simply false, because according to (4)(ii) $S(E^n)$ is solely determined by objective likelihoods and prior beliefs about hypotheses, but not by actual evidence. So if my prior for the statistical probability of a coin landing heads is uniformly distributed over all possible values, then my probability that it has landed heads as calculated by (4)(ii) will be 1/2, even if I have observed on which side the coin has landed – and this would be incoherent if the probability in (4)(ii) would express my actual degree of belief. (4)(ii) would only be correct for actual degrees of belief if we would replace the objective likelihood by the actual degrees of belief $B_C(E^n(a_1, \ldots, a_n) \mid H_i)$ – but then, as we have shown before, these quasi-likelihoods would no longer be objective and evidence-independent.

Support probabilities seem to be a 'third' kind of probability, but they are not a new primitive probability function, but are defined as a mixture of subjective prior probabilities and objective-statistical probabilities. A support probability is only entirely objective if the true hypothesis H_i, i.e., the true objective-statistical probability is known; otherwise it depends on one's subjective prior over the possible objective-statistical probabilities. But the important point is that a support probability does no longer depend on one's actual evidence – it is intersubjective. Conditional support probabilities S(A|B) intend to reflect the probability which B conveys to A independent of any specific background knowledge or evidence. In other words, neither the prior belief function B nor the support probability S depends on the system C of one's actual subjective beliefs or evidences, whence they are no longer relativized to a background belief system C.

Hawthorne emphasizes that support probabilities should not be conceived of as counterfactual subjective belief functions in the sense of Howson and Urbach (1993, pp. 404ff), i.e. not as one's counterfactual degree of belief in a contracted belief context in which one would not know the evidence. This seems right, because the contracted belief system may still contain many further pieces of evidence (which is illustrated by the example (2)(iii) above). However, this consideration still leaves open the possibility to consider support probabilities as hypothetical belief functions prior to all evidence and other informations. In this sense, Pearl (1988, p. 475) has suggested the *all-I-know*-interpretation, in which $S(H_i|E^n)$ expresses my hypothetical degree of belief in H_i given that all that I know is E^n (and the prior B(H) represents my hypothetical degree of belief in which I do not know anything at all).

Pearl's interpretation brings us back to Carnap's credibility function "Cred" (1971, pp. 21–23). This is a support probability interpreted as an 'absolute prior' probability, i.e. a degree of belief which one would have if she had no prior knowledge at all. This is, of course, an idealized notion, and support functions need not be interpreted in this way, but their important property is that they don't depend on any particular experience about particular individuals.[10] This has a fundamental consequence: support probabilities satisfy the axiom of *exchangeability* (as de Finetti called it) or the equivalent axiom of *symmetry* (as Carnap called it). An exchangeable (or symmetric) belief function is by definition one which is *invariant* w.r.t. arbitrary permutations of individual constants of the language, i.e., it holds that $B(A(a_1, \ldots, a_n)) = B(A(a_{\pi(1)}, \ldots, a_{\pi(n)}))$ for every bijective function $p{:}N\rightarrow N$ over the countable set of indices N of individual constants $\{a_i{:}i\in N\}$ (cf. Earman 1992, p. 89; Carnap 1971, pp. 117ff.). Exchangeable belief functions are thus independent from any particular experience; they assume that prior to all experience all individuals have the same probabilistic tendencies (whence they entail weak induction principles; cf. Kutschera 1972, pp. 74ff.; Earman 1992, p. 108). De Finetti's famous *representation theorem* says that a belief function B is exchangeable exactly if it is representable as an expectation (an average weighted by priors) of objective-statistical (Bernoullioan) probability functions. Line (4) (ii) is exactly such a definition of the degree of belief in a singular statement by mixtures of objective probabilities. In other words, exchangeable belief functions are nothing but support probabilities – and this is the link by which the framework of support probabilities and the older Carnapian theory of "logical" probabilities as symmetric apriori belief functions matches.

So far we have dealt only with general hypotheses H. For singular hypotheses F, e.g. predictions, the support probability is determined by their likelihoods relative to a sufficiently fine partition of hypotheses:

(6) For a singular sentence F all of whose individual constants appear in E_n: $S(F|E^n) = \Sigma_{1\leq i\leq m}\, P(F|E^n \wedge H_i)\cdot S(H_i|E^n)$, where $\{H_1, \ldots, H_m\}$ is the coarsest

10 Support probabilities are compatible with the assumption that the priors of hypotheses are dependent on previously made evidences E′ that are independent from the results of the experiments under consideration (cf. Hawthorne 2005, p. 305), i.e. $B(H) = B_{E'}(H)$. This assumption does not undermine the exchangeability of support probabilities because their indirect dependence on E′ is screened off by their dependence on the 'priors' $B_{E'}(H)$ that are independent from the individuals mentioned in E′ (via iteration of (4)). However, the assumption would imply that the value of S(H|E) does not only reflect probability-enhancing effects of E on H but also of E′ on H. One can avoid this distorting effect if B is understood to be 'absolutely prior' in Carnap's sense.

partition of possible hypotheses such that $P(F|E^n \wedge H_i)$ is objectively determined.[11]

If the true hypothesis H is known, (6) reduces to the conditional form of the statistical principal principle (5)(ii): $S(F|E^n) = P(F|E^n \wedge H)$.

To complete the picture, let us ask how support probabilities are related to actual degrees of belief. This is done by the principles of conditionalization. It is here where the already mentioned principle of total evidence comes into play as follows (my reconstruction essentially agrees with Hawthorne 2005, pp. 309–312, but simplifies it):

(7) *Strict conditionalization*: $B_C(A) = S(A|E_C)$, where E_C is the total evidence in the background system C (which is probabilistically relevant to A; see fn. 3).

(8) *Jeffrey-conditionalization*: $B_C(A) = \Sigma_{1\le i\le k} S(A|E_i)\cdot B_C(E_i)$, where $\{E_1,\ldots, E_k\}$ is a partition of (reasonably available) uncertain evidence-possibilities in background C such that each cell E_i entails the total certain evidence E_C.

Principle (7) is found in exactly the same way in Carnap (1971, p. 18) as the relation of the actual crendence function to the logical credibility function. (8) is the Jeffrey extension of (7) to actual evidences about which one is not certain.

To sum up – Bayesian statistics and confirmation theory need both subjective probabilities and objective (likelihood) probabilities. In our detailed reconstruction even four probabilities have been involved, though only two of them, objective-statistical probabilities and unconditional subjective probabilities (priors of hypotheses and actual degrees of belief in evidences) are primitive. The other two, support probabilities and actual degrees of belief in hypotheses, are derived.

11 For the uncountable partition of hypotheses $\{P(Gx|Fx)=r: r\in Reals\}$, an example would be: $S(Ga_{n+1}|Fa_{n+1}\wedge Fa_n\wedge Ga_n\wedge\ldots\wedge Fa_1\wedge Ga_1) = \int_r r\cdot d\,S(P(Gx|Fx)=r|Fa_n\wedge Ga_n \wedge\ldots\wedge Fa_1\wedge Ga_1)$.

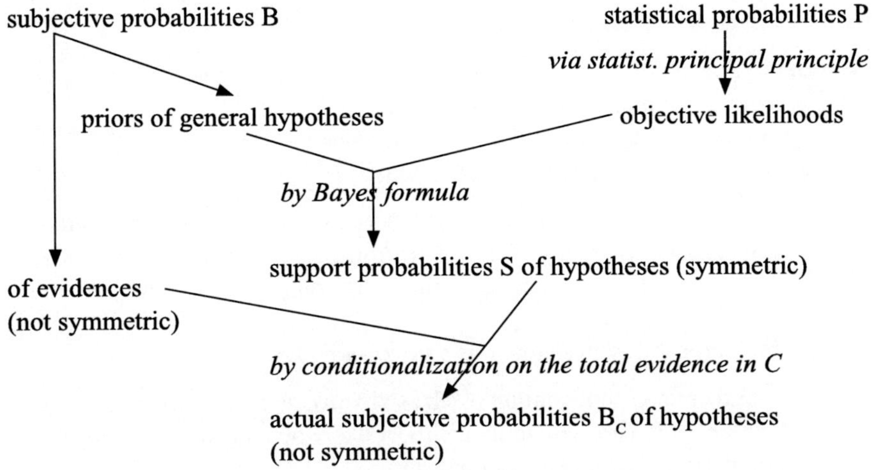

Fig. 5.1 The probabilistic framework of refined probabilism: two primitive and two derived probability functions

One may call this framework *refined probabilism* and it is illustrated in the diagram in Fig. 5.1. The relation of the results in this section with the Tweety and Nixon examples of the previous sections are as follows: the conditional probabilities in reasoning with exceptions represent in their generic formulation statistical probabilities and in their singular formulation support probabilities, while the unconditional probabilities represent actual degrees of beliefs. In the final section, we get back to our Tweety example.

5.4 Do Humans Reason with One Or Two Probability Functions? Tweety and Nixon Examples Put to Empirical Test

So far our reasoning was logical and normative. Now we ask the descriptive question whether in the intuitive reasoning of humans with uncertain information, one or two probability functions are involved. Will humans regard the probability assertions in the Tweety or Nixon examples as incoherent or not? And how will their judgment depend on the general versus singular formulation? Our plan is to test this by a series of experiments. So far, we have performed just a first explorative experiment. We presented 27 test persons[12] with probability assertions involving exceptions and asked the them

12 Test persons were paid 5 Euros for the test, which consisted of the six mentioned examples plus two minor variations of "Coh" and "Inc" that were inserted to avoid priming effects. 95% of the test persons were students; mean age was 26 years; 45% female. 50% reported they had one course in probability theory, 30% they had one course in logic, but we found no significant differences between the respective subgroups. Instructions were read loudly to the test persons; then they were asked, for each example: "Are the assertions of this example jointly contradictory, i.e., do they contradict each other?".

whether they think that these probability assertions are jointly contradictory or not. We confronted them with the following examples:

ESpecGen Exceptions with specificity (such as Tweety), generic formulation:
Tigers are most likely dangerous
Very young tigers are most likely not dangerous
This animal is most likely a very young tiger.

ESpecSing Exceptions with specificity (such as Tweety), singular formulation:
With high probability this animal is dangerous, given it is a lion.
With high probability this animal is not dangerous, given it is a very young lion.
With high probability this animal is a very young lion.

EConGen Exceptions with conflict (such as Nixon), generic formulation:
Animals which are able to fly are most likely oviparous (egg-laying).
Mammals are most likely not oviparous.
This animal is most likely a mammal which is able to fly.

EConSing Exceptions with conflict (such as Nixon), singular formulation:
With high probability, this animal is oviparous, given it is able to fly.
With high probability, this animal is not oviparous, given it is a mammal.
With high probability, this animal is a mammal which is able to fly.

To check whether the test persons had a minimal understanding of probability assertions, and also as a means of avoiding repetitions of similar task structures (priming effects), we inserted examples of the following sort:

Inc Clear cases of probabilistic incoherence such as:
It's highly probable that Peter is a student.
It's highly probable that Paul is a student.
It's highly probable that neither Peter nor Paul is a student.

Coh Clear cases of probabilistic coherence such as:
It's highly probable that Peter will travel to Berlin.
It's highly probable that Paul will travel to Berlin.
It's highly improbable that scientists have found water on Mars.

The results were as follows – "% incons." expresses the percentage of test persons who judged the example of the respective type as inconsistent.

Type of example	*% incons.*
Coh	0.05
Inc	0.95
ESpecGen	0.15
EspecSing	0.25

EConGen	0.45
EConSing	0.45

Although this was just a first experiment, the data tell us some interesting trends.

First: The clear cases of coherent and incoherence examples were judged correctly by the great majority of test persons. So, they indeed had a basic understanding of probability.

Second: Both the generic and the singular formulation of exceptions with specificity was regarded as consistent by the great majority. This indicates clearly that the test persons do indeed possess two distinct concepts of probability. The generic formulation, in which two distinct probability functions are linguistically encoded, was regarded by 85% of the test persons as coherent. However the fact that also in the singular formulation, in which no two distinct probabilities are linguistically indicated, 75% of the people regarded the situation as coherent seems to have only two plausible explanations: Either people's uncertain reasoning isn't probabilistically coherent at all – but this interpretation is already excluded by the results for the basic examples Coh and Inc. So the only remaining interpretation seems to be that our human mind works generally and automatically with these two distinct probability functions and people interpret conditional informations unconsciously in terms of generic statistical probabilities, and factual-categorial informations in terms of subjective-evidential degrees of belief.

Third: In the generic as well as in the singular formulation of examples with conflict the percentage of test persons which consider the example as incoherent is significantly higher but still less than 50%, namely around 45%. The fact that this result is completely independent from the generic vs. singular formulation, as well as the moderate percentage numbers, has in our view the following explanation: the 45% judgements of inconsistency do not reflect an incoherent probability function (people rather use also here two distinct probabilities as explained) – they rather reflect the qualitative conflict between the two opposite qualitative conclusions which are possible.

Our suggested explanations are prima facie and merely a first step. We plan a series of further experiments to increase the robustness and differentiatedness of these results. We also plan experiments about the evaluation of likelihood probabilities by test persons, in order to find out whether they are treated as evidence-independent or as evidence-dependent.

5.5 Conclusion

We have shown that the probabilistic reconstruction of reasoning with exceptions requires a distinction between two different probability functions, subjective-evidential probabilities and objective evidence-independent probabilities which are backed up by objective-statistical probabilities. If this is right, then the assumption of monistic Bayesianism, namely that the reconstruction of uncertain reasoning needs only one probability function, is wrong. Moreover, we have seen that also within the framework of standard Bayesian statistics and confirmation theory there exist strong reasons why these two probability functions have to be distinguished – the objectivity of likelihoods and the avoidance of the problem of old evidence. Drawing on Hawthorne (2005) we have sketched how with help of the statistical principal principle and Bayes rule, evidence-independent support probabilities can be defined from subjective priors of hypotheses and objective-statistical probabilities. These support probabilities have the characteristics needed to satisfy the conditions for convergence to objectivity with accumulating evidence. In the last section we turned from the normative to the factual domain: we presented a first experiment (of a series of planned experiments) which indicates that humans intuitive uncertain reasoning does indeed, if only unconsciously, involve two distinct probability functions. This finding is an important amendment to recent accounts of Bayesian reasoning in psychology such as Oaksford and Chater (2007).

Acknowledgements: Work on this paper was supported by the DFG-financed subproject P1 of the EuroCores LogiCCC project *The Logic of Causal and Probabilistic Reasoning in Uncertain Environments*. The experimental part of this paper was done by Matthias Unterhuber. For valuable discussions on the topic I am grateful to Matthias Unterhuber, Jim Hamthorne, Jon Williamson, Stephan Hartmann, James Joyce, Hannes Leitgeb, Horacio Arlo-Costa, Gernot Kleiter and Niki Pfeifer.

References

Jonathan Bennett (2003), *A Philosophical Guide to Conditionals*, Oxford: Clarendon Press.

Gerhard Brewka (1991a), *Nonmonotonic Reasoning. Logical Foundations of Common Sense.* Cambridge: Cambridge University Press.

Gerhard Brewka (1991b), "Cumulative Default Logic", in: *Artificial Intelligence* 50, pp. 183–205.

Rudolf Carnap (1950), *Logical Foundations of Probability*, Chicago: The University of Chicago Press.

Rudolf Carnap (1971), "Inductive Logic and Rational Decisions", in: Rudolf Carnap and Richard Jeffrey (Eds.), *Studies in Inductive Logic and Probability I.* Los Angeles: Univ. of Calif. Press.

John Earman (1992), *Bayes or Bust?* Cambridge/Mass.: MIT Press.

Ward Edwards, Harold Lindman and Leonard J. Savage (1963), "Bayesian Statistical Inference for Psychological Research", in: *Psychological Review* 70, pp. 193–242.

Jonathan St. Evans, Simon, J. Handley and David E. Over (2003), "Conditionals and Conditional Probability. *Journal of Experimental Psychology: Learning. Memory, and Cognition* 29/2, pp. 321–335.

Dov M. Gabbay et al. (Eds., 1994), *Handbook of Logic in Artificial Intelligence and Logic Programming, Vol. 3: Nonmonotonic Reasoning and Uncertain Reasoning*. Oxford: Clarendon Press.

Moises Goldszmidt and Judea Pearl (1996), "Qualitative Probabilities for Default Reasoning, Belief Revision and Causal Modeling", in: *Artificial Intelligence*, 84, pp. 57–112.

Alan Hájek (2008), "Arguments for (or against) Probabilism?", in: *British Journal for the Philosophy of Science* 59, pp. 793–819.

Joseph Halpern (2003), *Reasoning about Uncertainty*. Cambridge/Mass.: MIT Press.

James Hawthorne (2005), "*Degree-of-Belief* and *Degree-of-Support*: Why Bayesians Need Both Notions", in: *Mind* 114, pp. 277–320.

Collin Howson and Peter Urbach (1993), *Scientific Reasoning: The Bayesian Approach*. Chicago: Open Court (2nd ed.).

James M. Joyce (1998), "A Nonpragmatic Vindication of Probabilism", *Philosophy of Science* 65/4, pp. 575–603.

Franz von Kutschera (1972), *Wissenschaftstheorie, Bd. I und II*. München: W. Fink.

David Lewis (1976), "Probabilities of Conditionals and Conditional Probabilities", in: *The Philosophical Review* 85, pp. 297–315.

Vann MacGee (1989): "Conditional Probabilities and Compounds of Conditionals", in: *The Philosophical Review* 98, pp. 485–541.

Mike Oaksford and Nick Chater (2007), *Bayesian Rationality. The Probabilistic Approach to Human Reasoning*, Oxford: Oxford Univ. Press.

Klaus Oberauer and Oliver Wilhelm (2003), "The Meaning(s) of Conditionals: Conditional Probabilities, Mental Models, and Personal Utilities", in: *Journal of Experimental Psychology: Learning. Memory, and Cognition* 29/4, pp. 680–693.

Judea Pearl (1988), *Probabilistic Reasoning in Intelligent Systems*. Santa Mateo: Morgan Kaufmann.

Niki Pfeifer and Gernot Kleiter (2008), "The Conditional in Mental Probability Logic", in: Mike Oaksford (Ed.), *The Psychology of Conditionals*. Oxford: Oxford University Press.

John Pollock (1994), "Justification and Defeat", in: *Artificial Intelligence* 67, pp. 377–407.

Hans Reichenbach (1949), *The Theory of Probability*. Berkeley: University of California Press.

Raymond Reiter (1980), "A Logic for Default Reasoning", in: *Artificial Intelligence* 13, pp. 81–132.

Gerhard Schurz (2005), "Non-monotonic Reasoning from an Evolutionary Viewpoint", in: *Synthese* 146/1-2, pp. 37–51.

Gerhard Schurz (2007), "Human Conditional Reasoning Explained by Non-Monotonicity and Probability: An Evolutionary Account", in: Stella Vosniadou et al. (Eds.), *Proceedings of EuroCogSci07. The European Cognitive Science Conference 2007*, Lawrence Erlbaum Assoc., New York, 2007, pp. 628–633.

Institute of Philosophy
University of Düsseldorf
Universitätsstrasse 1
40225, Düsseldorf
Germany
schurz@phil.uni-duesseldorf.de

CHAPTER 6

DAVID ATKINSON AND JEANNE PEIJNENBURG

PLURALISM IN PROBABILISTIC JUSTIFICATION

6.1 INTRODUCTION

From Aristotle onwards, epistemic justification has been conceived as a form of inference. If a proposition E_n is epistemically justified by a proposition E_{n+1}, then according to the traditional view E_n is somehow inferred from E_{n+1}.

It took twenty-three centuries to modify this outlook. Today, many epistemologists construe epistemic justification in terms of probabilistic support rather than of inferential relations. In the modern view, E_n is epistemically justified by E_{n+1} if two requirements are fulfilled. First, E_{n+1} should probabilistically support E_n. By this we mean that the conditional probability of E_n, given E_{n+1}, exceeds the conditional probability of E_n, given not-E_{n+1}:

$$P(E_n|E_{n+1}) > P(E_n|\neg E_{n+1}) . \tag{6.1}$$

Second, the unconditional probability of $P(E_n)$ should not fall below some agreed threshold of acceptance.

This 'probabilistic turn' in epistemology opened the door to pluralism in epistemic justification. Imagine a sequence of propositions $E_0, E_1, E_2 \ldots$, such that E_0 is epistemically justified by E_1, which is epistemically justified by E_2, and so on. In 1956, when the probabilistic turn had not yet been fully made, Wilfrid Sellars still saw no more options than to construct this sequence as either a finite chain or a finite loop:

> One seems forced to choose between the picture of an elephant which rests on a tortoise (What supports the tortoise?) and the picture of a great Hegelian serpent of knowledge with its tail in its mouth (Where does it begin?). Neither will do.[1]

Present-day epistemologists, however, are not confined to these two possibilities. Thanks to the interpretation of epistemic justification as probabilistic support, they have at their disposal many different ways of reconstructing justificationary processes. A target proposition, E_0, can be probabilistically justified by a chain or by a loop, and both the chain and the loop can be finite or infinite.[2] Moreover,

1 Wilfrid Sellars, "Empiricism and the Philosophy of Mind", in: Herbert Feigl, Michael Scriven (Eds.), *The Foundations of Science and the Concepts of Psychology and Psychoanalysis*. Minneapolis: University of Minnesota Press 1966, pp. 253–329; p. 300.

2 The concept of an infinite loop might seem incoherent, but it is not. As we have shown elsewhere, an infinite loop differs in nontrivial ways from an infinite chain. See David

D. Dieks et al. (eds.), *Probabilities, Laws, and Structures*, The Philosophy of Science in a European Perspective 3, DOI 10.1007/978-94-007-3030-4_6,

in each of these four cases the conditional probabilities might be uniform, taking on the same values, or they might be nonuniform, differing throughout the chain or loop. This yields already eight different varieties of epistemic justification. In earlier papers we have discussed the four most intriguing ones – involving chains and loops of infinite size. We showed there that infinite chains and infinite loops can converge, yielding a unique and well-defined probability value for the target proposition E_0.

In the present paper we contribute to a pluralistic outlook by introducing even more possibilities for probabilistic justification. In contrast to the eight varieties above, all of which are one-dimensional, we will investigate probabilistic justification in more than one dimension. We shall concentrate again on structures of infinite size, and we show that many-dimensional networks can converge, too. Thus it makes sense to say that a target proposition, E_0, can be epistemically justified not only by an infinite one-dimensional chain or an infinite one-dimensional loop, but also by an infinite network of many dimensions.

We start, in Sect. 6.2, by recalling some facts about infinite, one-dimensional chains, where for convenience sake we restrict ourselves to chains that are uniform. In Sect. 6.3 we explain what happens when we go from a one-dimensional uniform chain to a two-dimensional uniform network. In Sect. 6.4, we contrast the properties of one-dimensional chains with those of two-dimensional networks. As we will see, there exists an intriguing difference between the two, which poses difficulties at an intuitive level. In Sect. 6.5 we indicate the relevance of this paper for disciplines outside epistemology and philosophy in general, by explaining an application of our analysis to genetics.

6.2 Infinite, Uniform Chains

Earlier we have shown that probabilistic epistemic chains of infinite length always converge. In the present section we summarize our findings, restricting ourselves for simplicity to uniform chains. However, the demonstration we give of convergence is different from earlier proofs, since we will now use fixed-point methods.[3]

The unconditional probabilities $P(E_n)$ and $P(E_{n+1})$ are related by the rule of total probability,

$$P(E_n) = P(E_n|E_{n+1})P(E_{n+1}) + P(E_n|\neg E_{n+1})P(\neg E_{n+1}). \tag{6.2}$$

As we have already indicated, we assume in this paper that the conditional probabilities are uniform, i.e. they are the same throughout the chain. However, it is important to keep in mind that the assumption of uniformity is not essential: the

---

Atkinson, Jeanne Peijnenburg, "Justification by Infinite Loops", in: *Notre Dame Journal of Formal Logic*, 51, 4, 2010, pp. 407–416.

3 Of the earlier proofs, the most general one can be found in Appendix A of David Atkinson, Jeanne Peijnenburg, "The Solvability of Probabilistic Regresses: A Reply to Frederik Herzberg", in: *Studia Logica*, 94, 3, 2010, pp. 347–353.

whole argument goes through without assuming uniformity, albeit in a somewhat more complicated form.

Under the assumption of uniformity, Eq. 6.2 may be rewritten in the form

$$P(E_n) = \beta + (\alpha - \beta)\, P(E_{n+1}), \tag{6.3}$$

where

$$\alpha = P(E_n|E_{n+1}) \quad \text{and} \quad \beta = P(E_n|\neg E_{n+1}).$$

Clearly $\alpha > \beta$ is equivalent to the condition of probabilistic support as expressed in (6.1).

Does the iteration (6.3) converge, giving a well-defined value for $P(E_0)$, $P(E_1)$, $P(E_2)$ and so on? If it does, then $P(E_n)$ and $P(E_{n+1})$ will have to be equal in the limit. Let us call this limiting value, if it exists, P_1^*. It is a fixed point of the iteration Eq. 6.3, i.e. it satisfies

$$P_1^* = \beta + (\alpha - \beta)\, P_1^*,$$

and this linear equation has the unique solution

$$P_1^* = \frac{\beta}{1 - \alpha + \beta}, \tag{6.4}$$

where we exclude $\alpha = 1$ (the case in which E_{n+1} entails E_n).

To show that the iteration (6.3) does indeed converge, we write $P(E_{n-1}) = \beta + (\alpha - \beta)\, P(E_n)$, which is (6.3) with $n - 1$ in place of n. Subtracting this from Eq. 6.3, we obtain

$$P(E_n) - P(E_{n-1}) = (\alpha - \beta)\,[P(E_{n+1}) - P(E_n)]\,.$$

Hence, by iteration,

$$\begin{aligned} P(E_n) - P(E_{n-1}) &= (\alpha - \beta)^2\,[P(E_{n+2}) - P(E_{n+1})] = \ldots \\ &= (\alpha - \beta)^s\,[P(E_{n+s}) - P(E_{n+s-1})]\,. \end{aligned} \tag{6.5}$$

We may take s to infinity on the right-hand side of (6.5), and since $(\alpha - \beta)^s$ tends to zero in this limit, it is clear that $P(E_n) - P(E_{n-1}) = 0$, i.e. $P(E_n) = P(E_{n-1})$ for all finite n. This shows that the iteration (6.3) converges. Indeed all the unconditional probabilities are equal to one another, and they are all equal to the fixed point, P_1^*.

6.3 Infinite, Uniform Networks

In this section we shall consider a two-dimensional probabilistic network, where a 'child' proposition is probabilistically justified by two 'parent' propositions, each of which is in turn probabilistically justified by two '(grand)parent' propositions, etc. So the network has a tree-like structure:

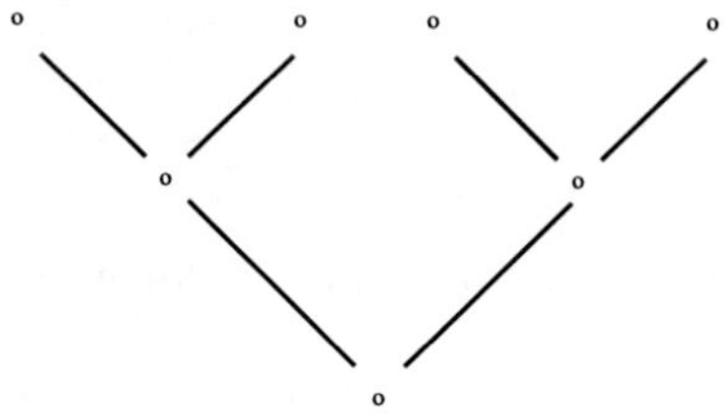

Figure 6.1: Justification in two dimensions

In the sequel, we shall often use 'child' and 'parent' for 'child proposition' and 'parent proposition'. We will talk about 'the probability that the child (parent) is true', or alternatively about 'the probability of the child (parent)'.

Much as in our treatment of the one-dimensional chain, we take it that a child proposition is justified by its parent propositions if two requirements are fulfilled. First, the parents must probabilistically support the child. By this we mean: the probability that the child is true, given that both its parents are true, is larger than the probability that the child is true, given that both parent propositions are false.[4] In symbols:

$$P(E_n|E_{n+1}\&E'_{n+1}) > P(E_n|\neg E_{n+1}\&\neg E'_{n+1}), \tag{6.6}$$

where E_n stands for the child, E_{n+1} for the one parent, and E'_{n+1} for the other. We will refer to the conditional probability in which both parents are true by α and that in which both parents are false by β, so (6.6) becomes:

$$\alpha > \beta.$$

The second requirement for justification is that the unconditional probability of the child proposition may not lie below a certain threshold of acceptance. The threshold could be 0.5, or 0.9, or even higher, dependent on the context of the case.

For simplicity, we make three further assumptions. To begin with, we assume that the parents are independent of one another, so the probability that both parents

4 Below we will consider the case in which one parent proposition is true while the other one is false. We will see that this situation has no bearing on the condition of probabilistic support.

are true equals the probability that one of them is true times the probability that the other one is also true:

$$P(E_{n+1}\&E'_{n+1}) = P(E_{n+1})P(E'_{n+1}) \,. \tag{6.7}$$

Moreover, we explicitly assume gender symmetry, i.e. both parent propositions have the same probability of being true, so $P(E_{n+1}) = P(E'_{n+1})$. Finally, we suppose that the conditional probabilities are the same throughout the whole two-dimensional structure. In other words, just like the linear chain in the previous section, our quadratic two-parent network in the present section is uniform. However, as was the case for the one-dimensional chain, the uniformity assumption is not essential. Nor do we need the assumptions of independence or of gender symmetry. Our argument can be made without these three assumptions, although we shall not show that here.

The unconditional probability that the child is true, $P(E_n)$, can be written in terms of the triple joint probabilities associated with two parent propositions as follows:

$$\begin{aligned} P(E_n) &= P(E_n\&E_{n+1}\&E'_{n+1}) + P(E_n\&E_{n+1}\&\neg E'_{n+1}) + \\ &\quad P(E_n\&\neg E_{n+1}\&E'_{n+1}) + P(E_n\&\neg E_{n+1}\&\neg E'_{n+1}). \end{aligned} \tag{6.8}$$

The first term on the right-hand side is the joint probability that the child proposition and both parent propositions are all true. This term can be written as

$$P(E_n\&E_{n+1}\&E'_{n+1}) = \alpha P(E_{n+1}\&E'_{n+1}), \tag{6.9}$$

where α, as we said, is now the conditional probability that E_n is true, given that both parents are true. Since the parents are independent of one another, see (6.7), and since we supposed that each parent has the same probability of being true, Eq. 6.9 can be written as:

$$P(E_n\&E_{n+1}\&E'_{n+1}) = \alpha P^2(E_{n+1}). \tag{6.10}$$

The last term on the right-hand side of (6.8) is the joint probability that the child proposition is true, and both parent propositions are false. It can be written as

$$\begin{aligned} P(E_n\&\neg E_{n+1}\&\neg E'_{n+1}) &= \beta P(\neg E_{n+1}\&\neg E'_{n+1}) \\ &= \beta P(\neg E_{n+1})P(\neg E'_{n+1}) \\ &= \beta\,[1 - P(E_{n+1})]^2, \end{aligned} \tag{6.11}$$

where β is the conditional probability that E_n is true, given that both parents are false.

The second and third terms on the right-hand side of (6.8) are the joint probabilities that the child proposition is true, when one parent is true, and one is false.

The terms can be written as

$$\begin{aligned} P(E_n \& E_{n+1} \& \neg E'_{n+1}) &= \gamma P(E_{n+1} \& \neg E'_{n+1}) \\ &= \gamma P(E_{n+1}) P(\neg E'_{n+1}) \\ &= \gamma P(E_{n+1})[1 - P(E_{n+1})], \end{aligned} \tag{6.12}$$

$$\begin{aligned} P(E_n \& \neg E_{n+1} \& E'_{n+1}) &= \delta P(\neg E_{n+1} \& E'_{n+1}) \\ &= \delta P(\neg E_{n+1}) P(E'_{n+1}) \\ &= \delta P(E_{n+1})[1 - P(E_{n+1})], \end{aligned} \tag{6.13}$$

where γ is the conditional probability that the child is true, given that only the first parent is true, and δ is the conditional probability that it is true, given that only the second parent is true.

On inserting the expressions (6.9), (6.11)–(6.13) into (6.8), we find, after rearrangement,

$$P(E_n) \quad = \quad \beta + 2(\varepsilon - \beta)\, P(E_{n+1}) + (\alpha + \beta - 2\varepsilon)\, P^2(E_{n+1}), \tag{6.14}$$

where ε is the average of the conditional probabilities that only one parent is true:

$$\varepsilon = \tfrac{1}{2}(\gamma + \delta).$$

In the special case that

$$\alpha + \beta = 2\varepsilon, \tag{6.15}$$

this relation takes on the linear form

$$P(E_n) = \beta + (\alpha - \beta) P(E_{n+1}),$$

just like the one-dimensional chain (6.3). We know from Eq. 6.5 that this sequence converges if $\alpha > \beta$ (we exclude the case $\alpha = 1$).

When the special equality (6.15) does *not* hold, Eq. 6.14 has two fixed points, namely the two solutions of the quadratic equation

$$P_2^* = \beta + 2(\varepsilon - \beta)\, P_2^* + (\alpha + \beta - 2\varepsilon)\, P_2^{*2}.$$

We show in the appendix that only one of these fixed points is attracting, namely

$$P_2^* = \frac{\beta + \frac{1}{2} - \varepsilon - \sqrt{\beta(1-\alpha) + (\varepsilon - \frac{1}{2})^2}}{\alpha + \beta - 2\varepsilon}. \tag{6.16}$$

As in the one-dimensional case, all the unconditional probabilities are the same, being equal to the fixed point, P_2^*.

6.4 Contrasting Chains and Networks

It is interesting to compare the properties of the linear one-parent chain of Sect. 6.2 with those of the quadratic two-parent net of Sect. 6.3. In the present section we will discuss first a similarity between the two structures, and then an intriguing difference.

The similarity concerns sufficient conditions for convergence in the two cases. As we have seen, the requirement of probabilistic support for the chain is $P(E_n|E_{n+1}) > P(E_n|\neg E_{n+1})$, i.e. the child E_n is supported by its parent E_{n+1}. For the net, on the other hand, the condition is $P(E_n|E_{n+1}\&E'_{n+1}) > P(E_n|\neg E_{n+1}\&\neg E'_{n+1})$, i.e. the child E_n is supported by both of its parents E_{n+1} and E'_{n+1}. In both cases, the condition turns out to be sufficient for convergence: it ensures convergence of not only the one-dimensional infinite iteration, but also of the two-dimensional infinite net. Note that, for the net, the conditional probability that the child is true, given that only one parent is true, has no relevance to convergence. The iteration has an attracting fixed point if $\alpha > \beta$; in the two-dimensional case, it does not matter how large or small ε is.

There is, however, a crucial difference between the chain and the net. This difference pertains to the situation in which β is zero. For the chain, as we have seen, $\beta = P(E_n|\neg E_{n+1})$; and the infinite, uniform chain leads to the fixed point (6.4):

$$P_1^* = \frac{\beta}{1-\alpha+\beta},$$

which clearly vanishes if $\beta = 0$ (assuming $\alpha < 1$). Thus if the child of a parent proposition that is false is *never* true then, after an infinite number of generations, the child proposition will certainly be false. This should not come as a surprise. After all, here the probabilistic justification of the target proposition E_0 by an infinite chain $E_1, E_2 \ldots$, and so on, is such that only the conditional probability $\alpha = P(E_n|E_{n+1})$ is positive, the conditional probability $\beta = P(E_n|\neg E_{n+1})$ being zero. Consequently, Eq. 6.2 becomes $P(E_n) = P(E_n|E_{n+1})P(E_{n+1})$, and so each link of the infinite chain contributes to the monotonic diminution of the value of $P(E_0)$, resulting finally in zero.[5]

5 This result on the one-dimensional infinite chain may shed light on the position of C.I. Lewis and Bertrand Russell, who both claimed that an infinite chain of probabilistic relations must lead to probability zero for the target proposition. This claim is incorrect as a general statement, as Hans Reichenbach pointed out, precisely because Lewis and Russell had forgotten the term in the rule of total probability, which corresponds to the probability that the child proposition is true when its parent is false (see Eq. 6.2 in the main text). However, as we can see now, Lewis and Russell were right in a very restricted situation. If in one dimension the probability of a child, given the falsity of its parent proposition, is always zero, then the probability of the target proposition will progressively decrease from link to link, tending to zero in the limit of an infinite chain. See C. I. Lewis, "The Given Element in Empirical Knowledge", in: *The Philosophical Review*, 61, 2, 1952, pp. 168–172; Hans Reichenbach, "Are Phenomenal Reports Absolutely Certain?", in: *The Philosophical Review*, 61, 2, 1952, pp. 147–159; Bertrand

The situation in two dimensions is entirely different. Now $\beta = P(E_n|\neg E_{n+1}\& \neg E'_{n+1})$; and the infinite, uniform net leads to the fixed point (6.16):

$$P_2^* = \frac{\beta + \frac{1}{2} - \varepsilon - \sqrt{\beta(1-\alpha) + (\varepsilon - \frac{1}{2})^2}}{\alpha + \beta - 2\varepsilon}.$$

In the case that β is zero, this formula reduces to

$$P_2^* = \frac{\frac{1}{2} - \varepsilon - |\frac{1}{2} - \varepsilon|}{\alpha - 2\varepsilon}. \qquad (6.17)$$

Notice that this is zero *only in the case that* $\varepsilon \leq \frac{1}{2}$. When $\varepsilon > \frac{1}{2}$, the expression (6.17) becomes

$$P_2^* = \frac{2\varepsilon - 1}{2\varepsilon - \alpha}. \qquad (6.18)$$

The interesting thing is that, if a child is false when both parents are false, then the unconditional probabilities $P(E_n)$ in the infinite net may, or may not be zero. It all depends on how probable it is that a child is true when *only one* of its parents is true. If this conditional probability is more than one half (that is, if the child is more likely to be true than false given that only one of the parents is true), then the unconditional probabilities $P(E_n)$ do not vanish. This is quite different from the one-dimensional situation, where $\beta = 0$ does imply that $P(E_n)$ vanishes. However, in order for the child to be justified, not only must ε be greater than one-half when $\beta = 0$, but also α must be large enough to ensure that the unconditional probability that the child proposition is true does not lie below the threshold of acceptance. Precisely why $\varepsilon = \frac{1}{2}$ marks the boundary between a zero and a nonzero unconditional probability is intuitively still unclear to us.

6.5 Relevance and Applications

Do the above exercises have any applications? Are the formalisms that we have developed of any philosophical relevance or utility in the outside world?

As far as the relevance to philosophy is concerned, we can be brief. In the introduction we already alluded to the venerable tradition concerning justification in epistemology; in particular justification related to an infinite regress is a subject that has been much discussed.[6] Most philosophers in the tradition took

Russell, *Human Knowledge: Its Scope and Limits*. London: George Allen and Unwin 1948; Jeanne Peijnenburg, David Atkinson, "Grounds and Limits: Reichenbach and Foundationalist Epistemology", in: *Synthese*, 181, 2011, pp. 113–124; Jeanne Peijnenburg, "Ineffectual Foundations", in: *Mind*, 119, 2010, pp. 1125–1133.

6 Cf. Bonjour: "Considerations with respect to the regress argument [are] perhaps the most crucial in the entire theory of knowledge", Laurence Bonjour, *The Structure of Empirical Knowledge*. Cambridge (Mass.): Harvard University Press, 1985; p. 18.

the view that argumentation to infinite regress shows that a certain position is absurd. Philosophers ranging from Zeno, Plato and Aristotle to Aquinas, Descartes, Leibniz, Hume and Kant have all used a *regressus ad infinitum* as a *regressus ad absurdum*: in their view, any argument that leads to an infinite regress is thereby vitiated. In our paper we show that this view is mistaken if an infinite regress is probabilistic in nature. We have explained this first for a one-dimensional chain of propositions, and then for a two-dimensional network.

As to the applications in the outside world, they are numerous, especially in view of the fact that our simplifying assumptions (namely uniformity, and, in two dimensions, gender symmetry and independence) can be relaxed without affecting the essential findings. Here we will restrict ourselves to one application, taken from the genetics of a population in which background conditions remain stable over time.

Consider the inheritance of a gender-specific genetic disorder in a human population, such as the tendency to prostate cancer in the male, or to breast cancer in the female. The probability that a child will develop the condition at some time in its life is different if the parent of the same gender has the complaint, or if that parent does not. If the relevant external conditions remain the same over time, the two conditional probabilities, α and β, will be uniform, that is, the same from generation to generation. The one-dimensional formalism of Sect. 6.2 is then applicable, and we conclude that the probability of disease, which we can equate to the relative frequency of its incidence in a large population, after any transient effects have died out, is given by the fixed point (6.4):

$$P_1^* = \frac{\beta}{1-\alpha+\beta}.$$

The values of α and β could be inferred from the statistics of two generations only, and one can then deduce the above relative frequency of incidence, which will be stable throughout the generations.

Our analysis of the two-dimensional case can be applied, for example, to the inheritance in a human population of albinism. The three conditional probabilities that a child will be normally coloured, namely α (if both parents are normally coloured), β (if neither parent is normally coloured, i.e. both are albinos), and ε (if one parent is normal and one is albino), can be estimated from the statistics of a large population. The relative frequency of normally coloured individuals in a large population is then given by the fixed point (6.16):

$$P_2^* = \frac{\beta+\frac{1}{2}-\varepsilon-\sqrt{\beta(1-\alpha)+(\varepsilon-\frac{1}{2})^2}}{\alpha+\beta-2\varepsilon}.$$

However, there is more to be said in this case.

To begin with, when two albinos mate, their children are nearly always albinos, i.e. they are almost never normally coloured, so $\beta = 0$ to a good approximation. Thus the situation obtains that we mentioned in Sect. 6.4.

Furthermore, our analysis of the two-dimensional case can be seen as a generalization of the famous Hardy-Weinberg rule in genetics.[7] The genetic fingerprint of every individual is given by his or her DNA, i.e. the double helix consisting of two strings of molecules called nucleotides. Sequences of nucleotides are grouped together to form genes. Many of these genes are such that they come in two possible forms or alleles, one of which stems from the father and one from the mother. Let us denote the one allele by the letter A, the other by a. So an individual's genetic make-up, as far as these particular genes are concerned, will be one of the following: AA, Aa, aA, aa. Now albinism arises from an allele, a, that is *recessive*: this indicates that only an individual with allele a in both strands of his or her DNA will be an albino. The allele A is called *dominant* because individuals carrying Aa, or aA are healthy, just like individuals carrying AA.

In a large population, suppose that the fraction of the recessive allele is q, and the fraction of the dominant allele is $p = 1 - q$. Then the albino fraction of the population, carrying aa, is q^2, while the healthy fraction that carries AA is p^2, and the healthy fraction that carries Aa or aA is $2pq$. Moreover, these fractions remain the same from generation to generation. This is the essence of the Hardy-Weinberg rule: it is based on the assumption of a theoretically infinite population and random mating.

In the Hardy-Weinberg model, it is possible to calculate the conditional probabilities in terms of q, the fraction of the recessive allele. We find

$$\alpha = 1 - \left(\frac{q}{1+q}\right)^2 \qquad \varepsilon = \frac{1}{1+q} \qquad \beta = 0, \tag{6.19}$$

where we suppress the details of the rather tedious calculation. The first thing to notice is that, if $0 < q < 1$, then ε is necessarily greater than one-half, and since $\beta = 0$ the formula (6.18) is applicable:

$$P_2^* = \frac{2\varepsilon - 1}{2\varepsilon - \alpha}.$$

With the values (6.19) we find $P_2^* = 1 - q^2$, which is clearly correct, since the albino fraction of the population, carrying two recessive alleles, aa, is q^2, and P_2^* is the complement of that.

When we take account of the fact that mutations from the recessive to the dominant allele are actually possible (very rarely), so that β is not quite zero, the general fixed-point formula (6.16) must be used instead of (6.18). This constitutes a modification of the Hardy-Weinberg model that can be readily handled by our methods.

7 G. H. Hardy, "Mendelian Proportions in a Mixed Population", in: *Science,* 28, 1908, pp. 49–50. Wilhelm Weinberg, "Über den Nachweis der Vererbung beim Menschen", in: *Jahreshefte des Vereins für vaterländische Naturkunde in Württemberg,* 64, 1908, pp. 368–382.

Appendix

When the special equality (6.15) does *not* hold, the quadratic iteration (6.14) can be put into canonical form by means of the substitution

$$q_n = (\alpha + \beta - 2\varepsilon)\, P(E_n) - \beta + \varepsilon\,. \qquad (6.20)$$

With this transformation, (6.14) becomes[8]

$$q_n = c + q_{n+1}^2\,, \qquad (6.21)$$

where

$$c = \varepsilon(1-\varepsilon) - \beta(1-\alpha).$$

The conditional probabilities, α, β and ε are all real numbers in the unit interval, so it follows that $c \leq \varepsilon(1-\varepsilon) = \frac{1}{4} - (\varepsilon - \frac{1}{2})^2 \leq \frac{1}{4}$. Further, since $\beta < \alpha$ (the condition of probabilistic support), $c \geq -\beta(1-\alpha) > -\alpha(1-\alpha) = -\frac{1}{4} + (\frac{1}{2} - \alpha)^2 \geq -\frac{1}{4}$. That is,

$$-\tfrac{1}{4} < c \leq \tfrac{1}{4}. \qquad (6.22)$$

Consider the fixed point,

$$q^* = \tfrac{1}{2} - \sqrt{\tfrac{1}{4} - c},$$

of the iteration (6.21). Via the inverse of the transformation Eq. 6.20, one can show that q^* corresponds to the fixed point P_2^* of Eq. 6.16.

To demonstrate that q^* is attracting, we change the variable from q_n to $s_n = q_n - q^*$, so that (6.21) becomes

$$s_n = s_{n+1}\left[1 - \sqrt{1-4c} + s_{n+1}\right]. \qquad (6.23)$$

If

$$\left|1 - \sqrt{1-4c}\right| < 1 \qquad (6.24)$$

and s_{n+1} is very small, we conclude that c^* is attracting. Indeed, since

$$s_n - s_{n+1} = (s_{n+1} - s_{n+2})\left[1 - \sqrt{1-4c} + s_{n+1} + s_{n+2}\right]$$

the mapping (6.23) is a contraction if $|s_n| \leq \rho$ and $\left|1 - \sqrt{1-4c} + 2\rho\right| < 1$. Hence if $|s_N| \leq \rho$ for very large N, and ρ satisfies the above contraction constraint, the iteration backwards to s_0 will be attracted to zero, that is to say q_0 will be attracted to q^*. The domain of attraction of the fixed point, q^*, is $-\frac{3}{4} < c < \frac{1}{4}$. Attraction is trivially guaranteed also when $c = \frac{1}{4}$. So absolute convergence is

8 Elsewhere we propose to trace the connection between the two-dimensional net and the Mandelbrot fractal. (Benoît Mandelbrot, *The Fractal Geometry of Nature.* New York: W.H. Freeman and Co 1982 – second printing with update).

assured if $|c| \leq \frac{1}{4}$; and we see from the inequalities (6.22) that this is consistent with the requirements $0 \leq \beta < \alpha < 1$ and $0 \leq \varepsilon \leq 1$.

Faculty of Philosophy
University of Groningen
Oude Boteringestraat 52
9712 GL Groningen
The Netherlands
d.atkinson@rug.nl
jeanne.peijnenburg@rug.nl

CHAPTER 7

JAN-WILLEM ROMEIJN, RENS VAN DE SCHOOT,
AND HERBERT HOIJTINK

ONE SIZE DOES NOT FIT ALL: PROPOSAL FOR A PRIOR-ADAPTED BIC

ABSTRACT

This paper presents a refinement of the Bayesian Information Criterion (BIC). While the original BIC selects models on the basis of complexity and fit, the so-called *prior-adapted* BIC allows us to choose among statistical models that differ on three scores: fit, complexity, and model size. The prior-adapted BIC can therefore accommodate comparisons among statistical models that differ only in the admissible parameter space, e.g., for choosing among models with different constraints on the parameters. The paper ends with an application of this idea to a well-known puzzle from the psychology of reasoning, the conjunction fallacy.

7.1 OVERVIEW

Statistical model selection concerns the choice among a set of statistical models. A model consists of a set of statistical hypotheses, where each hypothesis imposes a probability distribution over sample space. There are several ways of choosing among the models, leading to several so-called information criteria (ICs) that may regulate the choice. All of these are comparative: their absolute numerical values do not have meaning.

By way of introduction, we mention some of the dominant model selection tools. We may choose the model that, under relative entropy distance, brings us closest to the hypothesized true distribution. This distance is approximated by Akaike's information criterion or AIC (Akaike 1973; Stone 1977). The AIC can be interpreted naturally in terms of cross-validation and predictive success. The deviance information criterion, or DIC for short (Spiegelhalter et al. 2002), is based on a similar information-theoretic criterion. Like the AIC, it can be justified in terms of the predictive accuracy of the model, under a particular loss function to express accuracy. Alternatively, we may choose the model that allows us to capture the information contained in the data most efficiently, as proposed by model selection based on the so-called Minimum Description Length or MDL (Grunwald 2007; Balasubramanian 2005). Or finally, we may base our choice on the probability of the data averaged for the models at issue, called the marginal

D. Dieks et al. (eds.), *Probabilities, Laws, and Structures*, The Philosophy of Science in a European Perspective 3, DOI 10.1007/978-94-007-3030-4_7,

likelihoods. The Bayesian information criterion, or BIC (Schwarz 1978; Raftery 1995), approximates these likelihoods or, in other formulations, the resulting posterior probabilities over the models.

The selection criteria boil down to a trade-off between model complexity and model fit: we choose the model that optimizes the match between the data and the best fitting hypothesis within the model, but models that are more complex start with a handicap. One of the attractive features of the information criteria sketched above is that they are derived from first principles, and that the trade-off between fit and complexity shows up as a consequence of those starting points rather than being put in by hand. Moreover, it turns out that the trade-off is very similar for a number of different sets of first principles. It shows up in the approximations of entirely different notions, to wit, distance to the truth (AIC), predictive accuracy (DIC), minimum description length (MDL), and marginal likelihood (BIC). All these information criteria use the maximum likelihood of the model as a measure of fit, and for all these ICs the measure for complexity and the resulting penalty are related to the number of free parameters appearing in the statistical model.

The central problem of this paper is that this complexity measure is not effective if we want to compare models that do not differ in dimensionality, but merely in terms of the size of the admissible parameter space. There are numerous practical cases in which scientists are facing a choice between models differing in this way (Gelfand et al. 1992), e.g., when comparing models that impose order constraints on the parameters (Hoijtink et al. 2008; van de Schoot et al. 2010). Model selection criteria are in need of refinement if they are supposed to apply to such cases of model selection as well.

For the information criteria AIC and DIC, and for model selection based on the notion of minimum description length (MDL), adapted versions are available, or being developed, in which comparisons between models that differ in size can be accommodated (Anraku 1999; Balasubramanian 2005; van de Schoot et al. 2010). The idea is that models that admit a smaller range of parameter values are simpler, much like models that include a smaller absolute number of parameters. By reworking the derivations of the traditional model selection criteria, Anraku et al. and van de Schoot et al. arrive at respectively an AIC and a DIC that, next to model fit, express the complexity of the model in terms of dimensionality and in terms of the size of the admissible parameter space. The work of Rissanen (1996) and Balasubramanian (2005) goes even further. They identify terms that concern the inherent complexity of the model in virtue of model size as well as parametric complexity.

The object of the present paper is to accomplish something similar for the BIC: to reconsider its derivation in order to arrive at a *prior-adapted* BIC that expresses complexity in terms of both dimensionality and size, so that comparisons of the above type become possible. As may be expected from the strong similarity between MDL and BIC, we will arrive at an expression for complexity that is similar to the one that model selection based on MDL arrives at: it includes the relative size of the range of parameter values. However, differences emerge over the exact interpretation of the notion of size and the way in which it surfaces in the derivation. The results will give rise to a change in our understanding of statistical simplicity. Another novelty of this paper is in the application of model selection ideas to a well-known puzzle originally discussed by Kahneman et al. (1982): Linda the bank teller. The developments of the present paper suggest a particular take on this puzzle.

The setup of this paper is as follows. In Sect. 7.2 we review the BIC and clarify some interpretative issues surrounding it. In Sect. 7.3 we say exactly what models are at stake in the paper, and pinpoint the problem that is resolved in it. In Sect. 7.4 we propose the prior-adapted BIC as a solution. Then in Sect. 7.5 we interpret the additional term in the BIC as a particular kind of penalty, and we briefly contrast this penalty with the one featuring in MDL. In Sect. 7.6 we apply the result to the puzzle of Linda the bank teller. Section 7.7 concludes the paper.

7.2 Introduction to the BIC

We introduce the BIC and resolve an interpretative issue. This prepares for the application of the BIC to models with truncated priors.

7.2.1 BIC as Approximate Marginal Likelihood

Let $\langle W, \mathcal{F}, P \rangle$ be a probability space. We have W as a set of basic elements, often taken as possible worlds, and $\mathcal{F}$ an algebra over these worlds, often understood as a language. Sets of worlds are propositions in the language. Over the algebra we can define a probability function P, which takes the elements of the algebra, namely the propositions, as arguments. Let $\mathcal{D}$ be the algebra for the data samples D_n in which n indexes the number of observations in the sample. Let $\mathcal{H}$ the algebra based on a partition of statistical hypotheses H_θ, and $\mathcal{M}$ the algebra based on a partition of models M_i. The part of the algebra concerned with statistical theory is denoted $\mathcal{T} \subset \mathcal{M} \times \mathcal{H}$: it consists of a number of models, each including a distinct subset of statistical hypotheses. In the full algebra, each pair of model and hypothesis is associated with the full algebra of samples, so $\mathcal{F} = \mathcal{T} \times \mathcal{D}$.

We can define the hypotheses and models by means of so-called characteristic functions, c_θ and c_i, that assign a world w the value 1 if it belongs to the hypothesis H_θ and the model M_i respectively:

$$\begin{aligned} H_\theta &= \{w : c_\theta(w) = 1\}, \\ M_i &= \{w : c_\theta(w)\, c_i(w) = 1 \text{ and } \theta \in S_i\} \end{aligned}$$

A model M_i is connected to a collection of statistical hypotheses H_θ, labeled by parameters θ whose values are restricted to some set of values S_i. In the following it will be convenient to refer to the range of hypotheses associated with a model M_i:

$$\begin{aligned} R_i &= \{w : c_\theta(w) = 1 \text{ and } \theta \in S_i\} \\ &= \{H_\theta : \theta \in S_i\}. \end{aligned} \tag{7.1}$$

In this setup, a model covers a range of hypotheses as usual, but the hypotheses and ranges of hypotheses need not be strictly included in the models. They may intersect with any number of models.

The probability function P ranges over samples, hypotheses and models alike. But statistical hypotheses are a special type of set: they dictate the full probability assignment over the algebra of data samples associated with them. For every model with which the hypothesis intersects, we have a so-called likelihood of the hypothesis H_θ for the data D_n,

$$P(D_n|H_\theta \cap M_i) = f(\theta, D_n),$$

where f is some function of the sample D_n and the statistical parameters θ. Depending on the nature of the data D_n and the hypotheses H_θ, the function P will have to be a probability density function that can handle conditions with measure 0. To simplify the expositions below, we will assume that the observations are independent and identically distributed. Note that the model index i does not show up in the function f: the likelihoods of a hypothesis are independent of the model of which the hypothesis is part.

By the law of total probability we can now compute the marginal likelihood of the model, i.e., probability of the data D_n conditional on the model M_i:

$$P(D_n|M_i) = \int_{R_i} P(D_n|H_\theta \cap M_i)P(H_\theta|M_i)\, dH_\theta. \tag{7.2}$$

The marginal likelihood is the average of the likelihoods of the hypotheses in the model, weighted with the prior probability density over the hypotheses within the model. The marginal likelihood of a model is the central mathematical notion in this paper.

The idea behind the Bayesian information criterion (BIC) is that models are selected on the basis of their posterior probability, as determined by their marginal likelihood. The BIC is eventually an approximation of twice the negative logarithm of the marginal likelihood of the model:

$$-2\log P(D_n|M_i) \approx -2\log P(D_n|H_{\hat{\theta}} \cap M_i) + d_i \log(n) = BIC(M_i). \tag{7.3}$$

We choose the model for which the BIC is lowest. The expression $\hat{\theta}$, or more specifically $\hat{\theta}(D_n, M_i)$, signifies the maximum likelihood estimator, a function that maps the data D_n onto the hypothesis $H_\theta \in R_i$ for which $P(D_n|H_\theta)$ is maximal. So the term

$$P(D_n|H_{\hat{\theta}(D_n,M_i)} \cap M_i)$$

in the BIC is the likelihood for data D_n of the maximum likelihood estimate $H_{\hat{\theta}}$ within the model M_i, where this estimate is itself based on those data D_n. Loosely speaking, the expression $d_i \log(n)$ corrects for the fact that as an approximation of the marginal likelihood, the likelihood of the best estimate is too optimistic. The parameter d_i is the dimensionality of the model, i.e., the number of independent parameters in the model, and n is the number of independent observations in the data D_n. In the terms $P(D_n|H_{\hat{\theta}} \cap M_i)$ and $d_i \log(n)$, the BIC reflects both the fit and the complexity of the model.

7.2.2 *The Likelihoodist Information Criterion?*

Bayesian methods concern probability assignments to statistical hypotheses and not just to data samples. In the foregoing, hypotheses are assigned a probability while themselves being associated with a probability assignment over the data. One reason to call the above information criterion Bayesian is that it depends on such probability assignments over statistical hypotheses, $P(H_\theta|M_i)$. Another reason is that the marginal likelihoods enable us to compute the posterior probabilities over the models:

$$\frac{P(M_1|D_n)}{P(M_0|D_n)} = \frac{P(D_n|M_1)}{P(D_n|M_0)} \frac{P(M_1)}{P(M_0)}.$$

On the assumption of a uniform probability over candidate models, $P(M_1) = P(M_0)$, the likelihood ratio and the ratio of posteriors are equal. The logarithm of the ratio of posteriors is then approximated by the difference between the BICs of the models. Both interpretations of the BIC are viable.

This application of the BIC requires a specific setup of the probability space. Posterior model probabilities do not make much sense if the models under comparison are literally nested. Imagine that we had defined the following sets of hypotheses as our models:

$$\begin{aligned} R_0 &= \{H_\theta : \theta \in [0,1]\}, \\ R_1 &= \{H_\theta : \theta = 1\}. \end{aligned}$$

Then we have $R_1 \subset R_0$, in which case $P(R_0|D_n) > P(R_1|D_n)$ by definition, whatever the data. While we can still meaningfully compare the marginal likelihoods, a Bayesian comparison of posterior model probabilities for such sets of hypotheses is uninformative.

One possible response here is to replace R_0 by the set

$$R_2 = R_0 \setminus R_1 = \{H_\theta : \theta \in [0,1)\}.$$

The sets R_1 and R_2 are indeed disjunct, and because we have $P(R_1|R_0) = 0$ for any smooth probability density over R_0, the models R_0 and R_2 yield the same marginal likelihoods. However, in the setup of the probability space given above, we need not resort to redefining the models. To our mind, a more attractive response is to carefully define the models that are involved in the comparison, and to not confuse such models with sets of hypotheses. The sets H_θ may overlap with both model M_0 and model M_1.

While this makes posterior model probability a perfectly cogent notion, the BIC of a model remains primarily an approximation of marginal likelihood and not of the posterior of a model. For this reason, it is perhaps better to refer to the BIC as a likelihoodist information criterion, thus avoiding the interpretative problems with posterior model probability altogether. But here is not the place for proposing such terminological revisions. It suffices to note that the BIC is concerned with likelihoods as an expression of empirical support, and that a probabilistic comparison of seemingly overlapping models is not problematic, provided that we define our probability space cautiously.

7.3 Comparing Models with Truncated Priors

This paper concerns a particular application of the BIC, namely the comparison of models that do not differ in dimensionality but only in the admissible range of hypotheses. In such cases there is a problematic discrepancy between the BICs and the marginal likelihoods of the models.

7.3.1 *Truncated Priors*

The comparisons of this paper involve models whose ranges of hypotheses are nested, and whose priors only differ by a normalisation factor. Such comparisons are well-known from the practice of science, as for example in Klugkist et al. (2005).

By nested ranges of hypotheses, we mean that the range of hypotheses in model M_0 encompasses the range associated with model M_1, or equivalently, that the range of model M_1 is constrained relative to that of model M_0. Formally, we have $R_1 \subsetneq R_0$. The following encompassing and constrained ranges are a good example:

$$R_0 = \{H_\theta : \theta \in [0,1]\}, \tag{7.4}$$

$$R_1 = \{H_\theta : \theta \in [0, \tfrac{1}{2}]\} \tag{7.5}$$

Calling the associated models M_0 encompassing and M_1 constrained suggests that they are themselves nested, but in the present setup this is not so. Following Eq. 7.1, the models are disjunct sets. For $\theta \leq \frac{1}{2}$ the hypotheses H_θ intersect with both models, while for $\theta > \frac{1}{2}$ we have $H_\theta \subset M_0$. Hence we can meaningfully talk about posterior model probabilities as well as marginal likelihoods.

We tell apart encompassing and constrained models by the range of hypotheses R_i: the models are both part of a probability space with a single probability function P. Effectively, these models differ in the regions over which the prior is nonzero. The following abbreviation of the conditional probability function will be useful in the derivations below:

$$P_i(\cdot) = P(\cdot | M_i).$$

The probability density $P_i(H_\theta)dH_\theta$ is nonzero over $H_\theta \in R_i$ and zero everywhere else in $R_0 \setminus R_i$. For the likelihoods we simply have $P_i(D_n|H_\theta) = P(D_n|H_\theta)$ throughout $H_\theta \in R_0$.

We impose one further restriction to the class of model comparisons at stake in this paper: the prior P_1 is a truncated version of the prior P_0. Take any pair of encompassing and constrained models M_0 and M_1, and define a proper prior density $P_0(H_\theta)dH_\theta$ over the encompassing model M_0. This prior can be used to compute an associated prior over a constrained model M_1, the so-called truncated prior, as follows:

$$P_1(H_\theta)dH_\theta = \frac{1}{P_0(R_1)} P_0(H_\theta)dH_\theta,$$

where

$$P_0(R_1) = \int_{R_1} P_0(H_\theta)\, dH_\theta.$$

Within the domain that the two models have in common, the prior over the constrained model has the same functional form as the prior over the encompassing model. But it is normalized relative to the parameter space of the constrained model. Because of this normalisation, we have that $P(M_0) = P(M_1)$. The constraint is the only difference between them.

To illustrate the second restriction with the models of Eqs. 7.4 and 7.5, recall that $S_0 = [0, 1]$ and $S_1 = [0, \frac{1}{2}]$. A uniform prior density for the encompassing model is $P_0(H_\theta) = 1$. The corresponding truncated prior density for the constrained model is $P_1(H_\theta) = 2$ within $\theta = [0, \frac{1}{2}]$ and zero elsewhere.

7.3.2 *Peaked Likelihoods and Concentrated Posteriors*

We add two more restrictions on the model comparisons that are at stake in this paper, one concerning the posterior probability over the model M_0 and one concerning the likelihood function. Their import is that with increasing sample size, the posterior probability within model M_0 will collect around a unique maximum likelihood hypothesis $H_{\hat{\theta}}$.

First, we require that as the sample size n grows, the posterior probability in the encompassing model gets concentrated around the maximum likelihood point $\hat{\theta}$. Consider an environment

$$B(\hat{\theta}, r) = \left\{ H_\theta \in R_0 : |\theta - \hat{\theta}| < r \right\}$$

for some fixed $r > 0$. The model M_0 must be such that

$$\lim_{n \to \infty} P_0\left(B(\hat{\theta}, r) | D_n\right) = 1. \tag{7.6}$$

Note that r can be chosen arbitrarily small. The model M_0 is such that in the long run almost all probability will be collected inside this arbitrarily small environment around the maximum likelihood estimate. Importantly, if $H_{\hat{\theta}} \in R_1$, then the equivalence of the likelihood function and, up to normalisation, of the prior over the models M_0 and M_1 makes sure that the same limit statement holds for $P_1(B(\hat{\theta}, r)|D_n)$.

Second, for increasingly large sample size n we require that the likelihood function over the model M_0 is increasingly sharply peaked around $\hat{\theta}(D_n, M_0)$. Formally, we assume for all $\theta \neq \hat{\theta}$ that

$$\lim_{n \to \infty} \frac{P_0(D_n|H_\theta)}{P_0(D_n|H_{\hat{\theta}})} = 0. \tag{7.7}$$

This requirement may look superfluous, given that we have already assumed Eq. 7.6. But with a sufficiently non-smooth likelihood function it is possible to satisfy the requirement on the posterior for $B(\hat{\theta}, r)$ while maintaining an ever smaller patch of the parameter space outside $B(\hat{\theta}, r)$ at almost equally high likelihood. The requirement of Eq. 7.7 determines that such patches do not exist. In turn, this requirement does not entail that for growing sample size n the posterior probability gets concentrated around the maximum likelihood point: without the assumption of Eq. 7.6 the maximum likelihood might itself be an isolated peak. Relative to it, the peak around which all the posterior probability is collected may have negligible height.

We want to stress that the two requirements above are nothing out of the ordinary. Most models for independent and identically distributed trials for which the likelihood function

$$P_0(D_1|H_\theta) = f(\theta, D_1)$$

is sufficiently smooth will satisfy them. We rely on the blanket requirements because detailing the exact conditions on the asymptotic behaviour of the likelihood function presents us with an unnecessary detour.

7.3.3 *BIC for Truncated Priors*

In what follows we will be concerned with a problem in the application of the BIC to model comparisons of the above kind. In short, the problem is that the difference between the encompassing and constrained models always shows up in their marginal likelihoods, while their BIC is in some cases equal. Therefore, if the BIC is to accommodate the comparison of such models, it needs to be refined.

To explain the problem, we observe that the marginal likelihood of models that differ in terms of a truncated prior are intimately related. We start by rewriting the marginal likelihood of M_1:

$$\begin{aligned} P_1(D_n) &= \int_{R_1} P_1(H_\theta) P_1(D_n|H_\theta)\, dH_\theta \\ &= \frac{1}{P_0(R_1)} \int_{R_1} P_0(H_\theta) P_0(D_n|H_\theta)\, dH_\theta. \end{aligned}$$

So for both marginal likelihoods we can restrict attention to the probability function P_0. By Bayes' theorem the term appearing under the integration sign is

$$P_0(D_n|H_\theta) P_0(H_\theta)\, dH_\theta = P_0(D_n) P_0(H_\theta|D_n)\, dH_\theta.$$

The term $P_0(D_n)$ can be placed outside the scope of the integral, so the functional form of the terms within the scope is that of the posterior distribution over R_0. Note further that

$$\int_{R_1} P_0(H_\theta|D_n)\, dH_\theta = P_0(R_1|D_n),$$

so that we can derive

$$\frac{P(D_n|M_0)}{P(D_n|M_1)} = \frac{P_0(R_1)}{P_0(R_1|D_n)}. \tag{7.8}$$

We now consider this ratio of marginal likelihoods for large sample sizes, to see if its behaviour is matched by the behaviour of the BICs for the two models. We distinguish two regions in which the maximum likelihood estimate $\hat{\theta}(D_n, M_0)$ may be located, namely inside and outside the domain R_1.

We concentrate on the good news first. Say that the maximum likelihood estimate lies outside R_1, or more formally, $\hat{\theta}(D_n, M_0) \in S_0 \setminus S_1$. Following Eq. 7.6, we can always choose r such that $B(\hat{\theta}, r) \subset R_0 \setminus R_1$. From Eq. 7.6 and the fact that

$$P_0(R_1|D_n) < 1 - P\left(B(\hat{\theta}, r)|D_n\right)$$

for all n, we conclude that for increasing sample size n the ratio of marginal likelihoods of Eq. 7.8 is unbounded. Fortunately, the BIC replicates the behaviour of the marginal likelihoods in this case. From the requirement of Eq. 7.7 we have that the likelihood of the maximum likelihood hypothesis within M_1 is negligible in comparison to the likelihood of the maximum likelihood hypothesis within M_0:

$$\lim_{n\to\infty} \frac{P_1\left(D_n|H_{\hat{\theta}(D_n,M_1)}\right)}{P_0\left(D_n|H_{\hat{\theta}(D_n,M_0)}\right)} = 0$$

Moreover, the maximum likelihood term is dominant in both BICs: it grows with $O(n)$ while the complexity term $d_i \log(n)$ is the same for both models. So we can derive that $BIC(M_1) - BIC(M_0)$ tends to infinity for growing sample size n, thus matching the behaviour of the marginal likelihood.

But now consider the case in which the maximum likelihood estimate lies inside the domain R_1, that is, $\hat{\theta}(D_n, M_0) \in S_1$. We can always choose r such that $B(\hat{\theta}, r) \subset R_1$. From Eq. 7.6 and the fact that

$$P\left(B(\hat{\theta}, r)|D_n\right) < P_0(R_1|D_n)$$

we conclude that for increasing sample size n the ratio of marginal likelihoods of Eq. 7.8 tends to the value $P_0(R_1)$. We thus expect to see a difference of $2\log P_0(R_1)$ between the values of the BIC for the two models as well. But the BICs of the two models are exactly the same! The maximum likelihood terms are equal, because R_1 includes the maximum likelihood estimate $H_{\hat{\theta}}$. And the penalty terms are equal, because the encompassing and constrained models have an equal number of free parameters.

This is the problematic discrepancy between marginal likelihood and the BIC, alluded to at the beginning of this section. If the maximum likelihood $\hat{\theta}(D_n, M_0)$ lies inside the region R_1, the ratio of marginal likelihoods of encompassing and constrained models tends to the value $P_0(R_1)$. But this difference is not found back in the BICs of the models.

7.4 Prior-Adapted BIC

In the foregoing we derived how the marginal likelihoods for models with truncated priors behave, so we know what a more refined version of the BIC must look like. There may seem little point in an independent derivation of such a refined BIC, as a specialised approximation to something we already know about. Nevertheless, we will in the following closely scrutinize the original derivation of the BIC, based on Jeffreys (1961), Schwarz (1978), Kass and Wasserman (1992) and primarily Raftery (1995), and tweak this derivation in order to capture the effect of a truncated prior on the marginal likelihood. The gain of this is not so much that we thereby arrive at a new model selection tool. Something like that would require a much more extensive motivation. Rather it is that we can draw a parallel between the original BIC and the newly defined $PBIC$, and thus motivate a particular refinement of our notion of statistical simplicity.

7.4.1 Original Derivation

As spelled out in the foregoing, the BIC of a model M_i is an approximation of the marginal likelihood of the model M_i for data D_n. In the original derivation, as reproduced in Raftery (1995), it is shown that

$$\log P(D_n|M_i) = \log P(D_n|H_{\hat{\theta}} \cap M_i) - \frac{d_i}{2}\log(n) + \log P(H_{\hat{\theta}}|M_i) + \frac{d_i}{2}\log(2\pi) - \frac{1}{2}\log|I| + O(n^{-\frac{1}{2}}). \quad (7.9)$$

As before, d_i is the dimension of the model, n is the sample size, and $\hat{\theta}$ is shorthand for the maximum likelihood point $\hat{\theta}(D_n, M_i)$. The quantity $|I|$ denotes the determinant of the expected Fisher information matrix for a single observation D_1, evaluated at the maximum likelihood estimate $\hat{\theta}(D_n, M_i)$. The expression $O(n^k)$, finally, represents terms for which $\lim_{n\to\infty} O(n^k)n^{-k} = 1$.

The first term on the right hand side of Eq. 7.9 is of order $O(n)$, and the second term is of order $O(\log n)$. The next three terms in Eq. 7.9 are of order $O(1)$, and the last represents everything of order $O(\frac{1}{\sqrt{n}})$ or less. Removing the terms with order $O(1)$ or less and multiplying by -2 gives the BIC of Eq. 7.3. The terms of order $O(1)$ or less in Eq. 7.9 can be considered as an error of the estimation of $\log P(D_n|M_i)$. Arguably, the errors can be ignored because the first two terms will dominate the equation as n tends to infinity.

As shown in Kass and Wasserman (1992) and Raftery (1995), in some cases the terms of order $O(1)$ can be eliminated by the choice of a suitable prior. If we choose a distribution with mean $\hat{\theta}$ and variance matrix I^{-1}, we have that

$$\log P(H_{\hat{\theta}}|M_i) = \frac{1}{2}\log|I| - \frac{d_i}{2}\log(2\pi),$$

so that the terms of order $O(1)$ cancel each other out. Moreover, we have an independent motivation for this choice of prior. Roughly speaking, the variance matrix I^{-1} expresses that the prior contains the same amount of information as a single observation, while the mean $\hat{\theta}$ expresses that this information is in line with the average of the data set D_n. In other words, the prior expresses that we have little but adequate prior knowledge.

7.4.2 *Including the Prior in the BIC*

The key idea of the prior-adapted BIC is that this last step in the original derivation must be omitted. The effect of the truncated prior can be found back among the terms of order $O(1)$. Specifically, it can be identified in the prior probability density over the model. Recall that

$$P(H_\theta|M_1)dH_\theta = \frac{1}{P(R_1|M_0)}P(H_\theta|M_0)dH_\theta \tag{7.10}$$

for any value of $\theta \in R_1$ and therefore also for $\hat{\theta}$. We include terms of order $O(1)$ in the BIC, thus creating a *prior-adapted* BIC or $PBIC$ for short:

$$\begin{aligned} PBIC(M_i) \quad = \quad & -2\log P(D_n|H_{\hat{\theta}} \cap M_i) + d\log(n) \\ & -2\log P(H_{\hat{\theta}}|M_i) + d\log(2\pi) - \log|I|, \end{aligned} \tag{7.11}$$

where $P(H_{\hat{\theta}}|M_i)$ is the value of the density function at the point $H_{\hat{\theta}}$ within M_i. If we apply the $PBIC$ to the comparison of models with truncated priors, we recover the difference between the marginal likelihoods derived in the foregoing. To see this, note that

$$\log P(H_{\hat{\theta}}|M_1) = \log P(H_{\hat{\theta}}|M_0) - \log P(R_1|M_0)$$

by Eq. 7.10. As before, if $\hat{\theta} \in R_0 \setminus R_1$ then the first term completely dominates the comparison of the BICs of the models. But if we have that $\hat{\theta} \in R_1$, the first two terms in the $PBIC$ are equal for the encompassing and the constrained model. In that case the third term creates a difference of

$$PBIC(M_1) - PBIC(M_0) = 2\log P(R_1|M_0) < 0,$$

in accordance with the ratio of the marginal likelihoods of Eq. 7.8. By including terms of lower order in the approximation of the marginal likelihood, we thus recover the behaviour of the marginal likelihood.

The remaining task is to show that in applications of the $PBIC$ to model comparisons, the other terms of order $O(1)$ in Eq. 7.9 are the same for the two models. The term $d\log(2\pi)$ is clearly the same, as it only depends on the dimension which is equal for the encompassing and constrained model. We concentrate on the term $\log|I|$. It is the expectation value of how sharply the likelihood function is peaked around the maximum likelihood of one observation, or loosely speaking, it expresses how much information is contained in a single observation. Formally,

$$|I| = \det\left(-E\left[\frac{d^2P(D_1|H_\theta \cap M_i)}{dH_\theta^2}\,|_{\theta=\hat{\theta}}\right]\right),$$

where the expectation is taken over D_1 distributed according to $P(D_1|H_{\hat{\theta}} \cap M_i)$. Clearly, there will be no difference between the two models here, because they have the same likelihood function. Hence, in a comparison of the BIC of an encompassing and constrained model, only the term pertaining to the prior will differ. Retaining the terms of order $O(1)$ in the BIC thus resolves the discrepancy between the marginal likelihood and the BIC. We can use the original derivation of the BIC and get off one stop early to arrive at the $PBIC$.

A potential worry with this quick arrival at the $PBIC$ may be that the accuracy of the approximation of Eq. 7.9 is different for the encompassing and the constrained models, and that this introduces further errors of order $O(1)$ into the approximation. But that worry is unnecessary. Nothing in the accuracy of the approximation, in terms of the order of errors, hinges on the exact region of admissible parameter values.

7.4.3 More Details on the Error Terms

To substantiate the above claim, we now follow the original derivation of the BIC in more detail, and discuss possible differences between the encompassing and constrained model. This discussion is not self-contained but relies heavily on the derivation in Raftery (1995).

The derivation of the BIC employs the so-called Laplacian method for integrals on a Taylor expansion of the function

$$g(\theta) = \log[P(H_\theta|M_i)P(D_n|H_\theta \cap M_i)],$$

as it appears in the marginal likelihood. Its functional form is identical to that of the posterior distribution. This leads to

$$P(D|M_i) = \exp\left[g(\tilde{\theta})\right](2\pi)^{\frac{d_i}{2}}|A|^{-\frac{1}{2}} + O(n^{-1}) \qquad (7.12)$$

with $\tilde{\theta}$ the value where the function $g(\theta)$ is maximal, or in other words the mode of the posterior distribution, and

$$A = \frac{d^2 g}{dH_\theta^2} \,|_{\theta=\tilde{\theta}} \,.$$

Note that the value of $\tilde{\theta}$ may be different for the encompassing and the constrained model. However, the derivation itself is uncontroversially applicable to both models.

In the derivation it is then assumed that the mode of the posterior $\tilde{\theta}$ is close to the maximum likelihood estimator $\hat{\theta}$ for large n, so that $g(\tilde{\theta})$ can be approximated by $g(\hat{\theta})$. For the encompassing and the constrained model, this assumption is warranted by Eq. 7.6, which states that almost all posterior probability will eventually end up arbitrarily close to the maximum likelihood point. After taking the logarithm, we obtain

$$\begin{aligned} \log P(D|M_i) &= \log P(D_n|H_{\hat{\theta}} \cap M_i) + \log P(H_{\hat{\theta}}|M_i) \\ &\quad + \frac{d_i}{2}\log(2\pi) - \frac{1}{2}\log|A| + O(n^{-\frac{1}{2}}), \end{aligned}$$

thus recovering three terms in Eq. 7.9 from Eq. 7.12. The approximation introduces errors of order $O(n^{-\frac{1}{2}})$ for both the encompassing and constrained model, so there is no source for disagreement in this part of the derivation.

The eventual terms of Eq. 7.9, namely $-\frac{d_i}{2}\log(n) - \frac{1}{2}\log|I|$, result from the approximation

$$|A| = n^d |I| + O(n^{-\frac{1}{2}})$$

in Eq. 7.12. Here again, the evaluation at $\tilde{\theta}$ in the derivative A is replaced by the evaluation $\hat{\theta}$ in the derivative I, but we just argued that this is unproblematic. Apart from that, the approximation is based on two further assumptions. One is that the observations in D_n are independent and identically distributed, so that we can restrict attention to one observation. The other is that for large n, the second derivative of $g(\theta)$ is dominated by the likelihood factor, so that we can omit $P(H_\theta|M_i)$ from the derivative.

Both assumptions apply equally to the encompassing and the constrained model. The first was assumed all along. As for the second, recall that the priors of the encompassing and constrained models only differ by a constant factor, and that the likelihood functions of the models are equal. Therefore, if the assumption is indeed satisfied in the encompassing model, then it is also satisfied in the constrained model, and vice versa. Again, the approximation does not introduce any differences between the two models.

7.5 THE ROLE OF THE PRIOR

We conclude that the prior-adapted BIC adequately approximates comparisons of the marginal likelihood of encompassing and constrained models: it yields a difference of $2 \log P(M_1|M_0)$ in favour of the constrained model. As indicated before, we do not think this is enough motivation for using $PBIC$ across the board: for other model comparisons the terms of order $O(1)$ will differ in ways that have not been connected to natural aspects of model selection. In this section we merely argue that the term that shows up in comparisons between encompassing and constrained models has a very natural interpretation. It can lead us to redefine the notion of statistical simplicity accordingly.

Recall that the term $d_i \log(n)$ is often interpreted as a penalty for complexity, or conversely, as a term that affords simpler models a head start. The idea is that models with fewer parameters exclude particular statistical possibilities, therefore run more risk of failing to accommodate the data, and hence deserve pole position. We argue that the very same consideration applies to the constrained model M_1 when compared to the encompassing model M_0: the former deserves a head start because it excludes a number of statistical possibilities and hence runs more risk of being proved wrong. The statistical possibilities eliminated in a constrained model are of a different quantitative order than the possibilities eliminated by omitting a statistical parameter altogether, but constraints eliminate possibilities nonetheless.

Moreover, we believe that the head start of $2 \log P(R_1|M_0)$ is commensurate to the risk that model M_1 runs. Imagine that according to the prior over model M_0, the model M_1 occupies only a very small segment of the parameter space, meaning that $P(R_1|M_0)$ is very small. This means that, by the light of model M_0, the constraints of model M_1 rule out many statistical possibilities and hence that it runs a high risk of being proved wrong. In other words, the head start of model M_1 is proportional to the risk it runs by the light of model M_0.

This interpretation is in line with the model selection tool based on MDL, as derived and presented in Rissanen (1996), Myung et al. (2000) and Balasubramanian (2005). In the MDL quantity expressing the relative merits of models, we find very similar terms:

$$\begin{aligned} SC(M_i) &= -2 \log P(D_n|H_{\hat{\theta}} \cap M_i) + d \log n \\ &\quad +2 \log \left[\int_{R_i} |J(H_\theta)|^{\frac{1}{2}} dH_\theta \right] + \log \left[\frac{|I|}{|J(H_{\hat{\theta}})|} \right]. \end{aligned}$$

Specifically, the term $J(H_\theta)$ is a reparameterisation invariant prior over the space of hypotheses R_0. So the third term above effectively measures the volume of the range of hypotheses R_i, relative to the full range R_0. This term matches the term $2 \log P(R_i|M_0)$ in the expression of the $PBIC$. Similarly, the term involving the Fisher information matrix matches the term $\log |I|$ in the original BIC.

Such parallels are a far cry from showing some kind of equivalence between the prior-adapted BIC and the model selection tool deriving from MDL, or from bringing $PBIC$ up to the level of the MDL tool. For a more rigorous treatment of the relation between the approaches of MDL and BIC, we refer the reader to Grunwald (2007). For present purposes, we restrict attention to some salient differences between the two approaches.

One marked difference is in the interpretation of the Fisher information matrix. In the discussion of the $PBIC$ above this term is largely ignored, because it is the same for the models whose comparisons were the intended application of the $PBIC$, namely encompassing and constrained models. There are, however, numerous cases in which this term is relevant, and in fact very interesting. Very loosely speaking, the term expresses the sensitivity of the maximum likelihood estimate to small variations in the data. As argued in the MDL literature, higher sensitivity is penalised because it is associated with less reliable estimations, and accordingly with a more complex parameterisation over the space of hypotheses. Unfortunately, we must leave the exact role and interpretation of the Fisher information matrix in the $PBIC$ to future research.

In the context of the present paper, another difference between MDL and $PBIC$ merits more attention. Both model selection tools feature a term that expresses the size of the range of hypotheses within the model. But the nature of those terms is very different. In the MDL approach, the size term appears as the integral of an independently motivated reference prior over the model M_0. In the $PBIC$ approximation, by contrast, the size term derives from the way in which the priors over the encompassing and constrained models M_0 and M_1 are related. In turn, this relation between the priors is fixed for the reason that a comparison between the encompassing and constrained models should only hinge on the constraints. They make up the difference between the two models. Therefore, the comparison should not be affected by other differences between the priors than their normalisation.

This ties in with another difference between the two approaches, concerning the objectivity of the notion of size. In the MDL approach, the size term associated with a particular model is determined entirely by the independently motivated reference prior over M_0. The effective size of any range of hypotheses is determined by the density of statistically distinguishable hypotheses within that part. More precisely, the measure over R_0 is determined by the requirement that all equidistant hypotheses must be equally distinguishable throughout R_0, where the degree of distinguishability is equated with relative entropy. Against this, the size term in the $PBIC$, written $P(R_1|M_0)$, is essentially subjective because the prior over M_0 can be chosen at will. This has some repercussions for the interpretation of the size term in the $PBIC$. It expresses not size per se, objectively, but rather the subjectively perceived size of the constrained model M_1, as determined by the subjectively chosen prior over the model M_0.

7.6 Application to the Conjunction Fallacy

There is a nice parallel between the present discussion and a discussion in the psychology of reasoning on the so-called conjunction fallacy. In a nutshell, the problem is that in some cases people intuitively judge one statement more probable than some other statement that is logically entailed by it. A well-known example is the case of "Linda the Bank Teller": against the background of particular information about Linda, namely that she is 31 years old, single, outspoken, very bright, concerned with social issues, and so on, psychological subjects rank "Linda is a feminist bank teller" more probable than "Linda is a bank teller" (Kahneman et al. 1982). Needless to say, this response is incorrect because the proposition that Linda is a bank teller is logically entailed by her being a feminist bank teller.

The parallel between this puzzle and the model comparisons of the foregoing will be apparent. Much like "Linda is a feminist bank teller" cannot be more probable than "Linda is a bank teller" full stop, it seems that the constrained model M_1 cannot be more probable than the encompassing model M_0 (cf. Romeijn and van de Schoot (2008) for an earlier discussion of this parallel). Now one possible response is to resort to the probability space devised in the present paper and maintain that there is, in virtue of the parallel, no real puzzle in the case of Linda. Following this paper, we can provide a somewhat contrived semantics for the two propositions about Linda, such that they do not overlap.

We think that such a response misses the point. Puzzles like the one about Linda invite us to reconsider what the key components of our reasoning are, and how these components can be framed. They are not the objects, but rather the catalysts of investigation. So instead, we employ the parallel between the conjunction fallacy and the model comparisons of this paper to briefly explore a particular development in confirmation theory. We have no intention here to review the extensive literature on the conjunction fallacy or to suggest new solutions. We focus on one particular analysis of the fallacy, given by Crupi et al. (2008). They provide an attractive explanation of the experimental effects by suggesting that people do not respond by ranking the two propositions at stake on their respective probability, but rather on their confirmatory qualities. The proposition ranked first is the one that receives the strongest confirmation from the background knowledge provided.

The perspective of model selection by the $PBIC$ aligns with, and thereby reinforces the analysis of Tentori. The core of this analysis is that the propositions, e.g., those about Linda, are not compared on their probability, but on their likelihoods for the data about Linda. People rank Linda being a feminist bank teller, $F \cap B$, higher than her being a bank teller, B, in the light of data D because they compare $P(D|F \cap B)$ and $P(D|B)$ and judge the former higher than the latter. And this may be the case despite the fact that $F \cap B$ entails B. In our view, the parallel between the conjunction fallacy and the foregoing can be employed to give a further explanation of the relative sizes of the likelihoods. It seems not too far-fetched to portray the proposition B as a model, comprising of many different hypotheses concerning how Linda might be, each associated with its own probability for the data about Linda. In the same vein, the statement $F \cap B$ is a constrained model within B. The key point is that within this constrained model

B there are more hypotheses that assign the data about Linda a high probability, so that the marginal likelihood of B for the data about Linda is higher.

It may well be that, once in possession of a hammer, every problem looks like a nail. The benefits of this particular representation of the conjunction fallacy are perhaps minimal. Nevertheless, we think that the general idea of representing propositions as models is illuminating, and deserves further investigation. An illustration of its value is provided by Henderson et al. (2010), who represent scientific theories as models in order to illuminate issues in confirmation theory and reductionism by means of hierarchical Bayesian modelling. We think that this idea invites the use of model selection tools within probabilistic confirmation theory. It will be interesting to see if these tools can shed new light on old problems in confirmation theory, such as the seeming conflict between likeliness and loveliness (cf. Lipton 2004) and the interplay between accommodation and prediction (cf. Sober and Hitchcock 2004).

7.7 Conclusion

There are three cases when comparing encompassing to constrained models: the models differ in maximum likelihood, they differ in dimensionality, or they merely differ in the range of admissible hypotheses. In the first two of these cases the prior-adapted BIC, or $PBIC$ for short, boils down to the original BIC: relative to the contribution from the likelihood and dimension terms, the additional terms vanish. For the last case, however, none of the terms in the $PBIC$ differ except for the prior term. This term creates a difference between the models related to size, which allows us to choose between the models.

We have argued that the behaviour of the $PBIC$ is in line with the behaviour of the original BIC, and fitting for the comparison of encompassing and constrained models. In such a comparison, it replicates the behaviour of the marginal likelihood, which it is supposed to approximate, by returning a difference of $2\log P(R_1|M_0)$. Moreover, this term can be interpreted along the same lines as the term that involves the number of free parameters, $d_i \log(n)$. Both terms effect a head start for models that exclude statistical possibilities, and therefore run the risk of failing to accommodate the data. Finally, it was seen that the $PBIC$ is naturally aligned with model selection tools based on MDL. However, this is not enough to motivate replacing the BIC by the $PBIC$ across the board. There are many applications for which the behavior of the additional terms has not been interpreted or even investigated.

Instead, we submit that the benefit of deriving the $PBIC$ is conceptual. As argued, we can think of the prior term as pertaining to subjectively perceived model *size*. It is the natural development of the dimensionality of a model, which pertains to a size at a different order of magnitude. We have shown that, in trading simplicity against fit, we cannot act as if one notion of size fits all. Next to reduced dimensionality, a reduced model size gives a small but non-negligible contribution to the simplicity term in the model comparison. We believe that this adds to a clarification of the elusive concept of simplicity in model selection.

Acknowledgements: We thank audiences at the University of Kent and Düsseldorf for valuable discussions. Jan-Willem Romeijn's research is supported by The Netherlands Organization for Scientific Research (NWO-VENI-275-20-013), and by the Spanish Ministry of Science and Innovation (Research project FFI2008-1169). Rens van de Schoot's and Herbert Hoijtink's research is financed by The Netherlands Organization for Scientific Research (NWO-VICI-453-05-002).

References

Akaike, H. (1973). Information Theory and an Extension of the MaximumLikelihood Principle. In *2nd International Symposium on Information Theory*, B. N. Petrov and F. Csaki (Eds.), Akademiai Kiado, Budapest, pp. 267–281.

Anraku, K. (1999). An Information Criterion for Parameters under a Simple Order Restriction. *Journal of the Royal Statistical Society B*, 86, pp. 141–152.

Balasubramanian, V. (2005). MDL, Bayesian inference, and the geometry of the space of probability distributions. In *Advances in Minimum Description Length: Theory and Applications*, P. J. Grunwald et al. (Eds.), pp. 81–99. MIT Press, Boston.

Crupi, V., Fitelson, B. and Tentori, K. (2008). Probability, confirmation and the conjunction fallacy. *Thinking and Reasoning* 14, pp. 182–199.

Gelfand, A. E., Smith, A. F. M., and Lee, T. (1992). Bayesian analysis of constrained parameter and truncated data problems using Gibbs sampling. *Journal of the American Statistical Association*, 87, pp. 523–532.

Grunwald, P. (2007). *The Minimum Description Length Principle.* MIT press, Cambridge (MA).

Henderson, L., Goodman, N. D., Tenenbaum, J. B. and Woodward, J. F. (2010). The structure and dynamics of scientific theories: a hierarchical Bayesian perspective. *Philosophy of Science* 77(2), pp. 172–200.

Hoijtink, H., Klugkist, I., and Boelen, P. A. (2008). *Bayesian Evaluation of Informative Hypotheses*, Springer, New York.

Jeffreys, H. (1961). *Theory of Probability.* Oxford University Press, Oxford.

Kahneman, D., Slovic, P. and Tversky, A. (Eds.) (1982). *Judgment under Uncertainty: Heuristics and Biases.* Cambridge University Press, New York.

Kass, R. E. and Raftery A. E. (1995). Bayes Factors. *Journal of the American Statistical Association*, 90, pp. 773–795.

Kass, R. E., and Wasserman, L. (1992). *A Reference Bayesian Test for Nested Hypotheses with Large Samples.* Technical Report No. 567, Department of Statistics, Carnegie Mellon University.

Klugkist, I., Laudy, O. and Hoijtink, H. (2005). Inequality Constrained Analysis of Variance: A Bayesian Approach. *Psychological Methods* 10(4), pp. 477–493.

Lipton, P. (2004). Inference to the Best Explanation. Routledge, London.

Myung, J. et al. (2000). Counting probability distributions: Differential geometry and model selection. *Proceedings of the National Academy of Sciences* 97(21), pp. 11170–11175.

Raftery, A. E. (1995). Bayesian model selection in social research. *Sociological Methodology*, 25, pp. 111–163.

Rissanen, J. (1996). *IEEE Transactions of Information Theory*, 42, pp. 40–47.

Romeijn, J. W. and van de Schoot, R. (2008). A Philosophical Analysis of Bayesian model selection. In Hoijtink, H., Klugkist, I., and Boelen, P. A. (2008). *Bayesian Evaluation of Informative Hypotheses*, Springer, New York.

Schoot, R. van de, Hoijtink, H., Mulder, J., van Aken, M. A. G. Orobio de Castro, B., Meeus, W. and Romeijn, J. W. (2010a). Evaluating Expectations about Negative Emotional States of Aggressive Boys using Bayesian Model Selection. *Developmental Psychology*, in press.

Schoot, R. van de, Hoijtink, H., Brugman, D. and Romeijn, J. W. (2010b). A Prior Predictive Loss Function for the Evaluation of Inequality Constrained Hypotheses, manuscript under review.

Schwarz, G. (1978). Estimating the Dimension of a Model. *Annals of Statistics*, 6, pp. 461–464.

Silvapulle, M. J. and Sen, P. K. (2005). *Constrained Statistical Inference: Inequality, Order, and Shape Restrictions*, John Wiley, Hoboken (NJ).

Sober, E. and Hitchcock, C. (2004). Prediction Versus Accommodation and the Risk of Overfitting. *British Journal for the Philosophy of Science* 55, pp. 1–34.

Spiegelhalter, D. J., Best, N. G. Carlin, B. P. and van der Linde, A. (2002). Bayesian measures of model complexity and fit. *Journal of Royal Statistical Society B*, 64, pp. 583–639.

Stone, M. (1977). An Asymptotic Equivalence of Choice of Model by Cross-Validation and Akaike's Criterion. *Journal of the Royal Statistical Society B*, 39(1), pp. 44–47.

Faculty of Philosophy
University of Groningen
Oude Boteringestraat 52
9712 GL Groningen
The Netherlands
j.w.romeijn@rug.nl

Team B
Philosophy of the Natural and Life Sciences

Team D
Philosophy of the Physical Sciences

CHAPTER 8

MAURO DORATO

MATHEMATICAL BIOLOGY AND THE EXISTENCE OF BIOLOGICAL LAWS[1]

8.1 INTRODUCTION

An influential position in the philosophy of biology claims that there are no biological laws, since any apparently biological generalization is either too accidental, fact-like or contingent to be named a law, or is simply reducible to *physical* laws that regulate electrical and chemical interactions taking place between merely physical systems.[2]

In the following I will stress a neglected aspect of the debate that emerges directly from the growing importance of *mathematical models of biological phenomena*. My main aim is to defend, as well as reinforce, the view that there are indeed laws also in biology, and that their difference in stability, contingency or resilience with respect to physical laws is one of *degrees*, and not of *kind*.[3]

In order to reach this goal, in the next sections I will advance the following two arguments in favor of the existence of biological laws, both of which are meant to stress the similarity between physical and biological laws.

1. In physics we find an important distinction between laws of *succession* (holding between timelike-related or temporally successive events/facts) and laws of *coexistence* (holding between spacelike-related, coexisting events).[4] Examples of laws of coexistence are the Boyle-Charles law, relating pressure P and volume of gases V to their temperature T($PV=kT$), Ohm's law, relating resistance R to voltage V and intensity of current A ($V/A=R$), or the relation between the length and the period of a pendulum – $T=2\pi(L/g)^{1/2}$. While all of these laws relate events

1 Thanks to the editor Dennis Dieks for some helpful comments and suggestions.

2 See for one John Beatty, "The evolutionary contingency thesis", in: Gereon Wolters and John Lennox (Eds.), *Concepts, Theories and Rationality in the Biological Sciences*. Pittsburgh University Press 1995, pp. 45–81.

3 For a previous defense of this thesis, see Sandra Mitchell, *Unsimple Truths: Science, Complexity, and Policy*. Chicago: University of Chicago Press 2009. I hope to add new arguments so as to strengthen her view. For the idea of degrees of lawhood, see Marc Lange, "Laws, counterfactuals and degrees of lawhood", in: *Philosophy of Science*, 1999, pp. 243–267.

4 See Carl Hempel and Paul Oppenheim, "Studies in the logic of explanation", in: *Philosophy of Science* 15, 2, 1948, pp. 135–175, who contrast causal laws (of succession) with laws of coexistence. The difference between causal laws and laws of coexistence had been originally proposed by John S. Mill.

D. Dieks et al. (eds.), *Probabilities, Laws, and Structures*, The Philosophy of Science in a European Perspective 3, DOI 10.1007/978-94-007-3030-4_8,

or properties that are in some sense simultaneously existing, laws of succession instead describe the unfolding of physics systems in time.

Against the possibility of biological laws, it is typically argued that biological laws of evolution are either non-existent or just too complex to be formulated.[5] For the sake of the argument, let us suppose that this thesis is true.[6] It then follows that if we could prove that (i) in biology, unlike physics, there are also no laws of coexistence, or that (ii) such laws, if existent, are really all physical, we would have concluded against the existence of biological laws *tout court*. In Sect. 8.2, I will counter (i) and (ii) by discussing some examples of genuine biological laws of coexistence that I will refer to as *structural biological laws*.

2. Those who claim that there are no biological laws typically argue that lawlike-looking regularities in biology are either merely *mathematical* (and therefore a priori) *or purely physical*. In the former case, they are devoid of empirical content, in the latter they are empirical but not biological. The former claim has been put forward in particular by Brandom and Sober, and recently defended also by Okasha, by discussing examples like Price's equation, formulas in population genetics like Fisher's, or the simple Hardy-Weinberg's law in genetics.[7] Even though Sober does not think that this is an argument against the existence of laws in biology,[8] it clearly could be used in this way. What I will do in Sect. 8.3 is to counter this claim by citing some mathematical models that seem to be applicable to various biological entities, from cells to flocks of birds, and that are certainly *neither* tautologies nor interpretable just with entities or data models referring to the ontology of current physics.

Before discussing these two arguments in some more detail, however, it is important to clarify two methodological points raised by the issue I have been presenting so far.

---

5 By biological laws of succession I don't mean laws of law, but simply laws regulating the evolution of biological phenomena in time.

6 I don't think it is true, by the way, but I want to concede to the enemy of biological laws all the ground she needs.

7 Samir Okasha, *Evolution and the Levels of Selection*. Oxford: Oxford University Press 2006. By referring to Price's equation, Okasha writes: "though the equation is little more than a mathematical tautology …" *Ibid*, p. 3. Sober explains the Hardy-Weinberg's law with the properties of coin tossing. And then he adds "if we use the term mathematical tautology sufficiently loosely, then many of the generalizations in biology are tautologies" in: Elliott Sober, *Philosophy of Biology*. Oxford: Oxford University Press 1993, p. 72.

8 In Elliott Sober, "Two outbreaks of lawlessness in recent philosophy of biology", in: *Philosophy of Science* 64, 1997, p. S459, we read: "Fisher's theorem of natural selection says that the rate of increase in fitness in a population at a time equals the additive genetic variance in fitness at that time. When appropriately spelled out, it turns out to be a mathematical truth". And yet, he argues, a law need not be empirical but could also hold a priori.

(i) The first point is: when should we regard a regularity/law as biological or physical? In order to answer this first question, let me simply stipulate that a regularity/law can be regarded as biological (or physical) if it is formulated in the language of *current* biology (or physics). As long as a law contains notions or concepts that are regarded as belonging to current biology, we should consider it as biological, even if the notion in question were reducible to physics.[9] I will therefore completely ignore appeals to wholly vague and undefined *future and complete* physics or biology. After all, "in the long run", as Keynes would say, "we will all be dead", and what matters to us is to try to solve our problems relatively to our current state of knowledge.

(ii) The second point is the criterion of demarcation to be used to draw a distinction between genuine laws and merely accidental generalizations. Here I will appeal to counterfactuals, intentionally ignoring the difficulties raised by this criterion.[10] After all, such difficulties apply to physics as well as to biology, and it is not clear at all why the defenders of the existence of biological laws should solve them. Simply put, the main idea to be presupposed in the following is that while empirical generalizations do not hold counterfactuals, laws do. To repeat an oft-quoted example by Reichenbach, a generalization like "all gold cubes are smaller than one cubic kilometer", if true, is true accidentally, since the counterfactual "if x were a gold cube, it would be smaller than one cubic kilometer" does not hold, since no law prevents gold cubes from being larger that one cubic kilometer. On the contrary, given the laws of radioactive decay, "if x were a uranium cube, it would be smaller than one cubic kilometer" is true.

8.2 Laws of Coexistence in Biology

The reader will recall that in the previous section I posed the following two questions: (1) do we have laws of coexistence in biology? If so, (2) are they reducible to physical laws? I will now try to answer them in turn.

1. An important but often neglected source of biological laws might concern exactly laws of the "form", or of the structuring of biological space, in the tradition that spans from Goethe to Cuvier, and from D'Arcy Thompson to Thom and Gould and Lewontin. In this tradition, the permanence of forms or structures from one generation to another "is interpreted in relation to the pure game of three-dimensional space within which the constructive parameters of the organism are

9 Here I assume that reducibility does not entail elimination; and the case of thermodynamics is a clear exemplification of this claim: the reducibility of thermal physics to statistical mechanics does not entail that the properties that are typical of the former together with its laws disappear or are eliminated.

10 One of these is the smell of circularity raised by the criterion: one analyzes the notion of lawhood with counterfactuals but in order to know whether a counterfactual is true, one must already know which laws hold.

established."[11] In this sense the distinction, originating from physics,[12] between laws of coexistence and laws of succession would correspond in biology to the distinction between diachronic "laws of evolution" and "structural laws", the former related to *time*, and the latter constraining the structure of the *spatial* relationships between coexisting biological phenomena and entities.

The recent use of powerful computers has proved quite important to make us discover *structural* biological laws:

> Cardiovascular systems, respiratory systems, plant vascular systems, and insect tracheal tubes all exhibit the same continuously *branching structure* that increases or decreases in scale as a quarter power of body size.[13] (my emphasis)

This wide-scope biological regularity seems sufficient to allow us to respond positively to question (1): there are indeed biological laws of coexistence and they play a very important and generalized role. The following, natural question is whether they are reducible to physical laws which is our question (2).

2. The law of the quarter power mentioned in the quotation above is related to Kleiber's law, which connects the metabolic rate *R*, (*i.e.* the quantity of energy consumed in 1 s), to the dimensions of the animal, according to a precise ratio of proportionality, expressed by the cube of the fourth root of the organism's *body mass M*

$$R = (M)^{3/4} \tag{8.1}$$

For example, an animal c weighing one hundred times another animal m – $M_c = 100\,M_m$ – would have a metabolic rate that is only more or less *thirty* times greater.[14] This law is quite universal, as it holds from mitochondria, unicellular organisms to the largest animals (see Fig. 8.1), so that it definitely holds counterfactuals: if *a* were an animal, it would be related to its metabolism by the above relation.
It could be argued that in virtue of the criterion above, 1 counts as a *physical* law, because it only contains *physical* parameters ("the quantity of energy consumed in a second", "mass"). On the other hand, "metabolism" is typically applied in biological contexts, and "organism's mass" is after all a *physical* property of a

11 Barbara Continenza, and Elena Gagliasso, *Giochi aperti in biologia*. Milano: Franco Angeli, p. 67.

12 The principle of locality might induce one to think that physical laws of succession are more important than physical laws of coexistence, so that the latter somehow reduce to, or supervene on, the former. However, quantum non-separability and entanglement, even in the absence of action at a distance as in Bohm's interpretation, has rehabilitated the importance of laws of coexistence at a fundamental level.

13 J. Brown, G. West, B. Enquist, *Nature* CCLXXXIV, 1999, pp. 1607–1609. The work cited is taken from the website http://www.santafe.edu/sfi/publications/Bulletins/bulletin-summer97/feature.html. A later study published in Nature excluded plants from this generalization.

14 Brown and Enquist, work cited. Note that $M_c = (100)^{3/4}$ equals approximately 31 M_m.

biological entity. Laws of this kind are sort of mixed between physics and biology, and it should be no surprise that in many cases it is indeed difficult to conclude that a given nomic statement belongs to physics or biology. Consider "bridge" disciplines like *bio*physics or *bio*chemistry or molecular biology: any law in these fields cannot but "overlap" between the two disciplines. The existence of such an overlap, however, is good news for the defenders of biological laws, unless their enemies give them ground and retreat to the more limited claim that it is in purely biological domains that laws don't exist. Since this claim will be discussed in what follows, I can move on with my argument.

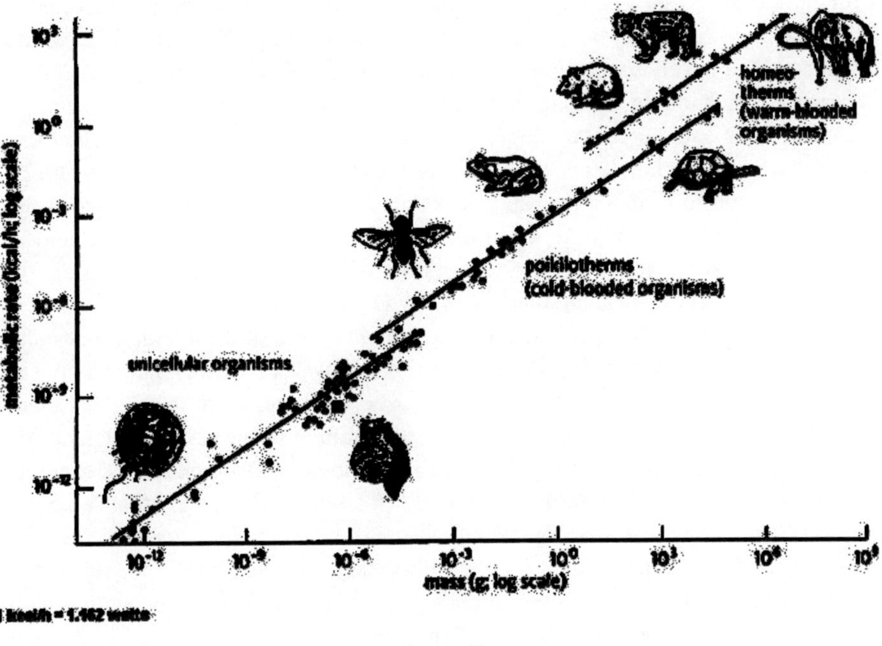

Fig. 8.1[15]

Interestingly, various hypotheses to explain this universal principle have been put forth since 1932. Lately, Kleiber's law has been derived, or explained, by a more profound law of coexistence, namely that the same *ramified model* – which refurnishes a vegetable or animal organism's vital fluids (lymph or blood) – fills the living organism's space like a *fractal*.[16] In a word, this type of ramified structure, which is essential to transport material to and from the cells, would be capable of explaining the existence of the otherwise mysterious proportionality between dimensions and the metabolic rate.

15 Taken from http://universe-review.ca/R10-35-metabolic.htm

16 Other geometrical considerations, involving the fixed percentage of the volume of the body that is occupied by the vessels, explain the presence of the cube in the formula above. The fractal law contributes only to the quarter power component. For more explanatory details, see http://universe-review.ca/R10-35-metabolic.htm.

The omnipresence of forms branching out like a "tree," and repeating themselves in different scales like fractals, can be explained by the fact that these structures optimize the transport of energy in all living species; as West, one of the authors of this theory expresses, "when it comes to energy transport systems, everything is a tree."[17]

While the key concepts entering Kleiber's law are somewhat mixed, the quotation above mentions "cardiovascular systems, respiratory systems, plant vascular systems, and insect tracheal tubes, all exhibiting the same continuously branching structure". We have seen that since all these notions are biological, the criterion for identifying a law as biological allows us to conclude that the fact that "all these structures have a tree-like shape" is a biological law. It could be noticed that it is implausible that a *physical* or "mixed", biophysical law like Kleiber's can be explained by a purely *biological*, structural law, exemplified by biological entities carrying life-sustaining fluids or, more in general, by entities that optimize energy transport. This could create evidence in favor of the view that also the fractal law is really a physical law. However, there is no violation of the causal closure of the physical world in this case, since it is the *shape* of the fractal that carries the explanatory role, and shape in a sense is an abstract, geometrical notion, and in another sense, when we consider it exemplified, is a spatial, topological property of biological entities. As such, the fractal law is a biological law.

The question of the relationship between such structural biological laws and evolutionary principles (or laws of succession, in my vocabulary) naturally poses itself at this point. I cannot enter this complex debate here, if not to note that there is a sense in which biological evolution is constrained by laws of coexistence of the kind we discussed above. On the other hand, however, against recent attempts at downplaying the role of natural selection,[18] it should be admitted that selection would obviously choose the organisms whose "forms" render more efficient the transport of those bodily fluids that are necessary for sustaining the life of the whole organism. In a word, if we could identify biological laws of succession with the family of models generated by the Principle of Natural Selection,[19] biological laws of coexistence and biological laws of succession could and should coexist peacefully, at least if we want to succeed in explaining the fact of evolution.

---

17 *Ibid.*

18 Jerry Fodor, Massimo Piattelli Palmarini, *What Darwin Got Wrong*. New York: Farrar, Straus and Giroux 2010.

19 For the view that the Principle of Natural Selection is really an abstract scheme to form more concrete models (like *F=ma*), see Mauro Dorato, *The Software of the Universe*. Aldershot: Ashgate 2005. For the view that the Principle of Natural Selection is to be understood within the semantic view of theories, see Marcel Weber, "Life in a physical world", in: F. Stadler, D. Dieks, W. Gonzales, S. Hartmann, T. Uebel and M. Weber (Eds.), *The Present Situation in the Philosophy of Science*. Dordrecht: Springer 2010, pp. 155–168.

In this respect, the tradition of the study of laws of the forms, if helped by the development of new mathematical models of the relevant phenomena, could help us to look at the sterile debate between selectionists and defenders of laws of the form in a whole new way. This claim will be illustrated in the next section, which will also provide evidence for the fact, too neglected by philosophers, that the development of a future "mathematics of living beings" will contribute much to both biology and mathematics.

8.3 Some Examples of Mathematical Models in Biology

The currently burgeoning field of mathematical biology can be regarded as providing crucial reasons to believe in the existence of biological laws. The argument for this claim is based on the following four premises, which presuppose a distinction between scientific laws (a defining feature of the model we use to represent the world) and what they purport to describe, namely lawmakers that I refer to as *natural* laws.

1. Scientific laws in physics are mainly dressed in *mathematical language*, a fact that is not an accidental feature, but rather an *indispensable* component of physics;

2. Mathematically formulated scientific laws in physics are part of the definition of the mathematical models of those natural phenomena (natural laws) that we intend to represent *via* the model itself;

3. The amazing effectiveness of mathematical models in *predicting* and *explaining* physical phenomena[20] can only be accounted for if there are natural laws in the physical world, laws that the models mentioned in 2. refer to or partially represent;

4. The three premises above apply also to biology, and guarantee the existence of biological laws rather than accidental generalizations if they do so in physics.

I take it that premise 1. is uncontroversial: since the modern times, it would be hard to do any physics without the abstract models of natural phenomena provided by mathematics. Premise 2. can also be granted: take for instance $ma = -kx$, which is Hooke's law; clearly, this statement also defines the main features of the corresponding abstract model, in the sense that anything that satisfies that law can be represented by the model of the harmonic oscillator.[21] Premise 3. is based on the claim that the existence of mathematical models that enable us to predict and explain physical phenomena *suffices* for the existence of physical laws. This premise

---

20 The claim that mathematics can be used also to *explain* physical phenomena is defended in Mauro Dorato and Laura Felline, "Structural explanation and scientific structuralism", in: A. Bokulich and P. Bokulich (Eds.), *Scientific Structuralism*. Boston Studies in Philosophy of Science: Springer 2011, pp. 161–176.

21 Ronald Giere, *Explaining Science*. Chicago: University of Chicago Press 1988.

is of course as controversial as is any realist claim based on inferences to the best explanation. Here I don't need to defend this premise explicitly, and actually I can take it for granted.[22] Note that 3. is sometimes accepted as being sufficient for the existence of physical regularities, and that here I could be content only with the conditional claim that *if* the inference works for physical laws *then*, in virtue of the analogy between physical and biological models of phenomena on which 4. is based, it also works for biological laws. A case study taken from a recent study of the collective behavior of starlings will, I hope, suffice to argue in favor of the analogy stated in 4.

8.4 Flocks of Starlings and Their Scale Invariant and Topologically-Dependent Interactions

Under the attack of a predator or even independently of it, flocks of starlings (*sturnus vulgaris*) can assume highly symmetrical and rapidly changing geometrical forms. These birds can synchronize their flight in such a way that one is led to think of the flock as a single, *super-individual organism*, whose parts always remain together in a strikingly coordinated fashion.

In the years 2006–2008, the Italian group of statistical physicists and biologists led by Giorgio Parisi has taken thousands of pictures of these birds (which some years ago had invaded parts of Rome with imaginable consequences …) in order to provide a precise empirical basis to study their collective behavior in three dimensions.[23] The guiding idea of the research program was that this empirical study, if suitably modeled, could be generalized to school of fishes, herd of mammals, flight of insects, etc. The scope and universality across the animal kingdom of these dynamical laws, if they could be found, would have been quite impressive.

The collective, cooperative behavior of the starlings is particularly important from an evolutionary point of view. Stragglers have a significantly larger probability of being attacked, while if the group remains together, each individual bird ends up being much safer.

The main question raised by this amazing collective behavior is, of course, how individual birds can remain in the group even when the latter, under attack by a predator changes significantly its form and density.[24] The biological qualitative

22 For a defence of the inference to the best explanation in realist contexts, see Stathis Psillos, *How Science Tracks Truth*. London: Routledge.

23 M. Ballerini, N. Cabibbo, R. Candelier, et al., "An empirical study of large, naturally occurring starling flocks: a benchmark in collective animal behaviour", in: *Animal Behaviour* 76, 1, 2008, pp. 201–215.

24 M. Ballerini, N. Cabibbo, R. Candelier, A. Cavagna, E. Cisbani, I. Giardina, V. Lecomte, A. Orlandi, G. Parisi, A. Procaccini, M. Viale, and V. Zdravkovic 'Interaction ruling animal collective behavior depends on topological rather than metric distance: Evidence from a field study', in: *Proc. National Academy of Science*, USA, 105, 2008, pp. 1232–1237.

laws that had been advanced so far presumed that the interaction among individuals decreased with the *metric* distance between any two birds, as in Newton's law of gravitation. However, this hypothesis would not explain the fact that even after density changes that are typical of starlings flight, the group continues to exist as such.

On the basis of models based on spin glasses and computerized vision, Parisi's group has advanced the new hypothesis that the birds' interaction depends not on *metric* distance (how many meters they are apart from each other) but on their *topological* distance, which is measured by *the number of birds separating each bird from the others with which it interacts*. This implies, for instance, that two interacting birds separated by ten meters and two birds that are one meter apart "attract" each other with the same "strength", independently of distance, since the number of intermediate birds in the two cases is the same.[25] This topological dependency – which I regard as a biological law, possibly interspecific and not just holding for sturnus vulgaris – allows cohesion to the flock even when the density changes. This hypothesis was tested with some simulations:

> Thanks to novel stereometric and computer vision techniques, we measured 3D individual birds positions in compact flocks of up to 2600 starlings ... whenever the inter-individual distance became larger than the metric range, interaction would vanish, cohesion would be lost, and stragglers would 'evaporate' from the aggregation. A topological interaction, on the opposite, is very robust, since its strength is the same at different densities.[26]

So the first species-specific law that we can express in this context, a law that can be expressed in a qualitative and quantitative way, is that *the interaction between starlings does not depend on metric distance but on topological distance*. According to our above specified criterion, this regularity is certainly purely biological. Does it hold counterfactuals, so that, in virtue of the criterion mentioned above, it counts as a law? Relatedly, can we generalize this law to other highly social species?

In order to answer these questions, it is appropriate to mention the fact that the mapping of the flight of the individual birds has shown an interesting *anisotropy*, which could be linked to the nervous system of the birds; this anisotropy means that it is more probable to find the neighboring birds on the side rather than in the direction of flight, and this holds up to six-seven individuals, since there is no interaction with the tenth nearest individual. Charlotte Hemelrijk, a theoretical biologist at Groningen, had found the same sort of anisotropy in school of fishes.[27]

The resilience of the flock against losing individual birds is a metaphor for the resilience of the following regularity: *starlings keep track of topological distance*

25 *Ibid.*

26 *Ibid.*

27 Toni Feder, "Statistical physics is for the bird", in: *Physics Today* 60, 28, p. 29.

by keeping track of 6/7 individuals against possible disturbing factors due to the presence of predators. I would add that the regularity in question is capable of holding counterfactual conditionals: "if *a* were a starling within a flock, it would adjust to changes of densities by keeping track of its 6/7 neighbors". Amazingly enough, the direct interaction with such a limited number of individuals is sufficient to spread correlation among a group that can be formed by thousands of birds!

In order to formulate another species-specific law that can generalize to other species, let me define the *correlation length* as the spatial length or spread of the behavioral correlation existing in a group, and the *interaction range* as the number of animals with which each animal is directly interacting: the former concept can be global, the latter is always local. An effective way to illustrate the difference between these two notions is using the example made by the authors of the research on the scale-free correlation of starlings flocks,[28] namely the "telephone game" played by *n* people. Suppose that each person in a group of *n* whispers a message to her neighbor and so on, and that there is no corruption of the message (no noise):

> The direct interaction range in this case is equal to one, while the correlation length, i.e. the number of individuals the phrase can travel before being corrupted, can be significantly larger than one, depending on how clearly the information is transmitted at each step.[29]

In the hypothesis of no noise, the whole group of *n* person is correlated (so that the correlation length in this example is *n*); of course, in more realistic examples, the information is always transmitted with some noise. We could note in passing that the possibility of sending the same (email) message to *n* people at once (interaction range = *n*) makes the correlation length grow exponentially in a very rapid time.

Cavagna et al. note furthermore that there are various ways to achieve order or correlation among social animals like starlings. One would be via a coordination of all birds' behavior with that of a *single* leader or of a few leaders; such a top-down method, however, would not be very efficient for the survival of birds. For example, if the leader did not notice the presence of a predator or of any other danger, the rigid rule of following the leader would not be of very much help, even if *all* birds, unlikely, had cognitive access to the remote position of the leader (flock can be made by numerous individuals). Furthermore, in this way any fluctuation in

---

28 Andrea Cavagna, Alessio Cimarelli, Irene Giardina, Giorgio Parisi, Raffaele Santagati, Fabio Stefanini, and Massimiliano Viale, "Scale free correlation in starlings flocks", in: *Proc. National Academy of Science*, 107, 26, Jun 29, 2010, pp. 11865–11870, available also on line at www.pnas.org/cgi/doi/10.1073/pnas.1005766107, p. 1.
29 *Ibid.*, p. 2.

the behavior of one bird would not be correlated to the behavior of another, unless the bird in question were the leader.[30]

A much more efficient way to get really cooperative and adaptive behavior is to avoid a centralized global order, but create a global correlation between all animals, a correlation that can be originally caused just by any one individual, the one, say, who notes the presence of a predator. If the change in direction of flight of this individual can rapidly influence all the flock via a few direct interactions between the single animals that is transferred to whole group, then the survival chances of each single animal will be enhanced, because no bird will be isolated. No part of the group can be separated from the rest, and the flock behaves like a critical system, capable of responding in a maximal way to a perturbation occurring to a single individual. With the words of our authors:

> For example, in bacteria the correlation length was found to be much smaller than the size of the swarm. In this case parts of the group that are separated by a distance larger than the correlation length are by definition independent from each other and therefore react independently to environmental perturbations. Hence, the finite scale of the correlation necessarily limits the collective response of the group. However, in some cases the correlation length may be as large as the entire group, no matter the group's size. When this happens we are in presence of scale-free correlations.[31]

The degree of global ordering in a flock is measured by the so-called polarization Φ,

$$\Phi = \frac{1}{N} \sum \frac{\vec{v}_i}{|v_i|}$$

where v_i is the velocity of bird i and N is the total number of birds within the flock (ibid.). Note that the fact that the polarization Φ is very close to 1 (birds fly parallel to each other) may be also considered to be an empirical, quantitative law, since also this statement holds counterfactuals.[32] Polarization is in fact a measure of the correlation of the animal's behavior, in the sense that when the correlation is, as in the case of starlings, close to 1, it is interpretable as the fact that the velocities of the birds are parallel, while when it is 0 "it means uncorrelated behavior, that is, non-parallel velocities."

30 *Ibid.*

31 *Ibid.*, p.1.

32 "Polarization is … a standard measure of global order in the study of collective animal behavior", since when the value is close to 1 it corresponds to parallel velocities, while when it is 0 is mean uncorrelated velocities", "Scale free," quoted, *ibid.*

8.5 Conclusion

The idea that in biology there are no laws (or event quantitative laws) seems to be simply due to a lack of imagination on our part, and to the fact that mathematical biology has not penetrated enough the community of philosophers of biology. So I conclude by quoting from an excellent, recent introduction to mathematical biology, which here I want to advertise, thereby signalling two interesting areas of research in mathematical biology, namely, population biology and ecology on the one hand, and phylogenetics and graph theory on the other.[33]

8.5.1 Population biology and ecology

The problems in population genetics and echology are similar to those illustrated in the case of the collective behavior of starlings, since they relate interaction between single members and collective, global properties. Imagine that a tree in an equally spaced orchard has a disease that, in analogy to the case of starlings, can be transmitted only to the nearest neighbors with a probability p. The problem is to calculate the probability that the correlation becomes scale-free, so that every tree in the forest becomes infected. Let $E(p)$ be the expected probability in question:

> Intuitively, if p is small, $E(p)$ should be small, and if p is large, $E(p)$ should be close to 100%. In fact, one can prove that $E(p)$ changes very rapidly from being small to being large as p passes through a small transition region around a particular critical probability p_c. One would expect p to decrease as the distance, d, between trees increases; farmers should choose d in such a way that p is less than the critical probability, in order to make $E(p)$ small. We see here a typical issue in ecological problems: how does behavior on the large scale (tree epidemic or not) depend on behavior at the small scale (the distance between trees).[34]

In this example scale-free correlations (epidemics among trees) depend on the existence of critical probabilities; it should be obvious how in this case, as in the previous one, the possibility of gathering empirical data allow us to make precise predictions about, say, the existence of scale-free correlations among individuals in a group (flocks, schools, trees in a forest, etc.).

8.5.2 Phylogenetics and graph theory

A connected graph with no cycles is called a tree. The tree has a vertex ρ, or root, and its vertices that have only one attached edge are called leaves. The problem consists in determining the trees that are consistent with our empirical and

---

33 Michael Reed, "Mathematical Biology", in: T. Gowers, J. Barrow-Green and I. Leader (Eds.), *The Princeton Companion to Mathematics*. Princeton University Press, pp. 837–848.

34 *Ibid.*, p. 845.

theoretical information about evolution.[35] Such *phylogenetics rooted trees* are used to select a particular empirical characteristic, say the number of teeth, and then define a function f from the leaves X, the set of current species, to the set of nonnegative integers. For a given leaf *x* (a species in X), one then let f(x) be the number of teeth of members of x.

It is characters such as these that are measured by biologists. In order to say something about evolutionary history, one would like to extend the definition of f from X to the larger set V of all the vertices in a phylogenetic tree. To do this, one specifies some rules for how characters can change as species evolve. A character is called convex if ... between any two species x and y with character value c there should be a path back in evolutionary history from x and forward again to y such that all the species in between have the same value c....A collection of characters is called compatible if there exists a phylogenetic tree on which they are all convex. Determining when this is the case and finding an algorithm for constructing such a tree (or a minimal such tree) is called the perfect phylogeny problem.[36]

The reader will excuse these long quotations. They have the purpose to allow me to conclude that it is by paying more attention to questions like these that a more thorough understanding of the relation physics and biology (and their nomic features) can be gained, a relation that is going to be deeper and deeper the more mathematics is becoming the common language of both. It seems fair to say that biology is becoming more and more, despite what is usually believed, a Galilean science, based as physics is "on sensible experiences and necessary demonstrations".[37]

Department of Philosophy
University of Rome 3
Via Ostiense 234
00144, Rome
Italy
dorato@uniroma3.it

35 *Ibid.*
36 *Ibid.*, p. 846.
37 See Stillman Drake, *Essays on Galileo and the History and Philosophy of Science*, vol. III, selected and introduced by N. Swerdlow and T. Levere, University of Toronto Press, p. 84.

Chapter 9

Federica Russo

On Empirical Generalisations

Abstract

Manipulationism holds that information about the results of interventions is of utmost importance for scientific practices such as causal assessment or explanation. Specifically, manipulation provides information about the stability, or invariance, of the (causal) relationship between (variables) X and Y: were we to wiggle the cause X, the effect Y would accordingly wiggle and, additionally, the relation between the two will not be disrupted. This sort of relationship between variables are called 'invariant empirical generalisations'. The paper focuses on questions about causal assessment and analyses the status of manipulation. It is argued that manipulationism is trapped in a dilemma. If manipulationism is read as providing a conceptual analysis of causation, then it fails to provide a story about the methods for causal assessment. If, instead, manipulationism is read as providing a method for causal assessment, then it is at an impasse concerning causal assessment in areas where manipulations are not performed. Empirical generalisations are then reassessed, in such a way that manipulation is not taken as methodologically fundamental. The paper concludes that manipulation is the appropriate tool for some scientific (experimental) contexts, but not for all.

9.1 Introduction

Manipulationist theorists, in slightly different ways, hold the view that information about the results of interventions is of utmost importance for scientific practices such as causal assessment or explanation.[1]

1 The main theoriser and partisan of the manipulationist (or interventionist) account is no doubt J. Woodward (see J. Woodward, "What is a mechanism? A counterfactual account", in: *Philosophy of Science* 69, 2002, pp. S366–S377; J. Woodward, *Making things happen: a theory of causal explanation*. Oxford: Oxford University Press 2003; J. Woodward, "Causation in biology: stability, specificity and the choice of levels of explanation", in: *Biology and Philosophy* 25, 2010, pp. 287–318. The approach has been also endorsed and used for various purposes by many other scholars, for instance M. Baumgartner, "Interventionist causal exclusion and non-reductive physicalism", in: *International Studies in the Philosophy of Science* 23, 2, 2009, pp. 161–178; S. Glennan, "Mechanisms, causes, and the layered model of the world", in: *Philosophy*

D. Dieks et al. (eds.), *Probabilities, Laws, and Structures*, The Philosophy of Science in a European Perspective 3, DOI 10.1007/978-94-007-3030-4_9,

Specifically, manipulation is meant to provide information concerning the invariance of the (causal) relationship between (variables) X and Y. This means that, in non-technical terms, were we to wiggle the putative cause X, the putative effect Y would accordingly wiggle and, additionally, the relation between the two will not be disrupted. This does not entail that wiggling X will necessarily make Y wiggle, but that, if it does, we will be interested in whether the relationship between X and Y is invariant in the sense sketched above. Such relationships are called *invariant empirical generalisations* and have the characteristic of being exploitable for explanation or for causal assessment. In this paper, I focus on questions related to causal assessment rather than explanation: I will focus on what makes empirical generalisations causal rather than with what makes them explanatory.

Section 9.2 presents the manipulationist account of empirical generalisations and makes it clear that *manipulation* is central for the account. The rest of the paper investigates the status of manipulation for questions of causal assessment. Section 9.3 argues that the manipulationist account is trapped in a dilemma. If the project is read as contributing to the conceptual analysis of causality, then it is at an impasse concerning the *methods* for causal assessment, i.e. no story about *how* to establish whether X causes Y is offered. If the project is read as contributing to the methodology of causality, then a second dilemma opens up. Strictly interpreted, manipulationism fails to offer methods for causal assessment in scientific areas where manipulations are not performed. Charitably interpreted, instead, manipulationism becomes so vague as to be an unilluminating – and even misleading – rationale underpinning causal reasoning in both experimental and nonexperimental contexts. In the light of the previous discussion, Sect. 9.4 reassesses empirical generalisations. The core of agreement with manipulationist theorists is that empirical generalisations are indeed *change*-relating relations and that for empirical generalisations to be causal they indeed have to be invariant, albeit in a sense that does not take manipulations as methodologically fundamental. The importance of the change-relating character of empirical generalisation has to do with the rationale underpinning causal reasoning: it is not manipulation but *variation* that does this job.

9.2 Manipulationist Empirical Generalisations

To understand the manipulationist project, we need to spell out the notions of (i) empirical generalisation, (ii) invariance, (iii) intervention, and the relations they stand with respect to each other.

---

and Phenomenological Research 81, 2, 2010, pp. 362–381; D. Hausman, "Causation, agency, and independence", in: *Philosophy of Science* 64, 4, 1997, pp. S15–S25. Supplement; D. Hausman and J. Woodward, "Manipulation and the causal Markov condition", in: *Philosophy of Science* 71, 5, 2004, pp. 846–856; C. K. Waters, "Causes that make a difference", in: *The Journal of Philosophy* CIV, 2007, pp. 55–579; J. Woodward and C. Hitchcock, "Explanatory generalizations, part I: A counterfactual account", in: *Noûs* 37, 1, 2003, pp. 1–24.

An *empirical generalisation* is a relation between variables that has the characteristic of being change-relating or variation-relating: changes in the putative causal-variable X are associated with changes in the putative effect-variable Y. Of course, the problem of distinguishing causal from spurious or accidental generalisations immediately arises. We could hit upon a change-relating relation that is accidental: an increased number of storks might be statistically associated with an increased number of births, but there is no causal link between these two variables. Or, a change-relating relation might be spurious: yellow fingers might be statistically associated with lung cancer but this is because they are effects of a common cause (cigarette smoking).

Change-relating relations have to show some invariability in order to be causal (or to be explanatory – this falls outside the scope of the paper). This requires manipulating the variables figuring in the relationship itself, and we will call the generalisation *invariant*, roughly speaking, if changing values of the cause-variable changes values of the effect-variable, and yet, the relationship between the cause and effect-variables is not disrupted. Invariant generalisations are then used to ask counterfactual questions about what would happen to the effect, had the cause been different.

However, in order to evaluate the effects of manipulations, not all counterfactuals will do. Relevant counterfactuals are those that describe outcomes of interventions. Consider an empirical generalisation between X and Y. An *intervention* I on X has to have three characteristics: (i) the change in the value of X is totally due to the intervention; (ii) the intervention will affect the value of Y, if at all, just through the change in the value of X; (iii) the intervention is not correlated with other possible causes of Y (other than X). Interventions establish whether changes in the cause will bring about changes in the effect, and yet the relation between the cause and the effect remains unaltered. If this is the case, then invariant empirical generalisations are in fact causal.

A number of examples from physics (e.g. the ideal gas law or Ohm's law) are discussed, inter alia, in Woodward.[2] Illustrations from biology are newer and less known. Consider Dawkins's fictious gene R.[3] When variant r is present, individuals have dyslexia and are unable to learn to read; when variant r′ is present, individuals can learn and read normally. The relation between the gene R and the ability to learn and read is not stable, however. In fact, differences in background conditions (e.g. schooling or culture) disrupt the relation between R and learning and reading. In other words, outcomes in learning and reading are not dependent on manipulations on the gene R. This has to be contrasted, instead, with invariant relationships involving other genes, for instance for eye colours or for external sexual characteristics.

2 See J. Woodward, *Making things happen: a theory of causal explanation*, *loc. cit.*

3 See J. Woodward "Causation in biology: stability, specificity and the choice of levels of explanation", *loc. cit.*

9.3 The Dilemma

The manipulationist account, I now argue, is caught in a dilemma. The dilemma arises because the manipulationist account can be given two readings: conceptual and methodological. *First*, according to the conceptual reading, the account aims to provide truth conditions for causal claims; the sought solution is that X causes Y if, and only if, manipulations on X would accordingly yield changes to Y. *Second*, according to the methodological reading, were manipulations on X to yield changes on Y, then we would be entitled to infer that X causes Y.[4]

Those two readings of manipulationisms lead to a dilemma. In the *first case* – that is if the project is conceptually read – manipulationism turns out to be unilluminating as to the methods to use for causal assessment. In the *second case* – that is if the project is methodologically read – then a second dilemma opens up: (a) *strictly* interpreted, methodological manipulationism is not in a position to offer a solution in domains where it is not possible to intervene (typically, the social sciences, but also astronomy); (b) *charitably* interpreted, the requirement of manipulation becomes so vague as to be not only unilluminating, but also misleading, as to the rationale underpinning causal reasoning.

Ultimately, the dilemma mirrors a more profound problem in the philosophy of causality: the relation between epistemology/methodology and metaphysics. Two remarks are in order. First, specific questions (epistemological/methodological and metaphysical) ought not to be conflated, and instead call for appropriate and distinct answers: we should not give a metaphysical answer to a methodological question, and vice-versa. Second, it is vital to investigate how metaphysical issues have a bearing on epistemological and methodological ones, and vice-versa. This can be done only insofar as different types of questions and of answers are kept distinct. With these caveats in mind, let us now analyse the horns of the dilemma.

9.3.1 Horn 1: Conceptual Manipulationism

According to the *conceptual reading* of manipulationism, X causes Y if, and only if, manipulations on X accordingly yield changes to Y. This, notice, amounts to giving truth conditions for causal claims, and consequently the project contributes to the analysis of the concept of causation. Manipulation is here the *concept* in terms of which causation is cashed out. Under this reading, manipulationism says what has to be true if X causes Y.

---

4 Another strong proponent of the quandary above is M. Strevens (see M. Strevens "Essay review of Woodward", in: *Making Things Happen. Philosophy and Phenomenological Research* 74, 2007, pp. 233–249 and M. Strevens "Comments on Woodward", in: *Making Things Happen. Philosophy and Phenomenological Research* 77, 2008, pp. 171–192.), who distinguishes between conceptual manipulationism and explanatory manipulationism. Given that I focus on questions of causal assessment rather than explanation, I prefer using the more general term 'methodological manipulationism'.

If this reading is correct, then manipulationism is unilluminating for the *methods* to establish whether in fact X causes Y. Nevertheless, it is desirable to have a conceptual analysis of causation that goes hand in hand with methodology. Once we know what a causal relation between X and Y amounts to, it helps a great deal to know *how* to find out what causes what. Conversely, if conceptual analysis and methodology are entirely disconnected, then our understanding and practice of causal inference are too fragmented to be successful. Many objections to standard accounts of causation (probabilistic, counterfactual, regularity, interventionist) stem from the fact that (i) epistemological, methodological and metaphysical questions are conflated and that (ii) most often the bearing of the epistemology/methodology side on the metaphysics side (and vice-versa) have not been thoroughly investigated.[5]

Thus, once we endorse the idea of having coherent (rather than disconnected) methodological and conceptual accounts of causation, then the only possible methodological candidate, under the conceptual reading of manipulationism, is a methodology based on manipulations. In this case, we have to investigate Horn 2 below, which discusses precisely methodological manipulationism.

It is worth noting that manipulationist theorists (and particularly Woodward) claim that the project is methodological rather than conceptual. Thus, Horn 2 below is prima facie more relevant. Nevertheless, the discussion of Horn 2 will reveal that the objections to the methodological reading do press the manipulationist theorist back into Horn 1, whence its relevance for our purposes. Yet, if escaping Horn 1 leads to Horn 2, and in turn, the branches of Horn 2 loop back into Horn 1, then it seems that the manipulationist is stuck between a rock and a hard place. But there is a way out: my reassessment of empirical generalisations offered in Sect. 9.4.

9.3.2 Horn 2: Methodological Manipulationism

According to the *methodological reading*, the perspective is reversed: were manipulations on X to yield changes on Y, *then* we would be entitled to infer that X causes Y. Manipulation is here a *method* to establish whether X causes Y. There is another dilemma opening up now. The requirement of manipulation can be either (a) *strictly* interpreted, or (b) *charitably* interpreted.

9.3.2.1 Horn (a): The Strict Interpretation

Strictly interpreted, manipulationism prescribes the following. In order to know whether X causes Y, perform an intervention on X, hold fixed anything else, and see what happens to Y.

The typical situation where this happens is the controlled experiment. Simply put, in a controlled experiment we compare results obtained from two groups: the experimental and the control group. Those are similar in all relevant respects,

5 See also N. Cartwright, *Hunting causes and using them: approaches in philosophy and economics*. Cambridge: Cambridge University Press 2007.

except for the putative cause, which undergoes manipulation in the experimental but not in the control group. The experimenter can then assess the influence of the putative cause X on the putative effect Y (i) by holding fixed any other possible influence of factors other than X, and (ii) by varying *only* the putative cause. Thus, in a controlled experiment, manipulation is indeed the key *tool* to establish causal relations.

Let me make clear what the *status* of manipulation is. Manipulation is a *tool* to get to know what causes what. It is also worth noting that the controlled experiment is here oversimplified and all the difficulties of experimental design are overlooked. Identifying the right or best groups (of people or any other type of units) to include in the experiment, choosing the right or best intervention to perform, and assessing the effects of such intervention are all far from being trivial and obvious tasks. Randomisation, *the* controlled experiment par excellence, is simple in *principle*, but not in *practice*. In practice, experimental design is a complex and delicate thing, having its own research tradition tracing back at least to the seminal work of Fisher.[6]

The problem, however, is that most situations in social science (and some exist in natural science too, e.g. astronomy) are *not* like controlled experiments. In observational contexts in social science we need methods to find out what causes what without resorting to manipulation. The problem isn't new. Early methodologists in social science recognised this difficulty already in the Sixties. For instance, Blalock,[7] trained in mathematics and physics, promptly admitted that well-designed experiments *could* allow the scientist to make causal inferences based on the outcomes of manipulations "with some degree of confidence" and "with a relatively small number of simplifying assumptions".[8] Blalock then noticed that this isn't the case when scientists have to deal just with *observational* data. The question is not whether *in principle* the same rules of inference can be applied, but how practical difficulties can be overcome in order to make reliable causal inferences on the basis of data coming from nonexperimental studies.

This is not to say that the social sciences do not perform interventions at all. Policy interventions, for instance, are indeed interventions, but the status of manipulation is here different than the one in controlled experiments discussed earlier. Policy interventions are based on a causal story, i.e. on valid empirical generalisations. The results of policy interventions may then lead us to further confirm or to question the validity of empirical generalisations. Thus, manipulation is not a tool to find what causes what. Instead, we manipulate *because* we know what causes what (to the best of our knowledge). In other words, manipulation is a

6 See R. A. Fisher, *The design of experiments*. Edinburgh: Oliver & Boyd 1935, 1st edition.

7 See H. M. Blalock, *Causal inferences in nonexperimental research*. The University of North Carolina Press 1961.

8 *Ibid.*, p. 4.

consequence of a causal story established (usually) in absence of interventions stricto sensu. Witness Birkland:

If the participants in policy making can at least approximate goal consensus, then the next thing they must do is to understand the causal theory that underlies the policy to be implemented. A causal theory is a theory about what causes the problem and what intervention (i.e. what policy response to the problem) would alleviate that problem. Without a good causal theory it is unlikely that a policy design will be able to deliver the desired outcome.[9]

The manipulationist will then rebut that, to find out what causes what, we don't have to *actually* intervene – ideal manipulations will do. In fact, the manipulationist thesis says that were we to intervene on the cause, the effect would accordingly change. Here are my replies to the objection.

First, some ideal interventions may not make any (physical) sense. For instance, imagining an intervention that would double the orbit of the moon (assuming Newtonian gravitational theory and mechanics) to see what would happen to the tides goes far beyond an ideal – in the sense of physically *possible* – intervention.[10] An intervention that is not physically possible – albeit ideal – must be *conceptual*. If we imagine moving the moon in a way that such and such changes on the tides will result, this spurs from our (already established) causal knowledge, but this is not *evidence* to establish a causal relation between the moon and the tides.

Consequently, it is reasonable to suspect that the manipulationist project (also) has a conceptual flavour. This suspicion is reinforced by claims such as "my aim is to give an account of the content or meaning of various locutions such as X causes Y".[11] However, in this way, the manipulationist is in a loop that sticks her back into Horn 1 discussed earlier.

Second, some other ideal interventions cannot be tested and, therefore, causally evaluated. For instance, Morgan and Winship supported the argument that attributes such as gender can be *ideally* manipulated thus:

[…] the counterfactual model could be used to motivate an attempt to estimate the average gain an employed black male working full time, full year would expect to capture if all prospective employers believed him to be white.[12]

However there is no way of testing such 'thought-experiments' against real data. This, again, raises the suspicion that manipulationism is (also) a thesis about the *meaning* of causation. This, again, brings the manipulationist back to Horn 1.

9 T. Birkland, *An introduction to the policy process: theories, concepts, and models of public policy making*. M.E. Sharpe 2010, third edition, p. 241.

10 See J. Woodward, *Making things happen: a theory of causal explanation, loc. cit.*, p. 131.

11 *Ibid.*, p. 38.

12 S. L. Morgan and C. Winship, *Counterfactuals and causal inference*. New York: Cambridge University Press 2007, p. 280.

Third, and most importantly, if the manipulationist stresses the counterfactual aspect of the thesis ('*were* we to intervene…'), then she is definitively providing a conceptual analysis of causation. The manipulationist is in fact stating what must be true about the relationship between X and Y, *if* X causes Y. 'What must be true' corresponds to providing the meaning – and thereby a conceptual analysis – of 'cause' in locutions such as 'X causes Y'. Choosing whether manipulation, or other notions, supplies the meaning of 'cause' may depend on our a priori intuitions about causation as well as on an analysis of the scientific practice – that's beyond the point at stake. The manipulationist theorist is indeed free to hold such a conceptual account, appealing to the best arguments she can produce. But the fact is, even if the conceptual story ('what must be true') is accepted, no methodological story ('how to know whether it is in fact true') is offered. Thus, the manipulationist is irreversibly brought back to Horn 1.

9.3.2.2 Horn (b): The Charitable Interpretation

Charitably interpreted, manipulationism does not prescribe that the *agent* intervenes to find out what causes what. If the agent cannot manipulate, *Nature* will do it for us. Thus, once Nature has manipulated for us, causal assessment is about evaluating changes, or variations, in the putative effect Y due to changes, or variations, in the putative cause X. This is roughly what happens, for instance, in 'natural experiments' in economics or epidemiology. In these observational contexts the assignment of treatment is done 'by Nature' rather than 'by the experimenter'. Two remarks are in order.

First, even if Nature can in principle manipulate (or randomise) for us, we need tools to find out *whether* Nature did in fact manipulate and, if so, whether the manipulation was effective. This means, eventually, that we have to establish what causes what in *nonexperimental* situations. This is exactly the kind of impasse that resulted from the strict reading of Horn (a) discussed earlier. Invariance under intervention (strictly interpreted) then turns out to be too strong a requirement for causal assessment of empirical generalisations in nonexperimental contexts. No wonder, then, that the need for a weaker version of invariance, i.e. *not* based on manipulation, come from the quarters of manipulationists themselves: this is Woodward's notions of weak invariance and of possible-cause generalisations to be discussed later in Sect. 9.4.

Second, the charitable reading of methodological manipulationism suggests that what is of utmost importance is to evaluate whether changes in the putative effect Y occur as a consequence of changes in the putative cause X. If this is correct, then manipulation cannot be interpreted as providing the rationale underpinning causal reasoning. Such interpretation is misleading and disingenuous. Let me elaborate this idea further.

If we let manipulation underpin causal reasoning, the risk is to create another 'gold standard', analogous to randomised clinical trials (RCTs) in evidence-based medicine. But RCTs aren't, by all means, the gold standard either. Criticisms of the alleged superiority of RCTs abound. Here is one. Thompson[13] is concerned that statistical methods *alone* cannot be a reliable tool for causal inference. More specifically, in his argument, the differences between trials in biomedical contexts and trials in agricultural settings – the origin of the Fisherian theory – are the key to understand why randomisation is by and large successful in the latter but not in the former.

The alleged superiority of manipulationist methods over observational ones is based on the idea that non-experimental models try to (actually, struggle to) reproduce the same methodology. Whence the widespread belief that experimental methods are intrinsically better than nonexperimental ones. This idea is questionable. Each scientific method – be it experimental or observational – has its own virtues (and weaknesses). Consequently, the goodness of a method has to be evaluated in the context in which it is used. If we cannot manipulate, it makes no sense to say that a controlled experiment would have been better than a cohort study. What does make sense is, for instance, questioning whether the chosen sampling method for the cohort study at stake was good or not in the given context. Methods are to be evaluated for the job they are supposed to do, not with respect to an alleged gold standard.

The motivation to have a rationale underpinning causal reasoning is to unify different methods under a principle that embraces them all. However, manipulation cannot do that – the impasse that ensued from the strict reading (Horn (a)) made the point. But there is indeed *one* rationale that unifies manipulationist and observational methods as *methods for causal inference*: this is the rationale of variation that I shall present next in Sect. 9.4.

9.4 Empirical Generalisations Reassessed

This section reassesses empirical generalisations. The arguments hereby presented are built upon the same formalism typically employed by manipulationist modellers: causal modelling (or structural modelling). *First*, I present causal modelling as the answer to the same methodological challenge identified by methodological manipulationism: how to find out what causes what. *Second*, I present the variational epistemology underpinning experimental and observational methods and I show how it works within causal modelling. *Third*, I develop a notion of invariance that does not necessarily require manipulation.

Let me make clear what the core of agreement with manipulationist theorists is. I do indeed share with them the idea that empirical generalisations are change-relating relations between variables of interest. I also share the idea that for empirical generalisations to be causal, they have to be invariant, albeit in a sense that

13 See R. P. Thompson, "Causality, theories, and medicine", in: P. Illari, F. Russo, and J. Williamson (Eds.), *Causality in the sciences*. Oxford University Press 2011, pp. 25–44.

I will specify later and that does *not* necessarily involve the notion of manipulation. Let me anticipate the importance of characterising empirical generalisations as *change-relating*: this aspect reflects the *variational* epistemology that underpins causal modelling. The full argument is given below.

9.4.1 Causal Modelling

Causal modelling (also, or alternatively, called structural modelling) constitutes the common ground for discussion with manipulationist theorists.[14] Causal modelling is a methodology the purpose of which is to establish what causes what in a given context. Causal modellers do so by specifying the data generating process (or mechanism) that accounts for the observations in the data set. There is no need to go into technical details, especially related to the statistical properties and tests of those models. The interested reader is directed to Mouchart and Russo; Mouchart et al.; Russo; Russo et al.; Wunsch; Wunsch et al., besides the well-known works of e.g. Pearl and Woodward.[15]

---

14 In the literature, causal and structural modelling are used interchangeably. I am not opposed to this practice, albeit a distinction between the two exists. 'Causal modelling' was introduced by methodologists such as Blalock in the Sixties, and covered different quantitative methods in social science. 'Structural (equation) modelling' is instead a term more familiar to econometricians, who intended to represent, with structural equations, the 'structure' of phenomena as prescribed by economic theory. The two terms can be legitimately used as synonyms insofar as causal models model mechanisms, that is causal structures (see F. Russo, "Correlational data, causal hypotheses, and validity", in: *Journal for General Philosophy of Science*, in press 2011a; "Explaining causal modelling. Or, what a causal model ought to explain", in: M. D'Agostino, G. Giorello, F. Laudisa, T. Pievani, and C. Sinigaglia (Eds.), *New Essays in Logic and Philosophy of Science*. SILF Series, London: College Publications 2011b, Volume I, pp. 347–361.).

15 M. Mouchart and F. Russo, "Causal explanation: recursive decompositions and mechanisms", in: P. Illari, F. Russo, and J. Williamson (Eds.), *Causality in the sciences*. Oxford University Press 2011, pp. 317–337; M. Mouchart, F. Russo and G. Wunsch, "Structural modelling, exogeneity and causality", in: H. Engelhardt and A. P. H-P Kohler (Eds.), *Causal analysis in population studies: concepts, methods, applications*. Dordrecht: Springer 2009, chapter 4, pp. 59–82; F. Russo, *Causality and causal modelling in the social sciences. Measuring variations*. Methodos Series, New York: Springer 2009; F. Russo, "Correlational data, causal hypotheses, and validity", *loc. cit.*; F. Russo, G. Wunsch, and M. Mouchart, "Inferring causality through counterfactuals in observational studies. Some epistemological issues", in: *Bullettin de Methodologie Sociologique – Bullettin of Sociological Methodology*, in press 2011; G. Wunsch, *Causal theory and causal modelling*. Leuven: Leuven University Press 1988; G. Wunsch, "Confounding and control", in: *Demographic Research* 16, 2007, pp. 15–35; G. Wunsch, F. Russo, and M. Mouchart, "Do we necessarily need longitudinal data to infer causal relations?", in: *Bulletin de Méthodologie Sociologique* 106, 1, 2010, pp. 1–14; J. Pearl, *Causality: models, reasoning, and inference*. Cambridge: Cambridge University Press 2000; J. Woodward, *Making things happen: a theory of causal explanation, loc. cit.*

Causal modelling can be schematically presented as step-wise methodology. The *first step* is to define the research question, the population of reference, and the causal context, broadly conceived. This includes taking into account well-established theories, comparative analyses, and preliminary analyses of data. Here is an example from social science in practice. Gaumé and Wunsch[16] investigate the determinants of self-rated health (i.e. of the individual's subjective perception of his/her own overall health). The first thing they do is to specify their research question and to define, consequently, the population of reference and the context: they analyse data related to Baltic countries in the Nineties, i.e. in a post-communist socio-political context.

On the basis of the outputs of step one, the *second step* is to give structure to the joint probability distribution of all the variables. This means 'breaking down' the joint probability distribution into smaller marginal and conditional components. This decomposition reflects the (recursive) structure among the variables.[17] This is also called, following Blalock, the 'conceptual model'. Gaumé and Wunsch,[18] simplifying things quite a lot, come up with a conceptual model where 'Self-rated health', the response variable (effect), directly depends on 'Education', 'Alcohol consumption', 'Locus of control', 'Psychological distress', and 'Physical health'. In their conceptual model, there are also indirect paths, for instance from 'Social support' to 'Self-rated health' via 'Psychological distress' – see Fig. 9.1.

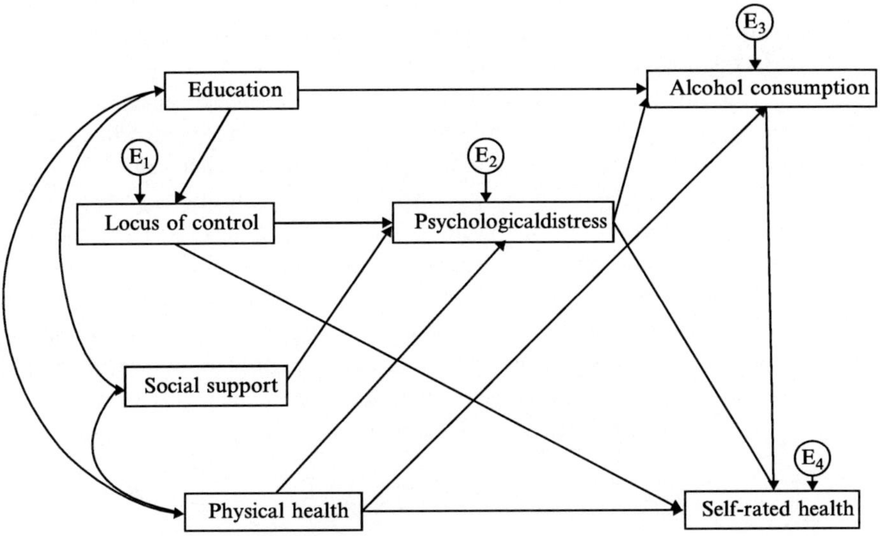

Fig. 9.1 Determinants of self-rated health in Baltic countries, 1994–1999

16 See C. Gaumé and G. Wunsch, "Self-rated health in the Baltic countries, 1994–1999", in: *European Journal of Population* 26, 4, 2010, pp. 435–457.

17 It is customarily assumed that causal models are recursive, that is feedback loops are not permitted. I do not enter here the debate on the plausibility of such assumption, nor the debate on the methods to deal with non-recursive models.

18 *Ibid.*

The *third step* is to translate a conceptual model into an operational model. This means choosing the variables that can be directly measured or proxies for them, choosing the statistical model and the methods of testing. Gaumé and Wunsch[19] fitted the model for four age groups (18–29, 30–44, 45–59, 60+), for both genders, for local individuals and for foreigners (mainly Russians). The authors evaluated the model through Bayesian structural equation modelling using a Monte Carlo Markov Chain procedure.

Once the operational model is in place, the *fourth step* consists in testing the model for invariance: *what* invariance is the matter of controversy with manipulationist theorists. I address this issue later.[20] In Gaumé and Wunsch,[21] the determinants taken into account (alcohol consumption, physical health, psychological health, psychological distress, education, locus of control, and social support) had a remarkable invariant impact on self-rated health across the different Baltic countries, across the time-frames analysed, across gender, ethnicity, or age group.

9.4.2 Variational Epistemology

In the previous section, I considered Horn (b) of the dilemma: if we grant a *charitable reading* to the manipulationist approach, it turns out that the rationale underpinning causal reasoning is misleading and disingenuous. In a nutshell, I am about to argue that the rationale underpinning causal reasoning – both in experimental and observational methods – lies in the notion of *variation*, not manipulation.[22]

To understand what a rationale is and does, we need a brief recap on the philosophy of causality. In the philosophy of causality, two broad areas of investigation may be distinguished: metaphysics and epistemology/methodology. The metaphysics of causality seeks to answer questions about what causality (or a cause) is. It is worth noting that conceptual analysis, in attempting to provide truth conditions for causal claims, or the ultimate content of various locutions such as ‘A causes B’, also contributes to answering questions akin to purely metaphysical ones. The epistemology and methodology of causality, instead, seek (i) to answer questions about how we know about causal relations and (ii) to develop or implement methods for discovery or confirmation of causal relations. It is worth noting that the border between epistemology and methodology is much more blurred than the border between metaphysics and epistemology-methodology.

19 *Ibid.*

20 Invariance is not the only test performed in causal models. Causal models also need to pass tests about goodness of fit or about exogeneity. I am just keeping the discussion focused on the controversy with manipulationist theories.

21 *Ibid.*

22 For a thorough discussion, see F. Russo, “The rationale of variation in methodological and evidential pluralism”, in: *Philosophica* 77, Special Issue on Causal Pluralism, 2006, pp. 97–124; F. Russo, *Causality and causal modelling in the social sciences. Measuring variations*, *loc. cit.*

The quest for a *rationale* of causality falls within the epistemology and methodology of causality and seeks to answer the following question. When we reason about cause-effect relations, what notion guides this reasoning? Is it regularity? Invariance? Production? Manipulation?

It is worth emphasising that rationale and truth conditions are not the same thing. A *rationale* is a principle, notion or concept underlying some decision, reasoning, modelling, or the like. *Truth conditions*, instead, are conditions under which a causal claim is true. According to manipulationist accounts, for instance, 'X causes Y is true' if, and only if, were we to manipulate X would yield changes to Y.

Let me now explain how the rationale of variation works. I give here just a taste of an argument from the causal modelling methodology presented above – the full argument, as well as other arguments, supporting the rationale of variation can be found in Russo.[23]

Causal modelling is regimented by a single rationale guiding model building and testing: the rationale of variation. For the sake of simplicity, consider the reduced form of a structural equation: $Y = \beta X + \varepsilon$, where Y is the putative effect, X the putative cause, β a parameter quantifying the effect of X on Y and ε represents the errors. The first question is whether the data set reveals meaningful co-variations between X and Y. If there are such meaningful co-*variations*, a second question arises: are those variations chancy or causal? In order to assess whether co-variations between X and Y are chancy or causal, we perform a number of tests, including (and perhaps most importantly) tests for invariance.

It is important to notice that the causal equation above can be given a *variational* and a *manipulationist* reading. However, whilst the former is more basic, the latter is derived. Let me explain further. At bottom, a structural equation is read as follows: variations in the putative cause X are accompanied by variations in the putative effect Y. How much Y varies in response to the variation in X is quantified by the parameter β. But this does not imply by all means that X has been manipulated. It could well be, as is typically the case in observational studies in social science, that statements about co-*variations* are based on calculated statistical correlations between the variables. The manipulationist reading is then derived from this basic variational reading as follows. In an experimental setting, manipulations on X make X varying, such that Y varies accordingly. In a controlled experiment, therefore, co-variations in X and Y are due to manipulations, unlike in observational studies.[24]

23 *Ibid.*

24 To be sure, there is also a counterfactual reading, which is, just like the manipulationist reading, derived from the basic variational one. Under the counterfactual reading, the equation says that were we to change X, Y would accordingly change. Notice that testing invariance under the counterfactual reading is far from being a trivial task. Some (e.g. S. Psillos, "A glimpse of the secret connexion: harmonising mechanisms with counterfactuals", in: *Perspectives on Science* 12, 3, 2004, pp. 288–319) have even come to the conclusion that the manipulationist account is, in this respect, parasitic on

This is all to say that variation not only guides causal reasoning in observational settings, but does so also in experimental ones. Notably, the variations we are interested in are exactly those due to the manipulations on the putative cause. In this sense, variation is a *pre*condition to other notions, notably to manipulation. This does not imply that there is no role left to manipulation, though. Manipulation is still a *tool* to find out what causes what, when it can be actually performed, but not always.

9.4.3 Invariance

The last issue to address is *what kind* of invariance is needed in order to establish whether change-relating generalisations are causal or not. Invariance, I argue, does not require interventions stricto sensu. This means that manipulation is not a necessary tool to establish what causes what. Instead, what is required in absence of manipulation is that the relation between the putative cause and effect(s) remains sufficiently stable across different partitions of the data set or across similar populations analysed in the same study.

In Gaumé and Wunsch,[25] no manipulation is performed. Causal assessment is instead made through testing the stability of the putative causal relationship across different 'portions' of the data set. The different 'portions' have to be carefully chosen. In fact, if we test invariance across sub-populations randomly sampled, we should indeed expect to find, approximately, the same values but with a larger confidence interval; consequently, this test wouldn't be terribly useful.[26] Instead, we should appropriately choose sub-populations, for instance considering different age strata, or different socio-demo-economic characteristics, or different geographical regions, or different time frames.

Invariance tests whether the relation between the cause-variable and the effect-variable(s) has some 'stability of occurrence': whether an empirical generalisation between X and Y is, in a given data set, 'regular enough'. Notice, however, that the kind of regularity hereby required is at variance with the 'traditional' Humean regularity. In fact, invariance is not a condition of regular succession of effect-events following cause-events, but a condition of constancy of the characteristics of the relation itself, notably of the causal parameters.[27]

So the manipulationists' requirement that empirical generalisations be invariant *under intervention* is, in nonexperimental contexts, pretty strong. The reason, as we have seen, is that in those cases we cannot intervene, and yet some form of invariance is required nonetheless. Some manipulationist theorists, apparently,

---

the existence of laws of Nature, which would justify why it is the case that Ohm's law turns out to invariant under (counterfactual) intervention.

25 C. Gaumé and G. Wunsch, *loc. cit.*

26 Thanks to Guillaume Wunsch for drawing my attention to this point.

27 For a discussion, see F. Russo, *Causality and causal modelling in the social sciences. Measuring variations*, *loc. cit*, ch. 4.

agree. According to Woodward,[28] there are 'possible-cause' generalisations that state, at bottom, that the presence of a type cause C raises the probability of an effect of type E. One example used by Woodward is 'Latent syphilis causes paresis'. These 'possible-cause' generalisations are exactly the kind of generalisations established by means of causal models, routinely used in the special sciences. Here, the invariance requirement is weakened[29]: 'weak invariance' is not to test whether the generalisation would remain stable were we to intervene, but whether the generalisation is stable across subpopulations.

To illustrate, Woodward[30] discusses a pioneer paper on the relations between smoking and lung cancer. Woodward notices that this paper was written in 1959, when detailed knowledge about the biochemical mechanism through which smoking produces cancer was still lacking. Thus, this study largely relies on epidemiological evidence – that is observational data – and only to a lesser extent on experimental studies of laboratory animals. Woodward then points out that the authors do not aim to formulate 'exceptionless generalisations' (i.e. laws); instead they establish a causal link between smoking and lung cancer because the relation turns out to be *invariant*. What kind of invariance? Exactly the kind of invariance discussed above: stability of the relationship across subpopulations. Let us read the passage:

> For example, the authors note that some association appears between smoking and lung cancer in every well-designed study on sufficiently large and representative populations with which they are familiar. There is evidence of a higher frequency of lung cancer among smokers than among nonsmokers, when potentially confounding variables are controlled for, among both men and women, among people of different genetic backgrounds, across different diets, different environments, and different socioeconomic conditions [...]. The precise level and quantitative details of the association do vary, for example, the incidence of lung cancer among smokers is higher in lower socioeconomic groups, but the fact that there is some association or other is stable or robust across a wide variety of different groups and background circumstances.[31]

The difference between the account of invariance hereby offered from the one of the manipulationists is that invariance is *not* counterfactually defined, *nor* does it necessarily involve manipulation. In making this move I am not claiming originality, as this kind of invariance is currently employed by practising scientists, and was indeed envisaged by Woodward. My point in these discussions is that non-counterfactual invariance, that is invariance *not* based on manipulation is methodologically more fundamental.

28 See J. Woodward, *Making things happen: a theory of causal explanation*, *loc. cit.*, ch. 5.8.

29 *Ibid.*, ch. 6.15 and ch. 7.8.

30 *Ibid.*

31 *Ibid.*, p. 312.

9.5 Conclusion

Manipulationist approaches hold the view that information about the outcomes of interventions is needed for a variety of scientific purposes, e.g. causal assessment or explanation. This paper narrowed down the scope to the role of manipulation for causal assessment. The solution of manipulationist theorists is that an empirical generalisation between X and Y is *causal* insofar as it is invariant under intervention.

I argued, however, that manipulationim is trapped in a dilemma. Manipulationism can in fact be read in two ways. First, if the project is given a conceptual reading, then it appears to be unilluminating from a methodological point of view (Horn 1). Second, if the project is given a methodological reading, then a second dilemma opens up (Horn 2). If methodological manipulationism is strictly interpreted (Horn (a)), then it fails to provide the methodology for observational studies. Or, if it is charitably interpreted (Horn (b)), then the requirement of manipulation becomes so vague and weak as to be not only unilluminating, but also misleading in providing a rationale for causal reasoning in either experimental or observational studies.

In the light of the previous discussion, I reassessed empirical generalisations. Empirical generalisations, I argued, are indeed *change*-relating (or, *variation*-relating) relations – this is the core of agreement with manipulationist theorists. This aspect of empirical generalisation is important because it mirrors the rationale underpinning both experimental and observational studies: it is *variation* – not manipulation – that guides causal reasoning. I also agree that for empirical generalisations to be causal, they have to be invariant. Yet, invariance need not take manipulation as methodologically fundamental.

The broad conclusion is that manipulation is *not* the building block of causal assessment. Manipulation is certainly a good tool, *when* it can be performed, but not always. In other words, there is still room for sound causal assessment in the absence of manipulation. Granted, it is no surprise that, ceteris paribus, manipulations give us higher confidence in causal assessment. But the ceteris paribus clause is important. Well-designed observational studies may deliver more reliable results than poorly designed controlled experiments.

Acknowledgements: I am very grateful to Lorenzo Casini, Brendan Clarke, Donald Gillies, Sebastian Mateiescu, and Jon Williamson for very helpful and stimulating comments. I am hugely indebted to Phyllis Illari. She discussed with me the structure and the contents thoroughly and several times. Of course, any mistakes or inaccuracies remain mine. Financial support from the British Academy is also gratefully acknowledged.

Department of Philosophy
University of Kent
Cornwallis North West
CT2 7NF, Canterbury
United Kingdom
f.russo@kent.ac.uk

CHAPTER 10

SEBASTIAN MATEIESCU

THE LIMITS OF *INTERVENTIONISM* – CAUSALITY IN THE SOCIAL SCIENCES

ABSTRACT

The paper confronts the interventionist theory of causation with one of its main competitors, namely the Causal Model. A dilemma raised against the former is analysed and some possible answers to this quandary are contemplated. Although the limits of interventionism are acknowledged, it is suggested these are not of a principle character. The strengths and the limits of Causal Model are also uncovered and its professed metaphysical neutrality is called into question. It is argued that this theory can not do its job without the need of the inference to the best explanation. The conclusion suggests the disagreement between the two theories lies in their different framing of the social ontology, respective of the ontology of the natural sciences.

10.1 INTRODUCTION

The interventionist theory of causation[1] is based on the principle that causal relationships are relationships which are relevant for manipulation and control. Roughly stated, given a cause C of an effect E, if I can manipulate C in an appropriate way then this should be a tool for modifying E itself. Thus, according to this interpretation, causal statements are analyzable by the means of intervening upon the putative causes and furthermore preserving invariant the relationship between the causal relata. This view has become increasingly popular in the last decades among philosophers and scientists, being used by econometricians, statisticians

1 James Woodward, *Making things happen: a theory of causal explanation*. Oxford: Oxford University Press 2003; James Woodward, "Causation and Manipulability", in: Edward N. Zalta (Ed.), *The Stanford Encyclopedia of Philosophy* (Winter 2008 Edition), online at http://plato.stanford.edu/archives/win2008/entries/causation-mani/; James Woodward, "Agency and Interventionist Theories", in: Helen Beebee, Christopher Hitchcock and Peter Menzies (Eds.), *The Oxford Handbook of Causation*. Oxford: Oxford University Press 2009, pp. 234–262.

D. Dieks et al. (eds.), *Probabilities, Laws, and Structures*, The Philosophy of Science in a European Perspective 3, DOI 10.1007/978-94-007-3030-4_10,

and more recently, by computer scientists.[2] Obviously, it must be that the vast application of this idea in so many causal contexts is also facilitated by the very structure of these states of affairs – they are open to interventions. It is tempting however to speculate about the application of this theory in domains where intervention is not an easy task, in social sciences, for instance. If interventionism will prove to be satisfactory enough in accounting for social causes, then it can be fairly considered the best candidate we have for a *general* theory of causation. If interventionism fails in this attempt, then it is worth seeing what the philosophical reasons for its limits are and what light they can shed upon the concept of causality instantiated in the social and natural sciences.

The aim of this paper is to make such an analysis of the virtues and limits of interventionism when applied to social sciences. I start by laying down a dilemma for the interventionist theory of causation due to Federica Russo's *On Empirical Generalisations*.[3] I then provide a detailed investigation of the structure and assumptions of this dilemma, which is to my knowledge, one of the strongest arguments against professing the validity of interventionism in the field of social sciences. I also provide three possible answers to the quandary raised by Russo (Sect. 10.3). I further draw on the features of the Causal Model, which is Russo's preferred theory of causation. Here I highlight the positive points of this interpretation and I submit the concept of invariance championed by this view to a separate analysis (Sect. 10.4). It will turn out from this detached examination that introducing new conceptual dimensions for the causal variables is the true counterpart of intervention in the Causal Model (Sect. 10.5). I however conclude this section with raising a serious objection to the professed metaphysical neutrality of the Causal Model. The final chapter allows me a short review of the investigation and suggests the root of disagreement between the two evocated theories may consist in their different views about ontology.

10.2 A Dilemma for Interventionism

Russo[4] proposes two possible readings of interventionism[5]: the *conceptual* reading, according to which interventionism provides an analysis for the notion of causation, and a *methodological* reading, which should supply us with a methodology for causal assessment. Under the reasonable assumption that a theory of causation should also tell us *how* to find out about what causes what, Russo[6] argues that interventionism fails to determine an adequate methodology for evaluating causation. Her strategy is to show first of all that the conceptual reading bears no methodological support (Horn 1 of the dilemma). Then, Russo argues the methodological

2 Judea Pearl, *Causality: models, reasoning, and inference*. Cambridge: Cambridge University Press 2000.

3 Federica Russo, "On empirical generalisations", in this volume.

4 *Ibid.*

5 Russo (*Ibid.*) uses 'manipulationism' for 'interventionism', but I prefer the latter one here.

6 *Ibid.*

reading performs no better when it comes to analyze causal relationships occurring in the social world (Horn 2 of the dilemma). As long as interventionism does not come up with an alternative version, it is deemed to be captive of a dilemma made of its theoretic claims (the two horns).

The conceptual interpretation of interventionism states that "C causes E *if*, and only if, manipulations on C accordingly yield changes to E".[7] Russo asserts this way of approaching interventionism affords grasping the meaning of causality. But this reading fails to provide us with a *method* for finding what actually causes what. Thus, the conceptual reading "amounts to giving identity conditions for causal claims, and consequently the project contributes to the analysis of the concept of causation … [Hence, this reading only] says what has to be true *if* C causes E".[8] Therefore, the conceptual reading remains silent when it comes to guide us in *finding* causes in the world. The methodological reading should be further investigated. We expect this later account will furnish a method for evaluating causal relationships. And indeed, under this construal we can assess causal claims by performing interventions in a specific manner: "were manipulations on C to yield changes in E, *then* we will be entitled to infer that C causes E".[9] Russo then goes for (a) a strict clarification of the methodological reading, respectively (b) a charitable interpretation. According to the strict rendering, interventionism urges to intervene on the putative cause C while keeping fixed anything else and then observe what happens to the putative effect E. The paradigm for this is the controlled experiment. Here the experimenter can check the change in the variable E by wiggling the putative cause C and isolating this causal relationship from any other causal path. Russo acknowledges the power of this method but she cast doubts on the validity of this view in analyzing social phenomena. It is a commonly accepted view that the social world resists to controlled experimentation and hence this turns the method professed by the methodological reading into something ineffective. Since we commonly cannot do experiments and intervene in the social realm in the way we do in a controlled experiment, interventionism faces a fatal limit here.[10] Russo further argues that the charitable reading of interventionism performs no better than the strict interpretation. The charitable version prescribes that it is *Nature* itself rather that the agent who manipulates the putative cause. Thus *Nature* 'intervenes' for us when we are not capable of doing this. And henceforth causal assessment reduces to registering appropriate variations in the putative effect when *Nature* changes the value of the putative causes. Nevertheless, Russo contends this is a valid manoeuvre for we still need a *method* to find out whether *Nature* indeed has manipulated for us or not. And it seems quite reasonable to contemplate that such a method should offer means of comparison between experimental and non-experimental cases again. But this, notice,

7 *Ibid.*, p. 3. I prefer here 'C causes E' instead of Russo's 'X causes Y'.

8 *Ibid.*

9 *Ibid.*, p. 4.

10 See Sect. 10.5 below for a method of 'intervening' in social sciences.

is exactly the kind of impasse that resulted from the strict reading of Horn [2](a) discussed earlier.[11] In conclusion, interventionism is lodged in a quandary. It either has to provide us with a reason for rejecting this dilemma or it must acknowledge its own limits in assessing causal claims. Since the second variant as advocated by Russo is explored in Sect. 10.4 below, let me now consider some possible replays to the proposed dilemma.

10.3 Some Remarks on the Dilemma

First, one should notice here that Russo does not claim that a third and different reading of interventionism is in principle forbidden. However, she correctly points out I think, that the interventionist proposal is to support a methodological interpretation for its own claims, at least in Woodward's[12] account. Consequently, interventionism should collapse directly on to Horn 2 of the dilemma, from which there is no way of escaping other than going back into Horn 1 and hence the dilemma.

Second, I take the challenge posed by the conceptual reading as based on a specific view of the semantics of the language. Specifically, it seems to me this reading hinges on the view that *reference* determines the *meaning* of a sentence. According to this position, the meaning of a sentence depends on the specific meanings of the words that compose it and the meaning of these words is in its turn set by the reference of these words. Consequently, if two different words, say 'Clark Kent' and 'Superman' have the same person as their referent, we expect they should have the same meaning. This is why I think Russo[13] considers that providing identity conditions for a statement is the same as determining its meaning.[14] However, for a *Fregean*, it is rather the *sense* or *content*[15] that determines the reference.[16] And it seems we have some good reasons to provide Woodward's account with such a Fregean reading. It may well be the case that for Woodward's theory, the *context* in which a clause is uttered contributes to the sense of such a statement. And why not allow for the *experimental setup* to be representative of the context in this theory? Continuing along these lines of thought it will eventually come up that although providing identity conditions refers primarily to setting out

---

11 Russo, "On empirical generalisations", *loc. cit.* See also the following section for this point.

12 Woodward, *Making things happen: a theory of causal explanation, loc. cit.*

13 Russo, "On empirical generalisations", *loc. cit.*

14 To preserve our example: stating that '*Clark Kent*' *is* '*Clark Kent*' and further that '*Clark Kent*' *is* '*Superman*' is the same with determining the meaning of 'Clark Kent' by the means of providing identity conditions for this name.

15 Also called 'proposition' in the theories of semantics. For details, see Devitt and Sterelny (Michael Devitt and Kim Sterelny, *Language and reality. An introduction to the philosophy of language*. Oxford: Blackwell Publishers 1999).

16 See *ibid.*

the meaning of a clause, this also bears a methodological import.[17] The reason for this is that identity conditions for a statement or expression must eventually ask for information about the context of utterance and implicitly about the details of the experimental setup.[18]

The supporter of interventionism is left here with the task of elaborating on the possible advantages of a different semantic view. No doubt, this is not an easy task. However, a different semantics still has to be completed with a *method* for finding what actually causes what. The interventionist also has to contemplate a way of refuting Russo's general assumption that methodological questions should be kept apart from metaphysical – and I add, semantic – issues.[19] The interventionist is then asked to show that the metaphysical/semantic, methodological and eventually, epistemological issues dealing with causality are inextricably linked together. The supporter of this view will eventually maintain that interventionism does justice to this situation by preserving the three ingredients as entangled. Thus, in contrast to Russo,[20] interventionism could profess the idea that conceptual reading is a necessary approach for elucidating the methods of identifying social or natural causes.[21]

Third, recall that Horn 2 of the dilemma swings between the distinction of experimental and non-experimental causal contexts. Russo however considers the domain of social science as the paradigm for non-experimental causal contexts. In social science we can at most refer to uncontrolled experiments, where generally not all the variables can be held under control and where causal influences others than those coming from the putative cause are also possible. Moreover, direct intervention in such an experimental context is rarely possible and it commonly

17 This is to show against Russo, that the conceptual reading of interventionism may comprise a methodological content.

18 Also notice that in many of Woodward's examples a detailed knowledge (similar I think with what the Causal Model refers to by 'background knowledge') of the causal *context* is supposed before elaborating on the causal relationship. See for this Woodward, "Making things happen: a theory of causal explanation", *loc. cit.*, especially chapters 1 and 2.

19 See Cartwright (Nancy Cartwright, *Hunting causes and using them: approaches in philosophy and economics*. Cambridge: Cambridge University Press 2007) for various reasons of why these three issues refer to different questions and hence why they should be kept separated when they are used in argumentation.

20 Russo, "On empirical generalisations", *loc. cit.*

21 My third remark here can be read along the same lines.

amounts to disrupt the underlying causal mechanism.[22] Woodward[23] however has made the important point that an intervention needs not be physically possible in order to assess a causal claim. The interventionist only needs to judge the situation *as if* an intervention had occurred. Nonetheless, Russo[24] rebuts this view by stating that a *possible intervention* is a conceptual tool. She forges the idea that ideal experiments are not physically possible and can not be tested against the data. So they must be relevant only at the conceptual level and this has to face Horn 1 of the dilemma. My opinion however is that, contrary to what Russo says, there are vast fields of social science which profess controlled experimentation, be it physically possible or just ideal. In Economics, for instance, researchers largely draw on ideal cases or simulations of real market situations and their methods prove to be highly relevant.[25] This shows that the applicability of interventionism to social science is not limited *in principle*. Moreover, although an intervention is not always possible in the field of social science, it may be the case that one can provide with proxies for such limits. It seldom happens that policy interventions yield adequate substitutes for interventions that are not actually possible. For instance, advertising for moderate fats consumption can be such a surrogate since there is no legal way of intervening in this context. Nevertheless, researchers keep doing studies on the impact of fats consumption upon one's state of health. This can lead to a better understanding of the social phenomenon and eventually provides the theoretical basis of proxies for intervention as it is in the case of advertising.[26]

Following the point mentioned above, I however expect Russo to push further the relevance of the distinction between experimental and non-experimental contexts. As a matter of fact, she accuses Woodward[27] of professing this same difference and hence she finds strong reasons for twisting interventionism on its own head.[28] Moreover, Russo points out that sneaking an eye into the actual practice of

---

22 See also Russo (Federica Russo, *Causality and causal modelling in the social sciences. Measuring variations*. Methodos Series. New York: Springer 2009) and Cartwright (Cartwright, *ibid.*) for this point. The same limits for intervening are equally valid in astronomy, either because of the distance between objects or because of practical reasons: one cannot manipulate the orbit of the moon (given Newtonian gravitation and mechanics) in order to see what happens to the tides (Woodward, *Making things happen: a theory of causal explanation, loc. cit.*, p. 131).

23 Woodward, *Making things happen: a theory of causal explanation, loc. cit.*

24 Russo, "On empirical generalisations", *loc. cit.*

25 See for this point, Wenceslao J. Gonzalez, "The role of experiments in the social sciences: the case of economics", in: Dov M. Gabbay, Paul Thagard and John Woods (gen. Eds.), *Handbook of the Philosophy of Science – General Philosophy of Science. Focal Issues*, Amsterdam: Elsevier, 2007, pp. 275–303.

26 For a similar point on this see Judea Pearl, "Review: Nancy Cartwright on hunting causes", in: *Economics and Philosophy* 26, 2010, pp. 69–94).

27 Woodward, *Making things happen: a theory of causal explanation, loc. cit.*

28 This is because Woodward (*Ibid.*, ch. 6 and 7) also accepts a weak form of interventionism, where intervening in order to analyse a causal claim becomes a secondary demand.

social scientists reveals that their main approach is to formulate causal hypotheses and then test them against the data. This amounts to focus on observational data which can do the job of verifying rather than finding tools on how to intervene in a given context. I don't feel fully persuaded by this type of reasoning. The issue at stake here is a matter of principle: whether the difference between experimental and non-experimental contexts puts any limits on interventionism. To invoke a *de facto* situation, for instance that social scientists in practice often use methods other than intervention may be illuminating but it begs the point.

Nevertheless, I do share the common intuition with the adversaries of interventionism that at least for many practical purposes other versions of causal assessment than interventionism may perform better. It seems indeed too strong a requirement to stress with interventionism that all non-experimental situations can be in principle translatable into (ideal) controlled experiments. As for Russo's advancement of a different theory of causality, one finds her intuition more than captivating. It thus becomes desirable to explore this new appraisal of causality.

10.4 The Causal Model

After pondering on the limits of interventionism, Russo passes to her preferred account of causation. She thus endorses the view that causality is best cached out in the framework of the Causal (or alternatively dubbed, Structural) Model.[29] Like interventionism, the Causal Model aims to formulate a method of evaluating causal claims. This paradigm aims to model given causal contexts in order to find out the mechanism that is responsible for the observed causal relationships. It is based on the fundamental idea that causation refers to variations among variables of interest. According to Russo,[30] the methodology of the Structural Model comprises four steps: (i) the *first* step is the formulation of a causal hypothesis, the delimiting the causal context and the selection of the population sample: "This includes taking into account well established theories, comparative analyses and preliminary analyses of data"[31]; (ii) the *second* step includes the building of a conceptual model by representing the situation by the means of relevant variables; also, a probability distribution is furnished among these variables in order to reflect their (recursive) structure; (iii) the *third* step is to supply the conceptual model with an operational interpretation: "This means choosing the variables that can be directly measured or proxies for them, choosing the statistical model and the methods for

29 See Russo (Federica Russo, *Causality and causal modelling in the social sciences. Measuring variations.* Methodos Series. New York: Springer 2009) for both a detailed presentation and historical description of the Causal Model.

30 Russo, "On empirical generalisations", *loc. cit.*

31 *Ibid.*, p. 7.

testing"[32]; (iv) the *fourth* step requires testing the stable character of the model or its *invariance*.[33] We saw that the invariance condition is one of the key features of interventionism as well. As advocated by interventionism, invariance certifies that the effect variable actually depends on the causal variable and not on other variables. Recall that in the interventionist framework this asked for an intervention in order to cancel out all causal paths on which the effect variable might depend. We expect the Causal Model to employ a different strategy for achieving this invariance condition. And indeed, Russo points out that interventionism and the Causal Model share the same intuition regarding the fundamental role of invariance but they differ in the way they settle this invariance stipulation.[34]

As I have already remarked, the Causal Model centers on the idea that causal claims are equivalent with claims about variations among variables of interest. Russo[35] highlights this point by stating that it is built in the very structure of the Causal Model to search for *variations* among the relevant variables. She exemplifies this by emphasizing that Causal Model usually reflects the dependencies among variables within the structure of (for convenience, linear) mathematical equations. A simple equation like $Y = \beta X + \varepsilon$[36] mirrors this situation as the parameter β quantifies the change in Y as due to a variation in X. The Causal Model therefore urges looking for meaningful co-*variations* that are observable in the data input and then to represent them in the variables of the model. When one hits upon these co-*variations*, one only has to check whether they are chancy or causal. The test of invariance as alleged by the supporters of the Causal Model provides us with the adequate tools for establishing the causal import of the detected meaningful variations. I postpone the analysis of the test of invariance for the next section, and for this section let me just add a few remarks on the virtues of the Causal Model. It is instructive to explore these qualities by comparing them with the claims made by interventionism.

The most important contribution of the Causal Model is the emphasis put on variations. Russo feels entitled to consider that the Causal Model takes on a *variational epistemology*, as this theory considers variation as the *unit of measure* when producing causal assessments. Manipulation plays only a secondary role, namely that of being a *constraint* imposed on the relevant variations. Russo rightly indicates I think that both in Woodward's theory, as within many interventionist

32 *Ibid.*, p. 8.

33 Russo exemplifies these stages with an example of Gaumé and Wunsch (Catherine Gaumé and Guillaume Wunsch, "Self-rated health in the Baltic countries, 1994/1999", in: *European Journal of Population* 26, 2010, pp. 435–457). See also Russo (*Causality and causal modelling in the social sciences. Measuring variations, loc. cit.*) for other similar examples from the practice of social science.

34 As will shortly become manifest, the two methods of establishing this invariance requirement reflect the true differences between interventionism and the Causal Model.

35 Russo, "On empirical generalisations", *loc. cit.*

36 Y represents here the effect variable, X is the causal variable and ε measures the error.

approaches, manipulation only comes in to fulfill restrictive goals. It delimits the scope of variations to those which are relevant for the causal context. Accordingly, when one searches for causes in the world he or she primarily looks for variations.

This amounts to saying that the guiding notion in analysis of causality is variation rather than intervention, which is I think another fundamental contribution of the Causal Model. In Russo's own phrasing, "... the rationale underpinning causal reasoning – both in experimental and observational methods – lies in the notion of *variation*, not manipulation".[37] Notice that equipping causality with a rationale is not the same as providing it with identity conditions: "A *rationale* is a principle, notion or concept underlying some decision, reasoning, modeling, or the like. *Identity conditions*, instead, are conditions under which a causal claim is true."[38] Stating such a rationale is one of the key epistemological features of a theory of.[39]

This allows me to emphasize a third positive aspect of the Causal Model. This theory fulfills the task of telling us what causation is and what methods we should use for evaluating causal claims, without immersing itself into deep metaphysical speculations. It is tempting to agree with the adherents to the Causal Model when they claim their view is metaphysically neutral. Although they do not deny the existence of causal mechanisms, they do not venture into speculations about the nature of these mechanisms, at least not from the beginning of the research. The Causal Model rather places the causal mechanism into a 'black box' and reasons about it by using only observational data.[40]

A fourth remark here regards the wide scope of the Causal Model. It seems this theory advocates a much more liberal philosophy than the interventionist account. Indeed, the Causal Model feels comfortable with allowing the implementation of different methods when it searches for what causes what. As long as one accepts the rationale of variation as the guiding principle of the methodology of causality, one is free to use intervention in cases of controlled experiments or just observational data in situations with uncontrollable causes. Once the variations are determined, it remains a personal option to decide what tool is most adequate for establishing the causal import of these variations. The interventionist would probably like to press this point further, as we have not said anything on how to establish the causal relevance of variations. We should come up with an explanation of how the stability or invariance of these chance-relating relations is established. Let us deal with this problem in the next chapter, in which I will also lay down some of my concerns about the theses of the Causal Model.

---

37 Russo, "On empirical generalisations", *loc. cit.*, p. 8.

38 *Ibid.*, p. 8.

39 *Ibid.*

40 See Russo, *Causality and causal modelling in the social sciences. Measuring variations*, *loc. cit.*

10.5 'Intervening' in the Causal Model

The strength of the Causal Model lies in its claim that in order to establish the invariance of a causal relationship one needs no intervention at all. What is then invariance and how to think about its actual settling? Recall the aforementioned study initiated by Gaumé and Wunsch[41] upon self-rated health in the population of the Baltic countries. The results of the study show that

> … the determinants taken into account (alcohol consumption, physical health, psychological health, psychological distress, education, locus of control, and social support) had a remarkable *stable impact* on self-rated health across the different Baltic countries, across the time-frames analysed, across gender, ethnicity, or age group [my emph., S. M.].[42]

So it is this 'remarkable stable impact' that accounts for the dependence of the effect upon its genuine causes. Why? Because it is hard to conceive of such a robust relationship as simply coming out of nowhere! It may happen to hit upon a chancy correlation but when it is one that shares in such a stable character, *it is reasonable* to think of it as a manifestation of an underlying causal mechanism. In this case, we all share the intuition that this invariant relationship *must be* there, somewhere in the 'world', especially when it is found to hold across different populations, in different social contexts. It must be strange indeed for a relationship to achieve a stable character and to lack the causal import. But let me remark in passing that expressions like 'it is reasonable' or 'it must be' suggest we encounter an inferential process here, which may be adduced as a fifth step of the methodology of the Causal Model. I will come back in a moment to the relevance of this issue for the present discussion.

It is now worth exploring the second part of the question that opened this section: how to establish the invariance of the causal relationship? What are the means by which one can achieve this? Contemplate Russo's answer:

> Causal assessment is instead made through testing the stability of the putative causal relationship across different 'portions' of the data set. The different 'portions' have to be carefully chosen. In fact, if we test invariance across sub-populations randomly sampled, we should indeed expect to find, approximately, the same values but with a larger confidence interval; consequently, this test wouldn't be terribly useful. Instead, we should appropriately choose sub-populations, for instance considering different age strata, or different socio-demo-economic characteristics, or different geographical regions, or different time frames.[43]

Therefore, breaking down the data input into more components and further registering the behavior of the causal hypothesis corresponds to evaluating its invariant

41 Gaumé and Wunsch, *loc. cit.*

42 Russo, "On empirical generalisations", *loc. cit.*, p. 8.

43 *Ibid.*, pp. 9–10.

character. If the hypothesis proves to be stable enough in the newly formed context, it is taken to signify a genuine causal relationship.

We are now in the position to look at the true *counterpart* of intervening in the Causal Model. If, as we have seen before, the interventionist prescribes to intervene in order to set the invariance of a causal connection, the Causal Model instead urges to *introduce* new *relevant* dimensions for the data set. If accordingly, no significant change is observed in the relationship that holds between the causal variables, than it *must be* that we have come across a true causal link. Russo[44] enforces this by highlighting that Gaumé and Wunsch[45] introduce new dimensions for the data set in their study: They talk about "different age strata, or different socio-demo-economic characteristics, or different geographical regions, or different time frames" and then check for the stability of the causal hypothesis in the new framework.[46] The idea is to compare the degree of stability of the causal hypothesis before and after the introduction of these new facets of the data input. Therefore, evaluating causal relationships across differences introduced here by gender, age strata, nationality and time frames allows the authors to assess the overall stability of the causal links, hence their invariance. Thus, in contrast with interventionism, the Causal Model does not directly manipulate the variables. Instead it induces significant variations in the causal context by fructifying the *multi-facet component* of the data set. Then it proceeds to compare the behavior of the causal hypotheses in the new and old framework. To restate, if no important changes occur, the invariance of the causal hypothesis is preserved and the causal import of the hypothesis is inferred.

Two short notes are needed here. I have previously used some qualifications for the causal variables. They were alleged to be 'relevant', 'important', etc. Obviously, one must call for an explanation here – the reason is that according to Russo, the adherent to the Causal Model largely employs these adjectives in his job of modeling the causal context to which Russo refers with the term 'background knowledge'. 'Background knowledge' consists of information about the causal context, that is information about how to select the causes, how to formulate the causal hypothesis, how similar hypotheses were treated in other studies or other theories, how comparable the populations are etc.[47] For instance, Gaumé and Wunsch[48] list at the beginning of their paper[49] what other studies in different contexts brought about the

44 *Ibid.*

45 Gaumé and Wunsch, *loc. cit.*

46 Russo (*Causality and causal modelling in the social sciences. Measuring variations, loc. cit.*) shows the same strategy is implemented by many other studies in social science, not all of them being limited to testing the 'subjective perception' of social actors as it is the case with the study of Gaumé and Wunsch (*loc. cit.*).

47 See Russo (*Causality and causal modelling in the social sciences. Measuring variations, loc. cit*) for a detailed account of the 'background knowledge'.

48 Gaumé and Wunsch, *loc. cit.*

49 *Ibid.*, pp. 436–438.

hypothesis they laid down. They also state that the population in the three Baltic countries is similar enough to allow comparisons.[50] Thus, 'background knowledge' is a necessary prerequisite for research under the Causal Model. Without the fundamental contribution of the 'background knowledge', the researcher is left with no clue of how to formulate the causal hypothesis.

This leads to my second remark here and eventually to my general concern with the Causal Model. The 'background knowledge' provides the researcher with sufficient information about how to formulate a plausible causal hypothesis. But according to Russo,[51] the Causal Model aims at testing this hypothesis against the data. If the test is passed,[52] then the causal hypothesis is supported. It only remains to demonstrate that the causal hypothesis represents a genuine causal relationship and not just a correlation among variables of interest. This also means one has to implement a test of invariance. Therefore, with Russo[53] we have a case of Hypothetico-Deductive Methodology (H-D): a causal hypothesis is initially formulated, some observational consequences are predicted on the base of this hypothesis and they are finally confronted with the data. Russo[54] draws on the crucial distinction between H-D and inductive methods: the causal hypothesis is tested by the data and not inferred from it. In contrast, inductivism prescribes performing an ampliative process whereby a causal statement is derived from the data. Russo moreover suggests that the label deductivism in H-D is misleading because "strictly speaking, there is no deduction going on. In causal modelling, hypothetico-deductivism does not involve deductions *strictu sensu*, but involves a weaker inferential step of 'drawing consequences' from the hypothesis".[55] All in all, the Causal Model should be taken as engaging a methodology for testing hypotheses rather than as a programme for producing knowledge by drawing inferences from the data. However, as I have already tried to suggest at the beginning of this section, the fourth step of the testing methodology seems to imply an important inferential process. Let us remember that the essence of this step is to check the invariance of the causal relationship by introducing new facets of the variables and observing the changes in the causal link. By noticing the stability of the causal hypothesis in the new framework, the researcher feels entitled to *infer* its invariance, hence its genuine causal character. This, I think, amounts to saying that H-D

50 See Russo (*Causality and causal modelling in the social sciences. Measuring variations, loc. cit.*) for a complete list of assumptions used in the Causal Model, other than the 'background knowledge': some of the most important are the direction of time, confounding, etc.

51 Russo, "On empirical generalisations", *loc. cit.*

52 It is assumed contra Popper that 'testing' rather than 'refuting' makes sense here.

53 Russo, *Causality and causal modelling in the social sciences. Measuring variations, loc. cit.*

54 *Ibid.*

55 *Ibid.*, p. 71.

is not as neutral from an inferential import as it would like to suggest.[56] If this is a sound remark, then a different but more interesting question opens up.

What exactly is the nature of this inference and toward what conclusion is it leading us? I think it is an inference from the data as the data set is broken up into its multi-facet component. Let us call to mind that in the study of Gaumé and Wunsch,[57] new features (age data, time frames etc.) of the data input – the populations of the Baltic countries – are taken into account in order to form new variations. Based on the observed stability of these new variations one infers their causal import. That is, one infers that *the best explanation* for the observed stability is the very fact of hitting on a genuine causal hypothesis. Although a hidden and unknown cause cannot in principle be ruled out here, it seems we have good reasons to trust our inference. Let me rephrase this as follows: we don't only share the feeling that 'it is reasonable' but we moreover think 'it must be' that the *only* explanation for the invariance of the causal hypothesis is its postulated causal content. Therefore, if my reading of the Causal Model is sound, this theory prescribes the *inference to the best explanation* (IBE) as the key tool for establishing the causal features of the hypothesis.[58]

The problems with this type of inference are well known and I refer the reader to Lipton[59] for an overview. Also, the use of IBE by the Causal Model contrasts with its acclaimed lack of commitment to the use of (inductive) inferential procedures. Moreover, when IBE is an integrating part of the steps of the methodology of testing, as it is the case here, it seems to flagrantly contradict with the central assumption of the Causal Model. As emphasized in section four above, the Causal Model assumes keeping apart the epistemological/methodological level from the metaphysical plane. However, the immersion of IBE – which is a procedure devoid of any empirical content – into the heart of the methodology of the Causal Model is nothing else than a metaphysical influence. This, I expect, will be used by the advocates of interventionism as a strong counterargument which parallels Russo's charge of mixing methodological questions with explanatory or metaphysical idiosyncrasies. The details of such a possible reply fall beyond the scope of this work.

---

56 One should not confuse this charge with a similar one of which Russo (*Causality and causal modelling in the social sciences. Measuring variations, loc. cit.*) is fairly aware: the use of the 'background knowledge' as a boundle of auxiliary hypotheses besides the causal hypothesis itself already suggests the need of an inferential process.

57 Gaumé and Wunsch, *loc. cit.*

58 More exactly, IBE grants the stability of variations and it is this stability which supports the causal import of the hypothesis. However, as long as a hidden or unknown cause can still be in place here, I think it is only the use of IBE which upholds this stability relationship.

59 Peter Lipton, "Inference to the best explanation", in: W. H. Newton-Smith (Ed.), *A Companion to the Philosophy of Science*. Blackwell 2001, pp. 184–194.

10.6 Conclusions

The goal of this paper was to contribute to the debate between interventionism and the Causal Model with some critical remarks. According to the Causal Model, interventionism proves itself to be ineffective for assessing causal claims in non-experimental frameworks, as it is the case with many causal contexts in social sciences. In my analysis of the dilemma raised against interventionism I tried to come up with a few possible answers to the problems laid down by Russo.[60] It firstly turned out that the interventionist can decide to choose for a different view of semantics than the one Russo[61] tacitly advanced. This will eventually open new possibilities for overcoming the lack of methodological import of the conceptual reading of interventionism.

Also, the distinction between experimental and non-experimental was challenged together with the acclaimed lack of use of thought experiments. It was argued, against Russo[62] that various fields of social science allow intervention in non-experimental contexts as well.

I however acknowledged that Russo's points do indeed press the reader to contemplate the Causal Model as a vigorous concurrent for the interventionist theory. Although I admitted the strong potential of the Causal Model, I called into question its metaphysical neutrality. I also concluded that the Causal Model can not do its job without the use of the inference to the best explanation.

Let me add that the method of settling the invariance in the Causal Model suggest a further important conclusion. The breaking down of the data set into its multi-facet component in order to obtain useful variations illustrates the conceptual complexity of this data. Of course, it may be the case that the decomposition will not be always possible,[63] but as Russo[64] argues, multiple studies in the social sciences perform it. I think this calls for a desideratum of conceptual complexity in the social science which, at least at a first sight, seems not to be necessary in the natural science. The methods within the latter domain rather aim at simplifying the situation for analysis and let us understand that the relevant information bears only one definite dimension. This method tends to give a unidirectional reading of the variables of interest because the information of interest lacks the complexity we encountered in the case of social phenomena. In natural science, the variables of interest are not susceptible of being broken-down into a multi-facet component once the causal context is sharply defined. Thus, there is only one conceptual *relevant* aspect echoed in the variables of interest and moreover, this key feature can be kept fixed when the variables are under control. This is also the reason why hidden

60 Russo, "On empirical generalisations", *loc. cit.*

61 *Ibid.*

62 *Ibid.*

63 See Russo (*Causality and causal modelling in the social sciences. Measuring variations, loc. cit.*) for a contemplation of this point.

64 *Ibid.*, ch. 1.

causal correlations can hardly make sense here. But this eventually justifies the need of an *interpretation* for the theory when *unexpected* correlations still come up, as it is for instance, with the (EPR) quantum correlations. The differences of the two approaches suggest a difference in the way they frame the data input, that is, in the way they regard the ontology. However, a comparative assessment of the ontology professed by social science and natural science is far beyond the scope of this essay.

Acknowledgements: I am very grateful to Prof. Dennis Dieks, Aura Nasta, Rosa Runhardt, Federica Russo and Valentina Spataru for very helpful discussions and suggestions. Of course, the entire responsibility for any mistakes or inaccuracies is mine. Also, special thanks to Prof. Dennis Dieks (Utrecht University, The Netherlands) for inviting me to attend the ESF-PSE Workshop "Points of Contact between the Philosophy of Physics and the Philosophy of Biology", London, UK 2010. Financial support from the University of Bucharest, grant POSDRU ID6827-6/1.5/12 is also acknowledged.

References

Nancy Cartwright, *Hunting causes and using them: approaches in philosophy and economics*. Cambridge: Cambridge University Press 2007.

Michael Devitt and Kim Sterelny, *Language and reality. An introduction to the philosophy of language*, Oxford: Blackwell Publishers 1999.

Catherine Gaumé and Guillaume Wunsch, "Self-rated health in the Baltic countries, 1994/1999", in: *European Journal of Population* 26, 2010, pp. 435–457.

Wenceslao J. Gonzalez, "The role of experiments in the social sciences: the case of economics", in: Dov M. Gabbay, Paul Thagard and John Woods (gen. Eds.), *Handbook of the Philosophy of Science – General Philosophy of Science. Focal Issues*, Amsterdam: Elsevier, 2007, pp. 275–303.

Peter Lipton, "Inference to the best explanation", in: W. H. Newton-Smith (Ed.), *A Companion to the Philosophy of Science*, Blackwell, 2001, pp. 184–194.

Judea Pearl, "Review: Nancy Cartwright on hunting causes", in: *Economics and Philosophy* 26, 2010, pp. 69–94.

Judea Pearl, *Causality: models, reasoning, and inference*. Cambridge: Cambridge University Press 2000.

Federica Russo, *Causality and causal modelling in the social sciences. Measuring variations*. Methodos Series. New York: Springer 2009.

Federica Russo, (2011). "On empirical generalisations". In this Volume.

James Woodward, "Agency and interventionist theories", in: Helen Beebee, Christopher Hitchcock and Peter Menzies (Eds.), *The Oxford Handbook of Causation*, Oxford: Oxford University Press 2009, pp. 234–62.

James Woodward, "Causation and manipulability", in: Edward N. Zalta (Ed.), *The Stanford Encyclopedia of Philosophy* (*Winter 2008 Edition*), URL=http://plato.stanford.edu/archives/win2008/entries/causation-mani/, accessed at 17.01.2011.

James Woodward, *Making things happen: a theory of causal explanation*. Oxford: Oxford University Press, 2003.

Department of Philosophy
University of Bucharest
Splaiul Independentei 204
060024, Bucharest
Romania
sebastian.mateiescu@gmail.com

CHAPTER 11

MICHAEL ESFELD

CAUSAL REALISM[1]

ABSTRACT

According to causal realism, causation is a fundamental feature of the world, consisting in the fact that the properties that there are in the world, including notably the fundamental physical ones, are dispositions or powers to produce certain effects. The paper presents arguments for this view from the metaphysics of properties and the philosophy of physics, pointing out how this view leads to a coherent ontology for both physics as well as biology and the special sciences in general.

11.1 INTRODUCTION

Causal realism is the view that causation is a real and fundamental feature of the world. That is to say, causation cannot be reduced to other features of the world, such as, for instance, certain patterns of regularities in the distribution of the fundamental physical properties. Causation consists in one event bringing about or producing another event, causation being a relation of production or bringing something into being.[2] I shall take events to be the relata of causal relations, without arguing for this claim in this paper, since this issue is not important for present purposes. More precisely, an event e_1, in virtue of instantiating a property F, brings about another event e_2, instantiating a property G. One can therefore characterize causal realism as the view that properties are powers. In short, Fs are the power to produce Gs. Saying that properties are powers means that it is essential for a property to exercise a certain causal role; that is what constitutes its identity. One can therefore characterize causal realism as the view that properties are causal in themselves. To abbreviate this view, I shall speak in terms of *causal properties*.

---

1 I'm grateful to Matthias Egg and Vincent Lam for comments on the draft of this paper.

2 See Ned Hall, "Two concepts of causation", in: J. Collins, N. Hall and L. A. Paul (Eds.), *Causation and counterfactuals*. Cambridge (Massachusetts): MIT Press 2004, pp. 225–276, for an analysis of the contrast between the production conception of causation and the regularity conception; the counterfactual analysis of causation is a sophisticated version of the regularity conception.

D. Dieks et al. (eds.), *Probabilities, Laws, and Structures*, The Philosophy of Science in a European Perspective 3, DOI 10.1007/978-94-007-3030-4_11,

To make the claim of this paper audacious, I shall defend the view that all properties that there are in the world are causal properties. The limiting clause "that there are in the world" is intended to leave open whether or not abstract mathematical objects exist: abstract mathematical objects, if they exist, do not cause anything, so that their properties are not powers. However, to the extent that properties are instantiated in the real, concrete world (by contrast to a – hypothetical – realm of abstract mathematical objects), it is essential for them to exercise a certain causal role. This is a sparse view of properties: it is not the case that for any predicate, there is a corresponding property in the world.

Properties, being causal in themselves and thus powers, are dispositions – more precisely, dispositions that manifest themselves in bringing about certain effects. Dispositions, thus conceived, are not Aristotelian potentialities, but real, actual properties. Furthermore, there is no question of dispositions in this sense requiring non-dispositional properties as their bases. If all properties are causal in themselves, being powers, then all properties are dispositions. The view defended in this paper hence coincides with the position known as dispositional monism or dispositional essentialism.[3] However, in claiming that all properties are dispositions, it is not intended to deny that properties are qualities. The view is rather this one: *in being certain qualities, properties are causal, namely powers to produce certain specific effects*.[4] Thus, for instance, in being a certain qualitative, fundamental physical property, charge is the power to create an electromagnetic field, manifesting itself in the attraction of opposite-charged and the repulsion of like-charged objects; and mass is a qualitative, fundamental physical property that is distinct from charge in being the power to create gravitational attraction (this example is meant to be a rough and ready illustration of this view of properties; an adequate scientific discussion would require much more details, and, notably, certain commitments in the interpretation of the relevant scientific theories).

The view of *causal properties* is both a metaphysical and an empirical position: it is a stance in the metaphysics of properties, and it is a claim about what is the best interpretation of the ontological commitments of our scientific theories. It is opposed to the view of *categorical properties*, that is, the view according to which properties are pure qualities, exercising a causal role only contingently, depending on the whole distribution of the fundamental physical properties in a given world and/or the laws of nature holding in a given world. That latter view also is both a metaphysical and an empirical position. As a metaphysical position, it is usually traced back to Hume's stance on causation and is today known as *Humean metaphysics*.[5] As an empirical position, it can be traced back to Russell's famous claim

3 See notably Alexander Bird, *Nature's metaphysics. Laws and properties*, Oxford: Oxford University Press 2007.

4 See Michael Esfeld and Christian Sachse, *Conservative reductionism*. New York: Routledge 2011, chapter 2.1, for a detailed exposition of this claim, drawing on John Heil, "Obituary. C. B. Martin", in: *Australasian Journal of Philosophy* 87, 2009, pp. 177–179.

5 See notably David Lewis, *Philosophical papers. Volume 2*. Oxford: Oxford University Press 1986, introduction, and David Lewis, "Ramseyan humility", in: D. Braddon-

that causation is a notion that has no place in the interpretation of contemporary physics.[6] For the sake of simplicity, I shall confront the view of causal properties only with the Humean view of categorical properties, thereby leaving out in particular views that invoke a commitment to universals and certain relations among universals in order to account for causation and laws[7]; the issue of a commitment to universals is not important for the arguments considered in this paper.

Accordingly, I shall mention the metaphysical argument for causal in contrast to Humean categorical properties in the next section, then move on to arguments from physics (Sects. 11.3 and 11.4) and finally consider the perspective for an account of the relationship between physics and the special sciences such as biology that this view offers (Sect. 11.5). Covering all these issues in a short paper means that I can only sketch out the main features of the central arguments here, providing the reader with some sort of an overview of the case for causal realism.[8]

11.2 The Metaphysical Argument for Causal Properties

The main metaphysical argument against the view of categorical properties is that this view is committed to quidditism. Accordingly, the main argument for the causal view of properties is that this view avoids any association with quidditism. If properties play a causal and nomological role only contingently, then their essence is independent of the causal relations in which they enter and the laws in which they figure. Their essence then is a pure quality, known as *quiddity*.[9] It is a primitive suchness, consisting in the simple fact of being such and such a quality, without that quality being tied to anything, notably not tied to certain causal or nomological relations. Consequently, it is not possible to have a cognitive access to the qualitative nature of the properties; that consequence is known as humility.[10]

The commitment to quiddities is objectionable, since it obliges one to recognize worlds as being qualitatively different, although they are indiscernible. Quidditism about properties is analogous to haecceitism about individuals. A haecceitistic difference between possible worlds is a difference that consists only in the fact that there are different individuals in two worlds, without there being any qualitative difference between the worlds in question. In other words, a haecceitistic difference is a difference between individuals which has the consequence

Mitchell and R. Nola (Eds.), *Conceptual analysis and philosophical naturalism*. Cambridge (Massachusetts): MIT Press 2009, pp. 203–222.

6 Bertrand Russell, "On the notion of cause", in: *Proceedings of the Aristotelian Society*, 13, 1912, pp. 1–26.

7 See notably David M. Armstrong, *What is a law of nature?* Cambridge: Cambridge University Press 1983.

8 For a detailed study, see Esfeld and Sachse, *loc. cit.*

9 See Robert Black, "Against quidditism", in: *Australasian Journal of Philosophy* 78, 2000, pp. 87–104.

10 See in particular Lewis, "Ramseyan humility", *loc. cit.*

that worlds have to be recognized as different, although they are indiscernible. If one maintains that the essence of properties is a primitive suchness (a quiddity), a similar consequence ensues: one is in this case committed to recognizing worlds as different that are identical with respect to all causal and nomological relations, but that differ in the purely qualitative essence of the properties that exist in them.

Thus, for instance, the property that exercises the charge role in the actual world can exercise the mass role in another possible world, since the qualitative nature of that property is on this conception not tied to any role that tokens of the type in question exercise in a given world. We can therefore conceive a swap of the roles that properties play in two possible worlds, such as the property *F* playing the charge role and the property *G* playing the mass role in world w_1, and *F* playing the mass role and *G* playing the charge role in w_2. The worlds w_1 and w_2 are indiscernible. Nonetheless, the friend of categorical properties is committed to recognizing w_1 and w_2 as two qualitatively different worlds. To put it in a nutshell, there is a qualitative difference between these two worlds that does not make any difference. Thus inflating the commitment to worlds is uncomfortable for any metaphysical position, and notably for a position that sees itself as being close to empiricism, as does Humean metaphysics.

The causal theory of properties avoids any association with quidditism by tying the essence of a property to its causal and thereby to its nomological role: instead of the essence of a property being a primitive suchness, the essence of a property is the power to enter into certain causal relations. Consequently, what the properties are manifests itself in the causal relations in which they figure (more precisely, the causal relations in which events stand in virtue of the properties that they instantiate). It is thus not possible to separate the properties from the causal relations.[11] The laws of nature supervene on the properties in revealing what properties can do in being certain powers.[12] Consequently, worlds that are indiscernible as regards the causal and nomological relations are one and the same world. Although being committed to objective modality by tying the essence of a property to a certain causal – and thereby a certain nomological – role, the causal view of properties thus is ontologically parsimonious.

---

11 See notably Sydney Shoemaker, "Causality and properties", in: P. van Inwagen (Ed.), *Time and cause*. Dordrecht: Reidel 1980, pp. 109–135; John Hawthorne, "Causal structuralism", in: *Philosophical Perspectives* 15, 2001, pp. 361–378; Alexander Bird, *Nature's metaphysics. Laws and properties*, *loc. cit.*; Anjan Chakravartty, *A metaphysics for scientific realism: knowing the unobservable*. Cambridge: Cambridge University Press 2007, chapters 3–5.

12 See e.g. Mauro Dorato, *The software of the universe. An introduction to the history and philosophy of laws of nature*. Aldershot: Ashgate 2005, chapter 4.

11.3 Structures and Causal Properties in Fundamental Physics

There is a so-called argument from science for the causal view of properties, drawing on the claim that the descriptions scientists give of the properties they acknowledge, including notably the properties they take to be fundamental, are causal descriptions, revealing what these properties can do in interactions. However, that argument is not cogent for two reasons. In the first place, without adding further premises, there is no valid inference from dispositional descriptions to an ontology of dispositional properties (that is, properties whose essence is a certain power or disposition). Such further premises are available; but they finally rely on the fact that the alternative view of properties, the categorical one, has to subscribe to metaphysical commitments such as the one to quiddities that do not serve any purpose for science, having notably no explanatory role, whereas the causal view of properties avoids any such free-floating commitments by identifying the essence of properties with their causal role.[13] In brief, due to the additional premises needed, the so-called argument from science does not make the case for the causal view of properties stronger than it is already as based on the mentioned metaphysical argument only.

Furthermore – and more importantly as far as the relationship between science and the causal view of properties is concerned –, the claim according to which the descriptions scientists give of the properties they acknowledge are causal descriptions is in dispute. One can maintain that at least as far as fundamental physics is concerned, the basic descriptions are structural rather than dispositional ones, drawing on certain symmetries rather than certain causal powers. More precisely and more generally speaking, one can associate these two types of descriptions with two different forms of or approaches to scientific realism. Entity realism, laying stress on experiments rather than theories, seeks for causal explanations of experimental results and commits itself to theoretical entities – such as e.g. electrons, or elementary particles in general – only insofar as these have the power, disposition or capacity to produce phenomena such as the ones observed in the experiments in question. Structural realism, by contrast, starts from the structure of scientific theories and maintains in its epistemic form (epistemic structural realism[14]) that there is a continuity of the structure of physical theories in the history of science; in its ontic form, going back to Ladyman,[15] structural realism maintains that structure is all there is in nature.

To strengthen the case for the causal view of properties, we should therefore, for the sake of the argument, base ourselves not on experiments and entity realism,

---

13 See Neil Edward Williams, "Dispositions and the argument from science", in: *Australasian Journal of Philosophy* 89, 2011, pp. 71–90.

14 John Worrall, "Structural realism: the best of two worlds?", in: *Dialectica* 43, 1989, pp. 99–124.

15 James Ladyman, "What is structural realism?", in: *Studies in History and Philosophy of Modern Science* 29, 1998, pp. 409–424.

but on theories and ontic structural realism (OSR). The first point that we can make in this context is to emphasize that nearly all the proponents of OSR conceive the structures to which they are committed in a non-Humean manner, namely as modal structures.[16] There is a clear reason for this commitment: it does not seem to make sense to conceive structures that are pure qualities. The identity of a structure obviously is constituted by its playing a certain nomological role. This is particularly evident when considering structures that are defined by certain symmetries.[17] Thus, it obviously does not make sense to conceive one and the same structure playing in one world, say, the role of the quantum structures of entanglement and in another world the role of the metrical-gravitational structures – as it does make sense in Humean metaphysics to conceive one and the same qualitative, intrinsic property to play the charge role in one world and the mass role in another world. The decisive question in this context therefore is this one: Is the nomological role that a structure plays also a causal role? Or is it a plausible move when it comes to structures to go for a separation between the nomological and the causal role – so that a structure necessarily plays a certain nomological role, the nomological role constituting its identity, but only contingently a causal role?

OSR is a realism with respect to the structure of a scientific theory. But this stance does not commit the ontic structural realist to Platonism about mathematical entities such as the mathematical structure of a fundamental physical theory. What the realist claims is that the mathematical structure of a fundamental physical theory *refers to* or *represents* something that there is in the world independently of our theories. In brief, the mathematical structure is a means of representation, and the point of OSR is the claim that what there is in the world, what the mathematical structure represents or refers to, is itself a structure, namely a physical structure.

To mention but one example, when one endorses a realist stance in the interpretation of quantum mechanics, one does not advocate Platonism with respect to mathematical entities such as the wavefunction (state vector) in a mathematical space; one maintains that these mathematical entities represent something that

16 See Steven French and James Ladyman, "Remodelling structural realism: quantum physics and the metaphysics of structure", in: *Synthese* 136, 2003, pp. 31–56; James Ladyman and Don Ross with David Spurrett and John Collier, *Every thing must go. Metaphysics naturalised*. Oxford: Oxford University Press 2007, chapters 2–4; Steven French, "The interdependence of structure, objects and dependence", in: *Synthese* 175, 2010, pp. 177–197, section 3; but see Georg Sparber, *Unorthodox Humeanism*. Frankfurt (Main): Ontos 2009 and Holger Lyre, "Humean perspectives on structural realism", in: F. Stadler (Ed.), *The present situation in the philosophy of science*. Dordrecht: Springer 2010, pp. 381–397 and Holger Lyre, "Structural invariants, structural kinds, structural laws", this volume 2011, for Humean versions of OSR.

17 See notably the "group structural realism" advocated by Bryan W. Roberts, "Group structural realism", in: *British Journal for the Philosophy of Science* 62, 2011, pp. 47–69.

there is in the world by contrast to being mere tools in calculating probabilities for measurement outcomes.[18] The realist in the interpretation of quantum mechanics therefore has the task to spell out what it is in the world that the quantum formalism refers to. Accordingly, the ontic structural realist has the charge to explain what a physical structure is in distinction to a mathematical structure that is employed as a means to represent what there is in the physical world, thereby replying to the widespread objection that OSR blurs the distinction between the mathematical and the physical.[19] Simply refusing to answer that question[20] is not acceptable.

In the context of a traditional metaphysics of universals and intrinsic properties, one can maintain that there are property types as universals, and that there are objects in the world that instantiate these property types. However, even if one is not an eliminativist about objects as is French,[21] but defends a moderate version of OSR that admits objects as that what stands in the relations in which the structures consist,[22] such a move is not available to the ontic structural realist in order to answer the question what distinguishes physical from mathematical structures: it presupposes the existence of objects as something that is primitively there to instantiate the mathematical structures, being ontologically distinct from the structures. But insofar as OSR is in the position to admit objects, it can recognize only what French[23] calls thin objects. More precisely, it can acknowledge objects only as that what stands in the relations that constitute the structures, the relations being the ways in which the objects are so that the objects do not have any existence or identity independently of the relations.[24]

Furthermore, the spatio-temporal criterion of existence, which traditional empiricism adopts, is not applicable in this context either. Four-dimensional space-time is itself a physical structure according to OSR.[25] Consequently, one cannot presuppose four-dimensional space-time as a background on the basis of which

18 See e.g. Tim Maudlin, "Can the world be only wavefunction?", in: S. Saunders, J. Barrett, A. Kent and D. Wallace (Eds.), *Many worlds? Everett, quantum theory, and reality*. Oxford: Oxford University Press 2010, pp. 121–143 for a clear statement in that sense in contrast to claims to the contrary, such as David Z. Albert, "Elementary quantum metaphysics", in: J. T. Cushing, A. Fine and S. Goldstein (Eds.), *Bohmian mechanics and quantum theory: an appraisal*. Dordrecht: Kluwer 1996, pp. 277–284.

19 See e.g. Tian Yu Cao, "Can we dissolve physical entities into mathematical structure?", in: *Synthese* 136, 2003, pp. 57–71 for that objection.

20 As do Ladyman and Ross, *loc. cit.*, p. 158.

21 French, *loc. cit.*

22 Michael Esfeld, "Quantum entanglement and a metaphysics of relations", in: *Studies in History and Philosophy of Modern Physics* 35, 2004, pp. 601–617; Michael Esfeld and Vincent Lam, "Moderate structural realism about space-time", in: *Synthese* 160, 2008, pp. 27–46.

23 French, *loc. cit.*

24 Michael Esfeld and Vincent Lam, "Ontic structural realism as a metaphysics of objects", in: A. Bokulich and P. Bokulich (Eds.), *Scientific structuralism*. Dordrecht: Springer 2011, pp. 143–159.

25 See e.g. Esfeld and Lam, "Moderate structural realism about space-time", *loc. cit.*

one could establish a distinction between physical and mathematical structures (the former, by contrast to the latter, existing in four-dimensional space-time). One could seek to counter that objection by simply stipulating that real physical structures have to be four-dimensional. But such a move would rule out a realist understanding of approaches like string theory in quantum gravity on purely *a priori* grounds, whereas such approaches have to be assessed according to their physical merits (or the lack of them). (Consider the historical case of the Kaluza-Klein theory: it is no cogent objection to realism about this theory that it considers the physical reality to be five-dimensional; but it is a knock-down objection against that theory that it gets the mass of the electron totally wrong).

Nonetheless, in order to answer the question how to distinguish physical from mathematical entities, the ontic structural realist can draw on another position that is widespread in traditional metaphysics, namely the causal criterion of existence, going back to the Eleatic stranger in Plato's *Sophist* (247e) and also known as Alexander's dictum: real physical structures distinguish themselves from their representations in terms of mathematical structures by being causally efficacious.[26] Concrete physical structures are first-order properties, too, namely first-order relations. They can be conceived as causal properties in the same manner as intrinsic properties: *in being certain qualitative physical structures, they are the power to bring about certain effects*. Structures can be causally efficacious in the same sense as intrinsic properties of events: as events can bring about effects in virtue of having certain intrinsic properties, they can bring about effects in virtue of standing in certain relations with each other so that it is the network of relations – that is, the structure as a whole – that is causally efficacious.[27] Furthermore, one thus accounts for the dynamics of physical systems: OSR is a proposal for an ontology of physical systems, but as such it is silent on their dynamical evolution.

Psillos[28] objects to the view of causal structures, in brief, that (a) it simply relocates the quidditism problem through its commitment to a holistic individuation of properties and that (b) it cannot show how structures are both abstract enough to be shareable by distinct physical systems and concrete enough to be part of the causal identity of physical systems. However, these are objections against the causal theory of properties in general. The adherent to causal OSR can draw on the resources in the literature on the metaphysics of properties to counter these objections – in particular follow the late Charlie Martin and John Heil in conceiving properties (including structures) as being qualitative and causal in one[29] and as

26 Michael Esfeld, "The modal nature of structures in ontic structural realism", in: *International Studies in the Philosophy of Science* 23, 2009, pp. 179–194.

27 See Esfeld and Sachse, *loc. cit.*, chapter 2, for details of such a metaphysics of causal structures.

28 Stathis Psillos, "Adding modality to ontic structural realism: an exploration and critique", in: E. Landry and D. Rickles (Eds.), *Structure, objects, and causality*. Dordrecht: Springer 2011.

29 Heil, "Obituary. C. B. Martin", *loc. cit.*

being tropes or modes,[30] thus acknowledging a perfect similarity among property (including structure) tokens as primitive.[31]

As this ongoing debate shows, the conception of causal structures is not the only game in town to answer the question what distinguishes real physical structures from their representation in terms of mathematical structures, to spell out the modal nature of structures in OSR and to account for the dynamics of physical systems on the basis of OSR. The reflection on these issues in the framework of OSR has just begun. But an answer to these questions is needed so that one can then engage in the business of assessing the options.

11.4 Causal Realism at Work in the Interpretation of Fundamental Physics

The arguments in the two preceding sections are rather abstract and general. In order to make a case for causal realism, one has to show in concrete terms how this interpretation applies to the current fundamental physical theories and what benefits one gets from doing so. A commitment to dispositions in the interpretation of quantum mechanics is usually linked with versions of quantum mechanics that recognize state reductions, leading from entanglement to something that comes at least close to classical physical properties with definite numerical values.[32] The theory of Ghirardi, Rimini and Weber[33] (GRW) is the most elaborate physical proposal in that respect. In the GRW framework, one can maintain that the structures of quantum entanglement are the disposition or the power to bring about classical properties through state reductions in the form of spontaneous localizations. Doing so answers a number of crucial questions in the interpretation of quantum mechanics: (a) it tells us what the properties of quantum systems are if there are no properties with definite numerical values, namely dispositions to bring about such properties, and these dispositions are real and actual properties (by contrast to mere potentialities); (b) it provides for a solution to the so-called measurement problem, without smuggling the notions of measurement interactions, measurement devices, or observers into the interpretation of a fundamental physical theory; (c) it yields the probabilities that we need to account for the quantum probabilities, namely objective, single case probabilities, by conceiving the dispositions for state reductions in the form of spontaneous localizations as propensities; (d) it provides for an account of the direction of time: processes of state reductions are

30 John Heil, *From an ontological point of view*. Oxford: Oxford University Press 2003, chapter 13.

31 See Esfeld and Sachse, *loc. cit.*, chapter 2, for details.

32 See Mauricio Suárez, "Quantum propensities", in: *Studies in History and Philosophy of Modern Physics* 38, 2007, pp. 418–438.

33 Gian Carlo Ghirardi, Alberto Rimini and Tullio Weber, "Unified dynamics for microscopic and macroscopic systems", in: *Physical Review D*, 34, 1986, pp. 470–491.

irreversible, thus singling out a direction of time; if these processes go back to entangled states as dispositions for state reductions, their irreversibility is explained by the relationship of dispositions and their manifestations being irreversible.[34]

Nonetheless, causal realism in the interpretation of quantum mechanics is not tied to realism with respect to state reductions. Regarding the quantum structures of entanglement as dispositions or powers also has certain benefits in the framework of the version of quantum mechanics that goes back to Everett,[35] recognizing no state reductions and taking the dynamics given by the Schrödinger equation to be the complete dynamics of quantum systems (and, by way of consequence, all physical systems). The claim then is that the structures of quantum entanglement are the disposition or the power to bring about a splitting of the universe into infinitely many branches through decoherence, the branches existing in parallel without interfering with each other; each of them appears like a domain of classical properties to an internal observer.

Notably the above mentioned points (a) and (d) apply also in this framework: decoherence and the splitting of the world into infinitely many branches is a fundamental, irreversible process, whereas the Schrödinger dynamics is time-reversal invariant. Conceiving entangled states as dispositions that manifest themselves spontaneously through decoherence and the splitting of the universe into infinitely many branches grounds that principled irreversibility. Furthermore, one has to provide an answer to the question of what entangled states are prior to the splitting of the universe into infinitely many branches. Simply drawing on the quantum formalism and proposing a realist attitude towards the wavefunction or state vector in configuration space does not answer that question, as pointed out in the preceding section. Conceiving entangled states as dispositions in the mentioned sense, by contrast, answers that question in setting out a clear ontology of what entangled states are objectively in the world, grounding the subsequent appearance of classical properties.

Again, conceiving entangled states as dispositions or powers may not be the only game in town. But work in the philosophy of physics has to be done in order to answer the mentioned ontological questions, instead of hiding oneself behind a mathematical formalism and passing what is de facto a realism with respect to mathematical entities for a realism with respect to the physical world.

Turning briefly to the other fundamental physical theory, general relativity, it seems at first glance that this theory suits well causal realism, since it abandons the view of space-time as a passive background structure, regarding instead the

34 See Mauro Dorato and Michael Esfeld, “GRW as an ontology of dispositions”, in: *Studies in History and Philosophy of Modern Physics* 41, 2010, pp. 41–49 for spelling out these points in detail, and see Mauro Dorato, “Dispositions, relational properties, and the quantum world”, in: M. Kistler and B. Gnassounou (Eds.), *Dispositions and causal powers*. Aldershot: Ashgate 2007, pp. 249–270 for dispositions in the interpretation of quantum mechanics in general.

35 Hugh Everett, “‘Relative state’ formulation of quantum mechanics”, in: *Reviews of Modern Physics* 29, 1957, pp. 454–462.

metrical field as a dynamical entity that accounts for the gravitational effects. It seems therefore that one can conceive the metrical properties of space-time points as dispositions or causal powers to bring about the gravitational effects.[36] But the case is not so clear: the gravitational effects are due to the movement of bodies along geodesics. One can therefore also argue that what seems to be gravitational effects are not effects that need a causal explanation, but is simply due to the geometry of curved space-time, not requiring a causal explanation in the same way as the inertial motion of a particle in Newtonian mechanics does not call for a causal explanation.[37] The case for causal realism in the philosophy of general relativity hangs on the ontology of the metrical field that one adopts, in other words, the stance that one takes in the traditional debate between substantivalism and relationalism cast in the framework of general relativity. Ultimately, the issue has to be settled in an ontology of quantum gravity.

In sum, causal realism can do a good ontological work in the framework of standard quantum mechanics with or without state reductions. The case of general relativity theory, however, depends on further parameters, such as the ontological stance that one adopts towards space-time (substantivalism or relationalism).

11.5 Causal Realism from Fundamental Physics to the Special Sciences

Assume that there are structures of quantum entanglement at the ontological ground floor which develop into classical properties that are correlated with each other in certain ways, or into the appearance of classical properties through the splitting of the universe into many branches. Assume furthermore that some of these classical properties build up local physical structures that distinguish themselves from their environment in bringing about certain effects as a whole – such as, for instance, a DNA sequence that produces a certain protein, or a brain that produces a certain behaviour of an organism. In conceiving the entangled states and, accordingly, such local physical structures as causal powers, causal realism provides for a unified ontology for fundamental physics as well as biology and the special sciences in general.[38]

---

36 See Alexander Bird, "Structural properties revisited", in: T. Handfield (Ed.), *Dispositions and causes*. Oxford: Oxford University Press 2009, pp. 215–241 and Andreas Bartels, "Modern essentialism and the problem of individuation of spacetime points", in: *Erkenntnis* 45, 1996, pp. 25–43, pp. 37–38, and Andreas Bartels, "Dispositions, laws, and spacetime", forthcoming in: *Philosophy of Science* 78, 2011 (Proceedings of the PSA conference 2010) – Bartels, however, voices also serious reservations about dispositional essentialism in this context.

37 See Vassilios Livanios, "Bird and the dispositional essentialist account of spatiotemporal relations", in: *Journal for General Philosophy of Science* 39, 2008, pp. 383–394, in particular pp. 389–390.

38 See Esfeld and Sachse, *loc. cit.*, for details.

On a Humean metaphysics of categorical properties, properties that are pure qualities and configurations of such properties can have a certain function in a given world and thus make true descriptions in dispositional, or functional terms; but there can be no functional properties, that is properties for which it is essential to exercise a certain causal role. However, on a widespread account of functions, namely the causal-dispositional one, the properties to which biology and the special sciences are committed are functional properties, consisting in exercising a certain causal role.[39] Notably the entire discussion of functionalism as the mainstream position in the philosophy of psychology and the social sciences is couched in terms of functional properties.

The advantage of causal realism is to be in the position to take the talk of functional properties literally: there really are functional properties in which biology and the special sciences in general trade out there in the world, for all the properties that there are in the world, down to the fundamental physical ones, are causal properties, being the disposition or the power to produce certain effects in being certain qualities. The commitment to causal properties in physics allows us to be realist about causal properties in the special sciences, and that commitment is a necessary condition for the latter realism: if there were no causal properties in physics, there would be no causal properties in the special sciences either (unless one were to maintain a dualism of free-floating properties of the special sciences). Taking for granted that the properties with which the special sciences deal supervene on the fundamental physical properties, if there is to be causation in the production sense in the domain of the special sciences, properties bringing about certain effects in virtue of their causal nature, there is causation in that sense in the fundamental physical domain, the supervenience base, as well. In other words, under the assumption of supervenience, if there is objective modality in the domain of the special sciences, there is objective modality also in the domain of fundamental physics.

Again, causal realism may not be the only game in town for a coherent ontology reaching from physics to biology and the special sciences in general. But, again, the task is to spell out such an ontology, and causal realism is one way to achieve that task. As any metaphysical position, causal realism has to be assessed on the basis of overall considerations, taking into account the metaphysics of properties, the philosophy of physics, and the philosophy of the special sciences.

Department of Philosophy
University of Lausanne
Quartier UNIL-Dorigny, Bâtiment Anthropole 4074
1015, Lausanne
Switzerland
Michael-Andreas.Esfeld@unil.ch

39 Robert Cummins, "Functional analysis". in: *Journal of Philosophy* 72, 1975, pp. 741–764.

CHAPTER 12

HOLGER LYRE

STRUCTURAL INVARIANTS, STRUCTURAL KINDS, STRUCTURAL LAWS

The paper has three parts. In the first part ExtOSR, an extended version of Ontic Structural Realism, will be introduced. ExtOSR considers structural properties as ontological primitives, where structural properties are understood as comprising both relational and structurally derived intrinsic properties or structure invariants. It is argued that ExtOSR is best suited to accommodate gauge symmetry invariants and zero value properties. In the second part, ExtOSR will be given a Humean shape by considering structures as categorical and global. It will be laid out how such structures serve to reconstruct non-essential structural kinds and laws. In the third part Humean structural realism will be defended against the threat of quidditism.

12.1 STRUCTURAL REALISM AND INTRINSICALITY: OSR EXTENDED

Many structural realists agree on two claims: they prefer ontic over epistemic versions of SR and they don't want to dismiss the idea of relata altogether. Therefore non-eliminative versions of ontic structural realism have become fashionable. They start from the idea that there are relations and relata, but that there is nothing more to the relata than the 'structural properties' in which they stand. But what are 'structural properties'? Are they all and only relations? Or must we allow for certain intrinsic properties as well? I do believe that, in order to cope with symmetry structures, one has to accept certain intrinsic features. The main reason is that symmetry structures come inevitably equipped with certain invariants under the symmetry. And symmetries and symmetry considerations play an eminent role in modern physics, notably as external spacetime structures and internal gauge symmetry structures. So SR proponents should take symmetry structure to be the most relevant structure of the world.

A symmetry of a domain D may be considered a set of one-to-one mappings of D onto itself, the symmetry transformations, such that the structure of D is preserved. The symmetry transformations form a group and exemplify equivalence relations (which lead to a partitioning of D into equivalence classes). From this we always get invariants under a given symmetry providing properties shared by all members of D. And insofar as such properties belong to any member of D

D. Dieks et al. (eds.), *Probabilities, Laws, and Structures*, The Philosophy of Science in a European Perspective 3, DOI 10.1007/978-94-007-3030-4_12,

irrespectively of the existence of other objects, they are 'intrinsic'. On the other hand, they do not suffice to individuate the members, since all members share the same invariant properties in a given domain. They are, in a still to be spelled out sense, 'parasitic' on the global structure. In my 2010 I call them "structurally derived intrinsic properties". They violate the strong Leibniz principle: as structure invariants they only serve to individuate domains, not entities.

Now consider non-eliminative OSR as a position characterized by the claim that there are relations and relata, but that there is nothing more to the relata than the structural properties in which they stand. We may then distinguish two versions:

- Simple OSR (SimpOSR): structural properties are only relational properties,
- Extended OSR (ExtOSR): structural properties are relational and structurally derived intrinsic properties (invariants of structure).

ExtOSR is the version favoured here (formerly labelled as "intermediate SR" in my 2010). In the taxonomy of Ainsworth,[1] ExtOSR is either a non-eliminativist OSR1 or close to OSR3, which takes relations and properties as ontological primitives, but objects as derived. And yet none of the categories really fits. The reason why Ainsworth's taxonomy seems to be transverse to ours is that it isn't fully exhaustive, which is why he discusses subcategories of all three versions that basically differ in the way objects are (re-) constructed. I will argue in favour of a modification of a bundle view of objects below.

In my 2010 paper[2] the Gedankenexperiment of a lone electron is introduced to show the differences between SimpOSR and ExtOSR. Under both eliminative and SimpOSR the lone electron cannot have a charge, since no other objects are left in virtue of which the electron's charge might be considered as relational. Under ExtOSR it is perfectly possible to allow, even in the trivial case of only one member in D, for the object to possess symmetry-invariant properties. I should emphasize that this Gedankenexperiment is exclusively meant to highlight the difference between SimpOSR and ExtOSR – it has a didactic value only. By no means do I claim, nor should ExtOSR proponents claim, that such a possible world is a nomologically possible world. Of course it isn't. It conflicts with QED and other fundamentally physical as well as operational assumptions. But it nevertheless highlights a metaphysical difference. An object may have its invariant properties according to the world's structure, the structure comes equipped with such properties. Moreover, such invariant properties should not be considered as relational to the structure, since this raises the problem of the possible Platonic existence of unexemplified structures. I take it that almost all OSR proponents of any stripe consider themselves to be *in re*-structuralists, not *ante rem*-Platonists. The world structure must therefore be an instantiated structure – instantiated by at least one member of D.

---

1 Peter M. Ainsworth, "What is ontic structural realism?", in: *Studies in History and Philosophy of Modern Physics* 41, 2010, pp. 50–57.

2 Holger Lyre, "Humean Perspectives on Structural Realism", in: Friedrich Stadler (Ed.), *The Present Situation in the Philosophy of Science*. Dordrecht: Springer 2010, pp. 381–397.

As far as I can see it, the most convincing reason from physics why we must take structure invariants seriously stems from gauge symmetries. I will give another argument, the argument of zero-value properties, below. It is obvious that the content of modern fundamental theories in physics is mainly given by symmetry structures. And in this respect, gauge theories figure as the most important case. But gauge symmetries are special in the sense that they are non-empirical symmetries. This means that gauge symmetry transformations possess no real instantiations, the physical content of gauge theories is carried all and only by the gauge symmetry invariants.[3] Such invariants are mathematically fully characterised (but not solely given) by the Casimir operators of the gauge groups (the Casimirs classify the multipletts and commute with the generators of the gauge Lie groups which correspond to the charges). We get mass and spin as Casimir operators of the Poincaré group and the various charges of the U(1)xSU(2) and SU(3) interaction groups. Hence, mass, spin, and charge (in the most general sense) are the most fundamental 'structurally derived intrinsic properties'. By focusing exclusively on relational properties, SimpOSR doesn't have the resources to take gauge theories into account, while ExtOSR apparently does.

But there's more. Elementary particle physics provides us with a taxonomy of the fundamental building blocks of the world. By characterizing particles via mass, spin, and charge, physicists regularly ascribe zero-value properties to particles. They will for instance say that the photon has zero mass or that the neutrino has an electric charge with value zero. As Balashov[4] points out, such zero values aren't merely absences of quantities or holes in being, they are considered to be as real as non-zero value properties. Balashov makes the following case:

> Suppose particle *a* is a bound state … of two particles … having non-zero quantities P+ and P-summing up to 0. … it is more reasonable to say that *a* has zero value of P … than to insist that it has no P at all. P-hood cannot simply disappear when combined with another P-hood in a productive way.

He calls this the *argument from composition*. But elementary particles aren't composites. We may, however, extend the argument by using parity and unification considerations to non-composite cases. P-hood may figure as part of the explanation of the generic behaviour of a particle in certain circumstances both in the case of $P \neq 0$ and $P = 0$. Conservation laws are the most important case of such explanations. We do for instance predict the behaviour of the yet undetected Higgs boson in part by the fact that it is assumed to have spin zero.

Consider also the well-known classification of elementary particles by means of the irreducible unitary representations of the Poincaré group.[5] The assumption

3 Holger Lyre, "Holism and structuralism in U(1) gauge theory", in: *Studies in History and Philosophy of Modern Physics* 35, 4, 2004, pp. 643–670.

4 Yuri Balashov, "Zero-value physical quantities", in: *Synthese* 119, 1999, pp. 253–286.

5 Cf. also my "Holism and structuralism in U(1) gauge theory", *loc. cit.*

behind it is that physical systems must possess relativistically invariant state spaces with the most elementary, irreducible representations possessing no invariant subspaces. And as we've already seen, the representations of the Poincaré group are mathematically fully characterized by its Casimir operators. This whole consideration affects all particles including the ones with zero mass, zero spin or both, since *all* particles are considered to be representations of the Poincaré group.

ExtOSR, I claim, naturally embraces the appearance of zero-value properties in fundamental physics by assuming that the world consists of a structure mainly given by the structure of the fundamental physical gauge groups (including the Poincaré group as a gauge group itself). Particles are instantiations of the world structure possessing all structurally invariant properties irrespective of whether the property value is zero or not. In what follows below I will show how this class of properties can also be accommodated from a non-dispositionalist point of view (pace Balashov[6] who argues otherwise).

Yet another commentary is necessary here. Recently, Roberts[7] has coined the term 'group structural realism' for the idea of identifying structure with the structure of symmetry groups. While on the one hand he acknowledges the fact that group structural realism has the advantage to provide us with a precise mathematical notion of structure, he on the other hand side diagnoses an, as he sees it, serious problem: the problem of an infinite regress of structures. Consider for instance the hierarchy that one can produce by ascending from a group *G* to *Aut G*, the group of all automorphisms of *G*, next to *Aut Aut G* and so on. But, as Roberts himself also acknowledges, the structural realist account "*perhaps most closest to the right attitude*" is to accept just the groups that are most naturally suggested by physics as the fundamental bottom of towers of structures. This is exactly the recipe I like to suggest here. While it is true that you can't easily read off your metaphysics from physics, one should nevertheless let physics be the main and solid guide in choosing the right metaphysics. And this in particular holds if we have an underdetermination in metaphysics which can be cured by physics! For this is just what Roberts does: construct an overblown and therefore underdetermined metaphysical hierarchy that can easily be cut back by physics as our primary guide.

Mention must finally be made that the present account is not bound to group structures. Surely, symmetry groups play a dominant role, but other structures come into play as well. The structural core of quantum theory is for instance given by the non-commutative algebra structure of the observables. It is, again, a physical, not a metaphysical question, what the fundamental structures in nature are.

---

6 Balashov, "Zero-value physical quantities", *loc. cit.*

7 Bryan W. Roberts, "Group structural realism", in: *British Journal for the Philosophy of Science 2010*, DOI: 10.1093/bjps/axq009

12.2 Humean Structural Realism: Structural Kinds and Structural Laws

After laying out ExtOSR as my favoured variant of structural realism, I shall now turn, for the rest of the paper, to the question of whether and in which sense I think structural realism can be combined with a Humean stance. I shall start with the issue of structural laws and then go over, in the third section, to defend non-modal categorical structures.

In its usual form, Humeanism rests on two basic features: first, the idea of an ultimate supervenience base – this is the reductionist spirit behind Humeanism. And, second, a quite rigorous scepticism about modalities – call this is the nominalist spirit. I shall focus on how this affects the Humeanist's view about properties and laws. Let's start with properties. From their nominalist inclinations it seems clear that Humeans will be non-dispositionalists and non-essentialists, that is they will favour categorical over dispositional (or modal) properties. It is of course not part of the Humean agenda that one must favour intrinsic over relational properties as in David Lewis' infamous doctrine of Humean supervenience. Moreover, Lewis' Humean supervenience is in glaring conflict with both quantum mechanics and gauge theories. In both types of theories non-local effects – EPR correlations on the one hand and holonomy effects on the other – suggest a stark violation of Humean supervenience: intrinsic properties of wholes do not supervene on intrinsic properties of their parts. By way of contrast, intrinsic properties of wholes may very well supervene on non-supervenient relations between the parts.

The natural supervenience base for structuralists consists, of course, of structures. Structures, in turn, seem to be "composed" out of relata and structural properties, being relations and structurally derived intrinsic properties. Leaving notorious questions of ontological priority for a moment aside, we can just say that the Humean structuralist shall consider non-modal, categorical structures as the proper supervenience base. And there are two aspects of such structures that are of interest here. There is on the one hand the aspect of categoricalism – this will be discussed below. On the other hand there's the aspect of such structures as being global entities. This is why the idea of structures as 'composites' must be taken with a grain of salt. If we think of the fundamental symmetry structures in physics, then we better conceptualize them as reflecting global regularity features of the world *in toto*. They neither are abstract mathematical *ante rem*-structures nor are they composed out of universals (as discussed by Psillos, forthcoming). They rather are concrete global and world-like *in re*-structures.

We may capture this characterisation by noticing that the usual metaphysical abstract/particular distinction must be complemented with a global/local distinction. We then arrive at the following matrix:

	concrete	*abstract*
local	particular	universal
global	*in re*-structures	*ante rem*-structures, universal structures, mathematical structures

Particulars or concreta are local and concrete entities, whereas universals are abstract. They are 'local' in the sense that they are instantiated by local exemplars. By way of contrast, structures aren't local, they are global or world entities. They may either be considered as abstract with mathematical structures as a prime example, but they may also be construed as concrete entities in the sense that they are directly given as elements of the spatiotemporal world *in toto*. This, I suggest, is the conception of structures that should be preferred by OSR proponents.

There are of course structuralists that, albeit coming close to the group theoretic considerations here, adopt an *ante rem* view of abstract structures.[8] Psillos,[9] following Bigelow and Pargeter,[10] discusses the pros and cons of the idea to construe structures as abstract entities or 'structural universals'. He diagnoses various difficulties of this view which can basically be traced back to the idea that such structural universals may have other universals as parts (as displayed in various cases of molecule configurations). I take this to indicate that we better refrain from characterizing structures as abstract. As far as I can see, however, none of Psillos' arguments speak against the possibility of structures as concrete elements of the world *in toto*. In considering structures as global or holistic entities the question of ontological priority of either relata or structural properties turns out as misguided. If talk of ontological priority makes sense at all, then structures as a whole should be prioritized. Ainsworth's[11] taxonomy should be supplemented by (at least) a fourth option (OSR4) which takes whole structures as basic and structural properties – relations and intrinsic invariants – as features of such structures. From them relata can be derived or reconstructed in the following sense: they are the placeholders between the relations and they are domain-wise individuated by the structural invariants which serve as structurally derived intrinsic properties of the relata.

8 E.g. Aharon Kantorovich, "Ontic Structuralism and the Symmetries of Particle Physics", in: *Journal for General Philosophy of Science* 40, 1, 2009, pp. 73–84; Tian-Yu Cao, *From Current Algebra Quantum Chromodynamics: A Case for Structural Realism*. Cambridge: Cambridge University Press 2010.

9 Stathis Psillos, "Adding Modality to Ontic Structuralism: An Exploration and Critique", in: Elaine Landry and Dean Rickles (Eds.), *Structure, Object, and Causality*. Dordrecht: Springer (forthcoming).

10 John Bigelow and Robert Pargeter, *Science and Necessity*. Cambridge: Cambridge University Press 1990.

11 Ainsworth, *loc. cit.*

This is also the reason why structuralism doesn't entirely collapse to variants of a bundle ontology. For instance, because of its nominalist spirit (and as will become clear in the next section), the ExtOSR version defended in this paper is close to an ontology of trope bundles. But bundles are usually construed as local. The picture I'd like to advocate is rather that the world consists of a global structure which can only approximately be reconstructed by a collection of more or less localizable objects. Another way of spelling out the worries about the 'local' is to say that structuralism seems to directly conflate with pointillisme – the doctrine that a physical theory's fundamental quantities are defined at spacetime points and represent intrinsic properties of point-sized objects located there (cf. Butterfield[12] as a forceful attack on pointillisme).

As we've seen, the structural invariants emphasized here provide us with properties that are shared by all members in a structure domain and thus serve to individuate such domains. In fact, they provide us with a concept of kinds – natural kinds. Generally speaking, natural kinds are human-independent groupings or orderings of particulars in nature. And it is one of the major tasks of science to reveal the kinds in nature, for if such kinds exist then we may expect our scientific explanations to become forceful precisely when they generalize over such kinds. But while it is straightforward to think of kinds as shared properties, the real problem is to understand what the reason for this 'sharing' is. Yet Humeanists usually don't provide an answer to this problem, since nine times out of ten they stick with a regularity view of laws. The orthodox account of natural kinds is therefore bound to essentialism, the view that there are essences in nature. Under such a view the shared properties that make up a kind are essential properties. Essential properties are modal properties in the sense that the particulars that possess them necessarily belong to the kind.

A rigorous Humean framework is incompatible with an essentialist conception of natural kinds. A remarkable feature of structural invariants, I claim, is however that they provide us with a non-modal Humean understanding of kinds without giving up the possibility of a further explanation for the universal sharing of certain features. For we may understand the perplexing empirical fact that, say, all electrons posses exactly the same property values for mass, spin and charge. These properties are in this sense universal properties. From the point of view of ExtOSR we may just trace this universality back to the globally built-in regularity of the world as possessing particular symmetry structures. That is to say we must not acquiesce the individual property-likeness of particular electrons as a brute fact of nature, as the traditional regularity view has it, but reduce it to the global world structure (e.g. particular gauge groups). Note that no necessity is involved in this conception since the global world structure itself is non-modal in the sense that it is a brute fact of nature that just this particular global structure exists. We

12 Jeremy Butterfield, "Against Pointillisme: A Call to Arms", in: Dennis Dieks, Wenceslao J. Gonzalez, Stephan Hartmann, Thomas Uebel and Marcel Weber (Eds.), *Explanation, Prediction, and Confirmation*. Dordrecht: Springer 2011.

have thus shifted the regularity one step further, from the level of local to global concrete entities. This conception of natural kinds might be dubbed a 'structural kinds' view. It is the conception of kinds offered by ExtOSR within a Humean framework.

To invoke structural kinds also means to invoke structural laws. For laws generalize over kinds. Structural laws, in turn, generalize over structural kinds. This is tantamount to say that structural laws just reflect the structures in nature. In the case of the fundamental physical structures the structural laws are essentially the mathematical equations that display the relevant symmetries. The symmetries are global built-in regularities of the world in the sense that other symmetries could exist as built-in instead. Obviously, this is in tune with a strict non-necessitarian conception of laws – and goes beyond structural realists' talk about 'modally informed' laws.[13] Moreover, and as I've pointed out in my 2010[14] the Humean structural laws view has the resources to overcome well-known problems of the orthodox regularity view such as non law-like regularities (by considering only global structures) and empty laws (by considering instantiated *in re*-structures only).

12.3 Humean Structural Realism: Categorical Structures

Humean SR sees structures as non-modal, categorical structures. They bring about nothing and constitute the Humean structuralist's supervenience base. They are "just there". Other structures could have been instantiated – or could be instantiated at any new moment in time (although more must be said, but cannot be said in this paper due to lack of space, about the temporal structure of the 4D world; see also the short remarks in the conclusion). As non-modal, categorical and determinate structures they should be taken as brute facts, ontologically irreducible, and primitive.

But there's a strong movement within structural realism to prefer modal or causal structures.[15] The perhaps most outspoken proponent of this movement is Michael Esfeld.[16] As he sees things, "the fundamental physical structures possess

13 Angelo Cei and Steven French, *Getting Away from Governance: A Structuralist Approach to Laws and Symmetries.* Preprint PhilSci-Archive 5462 (2010). – Though I'm of course very much in favour of the general tendency of this paper.

14 Lyre, "Humean Perspectives on Structural Realism", *loc. cit.*, sec. 22.

15 E.g. Anjan Chakravartty, *A Metaphysics for Scientific Realism: Knowing the Unobservable.* New York: Cambridge University Press 2007.

16 Michael Esfeld, "The modal nature of structures in ontic structural realism", in: *International Studies in the Philosophy of Science* 23, 2009, pp. 179–194; Michael Esfeld, *Causal realism.* This volume; Michael Esfeld and Vincent Lam, "Ontic structural realism as a metaphysics of objects", in: Alisa Bokulich and Peter Bokulich (Eds.): *Scientific structuralism.* Dordrecht: Springer 2011, pp. 143–159.

a causal essence, being powers".[17] Esfeld claims to overcome a couple of well-known difficulties connected to structural realism. The two most relevant problems are:

1. The mathematical/physical distinction of structures,
2. The problem of quiddities and humility.

Let's start with the mathematical/physical distinction (1). Esfeld believes that by assuming categorical structures Humean SR collapses to mathematical structuralism. He argues that while mathematical structures do not cause anything, real physical structures clearly distinguish themselves from mere mathematical structures in that they are causally efficacious.

I have two worries here. First, Esfeld raises the problem in such a way that it doesn't come out as a special problem of (particular versions of) structural realism, but of Humeanism or categoricalism in general. Any non-modal account of entities is affected by his kind of reasoning: causal efficacy cannot be accounted for by a Humean mosaic of non-dispositional properties but only by dispositional ones. But should we really consider this to be a knock-out argument against Humeanism? Dispositionalism will then become true by fiat. But you can't decide metaphysical debates like that. Humeans and non-Humeans agree that there are cause-effect regularities in our world. They disagree about the way how to conceptualize them metaphysically. It is of course true that we know about the various structures in physics by means of their causal efficacy. But this says nothing about the metaphysical conception of causation. Causal efficacy can very well be captured in regularist terms. Nothing in Esfeld's arguments enforces a metaphysically thick modal nature of structures.

But perhaps the real worry of Esfeld about structures with 'no causal contact' to the world lies elsewhere. In his 2009 he still sticks with SimpOSR and, hence, rejects any intrinsic properties.[18] Under this conception the question of how regularities of a pure microscopic web of relations hinge together with macroscopic causes and effects might indeed cause a certain uneasiness. However, the problem at this point is not the metaphysics of causation, but the notorious multirealizability of purely relationally individuated structures. By way of contrast, ExtOSR introduces structurally derived intrinsic properties to individuate structure domains. This provides us with an account to circumvent the problem of 'unintended domains'.[19] By introducing structure invariants the nature of the relations and relata in the structure is no longer completely indetermined. The idea is that in our experimental practice we are (more or less directly) acquainted with the intrinsic structure invariants. Hence, no multirealizability arises.

17 Esfeld, "The modal nature of structures in ontic structural realism", *loc. cit.*

18 Withdrawn by Esfeld and Lam, "Ontic structural realism as a metaphysics of objects", *loc. cit.*, sec. 8.4.

19 Cf. Lyre, "Humean Perspectives on Structural Realism", *loc. cit.*, sec. 12ff.

So let's go over to the second class of problems centred around quidditism and humility (2). As Esfeld[20] puts it:

> If the fundamental properties are categorical and intrinsic, then there are worlds that are different because they differ in the distribution of the intrinsic properties that are instantiated in them, although there is no difference in causal and nomological relations and thus no discernible difference between them. This position therefore implies quidditism and humility.

My first answer is that this is (again!) no special problem of SR, but of categoricalism in general. This seems to be granted by Esfeld: "what accounts for quidditism and humility is the categorical character of the fundamental properties, not their supposed intrinsic character".[21] In fact, he considers the threat of quidditism to be *the* master argument against categoricalism: only causal properties prevent from quiddities, because if all properties are causal and, hence, individuated by their causal profile only, then there's no room for extra quiddistic factors over and above the causal profile.

I certainly share Esfeld's worries about mysterious extra-metaphysical factors, which is what quiddities really are. But, I'm afraid, the antidote of causal properties isn't as strong as Esfeld wants it to be. This has rightly been pointed out by Psillos[22] by means of the following consideration. Suppose a world W1 in which two properties A and B work in tandem to produce a certain effect E but, taken individually, don't have any effect at all. Dispositionalism cannot distinguish W1 from a world W2 in which E is brought about by one single property. The metaphysical difference between W1 and W2 goes beyond causal roles. So this would be my second answer to (2): quidditism is the view that nomological roles do not supervene on properties, but nomological roles do not supervene on causal properties either!

So it seems that neither the dispositionalist nor the categoricalist can entirely get rid of any mysteriously hidden metaphysical factors. But clearly it would be neat if in particular the structural realist could obviate the threat of quidditism – at least to a certain extent. Well, I believe he can. This paves the way to a third answer to Esfeld's worries about quidditism. Here's a passage from Lewis who famously pronounced humility against quiddities:

> I accept quidditism. I reject haecceitism. Why the difference? It is not, I take it, a difference in prima facie plausibility. ... In both cases alike, however, we can feel an uncomfortable sense that we are positing distinctions without differences. [...] To reject haecceitism is to accept identity of qualitatively indiscernible worlds; to reject quidditism is to accept identity of structurally indiscernible worlds – that is, worlds that differ just by a permutation or replacement of properties.[...] It would be possible to combine my realism about possible worlds with anti-quidditism. I could simply insist that ... no property is ever instantiated

20 Esfeld, "The modal nature of structures in ontic structural realism", *loc. cit.*, p. 182.
21 *Ibid.*, p. 187.
22 Psillos, *loc. cit.*

in two different worlds. ... It could be for the sake of upholding identity of structurally indiscernible worlds, but I see no good reason for wanting to uphold that principle".[23]

While Lewis doesn't see a good reason for upholding the identity of structurally indiscernible worlds, structural realists certainly should. For structural realism is precisely the doctrine that is based on such a principle. We may even use the identity of structurally indiscernible worlds, or, shorter, the *identity of isomorphs*, to define SR and its major variants. While ESR is captured by the claim that the world is known up to structural isomorphs, OSR is the view that the world exists up to such isomorphs only. Surely, Lewis wouldn't be convinced by such a manoeuvre of simply 'quining quiddities' by means of the identity of isomorphs, since I've given no *metaphysical* reason to dismiss quiddities. Nevertheless, the identity of isomorphs is empirically supported to the extent to which structural realism is empirically supported by modern physics. So let me repeat the recipe already suggest at the end of Sect. 12.1: one should let physics be the main and solid guide in choosing the right metaphysics, particularly in cases of seemingly metaphysical excesses.

A fourth and final attempt to counter worry (2) is the following. Quidditism claims primitive suchness. It's the idea that a permuation of properties (or types) makes a difference. It follows that quidditism may also be understood as upholding the principle of trans-world property identity, since quiddities are instantiated at different possible worlds. This is quite in analogy to the traditional idea of universals as being instantiated in different spacetime regions. But why should one uphold such a principle? A Humeanist clearly wouldn't. Tropes as well as categorical structures violate trans-world property identity, since both tropes and categorical structures as (examples of) entities suited to constitute a proper Humean base are individuals. So neither tropes nor categorical structures are ever instantiated at two different worlds. While Esfeld's master argument wants to tell us that Humeanism implies quidditism, the contrary seems to be true: Humeanism virtually contradicts quidditism.

So neither is there any special problem with the distinction between the mathematical and the physical for Humean SR, nor does dispositionalism fare so much better in rejecting quidditism. But then the question must be raised what makes dispositionalism attractive at all. And here, as far as I can see, we should be quite reluctant. For the real problem with traditional dispositionalism is that it sticks with pointillism, the view that the fundamental quantities in physics are defined at local regions of spacetime. Structuralism, as we've seen, should however be construed as fundamentally holistic and conforming to globally defined entities, which is just what structures are. By way of contrast, the picture of local powers is a hopelessly outdated and naïve metaphysical picture of physics (to say the least).

23 David Lewis, "Ramseyan Humility", in: David Braddon-Mitchell and Robert Nola (Eds.), *Conceptual Analysis and Philosophical Naturalism*. Cambridge, MA: MIT Press 2009, pp. 209–210.

From this perspective, to combine dispositionalism with structuralism then means to try to combine two deeply opposing pictures, which in turn means that to prevent from quiddities by introducing mystic causal powers amounts to curing one metaphysical exaggeration with another one.

12.4 Conclusion

In this paper I've argued for an extended version of OSR, ExtOSR, that takes structural properties as ontological primitives, where structural properties are understood as comprising both relational and structurally derived intrinsic properties (structure invariants). ExtOSR is best suited to accommodate gauge invariants and zero value properties. I've then connected this with a Humean approach, in which one considers categorical and global structures to constitute the Humean supervenience base. As global entities the structures display a built-in global regularity and serve to understand the universality of the fundamental properties without invoking essences and, thus, providing us with a concept of non-essential structural kinds and laws.

The Humean position of structural realism just sketched avoids mysterious modal powers and ungrounded dispositions – but raises new questions, too. Of particular interest is most certainly how dynamics and temporal change fit into the overall picture of non-modal and global structures. A straightforward answer is that the group of temporal automorphisms of the state space is of course itself a structure. Another straightforward but at the same time extreme option would be to adopt something along the lines of either a perdurantist view or a block universe conception and take the entire four-dimensional world structure for granted. To whatever extent the structural realist wants to address these questions, it's definitely a topic that deserves further scrutinization.

Acknowledgement: Thanks to Kerry McKenzie for helpful remarks on an earlier draft of the paper.

References

Peter M. Ainsworth, "What is ontic structural realism?", in: *Studies in History and Philosophy of Modern Physics* 41, 2010, pp. 50–57.

Yuri Balashov, "Zero-value physical quantities", in: *Synthese* 119, 1999, pp. 253–286.

John Bigelow and Robert Pargeter, *Science and Necessity*. Cambridge: Cambridge University Press 1990.

Jeremy Butterfield, "Against Pointillisme: A Call to Arms", in: Dennis Dieks, Wenceslao J. Gonzalez, Stephan Hartmann, Thomas Uebel and Marcel Weber (Eds.): *Explanation, Prediction, and Confirmation*. Dordrecht: Springer 2011.

Tian-Yu Cao, *From Current Algebra Quantum Chromodynamics: A Case for Structural Realism*. Cambridge: Cambridge University Press 2010.

Angelo Cei and Steven French, *Getting Away from Governance: A Structuralist Approach to Laws and Symmetries*. Preprint PhilSci-Archive 5462 (2010).

Anjan Chakravartty, *A Metaphysics for Scientific Realism: Knowing the Unobservable*. New York: Cambridge University Press 2007.

Michael Esfeld, "The modal nature of structures in ontic structural realism", in: *International Studies in the Philosophy of Science* 23, 2009, pp. 179–194.

Michael Esfeld, "Causal realism." This volume.

Michael Esfeld and Vincent Lam, "Ontic structural realism as a metaphysics of objects", in: Alisa Bokulich and Peter Bokulich (Eds.), *Scientific Structuralism*. Dordrecht: Springer 2011, pp. 143–159.

Aharon Kantorovich, "Ontic Structuralism and the Symmetries of Particle Physics", in: *Journal for General Philosophy of Science* 40, 1, 2009, pp. 73–84.

David Lewis, "Ramseyan Humility", in: David Braddon-Mitchell and Robert Nola (Eds.): *Conceptual Analysis and Philosophical Naturalism*. Cambridge, MA: MIT Press 2009.

Holger Lyre, "Holism and structuralism in U(1) gauge theory", in: *Studies in History and Philosophy of Modern Physics* 35, 4, 2004, pp. 643–670.

Holger Lyre, "Humean Perspectives on Structural Realism", in: Friedrich Stadler (Ed.): *The Present Situation in the Philosophy of Science*. Dordrecht: Springer 2010, p. 381–397.

Stathis Psillos, "Adding Modality to Ontic Structuralism: An Exploration and Critique", in: Elaine Landry and Dean Rickles (Eds.), *Structure, Object, and Causality*. Dordrecht: Springer (forthcoming).

Bryan W. Roberts, "Group structural realism", in: *British Journal for the Philosophy of Science*, 2010, DOI: 10.1093/bjps/axq009.

Philosophy Department
University of Magdeburg
P.O. Box 4120
D-39016 Magdeburg
Germany
lyre@ovgu.de

Chapter 13

Paul Hoyningen-Huene

Santa's Gift of Structural Realism

As I am not a specialist in the subject matter of Holger Lyre's paper,[1] I had to learn from him that there are two different approaches to structural realism in the literature, the "Worrall-type" and the "French-Ladyman-type" approach, as he calls them.[2] The Worrall-type approach is a reaction to the difficulties that emerged in the defence of entity realism against various objections, especially against the well-known pessimistic meta-induction.[3] In contrast, the French-Ladyman-type approach tries "to present arguments from the sciences directly, more precisely from the way modern science, notably physics, informs us about the ontology of the world".[4] Thus, it seems to be the much more straightforward approach to structural realism, avoiding the detour of getting involved in a rather convoluted discussion. In the following, I shall present two critical remarks on this approach.

13.1

In contrast to the Worrall-type approach, the French-Ladyman-type approach decouples itself both from the history of science and from the history of philosophy in the following sense. It decouples itself from the history of science in considering contemporary theories of modern physics like quantum mechanics, quantum field theory, general relativity theory, quantum gravity, or gauge theories in their present state, like in a flash exposure, and not in their historical development. This, in itself, is clearly not illegitimate. However, this decoupling is connected with the decoupling from the history of philosophy in the sense that there, the pertinent

1 In fact, this is the kind of understatement possibly only appropriate in a talk at a UK university. I am absolutely ignorant of the technical details of structural realism when applied to real physical theories.

2 See Holger Lyre, "Humean Perspectives on Structural Realism", in: F. Stadler (Ed.), *The Present Situation in the Philosophy of Science*. Dordrecht: Springer 2010, pp. 381–397, p. 382; see also Holger Lyre, "Symmetrien, Strukturen, Realismus", in: M. Esfeld (Ed.), *Philosophie der Physik*. Berlin: Suhrkamp, in press.

3 As the starting point of the recent discussion, see John Worrall, "Structural Realism: The Best of Both Worlds?", in: D. Papineau (Ed.), *The Philosophy of Science*. Oxford: Oxford University Press 1996 [1989], pp. 139–165 (originally in *Dialectica* 43, 1989, pp. 99–124).

4 Lyre, "Humean Perspectives on Structural Realism", *loc. cit.*, p. 382.

D. Dieks et al. (eds.), *Probabilities, Laws, and Structures*, The Philosophy of Science in a European Perspective 3, DOI 10.1007/978-94-007-3030-4_13,

philosophical question has always been whether or not some form of realist interpretation of physical theories is legitimate in principle. This question gains a large part of its urgency from the history of science. By cutting themselves off from the history of science and the history of philosophy, this question disappears, and the structural realists can simply follow their "realist intuitions"[5] when interpreting modern physical theories.

With respect to following intuitions in philosophy, in many cases I am tempted to suggest that if someone has strong intuitions, he or she should urgently see a doctor.[6] As this is certainly not a very popular suggestion these days, I hasten to add that the sort of doctors I am thinking of include Dr Wittgenstein and Dr Ryle. As students and younger scholars of philosophy may be ignorant of them – depending on the geographical location of their education –, it may be worthwhile to be reminiscent of a strong philosophical motive common to both of them.[7] Certain hypotheses may appear extremely plausible simply because they are suggested by our language use; thus, they appear to us as very plausible, or even inevitable, "intuitions". This plausibility should, so the doctors go on, not be taken at face value and should not be followed uncritically. In our case, it is less language use that suggests realist intuitions but more our deeply entrenched and extremely successful everyday realism. Still, I believe that in philosophy the content of such intuitions should be subjected to careful logico-philosophical analysis. Roughly, the upshot is this: intuitions are there to be analysed, not to be trusted.[8]

Thus my first objection against the French-Ladyman-style structural realism is that in this approach, the critical philosophical question about the legitimacy of a realist interpretation of (modern) physics is not at all seriously considered but simply dismissed. Presupposing the principal adequacy of a (structural) realist interpretation of modern physics, the only remaining question is what the ontological structure is that fits a given theory best. I know that this is a highly demanding question, and the technically very sophisticated attempts to answer it certainly deserve great respect. Nevertheless, I dislike the tendency to not even consider the deeper, critical philosophical question about realism in general.

---

5 Lyre, oral communication, October 2010.

6 This is a variation on a quote attributed to the former German chancellor Helmut Schmidt: If you have visions, you should see a doctor. Given the current (February 2011) state of German political academic citation culture, I feel compelled to make explicit the source on which the above sentence is modeled.

7 See Gilbert Ryle, *The Concept of Mind*. London: Penguin 2000 [1949], and Ludwig Wittgenstein, *Philosophical Investigations*. Translated by G. E. M. Anscombe. Oxford: Blackwell 1958 [1953].

8 Holger assures me that the suspension of the foundational debate about realism in the French-Ladyman-type approach is only temporary as "you cannot do everything at the same time" (e-mail message, March 8th, 2011). Fine.

13.2

Surveying the discussion of structural realism by Holger Lyre, by Michael Esfeld[9] and others, I have been struck by the wealth of possible structures that were seen as candidates for a true ontology of physics. As the conference took place in December, I was reminded of wish lists one could submit to Santa Claus (and, if I remember correctly, in one of the earlier talks a picture of Santa Claus was presented, although in a different context). The abundance of possible structures gives rise to doubts whether it is really possible to identify the true (or most adequate) structure that would fulfil the program of structural realism. In other words, do we have enough constraints in order to single out the particular structure that can qualify as *the* structure of reality underlying a given theory? Indeed, in another paper Holger Lyre himself has discussed the question "Is Structural Underdetermination possible?".[10] This is a serious question indeed as, at least in the Worrall-type approach, structural realism is the strategy to circumvent the problem of theory underdetermination as it presents itself to entity realism. However, there seems to be a great variety of different types of structures that may be fitted to a particular theory. First of all, as the recent discussion has shown we have the choice between epistemic, ontic and semantic structural realism, and the second form of structural realism can be eliminative or non-eliminative,[11] resulting in four basic positions. As I am not an expert in the field, I cannot really tell which of the following alternatives can be combined with which of the basic positions mentioned (or even whether these alternatives produce more basic positions). At any rate, there are alternatives between positions with or without necessity, with or without causal powers, with or without Ramseyan humility, with or without (structurally derived) intrinsic properties, with or without dispositions, with or without relational properties, with or without categorical structures, with or without quiddities, and so on.[12] On top of these different metaphysical possibilities, there is the question whether the mathematical structures themselves are sufficiently unambiguously determined for some given theory.[13] Are we sure that a bunch of creative mathematicians would not come up with many more possibilities if one could sell this as an attractive task to them? Will in the end, after careful analysis, only one

9 This volume.

10 Holger Lyre, "Is Structural Unterdetermination Possible?", in: *Synthese*, 180: 235–247 (2011).

11 Lyre, "Humean Perspectives on Structural Realism", *loc. cit.*, pp. 382–383.

12 See Esfeld, this volume, Lyre, this volume, Lyre, "Humean Perspectives on Structural Realism", *loc. cit,.* and the rest of the literature.

13 I am grateful to Holger for making me aware that there is a relevant difference between the metaphysical and the mathematical aspect of the sought-after structure of the world (e-mail message, March 8th, 2011). Fixing the latter should be easier than fixing the former, and a permanent lack of consensus regarding the mathematical structures would place the whole program in jeopardy, as Holger concedes.

single structure fit a given theory, be it in the mathematical or the metaphysical sense? I doubt it. Will there be additional criteria like, for instance, simplicity, economy, elegance, or coherence that will lead us to those structures that truly represent Nature's own structures? I strongly doubt it – why should Nature be such that she complies with our ideas of simplicity or elegance? Perhaps she does – but do we know or could we know?

Acknowledgement: I wish to thank Holger Lyre for important critical remarks on an earlier version of this paper, Nils Hoppe for the final polishing of the English, and Marcel Weber for suggesting the sexy title of this paper replacing its boring predecessor.

Institute of Philosophy
Leibniz Universität Hannover
Im Moore 21
D-30167 Hannover
Germany
hoyningen@ww.uni-hannover.de

Chapter 14

Steven French

The Resilience of Laws and the Ephemerality of Objects: Can A Form of Structuralism be Extended to Biology?[1]

14.1 Introduction

Broadly defined, structuralism urges us to shift our claims of ontological priority, from objects to structures.[2] Historically it is a view that arose out of reflection on the nature of modern (that is, twentieth century) physics and in its most recent incarnations it is motivated by a) the attempt to capture what remains the same through radical theory change[3] and b) the implications of quantum theory for our conception of physical objects.[4] As broadly defined, it encompasses diverse understandings of the nature of structure and the relationship between structure and putative objects.[5] The question I wish to consider is whether this ontological shift can be extended into the biological domain.

---

1 I'd like to thank Marcel Weber for inviting me to give the presentation on which this paper is based and the audience of the ESF-PSE workshop "Points of Contact between the Philosophy of Physics and the Philosophy of Biology" in London, December 2010, for the excellent questions and general discussion. I'd also like to acknowledge the many helpful contributions from Angelo Cei, Kerry McKenzie and especially Alirio Rosales on various aspects of this attempt to extend structuralism.

2 S. French and J. Ladyman, "In Defence of Ontic Structural Realism", in: A. Bokulich and P. Bokulich (Eds.), *Scientific Structuralism*. Boston Studies in the Philosophy of Science: Springer 2011, pp. 25–42; J. Ladyman, "Structural Realism", in: *Stanford Encyclopaedia of Philosophy* 2009: http://plato.stanford.edu/entries/structural-realism/.

3 J. Worrall, "Structural Realism: The Best of Both Worlds?", in: *Dialectica* 43, 1989, pp. 99–124. Reprinted in: D. Papineau (Ed.), *The Philosophy of Science*, pp. 139–165. Oxford: Oxford University Press.

4 J. Ladyman, "What is Structural Realism?", in: *Studies in History and Philosophy of Science* 29, 1998, pp. 409–424.

5 M. Esfeld and V. Lam, "Moderate Structural Realism about Space–time", in: *Synthese* 160, 2008, pp. 27–46; S. French and J. Ladyman, "Remodelling Structural Realism: Quantum Physics and the Metaphysics of Structure", in: *Synthese* 136, 2003, pp. 31–56; J. Ladyman, "Structural Realism", *loc. cit.*; J. Ladyman and D. Ross, *Everything Must Go: Metaphysics Naturalized*. Oxford: Oxford University Press 2007.

D. Dieks et al. (eds.), *Probabilities, Laws, and Structures*, The Philosophy of Science in a European Perspective 3, DOI 10.1007/978-94-007-3030-4_14,

My considerations build on a previous paper[6] where I suggested that even in the absence of the sorts of robust laws that one finds in physics, biological models present a rich array of structures that the structural realist could invoke. I also touched on issues to do with the nature of biological objects that could perhaps motivate the above shift and my principal purpose in this essay is to take that discussion further by drawing on recent work on the heterogeneity of biological individuals.

Let me begin, however, by briefly reviewing the issue of laws and the structuralist conception of them.

14.2 Laws and the Lack Thereof

As realists, how should we read off our ontological commitments from theories? Here's one way: we focus on the relevant theoretical terms, such as 'electron', that feature in the successful explanations in which the theory figures, and understand such terms as referring to entities in the world. We conceive of these as objects that possess properties that are then inter-related via the laws of the theory. This presupposes commitment to what might be called an 'object-oriented' stance. Here's another way that some have argued is more natural: we begin with the laws and (crucially, in physics at least) the symmetries of the theory, and regard these as representing the way the world is. The relevant properties are then identified in terms of the role they play in these laws. And … we stop there and do not make the further move of taking these properties to be possessed by objects. This is the structuralist way of looking at things and we take the laws (and symmetries) as representing the structure of the world.

Now there is considerably more to be said, particularly about the nature and role of laws on this structuralist conception.[7] Note first of all, that the governance role of laws must be revised. How this understanding of laws as governing entered our conception of science is an interesting historical question but, depending on how one views properties, it fits nicely with an object-oriented metaphysics of science. On our 'second way' above, however, there are no objects, qua metaphysically robust entities, and so there is nothing for the laws to govern. Rather, the relevant relation is one of dependence, in the sense that properties depend on laws, since their identity is given by their nomic role. Secondly, how we conceive of the necessity of laws must also be understood differently. On the object-oriented view, we imagine possible worlds containing the same objects as this, the actual world, and then consider what relations between these objects continue to hold in such worlds. Those that do, count as, or feature in, the relevant laws which can then be regarded as (physically) necessary in the sense of holding in all (physically) possible worlds. This then allows us to distinguish between laws and accidental generalisations. On the structuralist view, we do not have the same fundamental

6 S. French, "Shifting to Structures in Physics and Biology: A Prophylactic for Promiscuous Realism", in: *Studies in History and Philosophy of Biological and Biomedical Sciences*, 42, 2011, pp. 164–173.

7 See A. Cei and S. French, "Getting Away from Governance: Laws, Symmetries and Objects", forthcoming.

metaphysical base consisting of some set of objects that we can hold constant between worlds. That feature of laws by which we can distinguish them from accidental generalisations has to be understood differently. Thus we might think of this feature in terms of the modal 'resilience' of laws, in the sense of remaining in force despite changes of circumstances. On the view adopted here,[8] this resilience is an inherent feature of laws, as elements of the structure of the world. And it is this resilience that gives laws their explanatory power – explaining why in every case, like charges repel for example. The explanation of this regularity – the reason why it obtains, and why it is, in a sense, unavoidable[9] – lies with the laws and, more profoundly perhaps, their inherently modal nature by which they have this resilience.

Now I have talked of 'resilience' here rather than necessity because the latter is associated with the above idea of 'holding in all possible worlds'. And this in turn is typically cashed out in terms of keeping a fundamental base of objects fixed in each such possible world and then showing that, given that, the laws of this, the actual world, will hold in all such worlds. From this perspective, the necessity of laws is a 'yes/no' matter as they either hold in all possible worlds or they do not. But this is not a perspective amenable to the structuralist, given its reliance on an object-oriented stance to begin with. Of course, there is still a tight metaphysical connection between laws and objects within the structuralist framework, but now it runs in the opposite direction: from laws and symmetries (as aspects of the structure of the world) in the fundamental base, via the relevant dependencies, to putative objects (such as elementary particles in physics). If the modality of the former is regarded as inherent, then 'necessity', as strictly conceived, is inapplicable; hence the use of the term 'resilience'. This then opens up some metaphysical space in which to consider laws in biology, or, rather, the supposed lack of them.

This feature is often highlighted as representing a major distinction between physics and biology and, in this context, as representing an equally major impediment to the extension of structuralism from the former to the latter. But this claimed lack of laws, and hence the distinction, rests on a characterisation of laws as necessary.

Consider, for example, Beatty's well-known 'Evolutionary Contingency Thesis':

All generalisations about the living world: (a) are just mathematical, physical, or chemical generalisations (or deductive consequences of mathematical, physical, or chemical

---

8 A. Cei and S. French, *loc. cit.*; but this is not the only option for the structuralist of course; see M. Esfeld, "The Modal Nature of Structures in Ontic Structural Realism", in: *International Studies in the Philosophy of Science* 23, 2009, pp. 179–194; H. Lyre, "Humean Perspectives on Structural Realism", in: F. Stadler (Ed.), *The Present Situation in the Philosophy of Science*. Springer 2010.

9 Cf. M. Lange, "Laws and Meta-Laws of Nature", in: *Studies in History and Philosophy of Modern Physics* 38, 2007, pp. 457–481, pp. 472–473.

generalisations plus initial conditions) or (b) are distinctively biological, in which case they describe contingent outcomes of evolution.[10]

If (a) is true, then biological laws 'reduce' (in whatever sense) to physical ones and there are no biological laws, per se (this obviously presupposes some form of reductionism and needs further argument to the effect that if *a* reduces to *b*, then *a* can be eliminated). If (b) is true, then the relevant generalisations are 'merely' contingent and thus cannot be necessary. (b) is certainly supported by the current conceptions of mutation and natural selection which imply that all biological regularities must be evolutionarily contingent. On that basis, they cannot express any natural necessity and hence cannot be laws, at least not on the standard understanding of the latter.[11] Thus, it follows that if either (a) or (b) is true, there are no biological laws.

Now one option would be to accept this thesis but insist that even though contingent, the relevant biological generalisations are still not 'mere' accidents in the way that, say, the claim that I have 67 pence in my pocket is. Thus one might argue that biological generalisations are fundamentally evolutionary, in the sense that under the effects of natural selection they themselves will evolve. In this sense, they cannot be said to hold in all possible worlds and thus cannot be deemed 'necessary'. If lawhood is tied to necessity, then such generalisations cannot be regarded as laws. However, given their role in biological theory, they cannot be dismissed as mere accidents like the claim about the contents of my pocket. They have more modal resilience than that. Perhaps then they could be taken to be laws in an inherently modal sense, where this is weaker than in the case of physical laws but still stronger and more resilient than mere accidents. Moreover, they are evolutionarily contingent in Beatty's sense. Putting these features together in the structuralist framework yields a form of 'contingent structuralism' in the sense that, unlike the case of physical structures where the structural realist typically maintains that scientific progress will lead us to *the* ultimate and fundamental structure of the world, biological structures would be temporally specific, changing in their fundamental nature under the impact of evolution.

Setting this option to one side, the standard view that there are no biological laws has been challenged by Mitchell.[12] She argues that this standard view assumes that natural necessity must be modeled on, or is taken to be isomorphic to, logical necessity.[13] But the crucial roles of laws – that they enable us to explain, predict, intervene and so on – can be captured without such an assumption. Indeed, what

10 J. Beatty, "The Evolutionary Contingency Thesis", in: G. Wolters and J. G. Lennox (Eds.), *Concepts, Theories, and Rationality in the Biological Sciences*. Pittsburgh: University of Pittsburgh Press 1995, pp. 45–81, pp. 46–47.

11 *Ibid.*, p. 52.

12 S. Mitchell, *Biological Complexity and Integrative Pluralism*. Cambridge University Press 2003.

13 *Ibid.*, p. 132.

characterizes laws as they feature in practice on her view is a degree of 'stability', in terms of which we can construct a kind of continuum[14]: at one end are those regularities the conditions of which are stable across space and time; at the other, are the accidental generalisations and somewhere in between are where most scientific laws are to be found. And even though biological generalisations might be located further towards the 'accidental' end of the continuum than the physical ones, this does not justify their dismissal as 'nonlaws'. Likewise Dorato, in his presentation at this workshop, argued that 'Biological laws differ from physical only in degree of stability and universality'. Such claims clearly mesh nicely with, and can be pressed into the service of, the kind of structuralist view I have sketched above, with 'resilience' equated with 'stability' and biological regularities regarded as features of the (evolutionarily contingent) biological structure of the world. It is this latter aspect that accounts for their (relative) resilience/stability and the way that aspect of their nature can explain why certain biological facts obtain.

Of course, one might complain that nevertheless there are fewer such laws in the biological domain than in physics, say, but this hardly seems strong grounds for blocking the extension of structuralism. Indeed, one can follow the advocates of the model-theoretic approach in responding to Beatty's arguments and look to the kinds of models and 'structures' in general that biology presents.[15] Not only have there been some useful discussions of biological models in recent years but their representation in terms of certain kinds of 'state spaces', for example, can be compared to similar representations in the case of physics. Thus on the biological side, we have the representation of, for example, Lotka-Volterra interspecific competition models in terms of state spaces or "phase portraits",[16] while in physics we find the representation of systems in terms of symplectic spaces, for example, as in the Hamiltonian formulation of classical mechanics.[17]

Nevertheless, we do not typically find the other feature of physical structures in biology, namely symmetries. Still, one can identify what might be called similarly 'high-level' features of biological structures. One such is Price's Equation, sometimes presented as representing 'The Algebra of Evolution'. Put simply, this states that

$$\Delta z = \mathrm{Cov}\,(w,z) + \mathrm{E}w(\Delta z)$$

where, Δz = change in average value of character from one generation to next; Cov (w,z) = covariance between fitness w and character (action of selection) and

14 *Ibid.*, p. 138.

15 Again, see S. French, "Shifting to Structures in Physics and Biology: A Prophylactic for Promiscuous Realism", *loc. cit.*

16 J. Odenbaugh, "Models", in: S. Sarkar and A. Plutynski (Eds.), *Blackwell Companion to the Philosophy of Biology*. Blackwell Press 2008.

17 J. North, "The "Structure" of Physics: A Case Study", in: *Journal of Philosophy* 106, 2009, pp. 57–88.

Ew(Δz) = fitness weighted average of transmission bias (difference between offspring and parents). Thus Price's equation separates the change in average value of character into two components, one due to the action of selection, and the other due to the difference between offspring and parents. There is a sense in which this offers a kind of 'meta-model' that represents the structure of selection in general.[18] Although obviously not a symmetry such as those we find in physics, this can nevertheless be regarded as a high-level feature of biological structure. As Rosales has emphasised, it is independent of objects, rests on no contingent biological assumptions and represents the modal, relational structure of the evolutionary process.[19] Just as the laws and symmetries of physics 'encode' the relevant possibilities, so Price's equation encodes how the average values of certain characters changes between generations in a given biological population.

There is of course more to say here and exploring the various features of both 'low-level' biological models and 'high-level' laws such as Price's equations will help reveal the extent to which structuralism can be extended into biology. Let me now turn to what I see as one of the principal motivations for dropping the object-oriented stance in philosophy of biology and adopting a form of structuralist ontology, namely the heterogeneity of biological objects.

14.3 The Fluidity and Heterogeneity of Biological Objects

As Dupré and O'Malley note, two of the implicit assumptions of biological ontology are that 'life' is organized in terms of the 'pivotal unit' of the individual organism and that such organisms constitute biological entities in a hierarchical manner.[20] Both of these assumptions can be challenged but they underpin the decomposition of biological organisms into individuals that are commonly taken to have the following fundamental characteristics: possessing three-dimensional spatial boundaries; bearing properties, acting as a causal agent.[21] Furthermore, biological individuals are generally taken to be countable and genetically homogenous (an assumption that forms part of what Dupré calls 'genomic essentialism').

However, there are well-known confounding cases that raise problems for this characterization.[22] Thus consider the case of the so-called "humungous fungus",

---

18 For a useful overview, see A. Gardner, "The Price Equation", in: *Current Biology* 18, 5, 2008, pp. 198–202; also S. Okasha, *Evolution and the Levels of Selection*. Oxford: Oxford University Press 2006, §1.2.

19 See A. Rosales, "The Metaphysics of Natural Selection: A Structural Approach", forthcoming, presented at the Annual Conference of the *BSPS* 2007.

20 J. Dupré and M. O'Malley, "Metagenomics and Biological Ontology", in: *Studies in History and Philosophy of the Biological and Biomedical Science* 28, 2007, pp. 834–846.

21 See R.A. Wilson, "The Biological Notion of Individual", in: *Stanford Encyclopaedia of Philosophy* 2007: http://plato.stanford.edu/entries/biology-individual/.

22 Some of these examples are taken from the papers and discussion at the symposium on

or *Armillaria ostoyae* which, in one case, covers an area of 9.65 square km. Previously thought to grow in distinct clusters, denoting individual fungi, researchers established through the genetic identity of these clusters that they were in fact manifestations of one contiguous organism that, as one commentator noted, "... challenges what we think of as an individual organism."[23] Or consider the case of the Pando trees in Utah, covering a area of 0.4 square km, all determined – again by virtue of having identical genetic markers – to be a clonal colony of a single 'Quaking Aspen'. In both cases, obvious problems to do with counting arise (how many 'trees' are there?) and at the very least force a liberal notion of biological individual to be adopted.

More acute problems for this notion arise with examples of symbiotes, such as that of a coral reef, which consists not just of the polyp plus calcite deposits but also zooanthellae algae that are required for photosynthesis. Another example is that of the Hawaiian bobtail squid, whose bioluminescence (evolved, presumably, as a defence mechanism against predators who hunt by observing shadows and decreases in overhead lighting levels) is due to bacteria that the squid ingests at night and which are then vented at the break of day, when the squid is hidden and inactive. The presence of the bacteria confers an evolutionary advantage on the squid and thus render the squid the individual that it is, from the evolutionary perspective, but they are, of course, not genetically the same as the squid, nor do they remain spatially contiguous with it.

Now there are two broad responses one can make to these kinds of examples: monistic and pluralistic, where the former attempts to construct a unitary account of biological individuals that can cover these cases and the latter abandons any such attempt and insists that there is no one such framework of biological individuality.

Thus, one option is to retain a form of monism while abandoning the genetic homogeneity assumption of biological individuality by shifting to a 'policing' based account. Thus, Pradeu offers an immunological approach to individuation which, he claims, moves away from the self/non-self distinction and is based on strong molecular discontinuity in antigenic patterns. A biological organism is then understood as a set of interconnected heterogeneous constituents, interacting with immune receptors.[24] This is an interesting line to take but concerns have been raised over its extension to plants, for example. Here cases can be given in which genetic heterogeneity is not appropriately policed.[25] One might also consider the

---

'Heterogeneous Individuals', at the *PSA 2010*, Montreal

23 USDA Forest Service, "Humongous Fungus a New Kind of Individual", in: *Science Daily* 27, March 2003.

24 T. Pradeu, "What is an Organism? An Immunological Answer", forthcoming.

25 E. Clarke, "Individuals and the Homogeneity Assumption", forthcoming, paper presented at the *PSA 2010*.

example of insect super-colonies where there is no conflict between colonies,[26] which revives the above issue of spatially extended individuals again.

Alternatively, one might try to maintain monism by adopting Wilson's 'tripartite account', according to which an organism is (a) a living agent, (b) that belongs to a reproductive lineage, some of whose members have the potential to possess an intergenerational life cycle, and (c) which has minimal functional autonomy.[27] Underlying this view is the assumption that organisms and the lineages they form have stable spatial and temporal boundaries but recent commentators have suggested if we pay attention to the microbial world as well as the macroscopic examples we are used to discussing, then rather than a 'tree' of life composed of such lineages, we have a 'web or network of life' in which the idea of stable and well-defined lineages begins to break down. Again, the example of symbiosis and indeed its pervasiveness suggests that lineages/individuals are fluid and ephemeral.[28]

Perhaps then one might be tempted by a pluralistic approach, as suggested by Dupré and O'Malley, who urge a shift from individual organismal lineages to the "overall evolutionary process in which diverse and diversifying metagenomics underlie the differentiation of interactions within evolving and diverging ecosystems."[29] Here they take the notion of the autonomous individual and argue that if it is applied consistently across the biological domain it actually breaks down, and rather than thinking of biological individuals in this way we should regard them as the product of multiple collaborations:

> To the extent that … individual autonomy requires just an individual life or life history, then it surely applies much more broadly than is generally intended by biological theorists. Countless non-cellular entities have individual life-histories, which they achieve through contributing to the lives and life-histories of the larger entities in which they collaborate, and this collaboration constitutes their claim to life. But – and this is our central point – no more and no less could be said of the claims to individual life histories of paradigmatic organisms such as animals or plants; unless, that is, we think of these as the collaborative focus of communities of entities from many different reproductive lineages.[30]

Such passages mesh nicely with Dupré's 'Promiscuous Realism', which holds that "there are countless legitimate, objectively grounded ways of classifying objects in the world."[31] and which underpins a metaphysics of 'radical ontological pluralism'.

26 See 'Ant mega-colony takes over world'; BBC, July 2009.

27 R. A. Wilson, *loc. cit.*

28 See for example, F. Bouchard, "Symbiosis, Lateral Function Transfer and the (Many) Saplings of Life", in: *Biol Philos* 25, 2010, pp. 623–641.

29 J. Dupré and M. O'Malley, "Metagenomics and Biological Ontology", *loc. cit.*

30 J. Dupré and M. O'Malley, "Varieties of Living Things: Life at the Intersection of Lineage and Metabolism", in: *Philosophy and Theory in Biology* 1, 2009, pp. 1–25, p. 15.

31 J. Dupré, *The Disorder of Things: Metaphysical Foundations of the Disunity of Science,* Cambridge MA: Harvard University Press 1993, p. 18.

In the metagenomic context it can be extended from kinds and classifications to objects and individuals, thus forming what we might call 'Promiscuous Individualism': there are countless, objectively grounded ways of individuating or, more generally, delineating biological objects and individuals. Here the obvious worry has to do with the extent to which we can legitimately call this a form of realism: if an object-oriented stance is assumed – as it typically is – so that biological theories are taken to represent or refer to biological objects, then pluralism will lead at best to a form of contextual reference or at worst to a kind of indeterminacy that may be incompatible with realism as typically understood.

Alternatively, we may eschew both monistic and pluralistic options, while retaining the above insights that power the latter and adopt the structuralist line. According to this, there are no biological objects (as metaphysically substantive entities), all there is, are biological structures, inter-related in various ways and causally informed. Putative objects – genes, individual organisms etc. – should be seen as dependent upon the appropriate structures ('nodes') and from the realist perspective, eliminable, or, at best, regarded as secondary in ontological priority. This then accommodates the 'fluidity' and 'ephemerality' of biological organisms, as evidence in the example of symbiotes. Furthermore, from this perspective, biological individuals are nothing more than abstractions from the more fundamental biological structure,[32] or can be viewed as no more than "... temporarily stable nexuses in the flow of upward and downward causal interaction".[33] This still allows for there to be appropriate 'units of selection', but such units are not to be conceptualised in object oriented terms. In particular, we can accommodate the view that,

> ... a gene is part of the genome that is a target for external (that is, cellular) manipulation of genome behaviour and, at the same time, carries resources through which the genome can influence processes in the cell more broadly.[34]

There are, of course, numerous issues to be tackled within this framework. Does the view of a biological object as a 'temporarily stable nexus' imply the elimination of objects (as elements in our metaphysics of biology – I am not suggesting the elimination of genes or organisms as phenomenologically grasped[35]) or can we hold a 'thin' notion of object, in the sense of one whose individuality is grounded

32 Cf. J. Dupré and M. O'Malley, "Metagenomics and Biological Ontology", *loc. cit.*

33 *Ibid.*, p. 842.

34 *Ibid.*

35 Can we envisage biological entities – squids, mushrooms, elephants even – that are not nexuses in biological structures? Given the evolutionary contingency noted earlier, it is hard to see how that possibility could arise in biological terms. Of course, one could imagine a possible world in which a squid just comes into being, between the stars, say, but just as in the case of the 'sparse' worlds containing 'lonely' objects (e.g. a single electron) so beloved by metaphysicians, one might be inclined to regard with some scepticism the claim that such worlds constitute genuine possibilities.

in structural terms?[36] Is the temporary stability of such objects sufficient for fitness to be associated with it? And can we articulate appropriate units of selection in such terms? Dupré and O'Malley suggest an ontology of processes, but can this be reduced to a form of structuralism if such processes are understood as temporal and evolving structures? And finally, if biological objects are viewed as '... temporarily stable nexuses in the flow of upward and downward causal interaction', what sense can we make of causation in the structuralist context?

These are all interesting issues (at least to me!) but here I shall briefly consider only the last.

14.4 Causation in Biology

One of the foremost concerns about structuralism as it has been expressed in the context of physics is that the reliance on mathematical structures to represent physical structure has blurred the distinction between the two. The obvious appeal to causal power as a way of re-establishing the ontological distinction[37] has run into the objections that the 'seat' of such power rests with the objects that structuralism eschews[38] and that difficulties arise when trying to articulate causal relationships within a structuralist framework.[39] Although there are moves available to the structuralist to respond to these concerns[40] – so, for example, one might insist that the 'seat' of causal power is the structure itself[41] – in the context of physics one can always fall back on the Russellian line that here there is little scope for any robust notion of causation in the first place.[42]

When it comes to biology, however, such a fall-back move is not so straightforward. In a useful review, Okasha argues that distinctive issues arise here that have no parallel in the physical sciences.[43] Thus he argues that Darwinian explanations are causal, but at the population-level, rather than singular. Insofar as natural selection is 'blind to the future' and genetic mutation is undirected, these explanations

36 See the discussion in S. French and J. Ladyman, "In Defence of Ontic Structural Realism", *loc. cit.*

37 S. French and J. Ladyman, "Remodelling Structural Realism: Quantum Physics and the Metaphysics of Structure", *loc. cit.*

38 A. Chakravartty, "The Structuralist Conception of Objects", in: *Philosophy of Science* 70, 2003, pp. 867–878.

39 S. Psillos, "*The* Structure, the *Whole* Structure and Nothing *But* the Structure?" in: *Philosophy of Science* 73, 2006, pp. 560–570.

40 S. French, "Structure as a Weapon of the Realist", in: *Proceedings of the Aristotelian Society*, 2006, pp. 1–19; S. French and J. Ladyman, "In Defence of Ontic Structural Realism", *loc. cit.*

41 S. French, "Structure as a Weapon of the Realist", *loc. cit.*

42 J. Ladyman and D. Ross, *Everything Must Go: Metaphysics Naturalized, loc. cit.*

43 S. Okasha, "Causation in Biology", in: H. Beebee, C. Hitchcock and P. Menzies (Eds.), *The Oxford Handbook of Causation.* Oxford University Press 2009, pp. 707–725.

certainly can be taken to have pushed teleology out of biology.[44] When it comes to genetics, matters are more nuanced. Here the distinction between singular and population-level causality is crucial as heritability analyses pertain only to the latter. In particular, such analyses "… tell us nothing about individuals."[45] Furthermore, the idea of the gene as sole causal locus has been undermined by the implicit relativity to background conditions.[46] Further challenges to the notion of the gene as the seat of causal power have also been posed by proponents of Developmental Systems Theory who advocate a form of 'causal democracy' (which brings to mind the 'nuclear democracy' of 1960s elementary particle physics). Here Okasha adopts a more cautious line, suggesting that genes might still play the more dominant causal role, although this is something that will be determined by further research.[47] And of course, even if that is granted, the structuralist can apply well-known pressure to the concept of the 'gene' and argue that even if this does play the dominant role in biological causation, it should not be understood in object-oriented terms.[48]

Again there is more to say here but the point I wish to emphasise is that talk of causal powers and associated causal loci *per se* does not represent a major obstacle to the structuralist. Even if one were entirely comfortable with such talk, one could follow Dupré and O'Malley and insist that these causal powers are derived from the interactions of individual components and are controlled and coordinated by the causal capacities of the 'metaorganism'. This sort of account seems entirely amenable to a structuralist metaphysics. Alternatively, one could acknowledge that causation is a kind of 'cluster' concept, under whose umbrella we find features such as the transmission of conserved quantities, temporal asymmetry, manipulability, being associated with certain kinds of counterfactuals and so on. Even at the level of the 'everyday' this cluster may start to pull apart under the force of counterexamples. And certainly in scientific domains only certain of these features, at best, apply: understanding causation in terms of the transmission of mass-energy, for example, may seem plausible in the context of Newtonian mechanics but it breaks down in General Relativity, where conservation of mass-energy does not apply. Likewise, establishing temporal asymmetry is famously problematic in the context of physics and here we can perhaps, at best, only say that a very 'thin' notion of causation holds, understood in terms of the relevant dependencies. Thus we may talk, loosely, of one charge 'causing' the acceleration of another charge, but what does all the work in understanding this relationship is the relevant law and from the structuralist perspective, it is that that is metaphysically basic and in terms of which the property of charge must be understood. It is the law – in this

44 *Ibid.*, pp. 719–720.

45 *Ibid.*, p. 722.

46 *Ibid.*, p. 721.

47 *Ibid.*, p. 724.

48 See S. French, "Shifting to Structures in Physics and Biology: A Prophylactic for Promiscuous Realism", *loc. cit.*

case and in the classical context, Coulomb's Law – that encodes the relevant dependencies that appear to hold between the instantiations of the property and that, at the phenomenological level, we loosely refer to as causal.

But once we move out of that domain, the possibility arises of 'thickening' our concept of causation in various ways. We might, for example, insist that for there to be causation there must be, in addition to those conditions corresponding to what are designated the 'cause' and the 'effect', a process connecting these conditions, where this actual process shares those features with the process that would have unfolded under ideal, 'stripped down' circumstances in which nothing else was happening and hence there could be no interference.[49] Such processes can be termed 'mechanisms'[50] and here one might draw upon mechanism based accounts of causation and explanation.[51] In particular, if such accounts were to drop or downplay any commitment to an object-oriented stance, possible connections can be established with various forms of structuralism. Thus McKay-Illari and Williamson[52] have noted that most characterisations of mechanisms can be broken down into two features: one that says something about what the component parts of the mechanism are, and another that says something about the activities of these parts. They advocate an interesting dual ontology with activities as well as entities – of which the parts of mechanisms are composed – in the fundamental base. Here consideration of putative asymmetries between activities and entities[53] mirrors to a considerable degree consideration of, again putative, asymmetries between objects and relations within the structuralist context. Indeed, a useful comparison could be drawn between McKay-Illari and Williamson's insistence that activities are not reducible to entities and that one needs both in one's ontology and certain forms of 'moderate' structural realism that set objects and relations ontologically on a par.[54] Or one could press further and argue that the kinds of examples that are typically given to establish the ontological fundamentality of entities are either 'toy' examples that do not match actual science or simply break down under further examination. Certainly, as I have sketched here, biological 'entities' seem to be much more fluid and ephemeral than might be initially supposed and there are grounds for shifting the ontological focus to the relevant activities and processes. Precisely

49 N. Hall, "Causation and the Sciences", in: S. French and J. Saatsi (Eds.), *The Continuum Companion to the Philosophy of Science.* Continuum Press 2011, pp. 96–199, p. 115.

50 *Ibid.*

51 See, for example, P. Machamer, L. Darden and C. Craver, "Thinking About Mechanisms", in: *Philosophy of Science* 67, 2000, pp. 1–25; for a useful critique, see S. Psillos, "The Idea of Mechanism", in: P. McKay-Illari, F. Russo and J. Williamson (Eds.), *Causality in the Sciences*. University of Oxford Press 2011.

52 P. McKay-Illari and J. Williamson, "In Defence of Activities", forthcoming.

53 *Ibid.*

54 M. Esfeld and V. Lam, *loc. cit.*

how these might be understood from the structuralist perspective requires further work, but there are clearly potentially fruitful avenues to explore.

14.5 Conclusion

It is a contingent fact of the recent history of the philosophy of science that structuralism in general, and the more well-known forms of structural realism in particular, have been developed using examples from physics. This has shaped these accounts in various ways but it would be a mistake to think that because of that, forms of structuralism could not be articulated within the biological context. The apparent obstacle of the lack of laws crumbles away under the appreciation that even in physics the standard connection between lawhood and necessity is not well-grounded. Adopting an understanding of laws in terms of their modal resilience allows one to accept certain biological regularities as law-like and there are models a-plenty to form the basis for a structuralist framework. Furthermore, the central claim of this paper is that there are good reasons for shifting one's ontological focus away from biological objects and towards something that is more fluid and contextual and, ultimately, structurally grounded. Causality can then be 'de-seated' and possible connections open up with activity-based accounts of biological processes. Certainly I would argue that the realist need not be promiscuous in this context, but can, and should, be a 'staid' structuralist instead. More importantly, given the theme of the workshop and this volume, this offers a useful framework for understanding the biology-physics inter-relationship in general.

Department of Philosophy
University of Leeds
LS2 9JT, Leeds
United Kingdom
S.R.D.French@leeds.ac.uk

CHAPTER 15

MICHELA MASSIMI

NATURAL KINDS, CONCEPTUAL CHANGE, AND THE DUCK-BILL PLATYPUS: LAPORTE ON INCOMMENSURABILITY

15.1 INTRODUCTION

In Chap. 5 of *Natural Kinds and Conceptual Change*[1] Joseph LaPorte defends the view that the meaning-change of natural-kind terms does not open the door to Kuhnian incommensurability and is compatible with scientific progress. LaPorte's strategy consists in disentangling meaning-change from theory-change, by contrast with proponents of the "'incommensurability thesis', who insist that conceptual change is marked by linguistic change".[2]

On LaPorte's view – as knowledge advances – kind terms, whose use was vague, get precisified as opposed to undergoing a conceptual shift that can make them vulnerable to incommensurability. He makes his case by arguing, on the one hand, that the descriptive theory of reference (traditionally considered the culprit of conceptual instability associated with meaning-change), does not necessarily lead to incommensurability; and, on the other hand, by accusing the Putnam-Kripke causal theory of reference (traditionally considered a weapon in the realist's arsenal against conceptual instability) of being useless in blocking incommensurability.

LaPorte endorses a descriptive theory of reference, according to which the reference of any kind term is fixed by a cluster of descriptions, including both essential properties (say, H_2O for the term 'water', or 'the clade that stems from ancestral group G' for the term 'Mammalia'), and superficial properties (say, 'colourless' for 'water', and 'live-bearing' for 'Mammalia'). Natural-kind terms get precisified whenever *ceteris paribus* – as knowledge progresses – one or more of either the essential or the superficial properties are dropped, added, or modified. Interestingly enough, LaPorte couples the descriptive theory of reference with a defence of the rigidity *de jure* of kind terms. Descriptions, albeit not having the same Kripkean status of names, are nonetheless rigid designators *de jure*, i.e. they rigidly designate by stipulation.

In Sect. 15.2, I review LaPorte's position. I argue that precisification – the way LaPorte defines it – does not cut any ice against Kuhn's incommensurability for two main reasons. First, to make a strong case for why precisification of kind terms

1 Joseph LaPorte, *Natural Kinds and Conceptual Change*. Cambridge: Cambridge University Press 2004.

2 LaPorte, *ibid.*, p. 112.

D. Dieks et al. (eds.), *Probabilities, Laws, and Structures*, The Philosophy of Science in a European Perspective 3, DOI 10.1007/978-94-007-3030-4_15,

differs from conceptual shift, LaPorte would need to make a principled distinction between descriptions capturing essential properties and descriptions capturing superficial properties. In the absence of such principled distinction, LaPorte's argument for precisification averting conceptual instability does not go through (Sect. 15.3). Second, far from addressing incommensurability, precisification violates a key Kuhnian requirement for translatability between scientific lexicons (Sect. 15.4). I conclude the paper with a brief report on the history of the duck-billed platypus as a biological counterexample to precisification (Sect. 15.5).

15.2 LaPorte on Meaning-Change, Incommensurability, and the Rigidity of Kind Terms

Kuhn's incommensurability has traditionally been regarded as challenging scientific progress. If main theoretical terms undergo a substantial change of meaning before and after a scientific revolution (say, "planet" before and after the Copernican revolution, or "species" before and after Darwin), to the point that there is no reference continuity across the revolutionary divide, how can the transition from the old to the new paradigm be considered scientific progress? It is this claim that LaPorte addresses in Chap. 5 of *Natural Kinds and Conceptual Change,* where he first addresses the issue of whether (1) the descriptive theory of reference is effectively responsible for the conceptual instability at work behind incommensurability, and (2) whether the Putnam-Kripke causal theory of reference is a solution to it. He gives negative answers to both questions. In the second half of the chapter, he advances the claim that linguistic change of the type he defends is in fact compatible with scientific progress and does not open the door to incommensurability. Let us briefly review each of these points.

LaPorte's attention concentrates on "conceptual" or "linguistic" incommensurability, famously championed by the late Kuhn[3] and intended as a form of untranslatability between scientific lexicons. This version of incommensurability affects kind terms undergoing meaning-change during a scientific revolution, and as such undermines referential continuity across theory-change. Traditionally, the culprit has been identified in the descriptive theory of reference, whereby the reference of a kind term is fixed by the descriptions associated with the term, so that when they change, so does also the reference of the term. It is this scenario that the causal theory of reference is meant to block.

3 Thomas S. Kuhn, "The Road Since Structure", in: A. Fine, M. Forbes, and L. Wessels (Eds.), *PSA 1990: Proceedings of the 1990 Biennial Meeting of the Philosophy of Science Association,* vol. 2 (East Lansing, MI: Philosophy of Science Association 1991), 3–13. Reprinted in Kuhn, *The Road since Structure. Philosophical Essays, 1970–1993, with an Autobiographical Interview.* Chicago: University of Chicago Press 2000, pp. 90–105.

But LaPorte warns us that "there are different versions of the description theory. Some lead to more radical changes in meaning and reference than others".[4] While Feyerabend's view implies some radical changes of meaning and reference, other versions of the theory (such as Kuhn's) lead to milder meaning-change that do not necessarily open the door to conceptual instability and reference discontinuity. LaPorte subscribes to the latter on the ground that it is compatible with his analysis of vague kind terms offered in Chaps. 3 and 4.

In these previous chapters, LaPorte defends the view that our use of kind terms is often vague and terms get precisified as knowledge grows. Precisification is different from the conceptual/linguistic shift at work in the incommensurability thesis. The conceptual/linguistic change responsible for incommensurability presupposes a clear (non-vague) use of kind terms before and after a revolution: since kind terms were (non-vaguely) used to refer to *different things* before and after a revolution, there is no reference continuity. On the other hand, precisification presupposes that our use of kind terms is *vague*: any kind term has an extension, an anti-extension and a boundary, whereby, as knowledge progresses, we learn how to refine the boundary without substantially modifying the extension of the old kind term. LaPorte gives the example of monotremes.[5] Before the discovery of platypus and echidna, speakers used to call 'mammal' whatever satisfied a cluster of descriptions including 'live-bearing', 'lactating', 'hairy' and so on. The anti-extension of 'mammal' consisted of whatever did not satisfy any of these descriptions. And the boundary included cases like the platypus and echidna, which although satisfying "enough of the descriptions" (or at least, "enough of the most important descriptions") nonetheless failed to satisfy others (such as "live-bearing"). In those cases, according to LaPorte, speakers took a decision to consider monotremes as mammals by simply dropping some of the previous descriptions, and retaining the rest of them. Precisification of vague terms always implies a decision, or better a

4 LaPorte, *loc. cit.*, p. 115.

5 "... proponents of the *cluster-of-descriptions* theory of reference...are not committed to any such drastic shifts of reference. Kuhn suggests such a theory in various places. Consider his example 'mammal'. On the cluster theory, speakers from centuries past would have referred by 'mammal' to whatever satisfies enough of the descriptions speakers associated with mammal, or enough of the most important descriptions. These would have included the descriptions 'live-bearing', 'lactating', 'hairy', and so on. Whatever possessed all of the descriptions clearly belonged to the extension, and whatever possessed none of the descriptions clearly failed to belong to the extension. But as Kuhn indicates, a group of organisms could meet some of the descriptions and not others: this allows for the discovery of borderline cases, too, which reveal vagueness in the term 'mammal' (...) the description 'live-bearing' was associated with the term 'mammal' before the discovery of the monotremes, but not after. But this is a gentler change of meaning than the one that Feyerabend recognises. Because the monotremes satisfy some but not all of the descriptions formerly associated with 'mammal', they were neither clearly in nor clearly out of the extension before speakers made a decision to call them 'mammals'. There is, therefore, some needed conceptual continuity, as well as change" (*Ibid.*, pp. 116–7).

stipulation, on behalf of speakers to drop, add, or modify one or more of the descriptions associated with kind terms. Interestingly enough, in LaPorte's account, this element of stipulation is meant to guarantee the rigidity of kind terms.

In Chap. 2, LaPorte defends Kripke's account of names as rigid designators, and goes beyond Kripke in offering a generous interpretation of rigidity, whereby kind terms rigidly designate kinds (including artificial kinds): they designate the same kind in all possible worlds. But kind terms do not rigidly designate their extensions, which can vary from world to world.[6] Under the Putnam-Kripke account, rigidity accomplishes another important job, namely that of getting a term hooked up to its referent, and putting meanings "out of the head". But LaPorte takes distance from Putnam and claims that this is instead the role of causal baptism, and Putnam conflated the roles of rigidity and causal baptism.[7] This is an important aspect of LaPorte's take on the Kripke-Putnam theory of reference: by decoupling rigidity from the causal theory of reference, he can go on to defend the view that kind terms are rigid designators while at the same time defending the descriptive theory of reference. But how can LaPorte defend the rigidity of descriptions such as 'H_2O' or 'the clade that stems from ancestral group G' or 'element with atomic number 79'? LaPorte claims that such descriptions are rigid *de jure*, not *de facto*. Namely, they are rigid by *stipulation.*[8]

A consequence of decoupling rigidity from the causal theory of reference (and endorsing descriptivism instead), is that – as meaning changes – the sentences where kind terms feature may be true at a given time and false at other times. The rigidity *de jure* of kind terms can guarantee that the term is still referring to the same abstract kind, although both its extension and the truth-maker of the sentence where the term features may change.

This is an important point because LaPorte can claim to defend some form of scientific progress, despite meaning-change. As knowledge advances, although later speakers accept as true different sentences from those accepted by earlier speakers, they can still communicate with each other because the use of the term was vague and the truth-makers of the sentences, where the term appeared, indeterminate.

Imagine a speaker before Darwin that denies that "New species arise by evolution" without distinguishing between Darwin-species and Hopkins-species, where Hopkins-species denote special creation species. According to LaPorte, although there is no accumulation of true sentences across the revolutionary divide, there is nonetheless accumulation of knowledge: "This speaker has progressed in understanding both what she formerly called 'species' and also what she now calls

6 *Ibid.*, p. 38.
7 *Ibid.*, p. 43.
8 *Ibid.*, p. 47.

'species'".[9] Precisification of kind terms with open texture[10] amounts then to a form of progress by accumulation of knowledge, not of sentences.

Putnam's causal theory of reference, on the other hand, is useless in blocking incommensurability because the causal baptism is performed by "speakers whose conceptual development is not yet sophisticated enough to allow the speakers to coin a term in such a way as to preclude the possibility of open texture, or vague application not yet recognised".[11] To sum up, LaPorte argues that Kuhn is right and Putnam wrong: meanings do change over time. The question is to what extent meaning-change amounts to precisification and whether precisification can avert conceptual instability. To defend this claim, LaPorte must show that

(i) His view of meaning-change clearly separates precisification from conceptual change;
(ii) And, it does not open the door to conceptual instability of the type at work behind Kuhnian incommensurability.

In this paper, I argue that LaPorte is not successful in showing either (i) or (ii). In particular, I argue that for LaPorte to make a strong case for why meaning-change does not open the door to conceptual instability of the type at work behind incommensurability (even in the Kuhnian, rather than Feyerabendian version) he would need stronger realist assumptions. He would need the very same Putnam's causal theory of reference that he disavows in favour of descriptivism. In Sect. 15.3, I address point (i) by claiming that without stronger realist assumptions, the difference between precisification and conceptual change becomes only a matter of degree. In Sect. 15.4, I argue against (ii): precisification is not sufficient to block conceptual instability of the Kuhnian type because, if anything, it violates a key Kuhnian requirement for translatability of kind terms. Section 15.5 concludes the essay by revisiting the story of the platypus. Far from being a case of precisification, I show how the complex process that led to the identification of the duck-bill platypus as an egg-laying mammal amounted to a genuine case of conceptual shift of the Kuhnian type.

15.3 Why LaPorte's View Does Not Cut Any Ice Against Kuhn's Incommensurability. Part I: Precisification Versus Conceptual Change

As we saw above, LaPorte's defence of scientific progress is based on the idea that meaning-change of natural-kind terms implies the precisification (as opposed to conceptual shift) of terms whose use was vague, or better, terms with open texture.

9 *Ibid.*, p. 132.

10 LaPorte defines open texture as follows: "Hidden vagueness in a word's application that is later exposed like this with more information is known as open texture" (*Ibid.*, p. 97).

11 *Ibid.*, p. 118.

Along lines similar to Bird,[12] I show that it is difficult to distinguish the precisification of a kind term from conceptual shift involving a change to its extension. In other words, meaning-change does not tell us that a kind term has been precisified any more than it has undergone a conceptual shift.

Consider again LaPorte's example of 'mammal'. He quotes Kuhn to make the point that the discovery of monotremes forced the precifisication of the kind term 'mammal', whereby some of the descriptions previously associated with the term (such as 'live-bearing') were dropped and replaced with others (i.e. 'egg-laying'). The use of the term 'mammal' was therefore vague before monotremes were encountered and scientists decided to include them as a borderline case in the extension of the term by refining some of the descriptions associated with it. But – on LaPorte's view – scientists might have decided otherwise, namely they might have decided to leave monotremes out of the extension of the term 'mammal' and they would not have been wrong in doing that, any more than we are right in having included them. After all, there was an element of stipulation in the precisification of the kind term 'mammal', as opposed to scientists discovering the true essence of mammals. This is what makes the kind term 'mammal' rigid *de jure*. As a result, the inclusion of monotremes in the extension of 'mammal' does not amount to conceptual shift of the Kuhnian type because it involved only peripheral changes to the boundary of the term, and not to its extension or anti-extension.

To make a strong case for why the inclusion of monotremes in the extension of 'mammal' is a case of precisification as opposed to conceptual shift, LaPorte needs a clear-cut definition of what the extension, anti-extension and boundary of this kind term, respectively, are. One could try to argue that perhaps the description 'live-bearing' was more peripheral to the term 'mammal', than other descriptions associated with the term, so that its removal amounted to a case of precisification, while the removal of other more essential descriptions amounted to a case of conceptual shift. For example, one could try to argue that the description 'clade that stems from ancestral group G' is more central to the term 'mammal' than superficial descriptions such as 'live-bearing', 'lactating' or 'hairy'.

But no such line of argument is open to LaPorte. To answer along these lines, one would need to distinguish between descriptions that capture the essential – genealogical or microstructural (depending on whether we are dealing with biological or chemical kinds) – properties of a kind, and descriptions that capture superficial observable properties, which is what the Putnam-Kripke theory does. Recall Putnam's[13] Twin Earth's story about water being XYZ in Twin Earth despite

12 Alexander Bird, "Discovering the essences of natural kinds", in: Helen Beebee and Nigel Sabbarton-Leary (Eds.), *The Semantics and Metaphysics of Natural Kinds*. Routledge 2009, pp. 125–136.

13 Hilary Putnam, "The meaning of 'meaning'", in: Keith Gunderson (Ed.), "Language, mind and knowledge", *Minnesota Studies in the Philosophy of Science* 7. Minneapolis: University of Minnesota Press 1975. Reprinted in Hilary Putnam, *Mind, language and reality. Philosophical papers*, Vol. 2. Cambridge: Cambridge University Press 1975,

having exactly all the same superficial characteristics of our water. By identifying the 'meaning' of meaning of natural-kind terms with the reference and its essential properties empirically discovered, and by relegating superficial properties to the stereotype of the term, Putnam's causal theory of reference would provide just what is needed to distinguish clearly among the extension, anti-extension, and boundary of any kind term.

But LaPorte cannot avail himself of the above line of argument, since he concedes that descriptions capturing the essential (microstructural or genealogical) properties and those capturing superficial ones are on a par: we cannot privilege the former on the ground that they pick out the essence of a kind. LaPorte defends this point profusely in Chaps. 3 and 4. For biological kind terms, he appeals to the existence of competing biological schools (evolutionary taxonomy and cladistics), and stresses that neither species nor higher taxa have essences captured by either classificatory method. For chemical kind terms, he revisits Putnam's example of the kind term 'jade' and argues that it provides a historical counterexample to Putnam.[14] Thus, LaPorte cannot appeal to a possible distinction between superficial and essential properties to draw a clear-cut distinction between extension, anti-extension, and boundary.

Here is a more promising line of argument that LaPorte might consider. If the extension of a term is given by a cluster of descriptions (including superficial and essential ones), then the anti-extension of the term includes whatever does not satisfy any of those descriptions; and the boundary is presumably the peripheral area of the extension at the intersection between extension and anti-extension, and partially overlapping with both. In the boundary, we would find items that share some of the descriptions associated with the extension but not others.

Presumably, precisification occurs when the boundary area overlaps as much as possible with the extension of the old kind term. Should the boundary area overlaps mostly with the anti-extension, i.e. should the items falling into the boundary area satisfy most of the descriptions associated with the anti-extension rather than with the extension, then including those items into the extension of the old kind term would not amount to a case of precisification, but instead to a case of conceptual shift. Presumably, on this view, conceptual shift would occur when the boundary area overlaps as much as possible with the anti-extension of the old kind term. Given the cluster of descriptions theory of reference, and given that superficial and essential descriptions are on a par, adding or dropping one or more of the descriptions becomes then a matter of degree: precisification and conceptual shift are the two ends of a continuum. So, against (i), precisification and conceptual shift are not clearly separated.

---

pp. 215–71.

14 See LaPorte, *loc. cit.*, p. 100. For a criticism of LaPorte's history of jade, see Ian Hacking, "The contingencies of ambiguity", in: *Analysis* 67, 2007, pp. 269–77.

Although there is no principled distinction between precisification and conceptual shift, one can still claim that adding monotremes to the extension of the term 'mammal' would amount to precisification, because the new kind term 'mammal' overlaps as much as possible with the extension of the old term and it simply includes few new referents that the old term did not include. But this mild definition of precisification does not cut any ice against Kuhn's incommensurability because, if anything, precisification violates a key Kuhnian requirement, the no overlap principle, to which I now turn.

15.4 Why LaPorte's View Does Not Cut Any Ice Against Kuhn's Incommensurability. Part II: Precisification And Translatability

The late Kuhn famously redefined incommensurability as untranslatability between scientific lexicons and introduced the no overlap principle as a key requirement for kind terms of any scientific lexicon to be translatable into another lexicon. The principle says that

> no two kind terms, no two terms with the kind label, may overlap in their referents unless they are related as species to genus. There are no dogs that are also cats, no gold rings that are also silver rings, and so on: that's what makes dogs, cats, silver, and gold each a kind.[15]

The principle precludes kind terms from being imported from one lexicon to another unless they are related as species to genus, i.e. unless the extension of an earlier kind term becomes a subset of the extension of a later kind term in the new lexicon. In any other case, where kind terms of the old and new lexicon are not a proper subset of one other, but they partially overlap, incommensurability as untranslatability arises. For example, the Copernican statement 'planets orbit the sun' cannot be translated into the Ptolemaic lexicon. The term 'planet' is a kind term in both lexicons, but the two overlap without either containing all the celestial bodies of the other, because a fundamental change has occurred in this taxonomic category during the transition from Ptolemaic to Copernican astronomy.[16]

Thus, although Kuhn would agree with LaPorte that for example Hopkins-speakers and Darwinian-speakers can communicate with each other and understand each other when using the term 'species', he would also insist that the process that allow them to understand each other is a form of bilingualism, not translation.[17] Each speaker on the two sides of the revolutionary divide would

15 Kuhn, "The Road since Structure. Philosophical Essays, 1970–1993, with an Autobiographical Interview", *loc. cit.*, p. 92

16 *Ibid.*, p. 94.

17 "To bridge the gap between communities would require adding to one lexicon a kind term that overlaps, shares a referent, with one that is already in place. It is that situation which the no-overlap principle precludes. Incommensurability thus becomes a sort

have to bear in mind what the term 'species' means in her own lexicon and in the other lexicon, in order to understand each other. The term cannot be imported from the Hopkins-lexicon to the Darwin-one at the cost of violating the no-overlap principle: Hopkins-'species' are not a proper subset of Darwin-'species', because given Darwinism, special creation species are not an option. Instead, the two terms partially overlap without the former including all the referents of the latter. For example, Hopkins-species would not include all post-Hopkins (post-1850s) discovered species and the not-yet-come-into-existence species resulting from speciation and evolutionary adaptation, which are instead included in the extension of Darwin-'species'. Translation would require a one-to-one mapping from the taxonomic categories and relationships of one lexicon to those of another lexicon at the cost of overlapping kind terms. It is this situation that the no-overlap principle precludes.

But it is precisely this overlapping of kind terms that precisification seems to require and imply. If precisification is defined as above, namely adding or dropping one or more of either superficial or essential descriptions so that items which were previously in the boundary get included in the extension of the old kind term and others which were included get dropped, then by definition precisification implies that the old kind term partially overlaps with the new kind term, without including all the referents of the other. By definition, precisification violates Kuhn's no overlap principle as a key requirement for translatability. Again, this does not mean that communication is impossible or that speakers on the two sides of the divide cannot understand each other. It is not communication or understanding that is jeopardised, but the very possibility of translation, which is precisely Kuhn's main point about incommensurability. Thus, claiming – as LaPorte does – that kind terms are vague and get precisified, far from addressing Kuhn's incommensurability, seems in fact to violate one of the key requirements for translation between lexicons.

Should we still regard the inclusion of monotremes in the extension of 'mammal' as a case of precisification, notwithstanding Kuhn's no overlap principle? After all, LaPorte's view about vagueness of kind terms seems to affect primarily boundary cases. I want to suggest that *even* in situations that look *prima facie* as boundary cases like the platypus, LaPorte's view does not apply. Indeed, if anything, the story of the platypus provides a biological counterexample to LaPorte on precisification and an illustration of how untranslatability arises when the no overlap principle is violated.

of untranslatability, localised to one or another area in which two lexical taxonomies differ. (…) Violations of those sorts do not bar intercommunity understanding. (…) But the process which permits understanding produces bilinguals, not translators, and bilingualism has a cost. The bilingual must always remember within which community discourse is occurring. The use of one taxonomy to make statements to someone who uses the other places communication at risk." (*Ibid.*, pp. 92–3).

15.5 The Story of the Duck-Bill Platypus. Or, Against Precisification

> "Little Dot clapped her hands. 'Oh, dear Kangaroo' she said 'do take me to see the Platypus!' There was nothing like that in my Noah's Ark'. 'I should say not!' remarked the Kangaroo. 'The animals in the Ark said they were each to be of its kind, and every sort of bird and beast refused to admit the Platypus, because it was of so many kinds; and at last Noah turned it out to swim for itself, because there was such a row. That's why the Platypus is so secluded'"
> From Ethel Pedley, *Dot and the Kangaroo.*[18]

As we saw in Sect. 15.2, LaPorte quotes Kuhn for the example of the egg-laying platypus as a borderline case, whose inclusion in the extension of the kind term 'mammal' would be a case in point for precisification of kind terms, as opposed to conceptual shift of the type at work behind incommensurability. In this section, I show how the real story of the platypus belies LaPorte's, and even Kuhn's own intuitions about monotremes. Far from being a case of precisification, the inclusion of monotremes in the category of Mammalia amounted to a genuine conceptual shift of the Kuhnian type, whose main actors *saw* the platypus in different incommensurable ways.

The platypus would be a case in point for precisification—according to LaPorte—because it satisfied most of the important descriptions for 'mammal' (namely, 'hairy' and 'lactating'), but not another key one, i.e. 'live-bearing', until speakers decided that an oviparous mammal was an acceptable option. But how can we make sense of LaPorte's suggestion that the platypus was neither clearly in nor clearly out of the extension of 'mammal', until such a decision was taken? How should we understand the notion of vagueness?

LaPorte gives us some helpful indications about vagueness in Chap. 4 on chemical kind terms, where he says that vagueness is exposed when a substance has *either* the right observable properties but the wrong microstructure, *or* the wrong observable properties and the right microstructure. In both cases, we are in the presence of a vague case. In what follows, I am going to work with LaPorte's helpful suggestion about vague cases in chemistry, and see what the biological counterpart would look like in the case of the platypus. Instead of (chemical) microstructure, we have genealogical descent for monotremes, i.e. their belonging to the 'clade that stems from ancestral group G', if we take that description as featuring in a theoretical identity statement for the term 'Mammalia'. Following LaPorte, I take as examples of observable properties for mammals the following: 'hairy', 'lactating' and 'live-bearing'.

---

18 Quoted from Ann Moyal, *Platypus*. Baltimore: Johns Hopkins University Press 2004, p. 200.

Thus, for the platypus to be a vague case of mammal, it would have to be the case that either it has the right observable properties and the wrong genealogical descent, *or* it has the wrong observable properties and the right genealogical descent. Interestingly enough, if we look at the story of the platypus, the confusion surrounding the first encounter with the *Ornithorhynchus paradoxus* (as Johann Blumenbach originally called it) originated precisely from such a split between the suspected genealogical descent and the observable properties of the duck-billed platypus. If we take the extension of the term 'mammal' to be fixed by a cluster of descriptions, both in the case of 'lactating' and 'live-bearing', there was a discrepancy between the available evidence and the underlying genealogical hypothesis. In both cases, the observable properties and the suspected genealogical descent came apart.

In the case of 'lactating', some naturalists wrongly identified the genealogical descent, despite the right observable properties of platypus' mammary glands. In the case of 'live-bearing', other naturalists rightly identified the platypus' genealogical descent from mammals, despite the wrong observable properties about platypus' eggs and reproductive system (which suggested that it was ovoviviparous like some lizards). Geoffrey St-Hilaire falls into the first group of naturalists, who could not accept the mammalian nature of the platypus despite evidence about mammary glands. Richard Owen and George Bennett belong to the second group of naturalists, who – despite the correct identification of the platypus as a mammal – could not entertain the idea of its oviparity.

Indeed, the real story of the platypus is an extraordinary example of the puzzling taxonomic classification of an animal, which defied well-established zoological standards and summed up aspects of different genera.[19] After the first encounter with specimens coming from New South Wales, Australia, in 1799, a debate began about the nature of the curious creature. Following Linnaeus' taxonomy, mammals were characterized by the presence of mammary glands and the suckling relation between mother and young. Moreover, all mammals were expected to give birth to live young, by contrast with oviparous (or egg-laying) animals such as birds and reptiles. Naturalists were baffled by the platypus. At the beginning of the nineteenth century, George Shaw placed it in the lowest Linnean order of Bruta; Blumenbach considered it as part of the family of anteaters and armadillos as a transitional form between mammals and birds.[20]

The leading British anatomist Everard Home thought it could not be a mammal because of the apparent absence of mammary glands, and advanced the hypothesis that it was ovoviviparous (hatching the young from an egg inside the mother's body) like lizards. It was not until 1833 that the British surgeon Richard Owen demonstrated that the platypus should be classed into the milk-producing

19 In what follows, I am drawing on Ann Moyal's, *loc. cit.*, excellent monograph on the history of the platypus.

20 *Ibid.*, p. 40.

order of Mammalia, on the basis of some new evidence of mammary glands found by the Lieutenant Lauderdale Maule stationed in New South Wales. Maule found a female and two young, whom he tried to keep alive feeding them with worms. When the female platypus died, Maule found that milk oozed through the fur on the stomach. However, the new evidence about the platypus' mammary glands was not universally accepted. The French savant Geoffrey St-Hilaire persistently refused to consider the glands as mammary glands because of the absence of nipples. He thought they were instead lubricating glands like those found in salamanders, and commented: "if those glands produce milk, let's see the butter!"[21]

On the other hand, there were naturalists that rightly identified the platypus as a mammal (thanks to the new evidence of mammary glands) but categorically refused the idea of an egg-laying mammal, in favor of the ovoviviparous option originally put forward by Home. The mystery surrounding the mode of reproduction of the platypus was caused by the elusive nature of the animal and scant evidence of eggs debris, which for more than seventy years made it impossible to identify the platypus' oviparous nature. Increasing evidence that the platypus laid eggs came in 1833 when the naturalist George Bennett, following local knowledge of Aborigines about the nesting burrows of the platypuses, captured and sent back to Richard Owen in England a specimen showing an egg in the uterus of a female Ornithorhynchus. Yet this discovery left open the question as to whether the platypus was oviparous or ovoviviparous. Owen and Bennett opted for the ovoviviparous option. And still in 1864, Owen refused to accept the idea of an egg-laying platypus when he heard the account of an Australian physician, whose recently captured platypus had laid two eggs. Owen dismissed the alleged 'eggshell' as excrement coated in urine salts. As Moyal put it:

> Sixty years into the platypus mystery, Owen was caught in a paradigm. It was a paradigm largely of his own making. With no other researchers challenging his opinion, the ovoviviparous generation of the *Ornithorhynchus* was judged to be an 'accepted truth'. (…) His research approach held a fatal flow: it failed to entertain and hence search out any evidence of laid eggs in the nesting burrows.[22]

It was not until August 1884 that evidence of a platypus laying an egg (and with a second egg in the mouth of the uterus) was finally found by William Caldwell, belonging to a new generation of researchers, who were not afraid of challenging Owen's orthodoxy. Caldwell's pioneering discovery of the platypus' eggs finally established beyond any doubt the peculiar nature of the platypus as an 'egg-laying' mammal, almost hundred years after the first encounter with the curious creature in 1799. It also opened the door to further important research. In the words of Ann

21 *Ibid.*, p. 58.

22 *Ibid.*, p. 128, 147.

Moyal, "the evidence from the Australian monotremes and marsupials would play a dynamic part in formulating a new theoretical framework for biology."[23]

Where does the platypus' story leave us? What should we make of LaPorte's claim that the inclusion of the platypus in the extension of the term 'mammal' is a case of precisification? In the light of the real history, we can conclude that the kind term 'mammal' was not vague. Naturalists worked with well-defined, non-vague criteria of what counted as a 'mammal', coming from Linnaeus, and the uncertainty as to whether the platypus was in fact a mammal was simply due to either a lack of sufficient evidence (given the platypus' elusive nature) or to evidence not recognized as such (because people worked under different conceptual frameworks). In the case of mammary glands, St-Hilaire was simply unable to identify them as such because of the absence of nipples. In the case of the missing egg, Owen and Bennett did not have the luck of Caldwell in finding a female platypus laying eggs on the spot, despite years of intensive research on nesting burrows.

Both in St-Hilaire's case and in the case of Owen and Bennett, the platypus was differently classified (as non-mammal and mammal, respectively) because scientists interpreted the available evidence about mammary glands in the light of different conceptual frameworks. Same goes for the available evidence of eggs debris. Owen could not accept reports of platypuses laying eggs because, in Moyal's words, he became trapped in the 'paradigm' of the ovoviviparous platypus that he had himself created. He just could not *see* the platypus as an oviparous mammal.

If anything, the story of the platypus shows how conceptually-driven naturalists' observations can be, and confirms the Hanson-Kuhn view that scientists *see* things differently before and after a revolution. It took almost hundred years for the duck-bill platypus to be recognized as a mammal in its own right. The process that led to the discovery (because it was a discovery made by Caldwell in 1884) of the platypus as an 'egg-laying' mammal was a genuine revolution, whose main actors endorsed different conceptual frameworks and fiercely battled for their views in the pages of respectable zoological journals. The platypus is an example of a genuine cross-cutting kind that defies Kuhn's no overlap principle and the hierarchy thesis behind it, namely the thesis that kind terms cannot be imported from one lexicon into another unless they are related as species to genus. The old kind term 'mammal' is *not* translatable into the new kind term (which now includes monotremes) because it is not a proper subset of it. 'Egg-laying' is *not* a defining feature of the new kind term 'mammal', somehow encompassing the old kind term, with its defining feature 'live-bearing', as a subset. Monotremes as egg-laying mammals are the exception that confirms the rule of live-bearing mammals, and show how mammals intersect reptiles (and birds) in the tree of life. As with any cross-cutting kind defying Kuhn's no overlap principle, translatability is at risk. Thus, far from being a case of precisification, the inclusion of monotremes

23 *Ibid.*, p. 47.

in the extension of the term 'mammal' amounted to a genuine conceptual shift between incommensurable paradigms.

15.6 Conclusion

In this paper I raised two problems for LaPorte. First, he does not make a strong case for a principled distinction between conceptual shift and precisification of kind terms. Second, his claim that meaning-change does not imply conceptual shift of the Kuhnian type relies on a definition of precisification that violates a key requirement for translatability between lexicons, namely the no overlap principle. To remedy both problems, and defend scientific progress, LaPorte would need more substantial realist assumptions. In particular, he would need to prove that there is accumulation of true sentences across scientific revolutions. After all, Kuhn himself admitted the possibility of communication across the revolutionary divide: he only insisted that communication requires the two communities to be bilingual not translators, with incommensurability intended as untranslatability between scientific lexicons. So, for LaPorte to cut any ice against Kuhn, he would need to prove that there is reference-continuity *and* accumulation of true sentences across the revolutionary divide. Precisification of kind terms *per se* does not deliver either, *pace* the rigidity of kind terms.

Acknowledgments: I am grateful to Hanne Andersen for helpful commentary on this paper at the LSE workshop on *Points of contact between philosophy of physics and philosophy of biology.* Special thanks to Marcel Weber for thought-provoking feedback on an earlier version of this paper.

References

Alexander Bird, "Discovering the essences of natural kinds", in: Helen Beebee and Nigel Sabbarton-Leary (Eds.), *The Semantics and Metaphysics of Natural Kinds.* Routledge 2009, pp. 125–136.

Ian Hacking, 'The contingencies of ambiguity", *Analysis* 67, 2007, pp. 269–77.

Thomas S. Kuhn, "The Road Since Structure', in: A. Fine, M. Forbes, and L. Wessels (Eds.) *PSA 1990: Proceedings of the 1990 Biennial Meeting of the Philosophy of Science Association,* vol. 2 (East Lansing, MI: Philosophy of Science Association 1991), 3–13. Reprinted in Kuhn (2000), 90–105.

Thomas S. Kuhn, *The Road since Structure. Philosophical Essays, 1970–1993, with an Autobiographical Interview.* Chicago: University of Chicago Press 2000.

Joseph LaPorte, *Natural Kinds and Conceptual Change.* Cambridge: Cambridge University Press 2004.

Ann Moyal, *Platypus.* Baltimore: Johns Hopkins University Press 2004.

Hilary Putnam, "The meaning of 'meaning'", in: Keith Gunderson (Ed.), 'Language, mind and knowledge', *Minnesota Studies in the Philosophy of Science* 7. Minneapolis: University of Minnesota Press 1975. Reprinted in Hilary Putnam, *Mind, language and reality. Philosophical papers*, Vol. 2. Cambridge: Cambridge University Press 1975, pp. 215–71.

Department of Science and Technology Studies
University College London
Gower Street
WC1E 6BT, London
United Kingdom
m.massimi@ucl.ac.uk

CHAPTER 16

THOMAS A.C. REYDON

ESSENTIALISM ABOUT KINDS: AN UNDEAD ISSUE IN THE PHILOSOPHIES OF PHYSICS AND BIOLOGY?

ABSTRACT

The consensus among philosophers of biology is that traditional forms of essentialism have no place in accounts of biological kinds and classification. Recently, however, several authors have attempted to resurrect essentialism about biological kinds, invoking various views of the nature of kind essences starting a new debate on what kind essentialism should be if it is to apply to biological kinds. In this paper I examine three contemporary forms of biological kind essentialism and conclude that the scope of philosophical work that these are able to do is quite limited.

16.1 INTRODUCTION

At least since the 1970s there has been a strong consensus among philosophers of biology that traditional forms of essentialism about kinds of biological entities, which assume for each kind a set of properties that are separately necessary and jointly sufficient for kind membership, have no role to play in accounts of biological kinds and classification. The main reason is that such forms of kind essentialism conflict with evolutionary theory, which is, after all, biology's core theoretical framework. Essentialism about biological kinds thus has long been a dead issue. In recent years, however, a number of authors attempted to resurrect essentialism about biological kinds, defending various views of the nature of kind essences, all different from traditional kind essentialism, and starting a new debate on what kind essentialism should be if it is to apply to biological kinds.

Philosophers of physics, in contrast, have not been much disturbed by the discussions on essentialism going on elsewhere. There seems to be no pronounced conflict between traditional kind essentialism and the central theories of physics and a comparatively straightforward, traditional essentialist view of kinds seems to fit well for kinds in the physical sciences. In addition, there does not seem to be a particular need to take recourse to essentialism in order to be able to make sense of kinds and classification in the physical sciences. Because of this, essentialism about kinds is not a big issue in the philosophy of physics: there, kind essentialism

D. Dieks et al. (eds.), *Probabilities, Laws, and Structures*, The Philosophy of Science in a European Perspective 3, DOI 10.1007/978-94-007-3030-4_16,

is nearly dead too. In contrast to philosophy of biology, however, this is not because kind essentialism is deeply problematic, but because it is unproblematic and, apparently, not particularly illuminating.

This situation gives rise to questions about the feasibility of essentialist accounts of scientific kinds, as well as the reasons for pursuing kind essentialism in general. For one, is the notion of 'essence' strictly necessary to reconstruct particular scientific practices involving kinds? If so, what work would essentialism do? I want to address these questions by examining three contemporary forms of biological kind essentialism, as essentialism is most controversial for biological kinds. I shall conclude that the scope of philosophical work that these are able to do is quite limited. This is not to say that kind essentialism could not be a viable position in the philosophies of physics or biology, though. It is just to say that the philosophies of physics and biology might be better off without it, as the costs of assuming kind essentialism probably outweigh the benefits.[1]

In Sect 16.2, I briefly explore what philosophical work kind essentialism could do, thus setting up a collection of motives for trying to resurrect kind essentialism. In Sect. 16.3, I turn to some recent attempts to resurrect kind essentialism for application to biological kinds and examine whether these can do the work that kind essentialism might be expected to do. The conclusion will be a negative one. As I won't say much about kinds in the physical sciences in these sections, Section 16.4 concludes by briefly addressing the question what prospects there are for an overarching kind essentialism, covering kinds in physics, biology and elsewhere. Again, my conclusion won't be positive.

16.2 What Work Could Kind Essentialism Do?

At least six tasks can be listed that kind essentialism is thought to be able to perform in contemporary philosophy. Taken together, these cover a considerable amount of philosophical work and would thus constitute a considerable motivation for attempting to resurrect kind essentialism.

1. Two roles for kind essentialism are deeply rooted in the philosophical tradition, tracing back at least to Locke's *Essay Concerning Human Understanding*. There, Locke defined real essences as "the very being of any thing, whereby it is, what it is",[2] his concern being with what *kind* of thing a given thing was. Accordingly, in the philosophical tradition kind essences are usually supposed to determine the identities of things as things of particular kinds, by specifying those properties a thing cannot lose without ceasing to belong to its kind.

2. The second traditional role for kind essentialism can also be illustrated by a quote from Locke. Immediately after he formulated what a real essence was,

1 See Ereshefsky for a similar conclusion (Marc Ereshefsky, "What's wrong with the new biological essentialism?", in: *Philosophy of Science* 77, 2010b, pp. 674–685, p. 675).

2 John Locke, *An Essay Concerning Human Understanding (Edited with an Introduction by Peter H. Nidditch)*. Oxford: Clarendon Press [1700] 1975, (Book III, Chap. III, §15).

Locke continued: "the real internal, but generally in Substances, unknown Constitution of Things, whereon their discoverable Qualities depend, may be called their *Essence*".[3] The idea is that members of a kind tend to exhibit the same (or at least highly similar) properties and behaviors, because they share an essence that makes them members of the same kind in the first place. Kind essences cause the observable properties and behaviors that are typical for the members of a kind and thus can be referred to in explanations of these typical properties and behaviors.

Following from the tradition, then, kind essentialism today is often conceived of as encompassing two independent claims[4]:

- All and only the members of a kind *K* have the kind essence associated with *K*. Having this kind essence is what makes things into *K*-things.
- The kind essence associated with *K* is responsible for the observable properties typically exhibited by the members of *K*.

While these claims reflect the two major traditional tasks of kind essentialism – fixing the kind identities of things and explaining the kind-specific observable properties of things –, kind essentialism seems to have an even broader potential for doing philosophical work.

3. Kind essentialism is sometimes invoked in explanations of everyday classificatory practices. Empirical studies in cognitive psychology have shown that both children[5] and adult people[6] tend to assume that things have intrinsic essences that make them into the particular kinds of things they are, a tendency that might be widespread because of the evolutionary advantage it confers on humans by providing a basis for inferences over kinds of things they encountered in their environment.[7] Essentialist thinking thus seems to come naturally to human beings, a claim that is usually known as "psychological essentialism".

Kind essentialism could serve to support psychological essentialism: people tending to be essentialists about kinds are confirmed, because kinds actually *do* have essences. However, psychological essentialism has been established in relation to "folk" classifications that generally do not tend to match the scientific classifications that apply to the same domain of reality. Thus, kind essentialism as a basis for psychological essentialism doesn't seem particularly relevant in the

---

3 *Ibid.*

4 Samir Okasha, "Darwinian metaphysics: Species and the question of essentialism", in: *Synthese* 131, 2002, pp. 191–213, p. 203; Marc Ereshefsky, "Species", in: Edward N. Zalta (Ed.), *The Stanford Encyclopedia of Philosophy*, 2010a, online at http://plato.stanford.edu/archives/spr2010/entries/species/, Section 2.1; Ereshefsky, "What's wrong with the new biological essentialism?", *loc.cit.*

5 Susan A. Gelman, *The Essential Child: Origins of Essentialism in Everyday Thought*. Oxford: Oxford University Press 2003.

6 Susan A. Gelman and Lawrence A. Hirschfeld, "How biological is essentialism?", in: Douglas L. Medin and Scott Atran (Eds.): *Folkbiology*. Cambridge (MA): MIT Press 1999, pp. 403–446.

7 H. Clark Barrett, "On the functional origins of essentialism", in: *Mind & Society* 3, 2001, pp. 1–30.

context of philosophy of science. Moreover, if kind essentialism obtains for "folk" kinds that do not match the relevant scientific kinds, kind essentialism about these scientific kinds seems to be unwarranted. Still, if psychological essentialism is a default position for children as well as adults, it should be expected that many scientists tend to conceive of the kinds they use in an essentialist manner, such that kind essentialism may do real work in accounting for those scientific classificatory practices in which scientists actually invoke essentialist principles.

4. In the essentialist tradition in the philosophy of language, tracing back to the work of among others Kripke and Putnam, kind essences are invoked to do semantic work. According to Kripke and Putnam, kind terms in everyday and scientific language refer to their kinds in the same way as the names of particular things refer to the things they refer to. Kind names, on their view, are linked to kinds in baptism events, in which a name is attached to a particular token entity or substance and is agreed to be used henceforth for all entities or substances that have the same (usually unknown) essence as the entity or substance involved in the baptism event. This view of how kind names refer also encompasses a view of the aims of science on which the discovery of kind essences is a task for scientific investigation.

5. Some authors take recourse to kind essences to ground the laws of nature.[8] Here, the idea is that the laws of nature describe how things by their natures are disposed to behave and thus follow from the natures of things. As all things of a particular kind share the same kind essence, they are disposed to similar behavior under similar circumstances, leading to certain laws to hold for all and only the members of that kind. Laws of nature thus are ontologically dependent on kind essences and kinds, not laws, are ontologically fundamental.

6. However, many areas of science don't seem to deal with laws of nature. This leads to another possible role for kind essentialism in philosophy of science. As Waters remarked, "once philosophers decided that biology lacked genuine laws, they seem to have lost interest in analyzing the empirical generalizations of the science. Meanwhile, biologists continue to generalize."[9] Thus, even if there are no "proper" laws in areas of science such as biology, there still is philosophical work to do in analyzing the generalizations that feature in biological reasoning. Recently, Devitt advanced this as the main reason for resurrecting kind essentialism:

> We group organisms together under what seem, at least, to be the names of species or other taxa and make generalizations about the morphology, physiology, and behavior of the members of these groups: about what they look like, about what they eat, about where they live,

---

8 E.g. Brian D. Ellis, *Scientific Essentialism*. Cambridge: Cambridge University Press 2001.

9 C. Kenneth Waters, "Causal regularities in the biological world of contingent distributions", in: *Biology and Philosophy* 13, 1998, pp. 5–36, p. 6.

about what they prey on and are prey to, about their signals, about their mating habits, and so on. … Generalizations of this kind demand an explanation.[10]

Here, kind essentialism might be invoked to account for the possibility of making stable generalizations suitable for use in scientific reasoning, most importantly in explanatory and predictive contexts. Whether or not there are biological laws of nature, it is uncontroversial that biology uses generalizations that are stable to various degrees in explanations and predictions of biological phenomena. In explaining why these generalizations hold lies a potential task for essentialism about biological kinds.

There seem, then, to be good reasons to try to resurrect kind essentialism. Kind essentialism has the potential to do metaphysical work (no. 1: determining the kind identities of things; no. 2: explaining the kind-specific observable characteristics of things; no. 5: providing a basis for the laws of nature), epistemological work (no. 3: explaining everyday classificatory practices; no. 6: supporting scientific generalizations, explanations and predictions) and semantic work (no. 4: providing a theory of reference for kind terms). But in how far are the kind essentialisms advocated in contemporary philosophy of biology able to fulfill this promise?

16.3 Kinds of Kind Essentialism

Let me now examine three types of essentialism advanced in the recent literature as fitting with how contemporary biology understands and explains living phenomena.[11] Each of these positions accepts that traditional kind essentialism fails for contemporary biology. Yet, each holds that some form of essentialism about biological kinds can and should be upheld, as it can do important philosophical work.

One reformulation of kind essentialism conceives of kind essences as relational instead of intrinsic. A few years ago, Okasha presented a relationally essentialist position with a focus on biological species of organisms; I shall illustrate what can be called "relational essentialism"[12] by examining Okasha's account.

According to Okasha, the principal arguments against essentialism about biological species only hold for essentialisms that conceive of species essences in terms of intrinsic properties of organisms.[13] Thinking of essences as relational

10 Michael Devitt, "Resurrecting biological essentialism", in: *Philosophy of Science* 75, 2008, pp. 344–382, 351–352.

11 I'll ignore a fourth form, Walsh's "developmental essentialism", as this position isn't intended as an account of biological kinds. (Denis M. Walsh, "Evolutionary essentialism", in: *British Journal for the Philosophy of Science* 57, 2006, pp. 425–448).

12 Ereshefsky, "What's wrong with the new biological essentialism", *loc. cit.*, p. 679.

13 Okasha, *loc. cit.*, p. 199.

could defuse them. Moreover, a closer look at the grounds on which biologists allocate organisms to species under the various available species concepts shows that there is a good case for relational essentialism about species: under interbreeding, phylogenetic and ecological species concepts, organisms are allocated to species on the basis of their mating relations to other organisms, their relations of ancestry and descent to other organisms, and their relations to the environments in which they live, respectively.[14]

Okasha's relational essentialism seems able to avoid the difficulties faced by traditional kind essentialism. In addition, it seems extensible to other kinds of biological kinds, in particular functionally defined kinds on an appropriately relational notion of biological function (on which an entity's function is conceived of in terms of its relations to other entities) or to the environment (for ecological kinds, for example). However, it still faces considerable problems. For one, while it is true that according to most species concepts an organism's relations to other organisms or to the environment fixes its species identity, it is unclear how such relations by themselves can fix kind membership.[15] A relation such as "is an offspring of", for instance, would place organisms of a present-day species in the same species with their distant ancestors, tracing all the way back to the origin of life on earth. One element of kind essentialism, however, is that an organism's kind essence completely fixes its kind identity.

Moreover, relational kind essentialism is unable to do some of the other work that kind essentialism promises to perform. As Okasha pointed out, the relational properties that under various species concepts fix the kind identities of organisms do not serve to explain their properties.[16] The traits of a given organism, after all, aren't *caused* by this organism's mating relations to other organisms, or by its ancestry (in any direct sense), or by its belonging to a particular branch on the Tree of Life. Thus, relational essentialism is only able to perform one of the two main tasks of kind essentialism.[17] Because of this, relational essences aren't able to support scientifically useful generalizations – let alone laws – either.[18] Furthermore, relational essences won't be able to explain everyday classificatory practices, as the essences that people tend to assume with respect to everyday kinds usually aren't relational essences.

What I shall call "historical essentialism" is a second kind of attempt to render kind essentialism compatible with evolutionary thinking. Paul Griffiths was one

14 *Ibid.*, p. 201.

15 Devitt, *loc. cit.*, p. 365. But see Ereshefsky, "What's wrong with the new biological essentialism", *loc. cit.*, pp. 680–682.

16 Okasha, *loc. cit.*, pp. 203–204.

17 Thus Ereshefsky concluded: "relational essentialism *is not* essentialism because it fails to satisfy a core aim of essentialism" (Ereshefsky, "What's wrong with the new biological essentialism?", *loc. cit.*, p. 683; emphasis added; cf. Ereshefsky, "Species", *loc. cit.*, Sec. 2.6).

18 Okasha, *loc. cit.*, pp. 208–209.

of the authors who prominently advocated historical essentialism and here I shall take his account to explicate the general approach.[19]

Griffiths' principal reason to try to resurrect kind essentialism about species was the perceived need to account for the role that reference to species plays in the formulation of scientifically useful generalizations (no. 6 in Sect. 16.3).[20] Species and other phylogenetically defined kinds, such as kinds of homologues, Griffiths argued, are such groups and, therefore, an account is needed that explicates what makes species and the like suitable to perform their generalization-grounding role in biological reasoning. According to Griffiths, the required account can be formulated in terms of kind essences on a suitably revised and loosened conception of what kind essences can be: "Any state of affairs that licences induction and explanation within a theoretical category is functioning as the essence of that category".[21]

In the case of biological species, as well as higher taxa and several other biological kinds, such essences can be found in the central Darwinian notion of common descent:

> Cladistic taxa and parts and processes defined by evolutionary homology have historical essences. Nothing that does not share the historical origin of the kind can be a member of the kind. ... Furthermore, cladistic taxa and parts and processes defined by evolutionary homology have no other essential properties.[22]

On this account, organisms, parts of organisms and biological processes are members of their kinds because of their ancestry: organisms are members of the same

---

19 Some authors see Griffiths' essentialism as a form of relational essentialism (e.g., Ereshefsky, "Species", *loc. cit.*, Sec. 2.6; Ereshefsky, "What's wrong with the new biological essentialism?", *loc. cit.*, p. 679), while others count Griffiths' and Okasha's positions as versions of the same position ("origin essentialism"; Olivier Rieppel, "New essentialism in biology", in: *Philosophy of Science* 77, 2010, pp. 662–672, p. 663). I'm not sure whether this is appropriate, though, as there is an important difference between Okasha's essentialism and Griffiths' essentialism: while according to Griffiths historical essences support generalizations about species and other taxa, according to Okasha relations cannot do this (Okasha, *loc. cit.*, pp. 208–209).

20 Paul E. Griffiths, "Squaring the circle: Natural kinds with historical essences", in: Robert A. Wilson (Ed.): *Species: New Interdisciplinary Essays*. Cambridge (MA): MIT Press 1999, pp. 209–228, 215–219.

21 *Ibid.*, p. 218. Griffiths here referred to Boyd's theory of Homeostatic Property Cluster kinds, which I'll discuss later, and embedded his account in Boyd's. However, I think Griffiths' account stands at some distance from Boyd's, as the former is framed in terms of historical origins and ancestry as essences, while the latter is framed in terms of causal mechanisms. It can thus be doubted whether Griffiths' account indeed constitutes a particular instantiation of Boyd's more general account, as Griffiths suggests (*Ibid.*, pp. 218–219). While I won't pursue this issue here, it constitutes my reason for treating Griffiths' account independently from Boyd's.

22 *Ibid.*, p. 219.

kind as their parents, and homologous organismal traits and processes are members of the same kinds as the traits with which they share a lineage of descent.[23]

Common descent also explains why organisms, parts and processes of the same kind resemble each other in such ways that stable, scientifically useful generalizations over them can be made. As Griffiths pointed out:

> The principle of heredity acts as a sort of inertial force, maintaining organisms in their existing form until some adaptive force acts to change that form. This *phylogenetic inertia* is what licences induction and explanation of a wide range of properties ... using kinds defined purely by common ancestry. If we observe a property in an organism, we are more likely to see it again in related organisms than in unrelated organisms. ... A hierarchical taxonomy based on strict phylogenetic principles will collect more of the correlations between characters ... than any other taxonomy we know how to construct.[24]

Thus, shared ancestry can be thought of as constituting the kind essences of biological species, higher taxa and many other biological kinds, because it performs the two principal tasks that kind essences should perform (see Sect. 16.2): to fix kind membership and to explain the observable properties typically exhibited by a kind's member entities.

Many biological kinds are defined by means of homology, so Griffiths' account should be broadly applicable to biological kinds. However, it is doubtful whether the majority of biological kinds are defined by homology. Many higher taxa are,[25] but lower taxa such as varieties, species and genera, are not. Thus, historical essentialism will not apply to the basic units of biological classification but only to more overarching units. Moreover, with respect to taxa defined by means of homologies, historical essentialism will apply only to the particular trait or traits that actually define the taxon under consideration. For the subphylum Vertebrata, for example, the fact that having a spinal column is an essential property of organisms of the taxon only supports the generalization that all Vertebrata have a spinal column. The historical essence of Vertebrata provides no foundation on which the possibility of making generalizations that hold for the members of the taxon can be extended beyond the particular essential traits of the kind. In addition, many biological kinds aren't defined by homology *alone*: consider, for example, the

---

23 For example, the radius bones in the arms of humans, in the fins of dolphins and in the wings of fruit bats are of the same overarching kind, because as discernible parts of organisms they all stand in the same line of descent from radius bones in limbs of a common ancestor (Thomas A.C. Reydon, "Gene names as proper names of individuals: An assessment", in: *British Journal for the Philosophy of Science* 60, 2009a, pp. 409–432, 420ff.).

24 Griffiths, *loc. cit.*, pp. 220–222; original italics.

25 For instance, Vertebrata is the taxon defined by a few traits shared by all and only vertebrates, including having a spinal column.

various kinds of genes, in the definitions of which functions play a role too next to descent.[26]

Furthermore, there are reasons to doubt whether phylogenetic inertia will be able to support species and other taxa as kinds over which stable generalizations are possible. In a general sense, phylogenetic inertia is the phenomenon that particular traits are conserved over long evolutionary time frames. Thus, a trait can be conserved over the entire lifetime of a species or higher taxon, from its origin in a speciation event up to its extinction, thus making it possible to generalize over the fact that all organisms of that species or higher taxon will posses this particular trait. However phylogenetic inertia is not a clearly defined notion in biology and is often taken as a phenomenon in need of an explanation rather than an explanatory factor by itself.[27] Griffiths grounded the occurrence of phylogenetic inertia in developmental causes, as developmental processes in organisms are inherited between ancestors and descendants and are sufficiently stable to cause the recurrent presence of similar traits in organisms related by descent.[28]

But there is no reason to assume that this stability will generally extend precisely over the lifetime of the species or higher taxon under consideration.[29] For some traits, the causal basis of phylogenetic inertia is such that the trait is only weakly conserved over a comparatively brief part of a species' lifetime. For other traits, the causal basis is such that the trait is strongly conserved far beyond the borders of species and higher taxa. The basic idea is easy to see: organisms of a species *S* that is a descendant of an ancestor species *A* share their descent with the organisms of species *A* and will thus possess some of the same developmental resources; therefore, it should not be surprising that some or many of the traits typical of *A*-organisms are conserved over the boundary between the two species and are also exhibited by many *S*-organisms. In sum: phylogenetic inertia may support generalizations pertaining to conserved traits, but the extent and the boundaries of these generalizations will not generally coincide with the extent and the boundaries of the kinds of organisms that biologists refer to when formulating their generalizations.

With respect to the other items of work for kind essentialism listed above, similar problems arise. While homologies may serve to fix the taxon identities of

---

26 Reydon, "Gene names as proper names of individuals: An assessment", *loc. cit.*

27 Thomas A.C. Reydon, "Generalizations and kinds in natural science: The case of species", in: *Studies in History and Philosophy of Biological and Biomedical Sciences* 37, 2006, pp. 230–255, 243–250.

28 Griffiths, *loc. cit.*, pp. 220–223. Griffiths's account here seems quite close to Walsh's account (see footnote 11). However, Walsh (Walsh, *loc. cit.*, p. 427) explicitly distanced his account from Griffiths' account. Indeed, Walsh's emphasis on the intrinsic nature of organisms and Griffiths' emphasis on common descent seem to yield essentialisms of which the differences outweigh the similarities. I won't pursue this issue here, though.

29 Reydon, "Generalizations and kinds in natural science: The case of species", *loc. cit.*

organisms under a cladistic view of biological classification, a view of homologies as kind essences will not yield explanations of the kind-typical properties of organisms (as the possession of a particular homologous trait does not explain the presence of most other traits of the organism), it will not yield a metaphysical grounding for any laws of nature and it will not do the semantic work that kind essences are often thought to do. Thus, there is reason to be skeptical about the applicability of historical essentialism to biological kinds in general and about the work that it can actually do in those cases in which it applies.

A third approach to kind essentialism in biology, which I here call "cluster essentialism", is based on Boyd's Homeostatic Property Cluster (HPC) theory.[30] As with historical essentialism, Boyd's account was intended to explain how the various kinds featuring in the special sciences were able to serve as the bases for scientifically useful generalizations. A prominent example in Boyd's work concerns biological taxa.

On HPC-theory, our ability to formulate explanatorily and predictively useful generalizations over biological species and other taxa becomes unsurprising once we note that the members of a taxon all exhibit largely similar properties due to the operation of largely the same causes, such as a common system of heredity and reproductive isolation between populations, shared developmental constraints, stabilizing selection in the same environment, etc.[31] Although typically there are no properties that *all* and *only* the member organisms of a taxon exhibit, there still are considerable similarities between the members of a taxon, the occurrence of which can be explained by taking recourse to among others the aforementioned factors. Accordingly, Boyd argued, biological taxa can be defined by the cluster of properties that are found to regularly, albeit not exceptionlessly, occur together in organisms of the same taxon in combination with the set of causal factors that underlie this clustering of properties.[32]

Because for a given species or other taxon there is no set of properties unique to and characteristic of all its members of that kind, the cluster of co-occurring properties cannot exhaustively define the taxon (if it could, a form of traditional kind essentialism would obtain). Accordingly, HPC-theory adds the set of causal factors that underlie this clustering to the definition and assumes the combination of these two elements to uniquely define the taxon as a kind. It conceives of this definition in an open-ended manner: no property is necessarily unique to one property cluster, no causal factor is necessarily unique to one set of homeostatic mechanisms, the property cluster of a kind may come to include new properties,

30 E.g., Richard N. Boyd, "Homeostasis, species, and higher taxa", in: Robert A. Wilson (Ed.): *Species: New Interdisciplinary Essays*. Cambridge (MA): MIT Press 1999, pp. 141–185; Richard N. Boyd, "Homeostasis, higher taxa, and monophyly", in: *Philosophy of Science* 77, 2010, pp. 686–701.

31 Boyd, "Homeostasis, species, and higher taxa", *loc. cit.*, p. 165.

32 Boyd calls these causal factors "homeostatic mechanisms", where he uses the term 'mechanism' in a very loose sense.

present properties may cease to be members, causal factors may begin or cease to operate, and there are no "core sets" of properties or underlying causal factors that all and only members of the corresponding kind exhibit or are affected by.

While there seems no particularly compelling reason to conceive of these two-part definitions of HPC-kinds as constituting the kind essences of taxa, Boyd himself suggested that his account should be seen as a form of kind essentialism:

> What is essential is that the kinds of successful scientific (and everyday) practice … must be understood as defined by a posteriori real essences that reflect the necessity of our deferring, in our classificatory practices, to facts about the causal structure of the world. … I'll argue that species (and, probably some higher taxa) do have defining, real essences, but that those essences are quite different from the ones anticipated in the tradition.[33]

While some authors take the causal factors (homeostatic mechanisms) underlying a kind as together making up the essence of that kind,[34] I believe that if one wishes to interpret HPC-theory in an essentialist manner, it is the set of clustering properties together with the set of underlying causal factors that should be seen as the essence of an HPC-kind on Boyd's account, because it is this combination that on Boyd's account constitutes the definition of a kind.

According to Boyd, the HPC-account of kinds applies widely to kinds in biology as well as in the other special sciences. Boyd himself suggested that species and other biological taxa are HPC-kinds and mentioned a diversity of other HPC-kinds, such as feudal economy, capitalist economy, monarchy, parliamentary democracy, the various religions, behaviorism,[35] and money.[36] Elsewhere, I have suggested that this broad scope of applicability constitutes both the strength and principal weakness of HPC-theory. My argument there was as follows.[37]

The scope of applicability, ranging from biological kinds to kinds of economic and political systems, is realized by conceiving of the defining essences of HPC-kinds in a non-traditional, open-ended manner. While this yields an account of kinds that is sufficiently flexible to accommodate all the various kinds that feature in the various special sciences, as well as more traditional natural kinds, precisely this flexibility causes a problem for HPC theory. Traditional essentialist accounts of kinds tell us which factors in nature determine the extensions of kind terms. If for a particular kind a kind essence in the traditional sense is identified, we immediately have a criterion for assessing whether or not a given entity is a member of that kind: does it exhibit all the properties deemed necessary and sufficient

33 Boyd, "Homeostasis, species, and higher taxa", *loc. cit.*, p. 146.

34 E.g., Griffiths, *loc. cit.*, p. 218.

35 Boyd, "Homeostasis, species, and higher taxa", *loc. cit.*, pp. 154–156.

36 Griffiths, *loc. cit.*, p. 218.

37 Thomas A.C. Reydon, "How to fix kind membership: A problem for HPC-theory and a solution", in: *Philosophy of Science* 57, 2009b, pp. 425–448. For additional criticisms of HPC-theory, see Ereshefsky, "What's wrong with the new biological essentialism?", *loc. cit.*

for membership in the kind? If one essential property is missing, the entity in question cannot be counted as a member of the kind. HPC-theory, however, fails to provide any such criteria. Even if we have identified all the elements in the property cluster and in the set of underlying causal factors for a given kind, we still have no criteria for determining the kind term's extension. The reason is the open-endedness of the defining essence of the kind: if one essential property is missing or one essential causal factor fails to operate, this does not say anything about whether or not the entity in question can be counted as a member of the kind. This can be seen particularly clearly for biological species, as species are subject to open-ended evolutionary change. Furthermore, in the case of a speciation event in which a new species branches off from its ancestor species, the member organisms of the two species will typically continue to be characterized by the same family of properties for quite some time and due to the operation of largely the same causal factors (cf. above). Thus, the combination of a property cluster and a set of underlying causal factors as a species' essence cannot serve to determine the boundaries of the species that it is supposed to define.

With respect to the philosophical work that kind essentialism might perform, this means that kind essences according to HPC-theory cannot serve to determine the kind identities of particular things as things of particular kinds (no. 1 in Sect. 16.2). Nor can they be invoked as explanations of the observable properties and behaviors that are typical for the members of a kind, as the causes of properties identified by HPC-theory aren't kind-specific but often extend beyond the kinds' boundaries or are limited to only a subgroup of a kind's members. By consequence, such kind essences will not be able to ground laws of nature or support scientific generalizations, explanations and predictions that pertain to particular kinds. Thus, notwithstanding its broad scope of applicability, the philosophical use of this form of kind essentialism appears to be quite limited.

16.4 Conclusion

Kind essentialism, it seems, is far from a dead issue in contemporary philosophy of biology. However, I do not think that this should be taken as implying that kind essentialism is a particularly promising position with respect to scientific kinds, i.e., an account of kinds that can do much important work to help us understand how science works.

For one, most kind essentialisms have only a very limited range of application, sometimes not even applying to most kinds in biological science (relational, historical and developmental essentialism), while one form of kind essentialism applies so widely that, if my arguments are correct, it becomes toothless (cluster essentialism). In addition, the various essentialisms discussed were aimed at different targets, none appearing able to perform all or even most of the tasks traditionally attributed to kind essentialism. As such, even if they are tenable, they will do only a limited amount of work in their specific domains. None of the essentialisms discussed above seem able to deliver on the great promise of kind essentialism.

Hence, there is reason to question the feasibility of essentialism about kinds in general. *If* some form of essentialism about biological kinds should turn out to be acceptable, this will not be of the sort that might apply to kinds in physics, because the essentialist positions are too much adapted to the requirements posed by biological science. Of the four positions discussed above, only cluster essentialism could be general enough to apply to physical kinds – but I have argued that there are general reasons for which cluster essentialism fails as an account of kinds. In addition, while traditional kind essentialism seems to work well for physical kinds, the essentialisms proposed for biological kinds break with the tradition on an important point, namely the traditional view that the essential properties associated with a kind should be exhibited by *all* and *only* the members of that kind. Thus, at most we will end up with a disunified account of scientific classification that could be called "local kind essentialism": one form of kind essentialism for physical kinds, another for biological kinds, and perhaps even several kind essentialisms for different domains of biology.

So, how much of an undead issue is kind essentialism? On a local level, that is when applied to kinds in one particular research context, some forms of kind essentialism may be alive and well. Globally, however, as a general account of kinds in biological science and beyond, kind essentialism does not show much promise with respect to being able to do important philosophical work. As a general thesis about kinds, then, kind essentialism is best left dead and buried.

References

H. Clark Barrett: "On the functional origins of essentialism", in: *Mind & Society* 3, 2001, pp. 1–30.

Richard N. Boyd: "Homeostasis, species, and higher taxa", in: Robert A. Wilson (Ed.): *Species: New Interdisciplinary Essays.* Cambridge (MA): MIT Press, 1999, pp. 141–185.

Richard N. Boyd: "Homeostasis, higher taxa, and monophyly", in: *Philosophy of Science* 77, 2010, pp. 686–701.

Michael Devitt: "Resurrecting biological essentialism", in: *Philosophy of Science* 75, 2008, pp. 344–382.

Brian D. Ellis: *Scientific Essentialism.* Cambridge: Cambridge University Press, 2001.

Marc Ereshefsky: "Species", in: Edward N. Zalta (Ed.): *The Stanford Encyclopedia of Philosophy (Spring 2010 Edition)*, 2010a, online at http://plato.stanford.edu/archives/spr2010/entries/species/.

Marc Ereshefsky: "What's wrong with the new biological essentialism?", in: *Philosophy of Science* 77, 2010b, pp. 674–685.

Susan A. Gelman: *The Essential Child: Origins of Essentialism in Everyday Thought.* Oxford: Oxford University Press, 2003.

Susan A. Gelman and Lawrence A. Hirschfeld: "How biological is essentialism?", in: Douglas L. Medin and Scott Atran (Eds): *Folkbiology.* Cambridge (MA): MIT Press, 1999, pp. 403–446.

Paul E. Griffiths: "Squaring the circle: Natural kinds with historical essences", in: Robert A. Wilson (Ed.): *Species: New Interdisciplinary Essays.* Cambridge (MA): MIT Press, 1999, pp. 209–228.

John Locke: *An Essay Concerning Human Understanding (Edited with and Introduction by Peter H. Nidditch).* Oxford: Clarendon Press, [1700] 1975.

Samir Okasha: "Darwinian metaphysics: Species and the question of essentialism", in: *Synthese* 131, 2002, pp. 191–213.

Thomas A. C. Reydon: "Generalizations and kinds in natural science: The case of species", in: *Studies in History and Philosophy of Biological and Biomedical Sciences* 37, 2006, pp. 230–255.

Thomas A. C. Reydon: "Gene names as proper names of individuals: An assessment", in: *British Journal for the Philosophy of Science* 60, 2009a, pp. 409–432.

Thomas A. C. Reydon: "How to fix kind membership: A problem for HPC-theory and a solution", in: *Philosophy of Science* 76, 2009b, pp. 724–736.

Olivier Rieppel: "New essentialism in biology", in: *Philosophy of Science* 77, 2010, pp. 662–672.

Denis M. Walsh: "Evolutionary essentialism", in: *British Journal for the Philosophy of Science* 57, 2006, pp. 425–448.

C. Kenneth Waters: "Causal regularities in the biological world of contingent distributions", *Biology and Philosophy* 13, 1998, pp. 5–36.

Institute of Philosophy & Center for Philosophy and Ethics of Science (ZEWW)
Leibniz Universität Hannover
Im Moore 21
30167 Hannover
Germany
reydon@ww.uni-hannover.de

CHAPTER 17

CHRISTIAN SACHSE

BIOLOGICAL LAWS AND KINDS WITHIN A CONSERVATIVE REDUCTIONIST FRAMEWORK

ABSTRACT

This paper argues for the existence of biological kinds and laws. After a general discussion of biological laws (Sect. 17.1), I shall outline a conservative reductionist approach towards biological property types (Sect. 17.2). Within this theoretical framework, it seems plausible to argue for biological laws (to a degree) and genuine biological natural kinds (Sect. 17.3).

17.1 BIOLOGICAL LAWS

John Beatty argues that biological generalizations are to some extent contingent and do not involve laws.[1] He construes the idea of laws as empirical generalizations without any exceptions (like "$\forall$x: if *F*x, then *G*x") and that contain a natural necessity; that are counterfactually robust.[2] Given this definition, he argues furthermore that biological generalizations that fit approximatively into the empirical and no exceptions framework are about genetically based traits that are subject to evolutionary forces. For instance, Mendel's first law or Hardy-Weinberg's law obtain only because of prior initial conditions that emerged *contingently* in the course of evolution, and could, thus, have been otherwise: "evolution can lead to different outcomes from the same starting point, even when the same selection pressures are operating."[3] Therefore, Beatty concludes that while empirical biological generalizations may correctly describe a causal relation over some period (from t_1 to t_2), they do not form laws in the sense that they are only true because

1 John Beatty, "What's wrong with the received view of evolutionary theory?", in: *Proceedings of the Biennial Meeting of the Philosophy of Science Association* Volume 2: Symposia and invited Papers, 1980, pp. 397–426; John Beatty, "The evolutionary contingency thesis", in: G. Wolters and J. Lennox (Eds.): *Concepts, theories, and rationality in the biological sciences: The second Pittsburgh-Konstanz Colloquium in the Philosophy of Science*. Pittsburgh: University of Pittsburgh Press, 1995, pp. 45–81.

2 Cf. Beatty, "The evolutionary contingency thesis", *loc. cit.*, p. 53, footnote 9.

3 Beatty, "The evolutionary contingency thesis", *loc. cit.*, p. 57.

D. Dieks et al. (eds.), *Probabilities, Laws, and Structures*, The Philosophy of Science in a European Perspective 3, DOI 10.1007/978-94-007-3030-4_17,

of some prior initial conditions *I* (that obtained at t_0). I shall come back to this argument later on.[4]

However, the principle of natural selection is a particular biological generalization. Here the argument from different circumstances, or from the contingency of evolutionary development, may not apply. Instead, all the circumstances we need for there to be natural selection consist in this: (a) *that* there are inheritable properties, which imply fitness differences; and (b) that both the inheritance mechanisms and the fitness differences may be physically realized in different ways. Whether this degree of generality is sufficient to avoid the contingency argument depend on a deeper discussion of contingency.[5] Let us suppose that it is. Still, according to Beatty, the principle of natural selection seems to have been defined so that it lacks *empirical* generalizability, and consequently does not count as a law, if fitness has been defined in a tautological way. This is the case if the fitness of an entity at t_1 is only determined by the evolutionary effects (e.g. number of descendants) it brings about at t_2. To put it differently, fitness differences can only be trivially linked to evolutionary changes by the principle of natural selection if we can define some former state of fitness upon which evolutionary changes work as the cause of present evolutionary changes.

However, one may argue that this tautology only exists at an epistemic level and can be theoretically avoided, following Rosenberg,[6] in distinguishing between the *operational* and the *conceptual* understanding of fitness.[7] Conceptually, we can understand the fitness contribution of a trait as its contribution to the organism's disposition to survive and its disposition to reproduce and both dispositions supervene locally on the physical properties of the organism and its environment.[8] The success (manifestation) of these dispositions to survive and reproduce depends on given environmental conditions, allowing us to attribute to a characteristic fitness function to any kind of organic trait (Fig. 17.1):

4 For a critique, see: Elliot Sober, "Two outbreaks of lawlessness in recent philosophy of biology", in: *Philosophy of Science*, 64, 1997, pp. S458–S467; Kenneth Waters, "Causal regularities in the biological world of contingent distributions", in: *Biology and Philosophy*, 13, 1998, pp. 5–36.

5 I sketch out one reply later on; for a more comprehensive discussion, see: Mauro Dorato, "Mathematical Biology and the Existence of Biological Laws", this issue.

6 Alexander Rosenberg, "Supervenience of biological concepts", in: *Philosophy of science* 45, 1978, pp. 368–386.

7 See Elliot Sober, *Philosophy of biology. Second Edition*. Boulder: Westview Press, 2000, ch. 3; Christopher Stephens, "Natural selection", in: M. Matthen and C. Stephens (Eds.), *Handbook of the philosophy of science. Philosophy of biology*. Amsterdam: Elsevier, 2007, pp. 111–127.

8 See furthermore: Marcel Weber, "Fitness made physical: The supervenience of biological concepts revisited", in: *Philosophy of Science* 63, 1996, pp. 411–431.

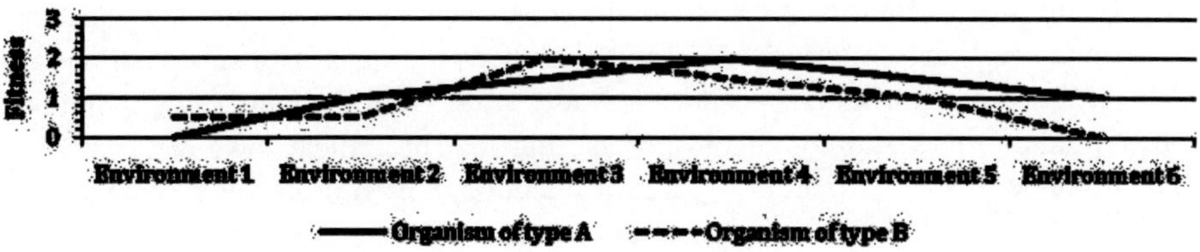

Fig. 17.1 *Fitness function*

Of course, the attribution of such fitness functions is rather difficult in practice. This, however, is an epistemic problem. It remains that fitness is ontologically determined by the dispositions to survive and reproduce, these dispositions are manifested under certain environmental conditions, and the principle of natural selection correctly registers the impact of fitness differences for evolutionary change. Thus, if we understand the principle of natural selection in this way, we can dispose of the non-empirical objection to it – for here it is surely an empirical effect on populations of organisms.

Following this reasoning, one could see no objection, following Rosenberg, to specifying the status of the principle of natural selection as a fundamental, non-derived law of physics[9]: the principle of natural selection is a fundamental law since it (a) can explain physical facts and (b) it cannot be derived from other physical laws because of the multiple realization of biological functions and thus of fitness (differences). In other words, the laws of physics need specific initial and boundary conditions to explain the distribution of the molecules (e.g. genes) at some later time, while the principle of natural selection can do so for infinitely many different initial conditions. This then suggests adding the principle of natural selection to the other fundamental physical laws.

Of course, as Rosenberg argues, if the principle of natural selection really is fundamental, then we can avoid any conflict with the principle of the completeness of physics by simply conjoining it to the physical laws.[10] However, at least in theory, there remains a categorial difference between the principle of natural selection and the (other) fundamental laws of physics that may seem, to a physicalist, like it calls for another act of reduction. If we decide not to adopt some kind of ontological property dualism (following Rosenberg's counsel), then we must say that the principle of natural selection and the (other) fundamental physical laws refer to the *same* properties, only in different manners. However, if this is our claim, we may question Rosenberg's argument from irreducibility to fundamentality. Furthermore, I argue later on that multiple realization does not actually present an obstacle to reducibility.

9 Alexander Rosenberg, *Darwinian reductionism. Or, how to stop worrying and love molecular biology*. Chicago: University of Chicago Press, 2006, ch. 6.

10 See also: Marcel Weber, "Review of Alexander Rosenberg, *Darwinian reductionism. Or, how to stop worrying and love molecular biology*", in: *Biology and Philosophy* 23, 2008, pp. 143–152.

In contrast to Beatty and Rosenberg, Sober wants to leave open the question of whether laws are empirical or *a priori*.[11] Understanding *a priori* propositions as laws if they are about causal processes, Sober argues that the way biologists build their models gives support to the proposition that biological laws are *a priori*. For instance, Fisher's theorem of natural selection, which proposes a mathematical proof, is a law, according to Sober, because it supports counterfactuals and describes causal and explanatory relations. More generally, if we accept Sober's construction of laws, and we accept that evolutionary processes are governed by biological laws, then we can conclude that evolution is lawful. Of course, Beatty's contingency argument is aimed at just these elements of Sober's argument. After all, any (empirical or *a priori*) biological law that has the general form "∀x: if *F*x, then *G*x" may be contingent on prior initial conditions *I*. However, this fact does not exclude reformulating the generalization in the form: "If *I* obtains at one time, then the generalization "∀x: if *F*x, then *G*x" will hold hereafter", from actually being contingent on anything.[12] Such reformulated non-contingent generalizations are laws since (a) they are about causal relations (between token of *F* and tokens of *G*) and (b) causation demands the existence of laws.

However these claims about laws are straightened out, one may still ask whether these laws aren't physical ones, at least in the last resort. After all, following the completeness claim, physics has the most detailed means to spell out the causal relations that lead to situations where, to take Beatty's examples, inheritance conforms to Mendel's first law. Furthermore, any naturalistic approach would suggest that the emergence of life, for whatever reason it happens, must ultimately reduce to physical law, from which is then derived the application of the principle of natural selection. On this reading it seems that Sober's reply to Beatty's contingency argument depends on the physical laws that have to be incorporated into his proposed reformulations. Therefore, biological laws are non-contingent only to the extent that they are in fact physical laws (or at least derivative from such laws). This suggests that reductionism gives us the only convincing reply to Beatty's contingency argument. Without reducing biological laws to the ones governing chemical and physical interaction between physical elements, we have no coherent account that allow us to conjoin the two ends of the theory of biological law: on the one end, the claim that biology is able to formulate *a priori* laws that support counterfactuals, which can be applied to causal relations concerning living things that give us scientific explanations; and on the other end, the claim that the truth of these laws supervenes on the truth of physical laws that are empirical ones. Moreover, we also confront, here, a problem quite similar to the tautology problem of fitness, intrinsic to the claim that *a priori* laws are mainly *operational* abstractions of physical laws that are genuine natural ones.

---

11 See: Sober, "Two outbreaks of lawlessness in recent philosophy of biology", *loc. cit.*

12 See: Sober, *Ibid.*

One could resolve these dilemmas by having recourse to a biological version of *ceteris paribus* laws, which – the claim would go – are genuine laws because biological laws differ from physical laws only *in degree* of their ceteris paribus type but not in kind.[13] To make a clear link to our previous discussion, this argument holds that laws do not have to be universal (contrary to the position of Beatty and Rosenberg) without necessarily adopting Sober's particular position on *a priori* laws. Still, following Beatty, there is a difference between biology and physics – and I spell out this difference within a reductionist framework in the next section, where I also keep in mind, following Rosenberg, to avoid any conflict with the completeness of physics and ontological reductionism. In addition, I take Sober's reply to Beatty's contingency argument for granted. Within this framework, I thus analyse here in more detail (a) the historical dimension of biology and (b) *ceteris paribus* clauses in biology. Then, given the decomposition of all laws, physical and biological, into *ceteris paribus* laws, we must show that the difference in degree in relation to physical laws is such that (c) these laws are distinctively *biological* ones.

(a) Biology is a diachronic discipline about biological events – for instance, speciation – that are unrepeatable in practice because of the differences between any biological organism. This means that there are no *types* of historical events, which disallows forming corresponding laws that take types as their object. However, physical theories like cosmology are also diachronic in the above given sense, in that they concern unrepeatable events. So, in comparing cosmology and biology, if we take it for granted that both refer to causal relations governed, in the last resort, by physical laws, then the difference in their objects appears to be more of a difference in degree of complexity than a difference in kind. To put it differently, it seems that the unrepeatable character of historic events *per se* does not exclude the existence of laws.[14] However, the question is not so much one of the historical dimension of biology but whether these are underlain by genuine biological laws, just as general relativity or quantum gravity underlies cosmology. In the next section, I will outline how this may work in biology.

(b) Biological laws are not universal since the existence of biological properties is spatiotemporally restricted. For instance, the principle of natural selection applies only to particular objects, living beings, and not to purely physical configurations. Biology always needs so-called *ceteris paribus* clauses in order to provide the applicability of its laws. Understanding *ceteris paribus* as "whenever the right

13 See: Dorato, "Ceteris paribus laws in physics and biology, or why there are biological laws", *loc. cit.*; Marc Lange, "Laws, counterfactuals, stability, and degrees of lawhood", in: *Philosophy of Science* 66, 1999, pp. 243–267; see also Marc Lange, *Laws & lawmakers*. Oxford: Oxford University Press, 2009.

14 See: Dorato, "Ceteris paribus laws in physics and biology, or why there are biological laws", *loc. cit.*

condition obtains"[15] (in distinction to "all other things being equal"[16]), one may then ask whether this feature really distinguishes biological laws from physical ones. The view that it doesn't mainly contains two parts. First of all, a *ceteris paribus* clause contains the right conditions and biology cannot specify them in its own terms. However, this seems to be an epistemic difficulty rather than a conclusive objection to a possible existence of biological laws. Second, of course, biological laws depend on initial conditions. However, this does not distinguish biological and physical laws, since initial conditions are required in physics as well.[17] The fact that adjustable parameters in the initial conditions may be much more numerous in biological laws than in physical is once again only a difference in degree.

(c) Following this line of argument, we still have to answer the question: what makes a law a distinctively biological one? After all, a complete *ceteris paribus* clause necessarily contains physical specifications. Still, a law may be called biological if it contains biological concepts that are irreducible to physics (or rather "irreplaceable" as I shall argue later on). And this seems to be the case, most philosophers agree, because of the multiple realization[18] of biological properties.

To conclude this section, it seems that if biological laws exist, they exist in the form of *ceteris paribus* laws. As I have argued, the view that biological laws differ only in degree to physical laws goes hand in hand with the irreducibility of biology due to multiple realization. In the following section, I will argue that this link is both unnecessary and moreover problematic. Multiple realization should not be seen as an irreducible impediment to reduction, nor should it be understood as an anti-reductionist argument. To the contrary, a conservative reductionist approach that embraces multiple realization as an *anti-eliminativist* argument gives us a stronger argument in favour of the existence of biological laws distinguished in degree from physical ones.

---

15 Nancy Cartwright, *How the laws of physics lie*. Oxford: Oxford University Press, 1983, p. 45 (taken from: Dorato, "Ceteris paribus laws in physics and biology, or why there are biological laws", *loc. cit.*).

16 See also: Stephen Schiffer, "Ceteris paribus laws", in: *Mind* 100, 1991, pp. 1–17; Jerry Fodor, "You can fool some of the people all the time, everything else being equal: Hedged laws and psychological explanations", in: *Mind* 100, 1991, pp. 19–34.

17 See: Dorato, "Mathematical Biology and the Existence of Biological Laws", *loc. cit.*; Mehmet Elgin, "There may be strict empirical laws in biology, after all", in: *Biology and Philosophy* 21, 2006, pp. 119–134.

18 See also: Lawrence Shapiro, "Multiple realizations", in: *The Journal of Philosophy* 97, 2000, pp. 635–654.

17.2 Conservative Reductionism

It is generally taken for granted that biological property tokens are identical with something physical.[19] Otherwise, at least one of the following widely accepted working hypotheses would be false: (1) biological properties supervene on complex configurations of physical properties[20]; and (2) physics is causally, nomologically and explanatorily complete with respect to biology[21]; and (3) biological properties are causally efficacious. Since, according to token-identity, biology and physics refer to the same entities, the problem of the autonomy of biology starts with explaining how their concepts, laws and explanations are related.

Let me start here with the argument that takes multiple realization to require an anti-reductionist stance, an argument that goes back to Fodor[22] and Putnam.[23] The principal point of the argument is that biological concepts cannot be bi-conditionally related to physical descriptions. They are not coextensive.[24] Therefore, biological functional explanations must constitute an autonomous and unifying explanatory level[25] (Fig. 17.2):

19 See among others: Michael Esfeld and Christian Sachse, *Conservative reductionism*. New York: Routledge, 2011, ch. 2.6; Jaegwon Kim, *Physicalism, or something near enough*. Princeton: Princeton University Press, 2005, ch. 2.

20 Rosenberg, "Supervenience of biological concepts", *loc. cit.*; Weber, "Fitness made physical: The supervenience of biological concepts revisited", *loc. cit.*

21 See: David Papineau, *Thinking about consciousness*. Oxford: Oxford University Press 2002, appendix.

22 Jerry A. Fodor, "Special sciences (or: The disunity of science as a working hypothesis)", in: *Synthese* 28, 1974, pp. 97–115.

23 Hilary Putnam, "The nature of mental states", in: H. Putnam, *Mind, language and reality. Philosophical papers. Volume 2*, Cambridge: Cambridge University Press 1975, pp. 429–440.

24 Note that natural selection is generally taken to be the reason why there is multiple realization of biological property types: the causal powers of a given physical configuration, realizing a biological property that is pertinent for selection, depends on the environmental conditions. See: David Papineau, *Philosophical naturalism*. Oxford: Blackwell 1993, p. 47; Alexander Rosenberg, "How is biological explanation possible?", in: *British Journal for the Philosophy of Science* 52, 2001, pp. 735–760.

25 See also: Philip Kitcher, "1953 and all that. A tale of two sciences", in: *Philosophical Review* 93, 1984, pp. 335–373.

Biological concepts (like "*B*") refer, because of functional similarities, in a homogenous way to biological property tokens that come under different physical descriptions (like "P_1" and "P_2"):

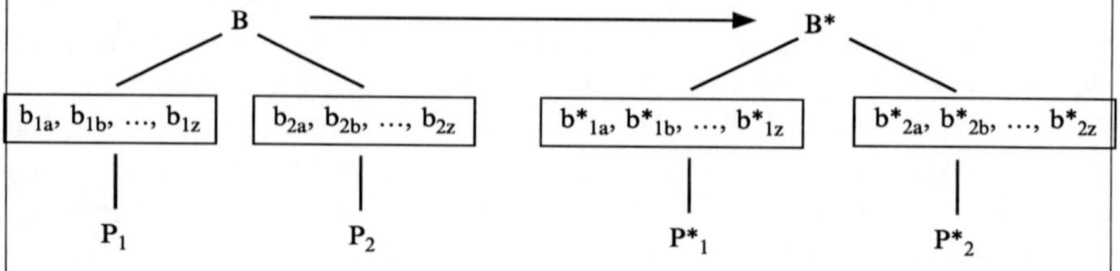

Fig. 17.2 Multiple realization

However, if no nomological coextension between physical and biological descriptions can be established, biological concepts would seem to not actually be about the *same* entities in the fine-grained sense, but are instead about *different* properties.[26] This then leads to a property dualism that contemporary anti-reductionists have tried to avoid, with its concomitant of making biological properties epiphenomenal. After all, it follows from token identity and the completeness of physics that for the biological property tokens b_{1a}, b_{1b}, ... b_{1z}, the fact of coming under a biological description *B* cannot signify some causal efficacy *beyond* what is spelled out by P_1 and, similarly, *B* applies as well to b_{2a}, b_{2b},... b_{2z} that are completely described by P_2. So, *B* cannot be something causal *in addition* to what physics tells us; *B* is either an abstraction or epiphenomenal. Epiphenomenalism implies eliminativism as regards the scientific quality of *B* (and of biology in general) since no causal explanation could be based on it. If we reject epiphenomenalism, then it *has to be theoretically possible* to construct biological concepts that are bi-conditionally related to physical descriptions. This then means to take a reductionist perspective that avoids epiphenomenalism and eliminativism as regards biological abstractions, which satisfy the following desiderata:

26 See: Michael Esfeld, "Causal properties and conservative reduction", in: *Philosophia naturalis* 47–48, 2010–2011, pp. 9–31; Michael Esfeld and Christian Sachse, "Theory reduction by means of functional sub-types", in: *International Studies in the Philosophy of Science* 21, 2007, pp. 1–17; Esfeld and Sachse, "Conservative reductionism", *loc. cit.*, ch. 5; Christian Sachse, *Reductionism in the philosophy of science*. Frankfurt: Ontos-Verlag 2007, ch. III.

1. Avoiding the conflict with the completeness of physics and ontological reductionism.
2. Biological concepts, laws and explanations are about causally efficacious property tokens ("Cau").
3. Biological concepts, laws and explanations are theoretically not replaceable ("¬Rep").

Fig. 17.3 *Minimal desiderata*

In order to combine "*Cau*" and "¬*Rep*" (Fig. 17.3), one has to consider multiple realization in more detail. According to it (as illustrated in Fig. 17.2), not everything that comes under *B* would also come under a single physical description P_1. Here, P_1 is a placeholder for a detailed homogeneous physical description that only applies to a subset of entities that come under *B*. However, if local physical structures coming under one concept *B* are described in terms of different physical concepts (like P_1 and P_2), then there is a difference in composition among their structures. Each of these physical concepts is about a minimal sufficient condition (realizer) to bring about the effects that define *B*, *ceteris paribus*. In order to get from structures coming under P_1 to structures coming under P_2, one has to substitute at least one of the necessary parts of the biological trait to bring about the effects in question with a part of another type. Any such replacement implies a systematic difference in the way in which these structures cause the effects that define *B*, which means that we cannot replace a local physical structure of type P_1 by a local physical structure of type P_2 (thus obtaining a different physical realizer of *B*) *without* making a causal difference.[27]

If the effects that define *B* can be brought about by two or more different configurations of physical properties (types of realizer), our claim is that we will still find a difference in the production of side effects that are systematically linked with the main effects in question over the entire trajectory of the trait's historical existence. Think of physically different genes[28] that all code for the same protein and thus come under one biological concept *B*. Such a case affords the possibility that different causal interactions with the physical environment within the cell will occur when these genes are transcribed and the proteins are synthesized. For any such difference in the causal sequence from the DNA transcription to the protein synthesis, there exists the possibility that the difference may become pertinent in particular environments[29] (see the illustration in Fig. 17.4, where the physically

27 See also: Kim, "Making sense of emergence", *loc. cit.* and Kim, "Physicalism, or something near enough", *loc. cit.*, p. 26.

28 Genes and functionally defined gene types should be generally understood as difference makers; see: C. Kenneth Waters, "Genes made molecular", in: *Philosophy of Science* 61, 1994, pp. 163–185; Kenneth Waters, "Causes that make a difference", in: *Journal of Philosophy* 104, 2007, pp. 551–579.

29 See also: Alexander Rosenberg, *Instrumental biology or the disunity of science.*

different genes differ in environment 1 and 6, but are alike in environment 2–5). Consequently, that difference can *in principle* also be considered in *functional* terms – terms *proper* to the biological domain to which *B* belongs.[30] The upshot of this argument is that more precise functional definitions may, in theory, account for different reaction norms (fitness functions), and thus, physical differences. Against this background, for the concept *B* (that is multiply realized by P_1 and P_2), it is possible to conceive two functional sub-types B_1 and B_2 taking different reaction norms into account:

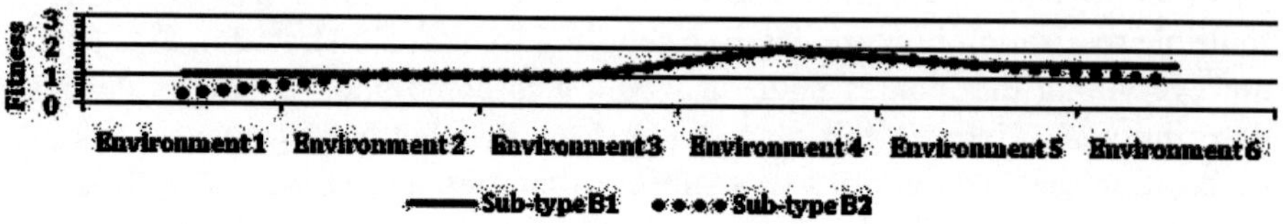

Fig. 17.4 *Fitness functions of sub-types*

For instance, consider a gene of *E. coli* whose expression is pertinent to the fitness function of the organism, and that is thus functionally defined in terms of biology. For instance, a genetic basis for cell-wall biosynthesis. Simplified, the gene tokens coming under *B* are defined by their characteristic expression of membrane proteins that are crucial for the cell growth of the bacterium before cell division, etc. Independently of our chosen level of genetic simplification, the gene tokens coming under *B* are identical with certain physical configurations (DNA sequences) that are described differently in terms of physics (by P_1 and P_2) since there are differences in the physical composition of the DNA sequences in question. Nonetheless, due to the redundancy of the genetic code, all these physically different DNA sequences code for proteins of the same type (or any other effect that is considered in the functional definition *B*). The crucial point here is that there are different physical paths to bring about the effect in *B* according to the physical differences between P_1 and P_2. These different ways to produce the effects (the proteins for instance) are systematically linked with possible side effects or reaction norms, as for instance differences in the speed or the accuracy of the protein production, of which we have more and more empirical evidences.[31] To sum up,

Chicago: University of Chicago Press, 1994, p. 32.

30 With regard to more fine-grained functional concepts of the special sciences, see also: William Bechtel and Jennifer Mundale, "Multiple realizability revisited: linking cognitive and neural states", in: *Philosophy of Science* 66, 1999, pp. 175–207.

31 See among many others: Michael Bulmer, "The selection-mutation-drift theory of synonymous codon usage", in: *Genetics* 129, 1991, pp. 897–907; Daniel L. Hartl, Etsuko Moriyama and Stanley Sawyer, "Selection intensity for codon bias", in: *Genetics* 138, 1994, pp. 227–234; Ulrich Gerland and Terence Hwa, "Evolutionary selection between

depending on variations in the environmental conditions, the optimality of certain DNA sequences over others can become selectively pertinent. This, then, should be taken into account in more precise functional definitions and explanations (see Fig. 17.5 below).[32]

By means of these sub-types we attain concepts of biology that are nomologically coextensive with physical concepts and thus make it possible to reduce biology to physical theories in a functional manner (if we assume multiple realization) in three steps (see also Fig. 17.5): (1) within an encompassing fundamental physical theory P, we construct the concepts P_1, P_2, etc. to capture the differences in composition among the local physical structures that are all described by the same concept B; (2) B is more precisely articulated by constructing functional sub-types B_1, B_2, etc. of B, each of which captures the systematic side effects linked to the different ways of producing the effects that define B. To put it differently, the sub-types are constructed from B in such a way that they are nomologically coextensive with the concepts P_1, P_2, etc., using the functional model of reduction; (3) B is reduced to P via B_1, B_2, etc. and P_1, P_2, etc. Reducing B (and thus biology) here means that starting from P, we can construct P_1, P_2, etc. and then deduce B_1, B_2, etc. from P_1, P_2, etc. given the nomological coextension. One derives B *by abstracting* from the conceptualization of the functional side effects contained in B_1, B_2, etc. given a environmental context in which the functional side effects are not manifested or are not pertinent to selection[33]:

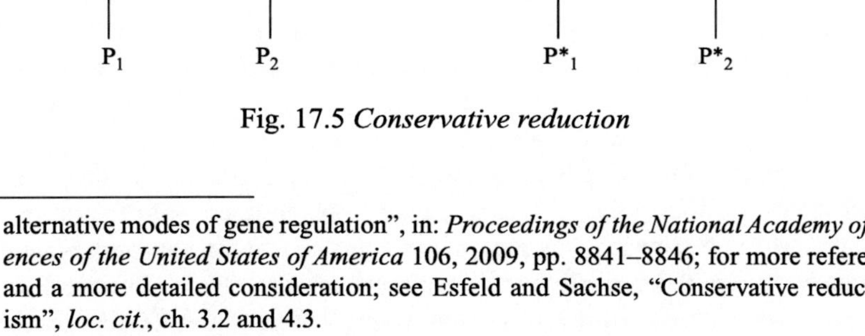

Fig. 17.5 *Conservative reduction*

alternative modes of gene regulation", in: *Proceedings of the National Academy of Sciences of the United States of America* 106, 2009, pp. 8841–8846; for more references and a more detailed consideration; see Esfeld and Sachse, "Conservative reductionism", *loc. cit.*, ch. 3.2 and 4.3.

32 See: Christian Sachse, "Conservative reduction of biology", in: *Philosophia naturalis* 48–49, 2010-11, pp. 33–65, for more details why sub-types are no longer multiply realizable and why sub-types and the original types have the same substantial meaning.

33 For more details, see: Esfeld and Sachse, "Conservative reductionism", *loc. cit.*, ch. 5; Sachse, "Conservative reduction of biology", *loc. cit.*

On the basis of the fundamental physical laws, one can construct laws in terms of P_1, P_2, etc. that refer to the properties on which biology focuses. From those laws, one can deduce biological laws in terms of B_1, B_2, etc., given the nomological coextension of these concepts. These sub-types and any laws and explanations that are based on them are not, then, about epiphenomena (thus vindicating "*Cau*"). Nonetheless, they were replaceable by physics because of nomological coextension (no vindication of "¬*Rep*"). However, one reaches the laws and explanations in terms of *B* by bracketing the conceptualization of the functional side effects that are represented in B_1, B_2, etc. Since the specification of the function of *B* is contained in each of its sub-types, the original and abstract concept *B* cannot be eliminated. The abstract laws of biology couched in terms of *B* are non-physical and not replaceable by physics in the sense that there is no single physical law having the same extension as any of these laws, vindicating "¬*Rep*" for *B*. When talking about complex objects such as e.g. genes, cells, or whole organisms, the physical concepts focus on the composition of these objects. Due to selection there are salient causal similarities among the affects produced as a whole by such complex objects, even though they differ in composition. So, the abstract concepts of biology possess a scientific quality in the sense of satisfying "*Cau*" *and* "¬*Rep*", figuring in law-like generalizations that capture something that is objectively there in the world. Furthermore, these concepts and law-like generalizations do not conflict with the completeness of physics and ontological reductionism, since the reductive method used to express them is based on the fundamental physical concepts and laws.

17.3 Perspectives

Conservative reductionism constitutes a plausible framework for biological laws and kinds. As regards the question of special biological laws, it is consistent with the claim that these exist as things different in degree from physical laws within the reductionist framework. We can specify the different degrees of lawhood in terms of different degrees of abstraction and generality. That is to say, biological generalizations may be, within their domain of application, law-like. The argument for this is stronger within a conservative reductionism because, as we have shown, it avoids conflicts with an ontological reductionism and the thesis of the completeness of physics that are usually held to be antithetical to the biological law claim. Moreover, by showing that the concepts constituting abstract biological generalizations are theoretically connectable via sub-types with physical descriptions and laws, we may formulate sub-type-laws that get their law-like character from physics *deductively*, on account of nomological coextension. From this move, the original biological generalizations can also be understood as inheriting their law-likeness, since they only abstract from certain functional details. Against this background, we can connect the principle of natural selection to physics by means of its application to specific units of selection, and thus confer on it its law-like character. Still, because of its extreme generality, the principle of natural selection is not replaceable by physics. Keep in mind that the kind and degree of abstraction is entirely a matter of the given and changing environmental conditions, and not on some theoretical protocol.

Against this background, one may consider the debate on biological taxa being natural kinds. Conservative reductionism supports a realist attitude with respect to biological kinds in the following general way: since the sub-types are nomologically coextensive with physical descriptions, it is possible to apply any argument in favour of (composed) physical kinds being natural ones to the biological sub-types as well. Thus, the more abstract biological concepts inherit their naturalness and counterfactual robustness from their sub-types, or, to put it differently, the reductionist framework makes explicit the hierarchical structure of a system of natural biological kinds that is *theoretically* achievable. Additionally, depending on environmental conditions, the abstract biological concepts such as biological taxa may not only be descriptive but also figure in biological laws and explanations. In this way, neither inheritance nor the biological sphere's systematic hierarchical structure contains, in the ideal case, any conventionalist aspect. This seems at least plausible for any kind of biological property type at a certain time.

However, things become more complicated as regards biological species that are evolving while in time, when physical natural kinds are not. Physical natural kinds are *perfectly similar* and can thus be rigidly designated, while biological kinds are at most *imperfect similar*. This difference suggests that we should deny any essence to the notion of the biological species. However, one may argue that imperfect similarities are sufficient for essence.[34] To show this, let us first consider the argument for the following claim: there is no principal difference whether we consider multiple realization of a type at one specific time or for a period of time. For instance, imagine an abstract concept B_{t1} such that it applies to any member of a species at t_1 and it can be conservatively reduced via its sub-types to physics. Look at that species at a later stage in evolution (at t_2) and imagine once again that an abstract concept B_{t2} applies to any member of a species and this concept can be conservatively reduced via its sub-types to physics as well. If we now compare both abstract concepts B_{t1} and B_{t2}, it is likely that they differ somehow and it is even more likely that their sub-types differ somehow since evolution has taken place. However, there is no principal objection to the view that both abstract concepts B_{t1} and B_{t2} may constitute themselves two sub-types for some more abstract concept that bring out salient characteristic similarities that figure in explanations. Call this a theoretical species concept that applies to B_{t1} and B_{t2}. Of course, common

34 See also: Kevin Lynch, "A multiple realization thesis for natural kinds", in: *European Journal of Philosophy*, DOI: 10.1111/j.1468-0378.2010.00420.x, 2010, pp. 1–18.

taxonomy may either satisfy these demands or not. But whenever it does, species concepts are natural ones and may theoretically figure in laws and explanations.

This then amounts to attributing essence to species. Say that the individuals of some species B differ physically and thus come under different physical descriptions P_1, P_2, etc. Applying the reductionist strategy, one may construct sub-types (B_1, B_2, etc.) of B that are nomologically coextensive with P_1, P_2, etc. Any attribution of essence to the constructed sub-types is justified since they are nomological coextensive with physical types (to which we generally attribute essences). Then, the species concept B can be understood as being nothing more than an abstraction from the essence *differences* of its sub-types. B spells out what all the individuals have essentially in common (similar to the functional similarity of biological types). The same reasoning is, as shown before, applicable to larger time scales. We may thus share some essence with humans of previous generations. However, since evolution continues, any particular essence may disappear one day. This then raises the question about the essence changing, or a speciation event.

Within the reductionist framework, speciation may be understood as arising when at least two sub-types (B_1 and B_2,) no longer share "enough" to come under the previously common species concept B. No longer sharing enough here means that functional (essence) differences that are spelled out in the sub-types become more important than their functional (essence) similarities. This poses no theoretical threat of conventionalism, since whether or not such situations emerge depends on the environmental conditions. Within the framework of conservative reductionism, our argument suggests that differences in essence (in combination with the given environmental conditions) constitute the starting point for whether the speciation event occurs or not. In other terms, phylogenesis during evolution does not depend on us but on the world and the underlying physical structures and changes that can be, in theory, considered in terms of sub-types and more abstract concepts. On that theoretical basis, rather descriptive classifications that mostly focus on a historical dimension like common ancestry are not impediments to the ahistorical construction of biological kinds with genuine essences that figure in genuine explanations.

Department of Philosophy
University of Lausanne
Quartier UNIL-Dorigny, Bâtiment Anthropole 4074
1015, Lausanne
Switzerland
christian.sachse@unil.ch

CHAPTER 18

MARIE I. KAISER

WHY IT IS TIME TO MOVE BEYOND NAGELIAN REDUCTION

18.1 INTRODUCTION

In this paper I argue that it is finally time to move beyond the Nagelian framework and to break new ground in thinking about epistemic reduction in biology. I will do so, not by simply repeating all the old objections that have been raised against Ernest Nagel's classical model of theory reduction.[1] Rather, I grant that a proponent of Nagel's approach can handle several of these problems but that, nevertheless, Nagel's general way of thinking about epistemic reduction in terms of theories and their logical relations is entirely inadequate with respect to what is going on in actual biological research practice.

I start with an overview of the long "success story" of the Nagelian account, which I think has not really found an ending yet (Sect. 18.2). Then I reveal the inadequacy of the Nagelian framework with respect to biology (Sect. 18.3) by arguing that Nagel focuses on the wrong relata of the relation of epistemic reduction (Sect. 18.3.2) and on the wrong kind of issues, namely on formal and not on substantive issues (Sect. 18.3.3). My argumentation is based on certain methodological assumptions about how to develop an adequate account of epistemic reduction (Sect. 18.3.1), which I specify by unfolding three criteria of adequacy that an account of epistemic reduction in biology must satisfy.

18.2 THE DOMINANCE OF THE NAGELIAN MODEL – A BRIEF HISTORY

The question about the reduction of the biological realm to, for instance, the physical realm is an old one. Reduction was an implicit topic of the mechanistic philosophy in the sixteenth and seventeenth century and it was controversially disputed in the debate about vitalism in the nineteenth and early twentieth century. In more recent years, when philosophy of biology emerged as a separate discipline in the 1960s/1970s the question whether biological theories can be reduced to molecular and in the end to physical theories was among the first issues disputed. Reductionism in biology became a central topic due to the impressive growth and development of molecular biology.

1 Cf. Ernest Nagel, *The Structure of Science. Problems in the Logic of Scientific Explanation*. London: Routledge 1961.

D. Dieks et al. (eds.), *Probabilities, Laws, and Structures*, The Philosophy of Science in a European Perspective 3, DOI 10.1007/978-94-007-3030-4_18,

Of particular interest was the question of whether classical genetics can be reduced to molecular biology. This special case was seen as a test case for the reduction of biology to physics in general.[2]

A few years before the debate about reduction in biology emerged, Nagel had published his *The Structure of Science*, in which he developed his *formal model of theory reduction*. In the spirit of logical empiricism, Nagel characterizes reduction as a deductive relation that holds between scientific theories, which he takes to be sets of law statements. In line with the deductive-nomological (D-N) model of explanation,[3] Nagel conceived reduction as a special case of explanation. For reduction to occur two conditions must be satisfied: the reduced theory has to be derived from the reducing theory ("*condition of derivability*"[4]). This presupposes that the reduced and the reducing theory either contain the same terms (in case of homogenous reduction) or that the former can be connected to the latter (in cases of heterogeneous reduction) via so called 'bridge laws' or, more neutrally, 'correspondence statements' ("*condition of connectability*"[5]). It should be acknowledged that Nagel contributed much more to the debate about reduction than this. For instance, he also proposed several non-formal conditions for distinguishing trivial from non-trivial cases of reduction,[6] discussed the issues of emergence[7] and "mechanistic explanation" in biology,[8] and identified different reasons why the whole can be more than the sum of its parts.[9] Nonetheless, the subsequent debate about Nagel's account focused on the *formal* conditions he identifies in his chapter on *theory* reduction. Although Nagel developed his formal model solely on basis of an example from physics (i.e. the reduction of thermodynamics to statistical mechanics), the early philosophers of biology considered it to be an adequate understanding of what epistemic reduction[10] in the sciences in general is and, thus, tried to apply it to biology.

2 Cf. Philip Kitcher, "1953 and All That: A Tale of Two Sciences", in: *Philosophical Review* 93, 1984, pp. 335–373 and Alexander Rosenberg, *The Structure of Biological Science*. Cambridge: Cambridge University Press 1985.

3 Cf. Carl Hempel, Paul Oppenheim, "Studies in the Logic of Explanation", in: *Philosophy of Science* 15, 2, 1948, pp. 135–175.

4 Nagel, *The Structure of Science*, p. 354.

5 *Ibid.*, p. 354.

6 Cf. *ibid.*, pp. 358–366.

7 Cf. *ibid.*, pp. 366–380.

8 Cf. *ibid.*, pp. 398–446.

9 Cf. *ibid.*, pp. 380–397.

10 With '*epistemic reduction*' I refer to the reduction of one body of knowledge (or parts of it like theories, explanations, methods, etc.) of a certain scientific discipline, e.g. biology, to another body of knowledge (or parts of it) of a different scientific discipline, e.g. physics. Epistemic reduction should be clearly distinguished from *ontological reduction*, which is the reduction of ontological entities of one kind (like objects, properties, facts, etc.), e.g. biological token objects, to ontological entities of another kind, e.g. physical token objects. In short, ontological reduction is a relation between things

It quickly became clear that Nagel's account not only had to face many general problems,[11] but that biology provides special obstacles for Nagelian reduction as well. In short: the objections were that neither the bridge laws that are needed to connect the terms of biological and physical theories nor the laws that constitute the units to be reduced, i.e. theories, are available in biology.[12] First, because evolution by natural selection is blind to structural differences with similar functions, most existing biological types of entities are *multiply realized* on the physical level.[13] For instance, the wings of different species of birds (let alone those of mammals and insects) vary strongly with respect to their structure and material composition although almost all of them share the same function, i.e., they enable their bearers to fly. The multiple realization of biological types makes it very difficult to establish those connections between the terms of biological theories (e.g. classical genetics) and physical or molecular theories (e.g. molecular biology) that are needed for reduction in the Nagelian sense. Second, another obstacle for a neat application of Nagel's model to biology is his assumption that theories are sets of law statements. The generalizations that can be found in biology (e.g. Mendel's laws of segregation and independent assortment) seem to be far away from describing laws of nature in the classical, strict sense. They typically have

in the world and epistemic reduction is a relation between parts of our knowledge about these things in the world. Nagelian theory reduction is a special case of epistemic reduction (other cases are explanatory and methodological reduction) because according to Nagel the relation of reduction holds between representational entities, i.e. theories. This is compatible with the claim that Nagel's regards bridge laws as stating identities or relations among extensions, i.e. as ontological links (although this is by no means clear, cf. for example Peter Fazekas, "Reconsidering the Role of Bridge Laws in Inter-Theoretical Reductions", in: *Erkenntnis* 71, 2009, pp. 303–322). Even if bridge laws are interpreted as stating ontological links, they are still linguistic entities (that represent relations that exist in the world) and not the relations in the world themselves.

11 For instance, Frederick Suppe, Ken Waters and others criticized the reliance of Nagel's account on a *syntactic view of theories* (cf. Frederick Suppe, *The Structure of Scientific Theories*. 2nd ed. Urbana: University of Illinois Press 1977 and Kenneth Waters, "Why the Antireductionist Consensus Won't Survive the Case of Classical Mendelian Genetics", in: *PSA 1990*, 1, 1990, pp. 125–139). Paul Feyerabend attacked Nagel's model by claiming the *incommensurability* of the meaning of the theoretical terms of the reduced and reducing theory (cf. Paul Feyerabend, "Explanation, Reduction and Empiricism", in: Herbert Feigl and Grover Maxwell (Eds.), *Scientific Explanation, Space, and Time*, Minneapolis: University of Minnesota Press 1962, pp. 28–97). Finally, Schaffner pointed out that in most cases of theory reduction the reduced theories first need to be *corrected* before they can be derived from the reducing theory (cf. Kenneth Schaffner, "Approaches to Reduction", in: *Philosophy of Science* 34, 1967, pp. 137–147 and "The Watson-Crick Model and Reductionism", in: *British Journal for the Philosophy of Science* 20, 1969, pp. 325–348).

12 Cf. for example Kitcher, *loc. cit.*

13 For a detailed elaboration of this point see, for instance, Alexander Rosenberg, "How Is Biological Explanation Possible?", in: *British Journal for Philosophy of Science* 52, 2001, pp. 735–760.

exceptions, are restricted in scope, and it can be argued that they are historically contingent.[14] This led many philosophers of biology to the conclusion: no laws in biology, hence, no cases of reduction in biology. The result was the formulation of the "*antireductionist consensus*".[15] About 20 years after the reductionism debate in the philosophy of biology had emerged it seemed as if everybody had become an antireductionist.[16] Even philosophers with strong reductionistic intuitions like Alexander Rosenberg gave up the hope that biology can be reduced to physics.[17]

It is important to note that during these 20 years and up to the 1990s the majority of philosophers took the obstacles with applying Nagel's model to biology to reveal the *non-existence* of reduction in this field and to support the *incorrectness* of reductionism in biology. Most of them did not choose the alternative option to question that Nagel's account is, in principle, the adequate way of thinking about reduction.[18] It was common practice to disagree about the details of the Nagelian model of theory reduction and to call for revisions. Many philosophers, most notably Kenneth Schaffner, tried to overcome the problems of Nagel's account by developing it further.[19] However, at that time hardly anybody questioned Nagel's

---

14 Cf. John Beatty, "The Evolutionary Contingency Thesis", in: Gereon Wolters, James Lennox (Eds.), *Concepts, Theories, and Rationality in the Biological Sciences*. Pittsburgh: University of Pittsburgh Press 1995, pp. 45–81.

15 Waters, "Why the Antireductionist Consensus Won't Survive the Case of Classical Mendelian Genetics", *loc. cit.*, p. 125. It is important to note that, contrary to the situation in the philosophy of mind, the reductionism debate in the philosophy of biology is a dispute about the frequency or possibility of *epistemic* reduction and not of *ontological* reduction. Ontological reductionism, at least in its weak version of a token-token physicalism, is the (often implicit) consensus in the philosophy of biology. However, this does not mean that it is impossible or fruitless to analyze ontological reduction or to dispute about ontological reductionism in biology. The epistemic questions are just taken to be more controversial than the ontological ones.

16 Notable exceptions are Ruse (Michael Ruse, "Reduction in Genetics", *PSA 1974*, 1976, pp. 633–651.) and Schaffner (cf. Kenneth Schaffner, *The Watson-Crick Model and Reductionism* and "Reductionism in Biology: Prospects and Problems", in: *PSA* 1974, 1976, pp. 613–632).

17 According to Rosenberg's view in the 1990s the impossibility of reductionism in biology inevitably leads to an instrumentalist interpretation of biological theorizing ("If reductionism is wrong, instrumentalism is right." Alexander Rosenberg, *Instrumental Biology or the Disunity of Science*. Chicago: University of Chicago Press 1994, p. 38) and to the abandonment of the unity of science above the level of physics. In the 2000s Rosenberg gave up this position again and became one of the few defenders of (epistemic) reductionism in biology.

18 Among the few exceptions were Wimsatt (cf. William Wimsatt, "Reductive Explanation: A Functional Account", in: *PSA 1974*, 1976, pp. 671–710) and Hull (David Hull, *Philosophy of Biological Science*. New Jersey: Prentice-Hall Inc. 1974).

19 Schaffner calls his revised version of Nagel's account 'general reduction-replacement model'. For a summary about how Schaffner supposes to cope with the problems of the Nagelian model see Kenneth Schaffner, *Discovery and Explanation in Biology and Medicine*. Chicago/London: University of Chicago Press 1993, chapter 9.

general way of thinking about reduction. In other words, most philosophers accepted the following two theses:

1. The adequate units of the relation of reduction are *theories* (whether they are conceived as sets of law statements or not, whether the theories need to be corrected before being reduced or not, and whether one adopts a syntactic view of theories or not).[20]
2. The relation of reduction is a relation of logical *derivation* (whether this means exact derivability or something weaker and whether the bridge laws that are necessary for the derivation are conceived as identity statements or not).[21]

The widespread acceptance of this general way of thinking about reduction in terms of theories and logical relations prevailed in the debate for a surprisingly long time. This is especially true for discussions that are not centered on but rather pick up the issue of reduction.[22] The most instructive example is Rosenberg, who nowadays explicitly argues for the need to abandon the Nagelian understanding of reduction[23] but, in the 1980s and 1990s, claimed that it "sounds suspicious to

---

20 Although some philosophers questioned the syntactic view of theories and called for a less formal alternative, up to the late 1990s almost nobody questioned the general thesis that *theories* are the adequate units of reduction. For instance, in his influential paper from 1990 Waters objected to Nagel's model of theory reduction but merely demanded the "reformulation of theoretical reduction" (Waters, *Why the Antireductionist Consensus Won't Survive the Case of Classical Mendelian Genetics*, p. 136). Nowadays Waters explicitly criticizes the concepts of "theoretical reduction" and "layer-cake antireduction" and the exclusive focus on *theoretical* developments in biology they imply (cf. Kenneth Waters, "Beyond Theoretical Reduction and Layer-Cake Antireduction: How DNA Retooled Genetics and Transformed Biological Practice", in: Michael Ruse (Ed.), *The Oxford Handbook of the Philosophy of Biology*. Oxford: Oxford University Press 2008, pp. 238–262).

21 At this point I want to emphasize that there, in fact, were a few philosophers of biology (most notably, David Hull, "Informal Aspects of Theory Reduction", in: *PSA 1974*, 1976, pp. 653–670 and Wimsatt, "Reductive Explanation: A Functional Account", *loc. cit.*) who early objected to this second thesis, i.e. Nagel's and Schaffner's presupposition that a model of theory reduction should focus on formal issues and reconstruct reduction as a relation of logical derivation.

22 One reason for the long survival of the Nagel-Schaffner model of theory reduction is that there was simply no popular alternative available, which could have replaced the thinking about reduction in terms of theories and logical relations. I think Wimsatt's functional analysis of reduction (cf. Wimsatt, "Reductive Explanation: A Functional Account", *loc. cit.*), which focuses on reductive explanations and mechanisms, had the potential to replace it but his account was, perhaps, not catchy and comprehensible enough.

23 Cf. Alexander Rosenberg, *Darwinian Reductionism. Or, How to Stop Worrying and Love Molecular Biology*. Cambridge: University of Chicago Press 2006, p. 40.

change the standards of reduction"[24] and conceived the alternative option of abandoning reductionism altogether as the "more reasonable"[25] option.

During the last 15 years more and more philosophers rejected the Nagel-Schaffner account and developed alternative ways of thinking about epistemic reduction in biology.[26] However, many opponents of the Nagelian approach do not put effort in elaborating an alternative view of epistemic reduction but rather argue for the *abandonment* of the focus on reduction altogether.[27] Despite these new developments, there clearly are philosophers, who adhere to the concept of reduction because they think it is an important conceptual tool for capturing many aspects of biological practice (or who think it is a philosophically interesting or fruitful concept one should not dismiss to easily). And many of these philosophers are far away from having given up thinking about reduction in terms of theories and logical derivation.[28]

---

24 Rosenberg, "The Structure of Biological Science", *loc. cit.*, p. 110.

25 Rosenberg, "Instrumental Biology or the Disunity of Science", *loc. cit.*, p. 22.

26 See e.g. Sahotra Sarkar, "Models of Reduction and Categories of Reductionism", in: *Synthese* 91, 1992, pp. 167–194; *Genetics and Reductionism*. Cambridge: Cambridge University Press 1998; *Molecular Models of Life. Philosophical Papers on Molecular Biology*. Cambridge: MIT Press 2005; William Wimsatt, *Reductive Explanation: A Functional Account*; *Re-Engineering Philosophy for Limited Beings: Piecewise Approximations to Reality*. Cambridge: Harvard University Press 2007; Rosenberg, *Darwinian Reductionism*; William Bechtel, *Discovering Cell Mechanisms. The Creation of Modern Cell Biology*. Cambridge: Cambridge University Press 2006; and *Mental Mechanisms. Philosophical Perspectives on Cognitive Neuroscience*. New York/London: Taylor and Francis Group 2008.

27 See e.g. Carl Craver, "Beyond Reduction: Mechanisms, Multifield Integration and the Unity of Neuroscience", in: *Studies in the History and Philosophy of Biological and Biomedical Sciences* 36, 2005: pp. 373–395; Carl Craver, *Explaining the Brain. Mechanisms and the Mosaic Unity of Neuroscience*. Oxford: Clarendon Press 2007; Sandra Mitchell, *Biological Complexity and Integrative Pluralism*. New York: Cambridge University Press 2003; Sandra Mitchell, *Unsimple Truths. Science, Complexity, and Policy*. Chicago/London: University of Chicago Press 2009; Sandra Mitchell and Michael Dietrich, "Integration without Unification: An Argument for Pluralism in the Biological Sciences", in: *American Naturalist* 168, 2006, pp. 73–79; Lindley Darden, "Relations Among Fields: Mendelian, Cytological and Molecular Mechanisms", in: *Studies in History and Philosophy of Biological and Biomedical Sciences* 36, 2005, pp. 357–371; Lindley Darden and Nancy Maull, "Interfield Theories", in: *Philosophy of Science* 44, 1977, pp. 43–64.

28 See e.g. Kenneth Schaffner, *Discovery and Explanation in Biology and Medicine*; "Reduction: The Cheshire Cat Problem and a Return to Roots", in: *Synthese* 151, 2006, pp. 377–402; John Bickle, *Psychoneural Reduction: The New Wave*. Cambridge: MIT Press 1998; John Bickle, *Philosophy and Neuroscience: A Ruthlessly Reductive Account*, Dordrecht: Kluwer Academic Publishers 2003; John Bickle, "Reducing Mind to Molecular Pathways: Explicating the Reductionism Implicit in Current Cellular and Molecular Neuroscience", in: *Synthese* 151, 2006, pp. 411–434; Ulrich Krohs, *Eine Theorie biologischer Theorien. Status und Gehalt von Funktionsaussagen und*

My main thesis in this paper is that it is finally time to leave Nagel's general way of thinking about reduction behind. However, I think this should not lead us to abandon the idea of reduction altogether. Rather, we should accompany authors like Sahotra Sarkar and William Wimsatt in their search for an adequate understanding of what epistemic reduction in biology *really* is. Thinking about reduction in terms of theories and the logical relation between statements has dominated the debate for too long. Instead of imposing an ill-fitting model on biology, we should develop a new account of epistemic reduction that "makes contact with real biology" (to use Rosenberg's words) and captures the *diversity* of reductive reasoning strategies present in current biological research practice.[29] Such an improved understanding will also disclose the importance as well as the limits of epistemic reduction in biology.

18.3 The Inadequacy of the Nagelian Account

In this section I do not want to echo the old criticism that has been put forward against Nagel's formal model of theory reduction in the early reductionism debate to reveal its general problems and its inapplicability to biology. This is the reason why my critique is focused on Nagel's *general way of thinking* about epistemic reduction (see Sect. 18.2) and abstract away from those details of Nagel's model that have turned out to be highly problematic. First, I grant that one could give up the concept of a *strict law* and adopt a more moderate account of what a scientific law is. For instance, one could allow laws to be ceteris paribus laws[30] or adopt the concept of a "pragmatic law".[31] This would allow one to claim that there exist genuine biological laws and, thus, that the relata for Nagelian reduction, namely

informationstheoretischen Modellen. Berlin: Springer 2004; Colin Klein, "Reduction Without Reductionism: A Defence of Nagel on Connectability", in: *The Philosophical Quarterly* 59, 2009, pp. 39–53; Foad Dizadji-Bahmani, Roman Frigg, Stephan Hartmann, "Who Is Afraid of Nagelian Reduction?", in: *Erkenntnis* 73, 2010, pp. 393–412; etc.

29 To be clear: This search for a new account of epistemic reduction cannot be the step of a desperate reductionist who seeks an understanding of reduction that allows him, finally, to defend reductionism in biology. One can speculate that this is exactly the way Rosenberg gets to his understanding of explanatory reduction, namely, that it allows him to defend *Darwinian Reductionism* (which is a specific version of explanatory reductionism). I think it is important to resist this temptation. An account of epistemic reduction should not reflect the wishes or ideals of philosophers. Rather, its search should be motivated by the aim to *understand* and *reconstruct* the various reductive reasoning practices characteristic for contemporary biological research.

30 Cf. for instance Marc Lange, *Natural Laws in Scientific Practice*. Oxford: Oxford University Press 2000.

31 Sandra Mitchell, "Pragmatic Laws", in: *Philosophy of Science* 64, 1997, pp. 468–479 and "Biological Complexity and Integrative Pluralism", *loc. cit.*

theories as sets of law statements, are available. Second, I admit that one could simply abandon Nagel's claim that theories must consist of law statements and allow each kind of general statements formulated in a formal language, i.e., first order logic. Third, one could go even further and abandon the "*syntactic view*" or "received view"[32] of theories and with it the requirement that theories must be formulated as statements in first-order logic. Alternatively, one could argue for a "semantic view"[33] of theories, according to which theories are families of models formalized in set theory. However, on closer inspection (see Sect. 18.3.2), this step turns out to be highly problematic as it leads the Nagelian model too far away from its core ideas. Forth, I allow the changes of the Nagelian model Schaffner made in his "general reduction-replacement model".[34] In line with Schaffner one could claim that reduction (in the revised Nagelian sense) also captures cases in which not the original theories themselves, but rather *corrected versions* of them are derived from each other. Finally, I grant that one can abandon the strong claim that bridge laws must be factual statements that express *identity relations* (though it is not at all clear whether Nagel holds this strong view[35]). Even if they are taken to be factual claims (and not e.g. mere stipulations, i.e. conventions) it is left open which ontological relation they express (for instance, mere correlations, necessary nomic connections, constitutional relations, identity relations, etc.[36]).

If a defender of the Nagelian account relinquishes all these problematic assumptions, what is left over is Nagel's general way of thinking about epistemic reduction, which can be characterized by the two theses introduced in the last section: first, the adequate units of the relation of reduction are *theories* and, second, the relation of reduction is a relation of logical *derivation*. My claim is that even this very moderate, thin version of the Nagelian account of reduction is deeply flawed. In the next sections I will reveal several reasons why it is inadequate to think about epistemic reduction in biology in terms of theories and the logical relations between them. The general line of my argument will be that a formal model of theory reduction does neither capture the *most important cases* of epistemic reduction in biology nor does it account for the *diversity* of reductive reasoning strategies present in current biological research practice. This leaves us with an account of epistemic reduction that reflects the ideals of some philosophers but

32 Frederick Suppe, "Understanding Scientific Theories: An Assessment of Developments, 1969–1998", in: *Philosophy of Science* 67, 2000, pp. 102–115, p. 102. See also Paul Thompson, *The Structure of Biological Theories*. Albany: State University of New York Press 1989.

33 Frederick Suppe, "The Structure of Scientific Theories", *loc. cit.* and *The Semantic Conception of Theories and Scientific Realism*. University of Illinois Press: Chicago 1989.

34 Cf. Schaffner, "Approaches to Reduction", *loc. cit.*; "The Watson-Crick Model and Reductionism", *loc. cit.*; "Reductionism in Biology: Prospects and Problems", *loc. cit.*; and "Discovery and Explanation in Biology and Medicine", *loc. cit.*

35 Cf. Nagel, "The Structure of Science", *loc. cit.*, pp. 354–358.

36 See also Dizadji-Bahmani et al., "Who Is Afraid of Nagelian Reduction?", *loc. cit.*

that is unconnected with real biological practice because it has no or at least a very restricted range of application. However, before I can move on it is necessary to make a few methodological clarifications.

18.3.1 How to Develop an Account of Epistemic Reduction

Why care about biological research practice in the first place? Why not stick to Nagel's formal model of theory reduction and view it as an *ideal* that does not need to be realized in biological practice? Schaffner, for instance, chooses this route and admits that theory reduction is "peripheral" to biological practice and should be regarded as a mere "regulative ideal".[37] I think that these two options – on the one hand, developing an account of epistemic reduction that captures actual biological practice and, on the other hand, analyzing epistemic reduction without caring about what epistemic reduction *in practice* is – are best seen as completely *different projects*. Those philosophers who want to understand what biologists actually do and how biological research practice really works will not be satisfied with a philosophical account that merely reflects the wishes or ideals of philosophers but does not capture what is really going on in biology itself. They will judge accounts of the second kind as *descriptively inadequate* and, probably, not continue thinking about them at all. Philosophers who pursue a project of the second type do not share the goal of capturing and understanding actual biological research practice but rather endorse other values of a philosophical account (for example, the fact that it captures certain philosophical or common sense intuitions, its suitability for a broader philosophical, for instance metaphysical, theory, its explanatory force, etc.). In the radical version of this kind of project descriptive adequacy is simply abandoned as a criterion of adequacy. The focus lies exclusively on analyzing reduction *in principle*. What characterizes reduction *in practice* is ignored.

However, looking at how philosophy of science is presently carried out reveals two points: first, although these two kinds of projects can be distinguished from each other they are, in fact, two end points of a *continuum* and, second, projects of the second type (at least in its radical version) are *rare*. Consider the first point. Since projects of the first type are philosophical projects they are more than *mere descriptions* of scientific practice. Rather they are actively pursued reconstructions that involve normative decisions of various kinds (in a broad sense, e.g. the choice of paradigmatic cases) and that can also result in normative claims about how science ideally should be carried out. On the other hand, only few philosophers, who pursue a project of the second type make claims about how science ideally should work without even having a quick glance at how science actually works. Thus, most projects are, in fact, located somewhere in the middle ground between

---

37 Schaffner, "Discovery and Explanation in Biology and Medicine", *loc. cit.*, pp. 508–513. In recent years even Schaffner has disavowed from his peripherality thesis and adopted a less spectacular view about reduction (cf. Schaffner, Reduction: The Cheshire Cat Problem and a Return to Roots).

the two extremes of the continuum. This leads us to the second point. Especially in philosophy of biology, most projects belong to (a moderate version of) the first type. Philosophers want to understand, for example, how the success and failure of explanation in different biological disciplines is in fact evaluated, why molecular research in various areas is as important as it is, which different roles models play in actual biological research practice, and how biologists *de facto* estimate the scope of biological generalizations. However, there are philosophers of science who are not primarily interested in capturing and understanding actual scientific practice. Their goal is to develop a view about science or about a specific element of science (like explanation, causation, confirmation, law, etc.) that is adequate because it captures certain philosophical intuitions, that is in line with a certain general philosophical picture or that has special explanatory force. But even the projects of this second kind are rarely pursued *without* relying (at least partially) on a view about how science really works and why it is actually successful. This is not surprising since it seems weird to make claims about how science ideally should work or certain elements of scientific practice like explanation and reduction should be understood without taking into account how science actually works and what scientific explanations and reductions *in fact* are. However, here I do not want to argue for this claim at length. Rather, I want to be explicit about where I stand and on basis of which criteria of adequacy I attack Nagel's general way of thinking about epistemic reduction.

My paper is concerned with the question whether Nagel's formal model of theory reduction is convincing if it is understood as a *project of the first kind*. Thus, the question is whether thinking about epistemic reduction in terms of theories and the logical relations between them "saves the phenomena (about the biological sciences)" (to borrow Bas van Fraassen's way of talking) and helps to understand what is going on in actual biological research practice. According to my view, there exist two *criteria of adequacy* on whose basis the quality of any philosophical account of epistemic reduction (pursued as a project of the first type) is judged:

A model of epistemic reduction should

1. capture and help to understand the cases of epistemic reduction that *actually occur* in current biological research practice, rather than focusing on epistemic reduction that can only be achieved in principle. In addition, it should
2. account for the *diversity* and *complexity* of the cases of epistemic reduction that are present in contemporary biology.

In the following sections I will argue why Nagel's general way of thinking about epistemic reduction in terms of theories and logical relations fails to satisfy these two criteria and, thus, should be assessed as inadequate to biology.

18.3.2 Theories as Relata of Reduction

One kind of objection that has been frequently put forward against Nagel's approach concerns the *non-existence* or *misrepresentation* of the relata of reduction. Nagel argues that the relation of reduction holds between theories, which he conceives as systems of statements, containing law statements and being formalized in first order logic.[38] In the subsequent discussion about the structure of scientific theories this view is referred to as the syntactic conception of theories. Nagel's account of what the relata of reduction are encounters several objections: first, it can be argued that the relata, i.e., theories containing law statements, do not exist since there are no strict laws in biology. Second, one can claim that Nagel misrepresents the relata of reduction because scientific theories in general and biological theories in particular do not satisfy the demands of the syntactic view. Rather, theories (in biology) are to be understood as families or sets of models meeting specific set-theoretic conditions.[39]

As stated at the beginning of Sect. 18.3, I am willing to allow several steps a proponent of Nagel's model could take in order to meet these objections and defend a modified version of the Nagelian account – at least if these modifications are carried out in a convincing manner. To counter the first objection, one could either argue for a more moderate conception of a 'law', according to which there exist genuine laws also in biology[40], or one could abandon Nagel's requirement that theories must contain law statements. However, it should be noted that the second option is highly problematic since Nagel conceives reduction to be a special case of explanation and explanation, according to the D-N model Nagel adopts, presupposes the availability of lawlike generalizations. Thus, it seems as if only the first option is accomplishable.

With respect to the second objection, a defender of Nagel's model of reduction could give up the syntactic view of scientific theories and adopt the alternative, semantic conception. The possibility of making this move is one reason why Foad Dizadji-Bahmani, Roman Frigg, and Stephan Hartmann want to convince us not to be afraid of Nagelian reduction anymore. The syntactic view is "unnecessary" to get Nagel's account "off the ground". We can replace first order logic "with any formal system that is strong enough to do what we need it to do".[41] They seem to

38 Cf. Ronald Giere, *Explaining Science: A Cognitive Approach*. Chicago: University of Chicago Press 1988. For details about Nagel's view of theories compare Chapter 5 "Experimental Laws and Theories" and 6 "The Cognitive Status of Theories" of his "The Structure of Science", *loc. cit.*

39 Many philosophers of biology have embraced this semantic view of theories, especially with respect to evolutionary biology. See John Beatty, "What's Wrong With the Received View of Evolutionary Theory?", in: *PSA* 1980, 2, 1981, pp. 397–426; Elisabeth Lloyd, *The Structure and Confirmation of Evolutionary Theory*. New York: Greenwood Press 1988; Thompson, *The Structure of Biological Theories*; and Peter Sloep, Wim Van der Steen, "The Nature of Evolutionary Theory: The Semantic Challenge", in: *Biology and Philosophy* 2, 1987, pp. 1–15; as well as the different responses to this paper in Biology and Philosophy Vol. 2, No. 1.

40 Cf. for instance Mitchell, "Pragmatic Laws and Biological Complexity and Integrative Pluralism", *loc. cit.*

41 Dizadji-Bahmani et al., "Who Is Afraid of Nagelian Reduction?", *loc. cit.*, p. 403.

have strong company on their side. John Bickle clings to the view that reduction is a relation between theories but argues for a semantic conception of theories.[42] Based on Clifford Hooker's approach to reduction[43] Bickle formulates his "new-wave account of intertheoretic reduction"[44] according to which the reduction of one theory T_R to another T_B requires the construction of an "image I_B of the set-theoretic structure of models of the reduced theory T_R within the set comprising reducing theory T_B".[45] The details of Bickle's "semantic" account of intertheoretic reduction are complex. However, what matters for the purpose of my paper is that Bickle explicitly contrasts his approach with the Nagelian idea of "characterizing intertheoretic reduction in terms of syntactic derivations".[46] But if theories are understood as sets of models satisfying certain set-theoretic conditions and no longer as sets of sentences in an axiomatized system of first order logic it is no longer clear what Nagel's condition of derivability amounts to. It even more seems as if the proponents of the semantic view must abandon the claim that it is logical derivation that connects the reduced to the reducing theory and are in need of a different specification of the reductive relation between theories (for instance, according to Bickle, in terms of 'analogy' or 'isomorphism' between the image I_B and the reduced theory T_R).[47] The alternative would be to adopt a very broad (and thus vague) notion of 'derivation' that also captures the relation between sets of models. But such a vague concept of derivation runs the risk that too much can be derived from something else and, hence, does not appear to be convincing.

The preceeding discussion reveals that the combination of an account of intertheoretic reduction with a semantic conception of theories takes us too far away from the core ideas of the Nagelian understanding of epistemic reduction (in particular, from the second thesis of Nagel's general way of thinking about reduction, i.e. that the relation of reduction is logical derivation).[48] This does not imply that the combination is untenable, but only that the resulting account is not "Nagelian" anymore. Hence, switching to the semantic view of theories in order to meet the

42 Cf. Bickle, "Psychoneural Reduction: The New Wave and Philosophy and Neuroscience: A Ruthlessly Reductive Account", *loc. cit.*

43 Cf. Clifford Hooker, "Towards a General Theory of Reduction. Part I: Historical and Scientific Setting. Part II: Identity in Reduction. Part III: Cross-Categorial Reduction", in: *Dialogue* 20, 1981, pp. 38–59, 201–236, 496–529.

44 Bickle, "Psychoneural Reduction: The New Wave", *loc. cit.*, p. 23.

45 Bickle, "Philosophy and Neuroscience: A Ruthlessly Reductive Account", *loc. cit.*, p. 27.

46 *Ibid.*

47 Cf. Bickle, "Psychoneural Reduction: The New Wave", *loc. cit.*

48 This claim is further confirmed by the fact that even explicit opponents of the Nagelian model of epistemic reduction adopt a semantic view of theories (see Carl Craver, "Structures of Scientific Theories", in: Peter Machamer, Michael Silberstein (Eds.), *The Blackwell Guide to the Philosophy of Science*. Malden/Oxford: Blackwell Publishers 2002, pp. 55–79; "Beyond Reduction: Mechanisms, Multifield Integration and the Unity of Neuroscience", *loc. cit.*; and "Explaining the Brain", *loc. cit.*).

second objection (i.e. the misrepresentation of the relata of reduction) is not an option for a proponent of Nagel's model of theory reduction.

Finally and most importantly, two further objections against Nagel's assumption that theories are the relata of reduction can be raised with respect to biology: first, biological research practice shows that, in general, theories are *not the only* and perhaps not the most *important* element of science. Second, biological practice reveals that for reduction, in particular, theories are *only peripherally important* since the most crucial and frequently occurring cases of epistemic reduction, i.e., reductive explanations, rarely involve fully explicated theories.

How could an opponent of the Nagelian account react to the first objection? As I have just argued, he must stick to the syntactic view of theories and, thus, is exposed to all the criticism that has been put forward against this conception. The overall tenor is: because the syntactic view focuses on theories as a whole and on their inferential structure it fails to capture what biological theories in fact are ("theories in the wild"[49]). For instance, it does not account for the diversity of representations of theories biologists actually use and which are neither restricted to first order logical predicates nor to linguistic representations at all (see e.g. Laura Perini's work on the importance of diagrams in biology[50]). Second, the syntactic conception focuses on full-established, static theories (context of justification) and lacks an account of the dynamics of biological theories (context of discovery), that is, how they are developed over time and which roles they play during that time.[51] Third, the syntactic view overestimates the role of theories by ignoring the important roles other epistemic units (such as models, descriptions of mechanisms, fragments of theories, etc.) play in the context of explanation, prediction, discovery, and manipulation in biology. The motivations for the development of the alternative, semantic conception of theories were to overcome these problems and to allow for the importance of models in scientific practice. I do not want to discuss the various versions of the semantic conception and the objections that can be put forward against them here. What is important for the topic of this paper is that even if a defender of the Nagelian model of reduction could adopt a semantic conception and could adjusts the notion of theories in a way that it is closer to what is going on in real biological research practice he still would not meet the first objection. Granted, theories (as sets of models) do occur in biological practice. However, theories are not the *only* and perhaps not the *most* important epistemic units in biology. To begin with, often not fully explicated theories as a whole, but rather *fragments* of theories, *individual models* and not

49 Craver, "Structures of Scientific Theories", *loc. cit.*, p. 65.

50 Cf. Laura Perini, "Explanation in Two Dimensions: Diagrams and Biological Explanation", in: *Biology and Philosophy* 20, 2005, pp. 257–269 and "Diagrams in Biology", in: *Knowledge Engineering Review*, forthcoming.

51 Cf. Lindley Darden, *Theory Change in Science: Strategies from Mendelian Genetics*. New York: Oxford University Press 1991 and Lloyd, "The Structure and Confirmation of Evolutionary Theory", *loc. cit.*

entire sets of models, and descriptions of particular *mechanisms* (if mechanistic models are understood as being parts of theories)[52] play important roles in explanation, prediction, discovery, and manipulation. Second, in biological practice there seem to exist many epistemic units that are relatively *independent* from theories and that, nevertheless, are crucial for the successful functioning of the biological sciences, for example, explanatory and investigative strategies,[53] semi-empirical rules,[54] mechanistic models,[55] to list only a few. Finally and as a further substantiation of the previous thesis, some authors have argued that scientific models *in general* are better conceived as being independent from theories, rather than being constitutive of them.[56]

The peripherality of theories to biological practice is even more apparent in the context of reduction. With respect to *diachronic* (or intralevel) reduction, Ken Waters has argued that a focus on theoretical developments fails to capture what is important for the successful transformation of biological disciplines, e.g., classical genetics.[57] He suggests that philosophers should direct their attention to the changes in the investigative practices of genetics instead. I argue that we should abandon the Nagelian focus on theories as the only or most important units of epistemic reduction also with respect to *synchronic* (or interlevel) reduction. Rather, we should concentrate our analysis on the most crucial and frequently occurring kind of epistemic reduction in biological research practice, namely *reductive explanations*. Part-whole explanations and mechanistic explanations, which are the paradigmatic cases of reductive explanations, have been strongly connected with reduction for a long time – not only by philosophers but also by biologists themselves.[58] However, individual explanations and the conditions that determine their reductive character have almost been neglected as a fruitful and *independent* subject of analysis so far. Granted, since Nagel took intertheoretic reduction to be a relation of explanation the debate about reduction has also been concerned with the issue of explanation. But discussions about explanation, which remain within the Nagelian framework, concentrate on the explanatory scope of theories

52 Cf. Craver, "Structures of Scientific Theories", *loc. cit.*

53 Cf. Waters, "Beyond Theoretical Reduction and Layer-Cake Antireduction: How DNA Retooled Genetics and Transformed Biological Practice", *loc. cit.*

54 Cf. Sarkar, "Models of Reduction and Categories of Reductionism", *loc. cit.*

55 Cf. Wimsatt, "Reductive Explanation: A Functional Account", *loc. cit.*, and Peter Machamer, Lindley Darden, Carl Craver, "Thinking about Mechanisms", in: *Philosophy of Science* 67, 2000, pp. 1–25.

56 See for instance Mary Morgan, Margaret Morrison, *Models as Mediators. Perspectives on Natural and Social Science*. Cambridge: Cambridge University Press 1999.

57 Cf. Waters, "Beyond Theoretical Reduction and Layer-Cake Antireduction: How DNA Retooled Genetics and Transformed Biological Practice", *loc. cit.*

58 To the philosophers belong, for instance, Wimsatt, "Reductive Explanation: A Functional Account", *loc. cit.*; Sarkar, "Models of Reduction and Categories of Reductionism", *loc. cit.*; and "Genetics and Reductionism", *loc. cit.*, and to the biologists Ernst Mayr, "The Limits of Reductionism", in: *Nature* 331, 1988, p. 475.

(e.g. on the question whether physical theories can be employed to explain certain biological theories or biological phenomena) or on reduction as a relation *between* explanations, i.e. between a higher-level explanation and a lower-level explanation of the same phenomenon.[59] Thus, they do *not* promote an understanding of what makes *individual* explanations *reductive*. Such an analysis would include the identification of the relata of reduction (roughly, explanandum and explanans) and the specification of the relation of reduction by analyzing the constraints on reductive explanations, that is, the various conditions on basis of which biologists evaluate the success and failure of the reductivity of explanations.[60] In Sect. 18.3.1 I claimed that an adequate account of epistemic reduction must capture and enlighten the cases of epistemic reduction that occur in *actual* biological research practice. According to this criterion of adequacy the fact that thinking about epistemic reduction in terms of theories and their logical relations does *not* yield an understanding of the reductive character of explanations is an important argument for the *inadequacy* of Nagel's general view of thinking about reduction. Thus, it seems fruitful to move beyond Nagelian reduction and shift the attention from theory reduction to reductive explanations.[61]

---

59 Cf. Rosenberg, "Darwinian Reductionism", *loc. cit.* Although Rosenberg explicitly abandons Nagel's model of theory reduction (*Ibid*., p. 40) his view of explanatory reductionism, nevertheless, remains closely connected to the Nagelian framework in a broad sense. For instance, he adheres to the view that laws are indispensable for explanation and his defense of explanatory reductionism is still centered on the question whether all biological phenomena can be explained with the resources of physical (or molecular) theory. See in particular Rosenberg "How Is Biological Explanation Possible?" and "Darwinian Reductionism", *loc. cit.*, ch. 4.

60 For an example of how such an analysis could look like see Marie I. Kaiser, "An Account of Explanatory Reduction in the Life Sciences", in: *History and Philosophy of the Life Sciences*, forthcoming, Sarkar, "Genetics and Reductionism", *loc. cit.*, and Andreas Hüttemann, Alan Love, "Aspects of Reductive Explanation in Biological Science: Intrinsicality, Fundamentality, and Temporality", in: *British Journal for Philosophy of Science*, forthcoming.

61 Although I am convinced that explanations are an especial fruitful subject of analysis I do not want to claim that giving an account of epistemic reduction by focusing on individual reductive explanations is the *only* possible way to analyze epistemic reduction in biology. Nor I want to argue that, on its own, it is *sufficient* to capture the diversity of reductive reasoning strategies present in current biology. Alternatively one could, for example, concentrate on methods (or investigative strategies) and try to specify what makes them reductive. This leaves room for the kind of *pluralism* Sarkar endorses: "There is no *a priori* reason to assume that all cases of reduction are so similar that they can all be captured by any single model of reduction." (Sarkar, "Models of Reduction and Categories of Reductionism", *loc. cit.*, p. 188). In this pluralistic picture also the Nagelian account of theory reduction could have its place – though it would be a *very small* place, as my discussion shows.

18.3.3 The Focus on Formal Issues

It is important to note that the criticism of Nagel's model of theory reduction (as well as the criticism of the syntactic view of theories and of the D-N model of explanation) is a part of a more general rejection of the logical empiricist's kind of philosophy, which focuses on *formal issues* (like the logical relations between sentences formalized in first order logic) and thereby ignores "substantive issues".[62] The attack against this kind of philosophy and the effort to replace it has a long history. However, as I pointed out in Sect. 18.2, in the context of reduction Nagel's formal model has shown a long persistency.

In this section I want to call attention to two issues: first, I argue that the logical empiricist's formal philosophy can only be rejected as a whole packet. Second, in addition to the criteria I have already identified (see Sect. 18.3.1) there is a third criterion of adequacy for an account of epistemic reduction, according to which a formal model of reduction like Nagel's comes away badly. Let us start with the first point. In his work on reductive explanations in genetics, Sarkar emphasizes that his analysis of what makes explanations reductive entails no commitment to a specific explication of what an explanation is (despite a few "basic assumptions"[63]). Rather, he tries to "keep the issues of reduction and explanation distinct" and identifies "*additional* criteria"[64] an explanation must satisfy in order to be a *reductive* explanation. I embrace Sarkar's goal not to conflate the question of what makes a representation (or a model) explanatory and, thus, distinguishes it from purely descriptive representations with the question of what makes an explanation reductive and distinguishes it from non-reductive explanation. I will come back to this issue when I present my third criterion of adequacy. What I think is important to note is that drawing this distinction between the issues of explanation and reduction and focusing exclusively on the latter does *not* guarantee that the provided account of epistemic reduction is *neutral* with respect to what the adequate model of explanation is. In fact, contrary to his own assertion, Sarkar's analysis cannot preserve the asserted neutrality. The reason is that Sarkar rejects Nagel's model of theory reduction because of its focus on "formal issues, [i.e.] the 'logical' form of reduction"[65] and wants to replace it with an analysis of "substantive issues" (i.e. what reductive explanations "assume about the world"), which he conceives to be "more interesting and important".[66] However, exactly this criticism seems to abolish the possibility of adopting a D-N (and I-S) model of explanation. At least, it seems to be very weird to reject Nagel's model because of its focus on formal issues, yet to adhere to the formal D-N model of

62 Sarkar, "Genetics and Reductionism", *loc. cit.*, p. 19.

63 *Ibid.*, p. 41.

64 *Ibid.*, p. 9.

65 *Ibid.*, p. 17.

66 *Ibid.*, p. 18f.

explanation that encounters very similar objections.[67] What has also become apparent in the discussion about the structure of scientific theories in Sect.18.3.2: if you reject Nagel's account of theory reduction because of its formal character and you want your whole philosophical position to be consistent you better get rid of the whole packet, including the D-N model of explanation and the syntactic view of theories.

Let us turn to the second issue. The discussion of Sarkar's approach revealed a *third criterion of adequacy* an account of epistemic reduction in biology must satisfy, namely to demarcate cases of epistemic reduction from cases where there is no reduction at all. With respect to reductive explanations (which a model of epistemic reduction must account for, see Sect. 18.3.2) this amounts to providing one or several *demarcation criteria* on basis of which reductive explanations clearly can be distinguished from explanations that are non-reductive. In sum:

A model of epistemic reduction should

3. *demarcate* reductive explanations from non-reductive explanations.

Nagel's formal model of theory reduction fails to meet this criterion since it equates explanation (of one theory by another) with reduction. As soon as a theory can be explained by and (according to the D-N model of explanation) thus be logically derived from another theory we have a case of theory reduction. What is important from Nagel's perspective is whether the two formal criteria, derivability and connectability, are satisfied or not.[68] But this does not endow us with resources to distinguish explanations of phenomena (types as well as tokens) that are *reductive* from those that are non-reductive. In order to draw this line of demarcation we need to refer to the relations that exist between the *things in the world* described in the explanandum and explanans in an explanation. For instance, we need to make claims of the kind that in many reductive explanations the entities referred to in the explanans are located on a *lower*, more *fundamental* level than (level fundamentality) or are *internal* to (internal fundamentality) the system whose behavior is to be explained.[69] Only thinking about epistemic reduction in a non-formal way directs our attention to these crucial substantive issues.

Finally, let me mention a related point. In so far as an analysis of the reductive character of biological explanations reveals that the reductivity of an explanation is not an "*all-or-nothing phenomenon*",[70] it succeeds much better than the Nagelian

67 This is why most of the early opponents of Nagel's model of theory reduction endorse a causal-mechanistic account of explanation (see e.g. Hull, "Philosophy of Biological Science", *loc. cit.*, and Wimsatt, "Reductive Explanation: A Functional Account, *loc. cit.*).

68 As I mentioned before, Nagel also proposed some non-formal conditions for theory reduction (cf. Nagel, "The Structure of Science", *loc. cit.*, pp. 358–366). However, these criteria help to distinguish trivial from non-trivial cases of theory reduction but they do not provide the demanded demarcation of reductive from non-reductive explanations.

69 Cf. Kaiser, "An Account of Explanatory Reduction in the Life Sciences", *loc. cit.*

70 Cf. Hüttemann, Love, "Aspects of Reductive Explanation in Biological Science: Intrinsicality, Fundamentality, and Temporality", *loc. cit.*

approach in capturing the diversity and complexity of epistemic reduction (second criterion of adequacy, see Sect. 18.3.1). According to Nagel, a theory can either be deductively derived from and thus be reduced to another theory or it cannot. In a specific case there are just two options: either reduction succeeds or it fails. Focusing on reductive explanation discloses that the situation in actual biological research practice is not as simple as suggested by Nagel's account. In fact, different respects in which an explanation can fail or succeed to be reductive need to be kept apart.[71] This important fact is obscured by Nagel's focus on theories and the logical relations between them.

18.4 Conclusion

Even if one grants that the proponents of the Nagelian model of theory reduction can handle several problems that have been raised in the past, Nagel's general way of thinking about epistemic reduction in terms of theories and focused on formal issues still remains inadequate with respect to what epistemic reduction in biology *really* is. In order to show this, I identified three criteria of adequacy and argued why Nagel's account fails to meet any of these criteria. First, it does not capture and enlighten those cases of epistemic reduction that are *most important* and *frequently occurring* in biological practice since it identifies relata, i.e., (fully-established) theories, that are not as important in biology as suggested, since it focuses on cases of epistemic reduction that are peripheral to biology, since it fails to account for the most crucial kind of epistemic reduction, i.e. reductive explanations, and since it focuses on formal issues and thereby ignores important substantive issues. Second, because of its restricted focus on formal issues and on theories, the Nagelian approach fails to account for the *diversity* of the cases of epistemic reduction that are present in contemporary biology as well as for the *complexity* of the conditions that determine the reductivity of biological explanations. Third, Nagel's account does not provide the recourses to *demarcate* reductive explanations from non-reductive explanations. All this strongly suggests that it is finally time to move beyond the Nagelian framework and break new ground in thinking about epistemic reduction.

Philosophisches Seminar
University of Köln
Richard-Strauß-Straße 2
50931, Köln
Germany
kaiser.m@uni-koeln.de

71 See also *ibid.*

Chapter 19

Charlotte Werndl

Probability, Indeterminism and Biological Processes

19.1 Introduction

Probability and indeterminism have always been core philosophical themes. Biology provides an interesting case study to explore these themes. First, biology is teeming with probabilities, and so a crucial question in the foundations of biology is how to understand these probabilities. Second, philosophers want to know whether the processes investigated by one of the major sciences – biology – are indeterministic.

This paper aims to contribute to understanding probability and indeterminism in biology. More specifically, Sect. 19.2 will provide the background for the paper. It will be argued that an omniscient being would not need the probabilities of evolutionary theory to make predictions about biological processes. However, despite this, one can still be a realist about evolutionary theory, and then the probabilities in evolutionary theory refer to real features of the world. This prompts the question of how to interpret biological probabilities which correspond to real features of the world but are in principle dispensable for predictive purposes. Sect. 19.3 will suggest three possible interpretations of such probabilities. The first interpretation is a propensity interpretation of kinds of systems. It will be argued that, contra Sober,[1] backward probabilities in biology do not present a problem for the propensity interpretation. The second interpretation is the frequency interpretation, and it will be argued that Millstein's[2] objection against this interpretation in evolutionary theory is beside the point. Finally, I will suggest Humean chances are a new interpretation of probability in evolutionary theory. Sect. 19.4 discusses Sansom's[3] argument that biological processes are indeterministic because probabilities in evolutionary theory refer to real features of the world. It will be argued that Sansom's argument is not conclusive, and that the question whether biological processes are deterministic or indeterministic is still with us.

1 Elliott Sober, "Evolutionary Theory and the Reality of Macro Probabilities", in: *Philosophy of Science*, Presidential Address 2004.

2 Roberta L. Millstein, "Interpretations of Probability in Evolutionary Theory", in: *Philosophy of Science* 70, 4, 2003, pp. 1317–1328.

3 Robert Sansom, "Why Evolution is Really Indeterministic", in: *Synthese* 136, 2, 2003, pp. 263–280.

D. Dieks et al. (eds.), *Probabilities, Laws, and Structures*, The Philosophy of Science in a European Perspective 3, DOI 10.1007/978-94-007-3030-4_19,

19.2 Realism, Indeterminism and Omniscient Beings

This section provides the background for the paper. First, the notions of realism, instrumentalism, determinism and indeterminism will be introduced. Then it will be explained that an omniscient being would not need the probabilities of evolutionary theory to make predictions about biological processes. It is argued that, despite this, one can still be a realist about evolutionary theory.

(Scientific) realism about a theory *T* is the idea that *T* corresponds to the world, i.e., *T* gives at least an approximately true description of the real-world processes falling under its scope. *Instrumentalism* relative to a theory *T* as understood in this paper is the negation of realism. Hence an instrumentalist about a theory *T* denies that *T* corresponds to the world. For what follows a definition of determinism for theories as well as for real-world processes is needed. A *theory T* is *deterministic* if and only if a state description of a system is always followed by the same history of transitions of state descriptions. A theory *T* is *indeterministic* if and only if it is not deterministic. A *process* is *deterministic* (concerning a specific set of kinds) if and only if a given state of a kind is always followed by the same history of transitions of states of kinds.[4] A process is *indeterministic* (concerning a specific set of kinds) if and only if it is not deterministic.[5]

Probabilities are of utmost importance in evolutionary theory, and the probabilistic character of evolutionary theory is widely accepted.[6] An example is the concept of fitness of an organism in an environment (see Sect. 19.3 for other examples of probabilities in evolutionary theory). Since one wants to allow that in unusual circumstances less fit organisms have more offspring than fitter ones, fitness of an organism[7] is captured by means of the probability to have a certain level of reproductive success.[8]

An omniscient being would not need the probabilities of evolutionary theory to make predictions about biological processes. The next two paragraphs will explain why this is so. In essence, this is a consequence of the fact that evolutionary theory ignores certain details and factors. For example, evolutionary theory does not include detailed models of flashes of lightning (because which organisms will

4 Jeremy Butterfield, "Determinism and Indeterminism", in: *Routledge Encyclopaedia of Philosophy Online* 2005; John Earman, *A Primer on Determinism*. Dordrecht: D. Reidel Publishing 1986.

5 What I call "determinism" of theories and processes is also sometimes called "future determinism". This is to highlight that it is not required that any state is also always preceded by the same history of transitions of states (see Earman, *loc. cit.*, pp. 13–14).

6 Sansom, *loc. cit.*, pp. 268–269.

7 Robert N. Brandon, *Adaptation and Environment*. Princeton: Princeton University Press 1990, p. 15.

8 How to define or measure fitness exactly turns out to be tricky. For an organism in a specific environment it is not enough to consider the expected value of offspring number, sometimes also the variance and other measures need to be taken into consideration (cf. Brandon, *ibid.*, p. 20).

struck by lightning is random – i.e., not related to their actual traits). Another example is that the exact location at each point of time of a chimpanzee in a forest does not appear as a variable in evolutionary theory (because this location is not correlated to reproductive success). Now consider models which correctly describe biological processes in all their details at the level of macrophysics. These macro-physical models will be very different from models in evolutionary theory because the former include details and factors which are ignored by the latter. For example, such macro-physical models include a description of flashes of lightening and they include variables for the exact location of chimpanzees.

Are these macro-physical models of biological processes deterministic or indeterministic? This is a matter of debate. Rosenberg[9] argues that they are deterministic. Abrams[10] and Graves et al.[11] claim that these models are "nearly deterministic". What is meant by this is that they are indeterministic because quantum mechanical probabilities can percolate up and quantum mechanics is indeterministic.[12] However, because macro-physical objects consist of many particles, the probabilities at the macro-level are very close to zero or one. Others such as Millstein[13] and Weber[14] argue that we do not know enough about the role of quantum

9 Alexander Rosenberg, *Instrumental Biology or the Disunity of Science.* Chicago: The University of Chicago Press 1994.

10 Marshall Abrams, "Fitness and Propensity's Annulment?", in: *Biology and Philosophy* 22, 1, 2007, pp. 115–130.

11 Leslie Graves, Barbara L. Horan and Alexander Rosenberg, "Is Indeterminism the Source of the Probabilistic Character of Evolutionary Theory?", in: *Philosophy of Science* 66, 1999, 1, pp. 140–157. See also Alexander Rosenberg, "Discussion Note: Indeterminism, Probability, and Randomness in Evolutionary Theory", in: *Philosophy of Science* 68, 4, 2001, pp. 536–544.

12 These positions and generally philosophers of biology take it to be uncontroversial that quantum theory is indeterministic (see Abrams "Fitness and Propensity's Annulment?", *loc. cit.*, pp. 119–121; Graves et al., *ibid.*, pp. 144–145; Rosenberg, "Discussion Note: Indeterminism, Probability, and Randomness in Evolutionary Theory", *loc. cit.*, pp. 537–538; Sansom, *loc. cit.*, p. 267). However, this is questionable. As generally agreed in philosophy of physics, there are coherent deterministic interpretations of quantum theory and "the alleged indeterminism of quantum theory is very controversial: it enters, if at all, only in quantum theory's account of measurement processes, an account which remains the most controversial part of the theory" (Butterfield, *loc. cit.*). Similarly, it is often simply assumed that macrophysics is deterministic (e.g. Graves et al., *ibid.*, p. 145; Rosenberg "Discussion Note: Indeterminism, Probability, and Randomness in Evolutionary Theory", *loc. cit.*, p. 537). Yet, research in philosophy of physics has shown that it is unclear whether macrophysics is deterministic (see Earman, *loc. cit.*, Chapter III). These assumptions are questionable, but they will not matter for what follows.

13 Roberta L. Millstein, "Is the Evolutionary Process Deterministic or Indeterministic? An Argument for Agnosticism", Presented at the Biennial Meeting of the Philosophy of Science Association Vancouver, Canada, 2000.

14 Marcel Weber, "Indeterminism in Neurobiology", in: *Philosophy of Science (Proceedings)* 71, 2005, pp. 663–674.

events at the macroscopic level and hence should remain agnostic: these models could be deterministic or indeterministic with probabilities very close to zero or one. The upshot is that even if there are nontrivial probabilities for macro-physical models of biological processes, they are different from those probabilities figuring in evolutionary theory. Consequently, *evolutionary theory appeals to probabilities which at least partly arise from ignoring certain details and factors. Hence an omniscient being would not have to rely on the probabilities of evolutionary theory to make predictions about biological processes*. If the world at the macro-physical level is deterministic, an omniscient being could appeal to a deterministic theory to predict biological processes. If the world at the macro-physical level is indeterministic, the omniscient being could appeal to a very different indeterministic theory (with probabilities close to zero and one) to predict biological processes.

Does this have any implications about whether one should be a realist or instrumentalist about evolutionary theory? Rosenberg thinks so. Because an omniscient being would not need evolutionary theory, he argues that "[t]his makes our actual theory of natural selection more of a useful instrument than a set of propositions about the world independent of our beliefs about it".[15] So Rosenberg argues that because an omniscient being would not need evolutionary theory, this implies instrumentalism about evolutionary theory.

Weber[16] disagrees with Rosenberg. He points out that:

> A theory may be dispensable in the sense that an omniscient being would be able to understand the phenomena in question at a deeper level, but it is still possible that this theory correctly represents some aspects of reality. To put it differently, a theory may be indispensable merely for *pragmatic* reasons i.e., for reasons which have to do with our cognitive abilities, but still be open to a realist interpretation. The fact that a theory falls short of giving us a *complete* account of some complex causal processes does not imply that this theory has no representational content whatsoever. A scientific realist is not committed to the thesis that even our best scientific theories provide complete descriptions of reality.[17]

In my opinion, Rosenberg is in principle right that the dispensability of evolutionary theory for an omniscient being can lead to the rejection of realism about evolutionary theory. However, this is only the case when one endorses an *extremely strong version of realism,* viz. a realism which demands that theories should match reality to such a high degree that an omniscient being could not use another theory to predict the processes in question. Weber correctly points out that *such a strong version of realism is hard to swallow.*[18] *Hence one can be a realist about evolu-*

15 Rosenberg, "Instrumental Biology or the Disunity of Science", *loc. cit.*, p. 83.

16 Marcel Weber, "Determinism, Realism, and Probability in Evolutionary Theory", in: *Philosophy of Science (Proceedings)* 68, 2001, pp. 213–224.

17 Weber, "Determinism, Realism, and Probability in Evolutionary Theory", *loc. cit.*, p. 217, original emphasis.

18 For an example of a kind of scientific realism that does not demand that our best scientific theories provide complete descriptions of reality, see Kenneth C. Waters,

tionary theory even if an omniscient being would not have to rely on evolutionary theory to predict biological processes. To give an example, assume that Newtonian mechanics truly describes the world. Then, according to Rosenberg's argument, it would follow that one cannot be a realist about statistical mechanics. Yet, most physicists and philosophers contend that it is possible to be a realist about statistical mechanics: statistical mechanics correctly represents certain features of systems even if these systems can be described in more detail at the microscopic level by Newtonian mechanics.[19]

To conclude, one can still be a realist about evolutionary theory even it this theory is dispensable for an omniscient being for predictive purposes. Many biologists and philosophers of biology are realists in such a sense, and then the interesting question arises of how to interpret the probabilities figuring in evolutionary theory. Because of realism, these probabilities are *ontic* in the sense that they refer to real feature of the world.[20] Yet, an omniscient being would not need these probabilities to make predictions (because an omniscient being could use a more fine-grained theory which is either deterministic or invokes probabilities different from evolutionary theory). So the task is to find interpretations of ontic probabilities which could in principle be eliminated for predictive purposes.

19.3 Interpretations of Ontic Probabilities in Evolutionary Theory

This section will discuss three possible interpretations of ontic probabilities in *evolutionary theory* consistent with the claim that the probabilities are in principle dispensable for predictive purposes, namely a propensity interpretation of kinds of systems (Sect. 19.3.1), the frequency interpretation (Sect. 19.3.2) and Humean chances (Sect. 19.3.3). It is worth pointing out that also in *several other contexts* scientists and philosophers talk about ontic probabilities which are in principle dispensable for predictive purposes. Examples are setups where the world is supposed to be deterministic at a more fundamental level, such as the probabilities in statistical mechanics or the probabilities arising from coin tosses, roulette wheels and similar processes.[21]

"Tempered Realism About the Force of Selection", in: *Philosophy of Science* 58, 4, 1991, pp. 553–573.

19 Roman Frigg, "A Field Guide to Recent Work on the Foundations of Statistical Mechanics", in: Dean Rickles (Ed.), *The Ashgate Companion to Contemporary Philosophy of Physics*. London: Ashgate 2008, pp. 99–196.

20 Hugh Mellor, *Probability: A Philosophical Introduction*. Cambridge: Cambridge University Press 2005. Mellor calls these probabilities "chances". I prefer the term "ontic" because some philosophers think that the term "chance" should only be used to refer to probabilities in an indeterministic world.

21 Frigg, "A Field Guide to Recent Work on the Foundations of Statistical Mechanics",

Millstein[22] already proposes two versions of the propensity account as possible interpretations of probability consistent with both determinism and indeterminism (hence these interpretations are consistent with the claim that probabilities are in principle dispensable for predictive purposes). The discussion of this paper differs in four respects. First, two interpretations are suggested which were not suggested by Millstein. In particular, I propose Humean chances as a possible interpretation of biological probabilities, and to the best of my knowledge, Humean chances have not previously been suggested as an interpretation of probabilities in evolutionary theory. Second, Sober's[23] objection to the propensity interpretation based on backward probabilities in biology is examined and dismissed; this objection has not been discussed by Millstein. Third, as outlined below, I disagree with Millstein's argument against frequency interpretations in evolutionary theory. Fourth, Millstein[24] proposes an interpretation based on Giere's single-case propensity interpretation. Single-case propensities provide an interpretation of probabilities that are not in principle dispensable for predictive purposes.[25] Hence this interpretation cannot be applied to probabilities as they arise in evolutionary theory. Yet Giere[26] suggests, and Millstein follows him in this, that from a pragmatic perspective his interpretation of probability can also be applied to probabilities that are in principle dispensable but behave like if there were not dispensable. However, if one makes this pragmatic move, one does not understand what probabilities are, and one cannot say that probabilities really exist. Consequently, I do not think that interpreting Giere's account pragmatically leads to a satisfying interpretation of probabilities which are in principle dispensable for predictive purposes.

19.3.1 Propensity Interpretation

The three interpretations of ontic probabilities will now be presented. The first interpretation is a version of the *propensity interpretation*, namely what Millstein calls a "propensity interpretation that views propensities as adhering to *kinds* or classes".[27] According to this interpretation, what one means by saying that a kind of system has a certain probability to change or to remain in a specific state is that

loc. cit.; Mellor, *Ibid.*, p. 55.

22 Roberta L. Millstein, "Interpretations of Probability in Evolutionary Theory", in: *Philosophy of Science* 70, 4, 2003, pp. 1317–1328.

23 Sober, "Evolutionary Theory and the Reality of Macro Probabilities", *loc. cit.*

24 Millstein, "Interpretations of Probability in Evolutionary Theory", *loc. cit.*, pp. 1322–1324.

25 Ronald N. Giere, "Objective Single-Case Probabilities and the Foundations of Statistics", in: Patrick Suppes, Leon Henkin, Grigore Moisil and Athanase Joja (Eds.), *Logic, Methodology, and the Philosophy of Science*. North Holland: Amerikan Elsevier, 1973, pp. 467–483.

26 Giere, *ibid.*, p. 481.

27 Millstein, "Interpretations of Probability in Evolutionary Theory", *loc. cit.*, p. 1324, original emphasis.

it has a disposition to produce specific long-run frequencies. Here the question emerges to what *kind of kind of systems* propensities should be attributed. Millstein argues that for probabilities in evolutionary theory a kind is specified by the causal factors that influence population level processes, ignoring details particular to one population such as the relative locations of organisms within the environment. For our purposes it is important that since this interpretation attributes a propensity to a kind of system, the probabilities are in principle dispensable for predictive purposes. Besides, according to this interpretation, the probabilities are ontic because they correspond to features of kinds of systems.

Like all the major interpretations of probability, propensity interpretations are controversial.[28] The main concerns are to explain what exactly a propensity is, and whether one can accept that a propensity, which is a very peculiar sort of entity, type of causation or property, is a part of the world. These problems are serious. Yet, in my opinion, they do not imply that the propensity interpretation is doomed to failure but rather call for further clarification or research. For Sober the main problem of the propensity interpretation in evolutionary theory is Humphrey's paradox, viz. that the propensity interpretation cannot make sense of *backward probabilities* as they appear, for example, in coalescence theory.[29] I will now argue that these backward probabilities do not present a problem.

Coalescence theory gives probabilities of how long ago the most recent ancestor of two organisms existed. A simple model of coalescence theory is as follows: the population number is constant, i.e., there are N organisms in each generation; the likelihood that an organism is a parent of an organism in the next generation is $1/N$; and the parents of the organisms in a generation are probabilistically independent. Under these assumptions, the probability that the first two organisms of a generation share a parent is $1/N$, and the probability that the most recent common ancestor existed t generations in the past is $(1-1/N)^{t-1}(1/N)$. These probabilities are backward probabilities in the sense that the question is whether for two organisms which live *now* the most recent common ancestor existed t generations in the *past*. For such backward probabilities the worry is that there are no nontrivial propensities: the linage of the two organisms is determined. Hence the two organisms either have or do not have the most recent common ancestor t generations in the past.

For single-case propensity interpretations such as Giere's interpretation discussed above, this might present a problem. However, there is no problem for the propensity interpretation in evolutionary theory, which appeals to kinds of systems. For the simple model of coalescence theory, consider the kind of system where there are N organisms at the start and the organisms reproduce over t generations. Then the probability that the first two organisms have the most recent ancestor t generations in the past is the *propensity of this kind of system* to produce a first and

28 Colin Howson and Peter Urbach, *Scientific Reasoning, the Bayesian Approach*. Peru/Illinois: Open Court 1996, pp. 338–351; Mellor, *loc. cit.*, Section 4.

29 Sober, "Evolutionary Theory and the Reality of Macro Probabilities", *loc. cit.*

a second organism in the *t*-th generation which have their most recent common ancestor *t* generations in the past. Thus, there is nothing like a backward propensity here. For each run of the system the first two organisms either have or do not have their most recent ancestor *t* generations in the past. Yet this is entirely compatible with a nontrivial propensity of a system to produce organisms that have their most common ancestor *t* generations in the past.[30] To conclude, *backward probabilities do not represent a problem for propensity theories in evolutionary theory, which appeal to kinds of systems*.

19.3.2 Frequency Interpretation

The second interpretation is the *frequency interpretation*. According to the most widely accepted version, the probability is the frequency of a hypothetical infinite sequence of trials. In our context it is important to note that, according to the frequency interpretation, probabilities are ontic because the frequencies correspond to real features of the world. Furthermore, because the notion of a frequency applies to sequences of outcomes, the probabilities are in principle dispensable for predictive purposes.

Frequentists are confronted with difficult questions.[31] A serious worry is that the frequency interpretation overstates the relation of probabilities to frequencies. As treated in the mathematical field of probability theory, a probability can also lead to an infinite sequence of outcomes where the frequency of the sequence differs from the probability. For instance, a fair coin can land heads each time in an infinite run of tosses (though this sequence has probability zero). It is plausible to demand that interpretations of probability should allow for this too, but the frequency interpretation does not.[32] There is no way out of this by postulating that the probability for an infinite sequence to yield the correct frequency is one. Clearly, this would be circular because probability would be defined by referring to probability. Another problem for hypothetical limiting frequentists is to explain what exactly fixes the outcomes of hypothetical infinite sequences, why counterfactual frequencies are determinate, and why they agree with the probability.[33] Furthermore, what can happen more or less frequently is not that a single experiment yields an outcome but that members of some class of experiments yield an outcome. This class is called a reference class, and frequentists have to answer the

30 This solution to Humphrey's paradox in evolutionary theory is similar to the solution proposed by Gillies and Mc Curdy. See Donald Gillies, "Varieties of Propensities", in: *The British Journal for the Philosophy of Science* 51, 4, 2000, pp. 807–835; Christoper S. I. McCurdy, "Humphreys's Paradox and the Interpretation of Inverse Conditional Propensities", in: *Synthese* 108, 1, 1996, pp. 105–125.

31 Howson and Urbach, *loc. cit.*, pp. 319–337; Mellor, *loc. cit.,* Section 3.

32 Sober, "Evolutionary Theory and the Reality of Macro Probabilities", *loc. cit.*

33 Marshall Abrams, "Infinite Populations and Counterfactual Frequencies in Evolutionary Theory", in: *Studies in History and Philosophy of the Biological and Biomedical Sciences* 37, 2, 2006, pp. 256–268; Mellor, *loc. cit.,* Section 3.

question of what constitutes a reference class. That this problem is difficult is illustrated by the fact that it is easy to change the order of an infinite sequence such that the frequency changes. Thus an answer to the reference class problem also needs to explain why only a certain order of experiments is allowed and others are not allowed.

In conclusion, the frequency interpretation faces serious problems. In my opinion, they do not imply that the frequency interpretation is doomed to failure. Yet, some of the problems seem hard to solve, and further work is needed to make progress on these problems. Millstein[34] has argued that the frequency interpretation is of no use in evolutionary theory because it faces an insurmountable problem involving the change of frequencies. I will now argue that Millstein's objection is misguided.

Millstein's argument starts from considering random drift – a process where physical differences between organisms are causally irrelevant to differences in reproductive success. A simple model of drift is as follows[35]: suppose that the population size is a constant N with $2N$ alleles and that there are i alleles of type A. Further, suppose that the number of alleles of type A in the next generation is the sum of $2N$ independent Bernoulli variables where the probability for an allele of type A is $i/2N$ (the ratio of allele A in the current population). Then the probability that the population will go from i alleles of type A to j alleles of type A is[36]:

$$p_{ij}=\frac{2N}{(2N\text{-}j)!j!}\left(\frac{i}{2N}\right)^{j}\left(1\text{-}\frac{i}{2N}\right)^{2N\text{-}j}.$$

Clearly, this implies that when drift occurs over a number of generations, the ratio of alleles of type A can fluctuate from generation to generation, especially in small populations. Any interpretation of probability in evolutionary theory has to be able to successfully interpret these probabilities. Millstein argues that these probabilities cannot be interpreted as frequencies because "frequencies may increase, decrease, or remain constant. In an ensemble of populations, eventually each population undergoing drift will go to fixation for one of the types, but which type cannot be predicted".[37]

However, *Millstein's worries are unjustified.* All the frequency interpretation says for the simple model of drift is that if, again and again, one considers a population with $2N$ alleles and i alleles of type A, the frequency that such a population will go to j alleles of type A is p_{ij}. This is entirely consistent with the fact that the ratio of alleles of type A and the transition probabilities can change from one gen-

34 Millstein, "Interpretations of Probability in Evolutionary Theory", *loc. cit.*, p. 1322.

35 Jonathan Roughgarden, *Theory of Population Genetics and Evolutionary Ecology: An Introduction*. Upper Saddle River: Prentice Hall 1996, pp. 65–66.

36 This equation is a correction of Millstein's equation, where there is a typo.

37 Millstein, "Interpretations of Probability in Evolutionary Theory", *loc. cit.,* p. 1322.

eration to the next and that populations will go to fixation for one of the types. The point is that for a given reference class the frequencies and hence the probabilities are well defined. If the number of alleles of type A changes in one generation from i to k (i□k), then also the probabilities p_{ij} and p_{kj} will be different. However, far from being a problem, this is as it should be because p_{ij} and p_{kj} are the probabilities corresponding to different reference classes.

My argument can be illustrated with a more familiar example. Suppose that at time t_0 a ball is drawn randomly from an urn with six red and six black balls (probability for red 1/2, probability for black 1/2), at time t_1 a ball is drawn randomly from an urn with 3 red and 1 black balls (probability for red 3/4, probability for black 1/4), at time t_2 a ball is drawn randomly from an urn with two red and three black balls (probability for red 2/5, probability for black 3/5), and so on. Millstein's argument would amount to the claim that the frequency interpretation cannot make sense of the probability to draw a red ball from an urn because the proportion of red balls changes with time. This seems misguided. The probabilities change with time because they correspond to different reference classes. This is as it should be and is unproblematic since for a given reference class the probability is well defined.

19.3.3 Humean Chances

As a third interpretation I want to suggest *Humean chances* as recently endorsed by Frigg and Hoefer as a new interpretation of probabilities in evolutionary theory.[38] The Humean mosaic is the collection of all events that actually happen at all times. (Here Frigg and Hoefer make the assumption of ontological pluralism, i.e., entities at different levels of the world, and not only the entities at the most fundamental level, are real.) Humean chances supervene on the Humean mosaic. More specifically, imagine all possible systems of probability rules about events in the Humean mosaic. There will be a best system in the sense that the probability rules of this system can best account for the Humean mosaic in terms of simplicity, strength and fit. The strength of a system of rules is measured by its scope to account for large parts of the Humean mosaic, and fit is measured in terms of closeness to actual frequencies. Then Humean chances are the numbers that are assigned to events by the probability rules of this best system. The reason why the best system contains rules about macro-processes, such as the processes involving the kinds postulated by evolutionary theory, is simplicity in derivation: even if it were the case that the facts about macro-processes could be derived from fundamental physics, "it is hugely costly to start from first principles every time you

38 Roman Frigg and Carl Hoefer, "Determinism and Chance from a Humean Perspective", in: Dennis Dieks, Wenceslao Gonzalez, Stephan Hartmann, Marcel Weber, Friedrich Stadler and Thomas Uebel (Eds.), *The Present Situation in the Philosophy of Science*. Berlin: Springer 2010, pp. 351–372; Carl Hoefer, "The Third Way on Objective Probability: A Sceptic's Guide to Objective Chance", in: *Mind* 116, 463, 1007, pp. 549–596.

want to make a prediction about the behaviour of a roulette wheel. So the system becomes simpler in that sense if we write in rules about macro objects".[39]

Proponents of Humean chances are confronted with the difficult question of how to characterise simplicity, strength and fit in detail. Providing a detailed account of simplicity, strength and fit is crucial because otherwise it remains vague and unclear what probabilities really are. For our purposes it is important to note that because Humean chances are facts entailed by actual events in the world, probabilities, thus understood, correspond to real features of the world. Furthermore, Humean chances as described above differ from Lewis's original proposal in that laws and chances are not analysed together, which implies that the interpretation presented here can also apply to probabilities which are in principle dispensable for predictive purposes.[40] Indeed, Frigg and Hoefer's main concern is to argue for Humean chances as an account of ontic probabilities in deterministic worlds. In particular, they defend Humean chances as an interpretation of probability in statistical mechanics and as an interpretation of the probabilities associated with deterministic processes such as coin tossing and the spinning of roulette wheels.

In sum, *propensities of kinds of systems, frequencies and Humean chances are possible interpretations of probabilities in evolutionary theory in the sense that the probabilities are ontic and can in principle be eliminated for predictive purposes.*

19.4 Criticism of Sansom's Claim That Biological Processes Are Indeterministic

Because probabilities are ontic in evolutionary theory, Sansom[41] concludes that biological processes are really indeterministic. This section will argue that Sansom's argument is inconclusive. First of all, Sansom's argument needs to be introduced in more detail. Sansom distinguishes between two kinds of realism, which he regards as the only two versions of realism worthy of further consideration: innocent pluralism and monorealism. *Innocent pluralism* asserts that different theories describing the same part of the world at different levels can be true and that no level of the world is privileged.[42] On this view, for instance, the same part of the world can be adequately described by quantum theory and macrophysics. *Monorealism* holds that the world is truly described by only one theory. For example, some physicists and philosophers have contended that quantum theory is the only theory capturing reality.

39 Frigg and Hoefer, *loc. cit.*, p. 21.

40 Hoefer, *loc. cit.*, pp. 558–560.

41 Sansom, *loc. cit.*

42 Sansom introduces this concept by alluding to the presentation of this view by Sober – see Elliott Sober, *The Nature of Selection.* Cambridge/MA: MIT Press 1984.

Imagine an innocent pluralist who thinks that quantum theory and macrophysics truly describe the world watching a ball rolling across a table. Then, assuming that macrophysics is deterministic and that quantum theory is indeterministic, from the innocent pluralist's point of view the process is indeterministic relative to quantum theory and deterministic relative to macrophysics. Consequently, as Sansom correctly remarks, an innocent pluralist has to accept the "*relativity of determinism*", namely that the world is neither merely deterministic nor indeterministic, but that whether or not determinism is true is relative to the kinds under consideration.

Sansom argues for realism about evolutionary theory and innocent pluralism by referring to Geach's[43] view of relative identity. Because processes are indeterministic relative to the kinds posited by evolutionary theory, Sansom concludes that biological processes are really indeterministic.

Sansom is right that processes are indeterministic relative to the kinds posited by evolutionary theory. However, the question arises *why one should exclusively focus on the kinds posited by evolutionary theory*. To understand this point, a comparison with physics will help. For an innocent pluralist there are many physical realities – the processes relative to quantum-mechanical kinds, the processes relative to the kinds posited by general relativity theory, the processes relative to statistical-mechanical kinds etc. Now suppose that in biology there are also two realities: processes involving life relative to the kinds posited by evolutionary theory and processes involving life relative to macro-physical kinds. Relative to the macro-physical kinds the processes might be deterministic. Then the question whether biological processes are deterministic has no clear answer for an innocent pluralist: biological processes are indeterministic relative to the kinds posited by evolutionary theory and deterministic relative to the macro-physical kinds.

Sansom's concern are the biological realities as considered by biologists and philosophers of biology.[44] He simply assumes and does not provide any argument for the exclusive focus on the biological reality of the processes relative to the kinds posited by evolutionary theory. Is there no need to justify this assumption because it is uncontroversial that there is only one biological reality, viz. the processes involving life relative to the kinds posited by evolutionary theory? This is *not* so. The extant literature speaks at least about two biological realities: namely, about a biological reality of the processes involving life relative to the kinds posited by evolutionary theory, and about another biological reality of the processes involving life relative to macro-physical kinds. Important for our purpose is that the latter is standardly referred to as a biological reality.[45] Indeed, there is a lively debate

43 Peter Geach, "Ontological Relativity and Relative Identity", in: Milton K. Munitz (Ed.), *Logic and Ontology*. New York: New York University Press 1973, pp.287–302.

44 Clearly, Sansom cannot arbitrarily decide what to call "biological reality" because this would render his argument uninteresting.

45 Abrams, "Fitness and Propensity's Annulment?", *loc. cit.;* Millstein, "Interpretations of Probability in Evolutionary Theory",, *loc. cit.;* Millstein, "Is the Evolutionary Proc-

in the philosophy of biology about the question whether determinism holds true for the biological reality of the processes involving life relative to macro-physical kinds. As already mentioned in Sect. 19.2, Rosenberg[46] argues that this biological reality is deterministic. Abrams[47] and Graves et al.[48] claim that it is indeterministic but that all probabilities are very close to zero and one. Others such as Millstein[49] and Weber[50] argue that we do not know enough about the role of quantum events at the macroscopic level and hence should remain agnostic about whether or not this biological reality is deterministic.

To conclude, *Sansom simply assumes that "biological reality" refers to the processes relative to the kinds posited by evolutionary theory, but this assumption is not justified*. The extant literature speaks at least about two biological realities – processes involving life relative to the kinds posited by evolutionary theory and processes involving life relative to macro-physical kinds. Consequently, for an innocent pluralist the question whether biological processes are deterministic has to broken up into (at least) two subquestions: Are processes involving life deterministic relative to the kinds posited by evolutionary theory? Are processes involving life deterministic relative to macro-physical kinds? Hence for Sansom's argument to be tenable, he would need to show that biological processes are indeterministic relative to these two sets of kinds. However, he has not shown that processes involving life are indeterministic relative to macro-physical kinds. And, as illustrated by the debate in philosophy of biology,[51] the question whether biological processes are deterministic relative to macro-physical kinds is controversial and has no easy answer. *Consequently, Sansom's argument that biological processes are really indeterministic (for an innocent pluralist) does not succeed.*

---

ess Deterministic or Indeterministic?", *loc. cit.;* Rosenberg, "Instrumental Biology or the Disunity of Science", *loc. cit.;* Rosenberg, "Discussion Note: Indeterminism, Probability, and Randomness in Evolutionary Theory", *loc. cit.*

46 Rosenberg, "Instrumental Biology or the Disunity of Science", *loc. cit.*

47 Abrams, "Fitness and Propensity's Annulment?", *loc. cit.*

48 Graves et al., "Is Indeterminism the Source of the Probabilistic Character of Evolutionary Theory?", *loc. cit.*

49 Millstein, "Is the Evolutionary Process Deterministic or Indeterministic?", *loc. cit.*

50 Weber, "Indeterminism in Neurobiology", *loc. cit.*

51 Robert N. Brandon and Scott Carson, "The Indeterministic Character of Evolutionary Theory: No 'No Hidden Variables Proof' but Not Room for Determinism Either", in: *Philosophy of Science* 63, 3, 1996, pp. 315–337; Graves et al. "Is Indeterminism the Source of the Probabilistic Character of Evolutionary Theory?", *loc. cit.*; Millstein, "Is the Evolutionary Process Deterministic or Indeterministic?", *loc. cit.*

19.5 Conclusion

Probability and indeterminism have always been central philosophical themes. This paper contributed to understanding these themes by investigating probability and indeterminism in biology.

The starting point was the following argument: an omniscient being would not need the probabilities of evolutionary theory to make predictions. Despite this, one can still be a realist about evolutionary theory. For a realist about evolutionary theory the probabilities are ontic, i.e., they refer to real features of the world. This prompted the question of how to understand probabilities which are ontic but which are in principle dispensable for predictive purposes.

The contribution of the paper to this question was to suggest three possible interpretations of such probabilities in evolutionary theory. The first interpretation was a propensity interpretation of kinds of systems. Since this interpretation attributes a propensity to kinds of system, the probabilities are ontic and are in principle dispensable for predictive purposes. Sober's objection that propensity theories cannot deal with backward probabilities in biology was discussed. By investigating backward probabilities in coalescence theory, I concluded that backward probabilities are unproblematic because they can be understood as propensities of kinds of systems. The second interpretation was the frequency interpretation. Since a frequency applies to a sequence of outcomes, the probabilities are ontic and are in principle dispensable for predictive purposes. I examined Millstein's objection that in the case of drift frequencies often change, implying that biological probabilities cannot be interpreted as frequencies. I argued that this objection is beside the point because it is normal that there are different frequencies for different reference classes. Third, I suggested Humean chances as a new interpretation of probability in biology. Humean chances are the numbers assigned to events by the probabilities rules of the best system (the best system is identified by the probability rules that can best account for the collection of all actual events in terms of simplicity, strength and closeness to frequencies). Humean chances are ontic because they are facts entailed by all actual events. Furthermore, because of simplicity of derivation, probabilities are also assigned to macro-processes, and hence Humean chances are in principle dispensable for predictive purposes. All three interpretations suffer from problems, and further research is required to tackle them. Yet they at least show us three possible ways of understanding ontic probabilities in evolutionary theory.

Finally, I criticised Sansom's claim that biological processes are really indeterministic. Sansom is a realist about evolutionary theory and subscribes to the view that different theories describing the same part of the world at different levels can be true. Because processes are indeterministic relative to the kinds posited by evolutionary theory, Sansom concludes that biological processes are indeterministic. Sansom's argument presupposes that "biological reality" refers to the processes relative to the kinds posited by evolutionary theory. However, this assumption is not justified. The extant literature in biology and philosophy is concerned with at least two biological realities – processes involving life relative to the kinds posited by evolutionary theory and processes involving life relative to macro-physical kinds. Consequently, Sansom's argument that evolution is really indeterministic is not conclusive. The problem whether biological processes are deterministic or indeterministic is still with us.

Department of Philosophy, Logic and Scientific Method
London School of Economics and Political Science
Houghton Street
WC2A 2AE, London
United Kingdom
c.s.werndl@lse.ac.uk

CHAPTER 20

BENGT AUTZEN[1]

BAYESIANISM, CONVERGENCE AND MOLECULAR PHYLOGENETICS

ABSTRACT

Bayesian methods are very popular in molecular phylogenetics. At the same time there is concern among biologists and philosophers regarding the properties of this methodology. In particular, there is concern about the lack of objectivity of evidential statements in Bayesian confirmation theory due to the role of prior probabilities. One standard reply to be found in the Bayesian literature is that as data size grows larger differences in prior probability assignments will "wash out" and there will be convergence of opinion among different agents. This paper puts the "washing out of priors" argument to the test in the context of phylogenetic inference. I argue that the role of nuisance parameters in molecular phylogenetics prevents the application of convergence arguments typically found in the literature on Bayesianism.

20.1 INTRODUCTION

Bayesian methods are used widely across scientific disciplines. For instance, in cosmology Bayesian methods have gained considerable popularity due to the increase of cosmological data in combination with modern computational power.[2] Similarly, the dramatic growth in molecular sequence data in biology has lead to a heightened interest in Bayesian statistical techniques for the purpose of phylogenetic inference.[3] Besides its popularity in science, Bayesianism provides the most popular account of evidence and confirmation in the philosophy of science.[4]

---

1 I would like to thank Eric Raidl and Marcel Weber for their comments on earlier drafts of this paper as well as the participants of the ESF workshop 'Points of Contacts between the Philosophy of Physics and the Philosophy of Biology' for helpful discussion.

2 Roberto Trotta, "Bayes in the Sky: Bayesian Inference and Model Selection in Cosmology", in: *Contemporary Physics* 49, 2008, pp. 71–104.

3 Mark Holder and Paul Lewis, "Phylogeny Estimation: Traditional and Bayesian Approaches", in: *Nature Reviews Genetics* 4, 2003, pp. 275–284.

4 Luc Bovens and Stephan Hartmann, *Bayesian Epistemology*. Oxford: Oxford University Press 2003.

D. Dieks et al. (eds.), *Probabilities, Laws, and Structures*, The Philosophy of Science in a European Perspective 3, DOI 10.1007/978-94-007-3030-4_20,

Tracing back to Carnap,[5] there are two distinct notions of confirmation suggested by Bayesian epistemologists. According to the 'relative' notion of confirmation, D confirms H if and only if $P(H/D) > P(H)$. In contrast, the 'absolute' notion of confirmation asserts that D confirms H if and only if $P(H/D) > k$, where k denotes some threshold of high probability. Typically it is assumed that k equals 1/2.[6] Bayesian epistemologists evaluate the posterior probability of a hypothesis $P(H/D)$ by means of Bayes' theorem which states that the conditional probability of a hypothesis H given some data D equals the product of the likelihood of the hypothesis and the prior probability of the hypothesis divided by the prior probability of the data, that is,

$$P(H|D) = \frac{P(D|H) * P(H)}{P(D)}$$

Both Bayesian accounts of confirmation require the assignment of prior probabilities. The question of how to assign these prior probabilities is referred to as the 'problem of the priors' in the Bayesian literature. Following Earman[7] there are two broad strategies of addressing the problem of the priors. The first strategy is to constrain the priors. While all Bayesian epistemologists agree that prior degrees of belief should satisfy the probability calculus, there is a long-lasting debate about imposing further constraints on prior probability distributions. Examples of this strategy include the principle of indifference,[8] the principle of maximum entropy,[9] or the Principal Principle.[10] The second strategy asserts that the numerical values of prior probabilities do not matter as long as the amount of data analysed is sufficiently large. It is argued that as data size grows larger differences in prior probability assignments will "wash out" and there will be convergence of opinion among different agents.

This paper sets the strategy of constraining prior probabilities aside and focus only on the 'washing out of priors' strategy. The reason is that if the 'washing out of priors' argument works in the phylogenetic context, we can avoid the difficult task of justifying a particular way of constraining prior probabilities in this domain.

5 Rudolf Carnap, *The Logical Foundations of Probability*. 2nd edition. Chicago: Chicago University Press 1967.

6 See, for instance, Peter Achinstein, *The Book of Evidence*. New York: Oxford University Press 2001, p. 46.

7 John Earman, *Bayes or Bust: A Critical Examination of Bayesian Confirmation Theory*. Cambridge, Mass.: MIT Press 1992, p. 139.

8 John Maynard Keynes, *A Treatise on Probability*. New York: MacMillan 1921.

9 Edwin Thompson Jaynes, *Papers on Probability, Statistics, and Statistical Physics*. Roger Rosenkrantz (Ed.), Dordrecht: Reidel 1981.

10 David Lewis, "A Subjectivist's Guide to Objective Chance", in: Richard Jeffrey (Ed.), *Studies in Inductive Logic and Probability Vol II*. Berkeley: University of California Press 1980, pp. 263–293.

This view parallels Hesse's position who writes that convergence of opinion theorems "if relevant to scientific inference in general, would be a very powerful aid to confirmation theory, for it would discharge us from discussing the details of the initial probability distribution".[11]

Without constraining prior probabilities the Bayesian approach raises the question of how it can make sense of the idea that scientific methodology requires that there is some interpersonal agreement on how the data bear on the hypotheses under consideration. And without such an agreement the idea of the objectivity of science seems futile. In order to counter the charge that confirmation statements are entirely down to the subjective beliefs of individual researchers, Bayesians typically invoke the argument that there will be a convergence of opinions (as well as convergence to the truth if the truth is part of the considered hypotheses) from widely differing initial opinions as the amount of available data grows larger. Earman[12] summarizes what he calls the "Bayesian folklore" of 'washing out of priors' as follows:

Differences in prior probabilities do not matter much, at least not in the long run; for (as the story goes) as more and more evidence accumulates, these differences wash out in the sense that the posterior probabilities merge, typically because they all converge to 1 on the true hypothesis.

Historically, Bayesian philosophers and statisticians alike have endorsed the idea of 'washing out of priors'. Here are two examples from the literature illustrating this line of reasoning. Edwards, Lindman, and Savage[13] write:

Although your initial opinion about future behaviour of a coin may differ radically from your neighbour's, your opinion and his will ordinarily be transformed by application of Bayes' theorem to the results of a long sequence of experimental flips as to become nearly indistinguishable.

Suppes[14] argues in a similar vein when he writes:

It is of fundamental importance to any deep appreciation of the Bayesian viewpoint to realize the particular form of the prior distribution expressing beliefs held before the experiment is conducted is not a crucial matter [...] For the Bayesian, concerned as he is to deal with the real world of ordinary and scientific experience, the existence of a systematic method for reaching agreement is important [...] The well-designed experiment is one that

11 Mary Hesse, *The Structure of Scientific Inference*. Berkeley: University of California Press 1974, p. 116.

12 Earman, *loc. cit.*, p. 141.

13 William Henry Edwards, Harold Lindman, and Leonard Savage, "Bayesian Statistical Inference for Psychological Research", in: *Psychological Review* 70, 1963, pp. 193–242.

14 Patrick Suppes, "A Bayesian Approach to the Paradoxes of Confirmation", in: Jaakko Hintikka and Patrick Suppes (Eds.), *Aspects of Inductive Logic*. Amsterdam: North-Holland 1966, p. 204.

will swamp divergent prior distributions with the clarity and sharpness of its results, and thereby render insignificant the diversity of prior opinion.

Underlying the 'washing out of priors' reasoning are mathematical convergence to the truth theorems, such as, Savage's theorem[15] or Gaifman and Snir's theorem[16] among others. While convergence theorems have been discussed on general grounds in the philosophical literature,[17] it seems worthwhile to put these theorems to the test in particular cases of scientific inference. The reasons for this are twofold. For one thing, it is of interest to the scientist whether these theorems provide some comfort when confronted with the subjectivity objection raised against the Bayesian methodology. For another, it might demonstrate to the philosopher which assumptions of these theorems actually conflict with scientific practice.

This paper examines the application of two convergence theorems (i.e., Savage's theorem and Gaifman and Snir's theorem) to molecular phylogenetics. I argue that the role of auxiliary assumptions in Bayesian phylogenetic inference prevents the applications of these theorems. More specifically, the structure of this paper is as follows. Section 20.2 gives a brief introduction to Bayesian phylogenetics. Section 20.3 discusses the application of Savage's theorem (Sect.20 3.1) and Gaifman and Snir's theorem (Sect. 20.3.2) to molecular phylogenetics.

20.2 The Bayesian Approach to Phylogenetic Inference

Phylogenetics is the field of biology that seeks to reconstruct phylogenetic trees out of molecular data (e.g., amino acid or nucleotide sequences) or morphological data. A phylogenetic tree is a graphical representation of the genealogical relationship among species, among genes, among populations or among individuals. For instance, consider the three species chimpanzee, human and gorilla. What is the evolutionary relationship among these species? The following phylogenetic tree or, more precisely, 'tree topology'[18] T_1 asserts that chimpanzees (A) are more closely related to humans (B) than to gorillas (C).[19]

15 Leonard Savage, *The Foundations of Statistics*. New York: Dover 1972.

16 Haim Gaifman and Marc Snir, "Probabilities Over Rich Languages, Testing and Randomness", in: *Journal of Symbolic Logic* 47, 1982, pp. 495–548.

17 See, for instance, Hesse, *loc. cit.*, Earman, *loc. cit.* and Clark Glymour, *Theory and Evidence*. Princeton: Princeton University Press 1980.

18 A 'tree topology' is a branching diagram with labels at the tips of the tree.

19 It is often practical to refer to a phylogenetic tree in terms of the 'Newick format'. Written in Newick format, the tree topology T_1 reads ((A, B), C).

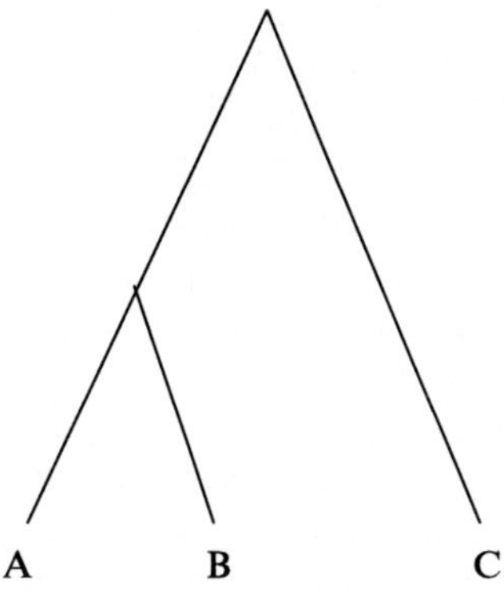

Tree T_1

Fig. 20.1 A rooted, binary tree

Biologists call the edges connecting two nodes of the tree the 'branches' of the tree.

A variety of methods for inferring phylogenetic trees have been proposed in the biological literature. This paper deals with the Bayesian approach to phylogenetic inference. A Bayesian phylogenetic analysis requires the calculation of the posterior probability distribution of phylogenetic trees by means of Bayes' theorem. In the phylogenetic context Bayes' theorem states that the conditional probability of a tree topology T_i given some molecular sequence data D equals the product of the likelihood of the tree topology and the prior probability of the tree topology divided by the prior probability of the data:

$$P(T_i|D) = \frac{P(D|T_i) * P(T_i)}{P(D)}$$

Let us assume that there are m possible tree topologies. We can then calculate the probability of the data $P(D)$ by means of the law of total probability:

$$P(D) = \sum_{i=1}^{m} P(D|T_i) * P(T_i) .$$

The probability of the data D given a tree topology T_i (i.e., the likelihood $P(D/T_i)$) is not determined without invoking auxiliary assumptions regarding the evolutionary process at the molecular level. This is just a token of Duhem's problem, that is, the problem that scientific hypotheses typically do not deductively entail observable consequences. Consider again tree topology T_1 = ((A, B), C). If this tree is supposed to confer probabilities on observed distributions of molecular sequence data for the three species A, B and C, then values must be specified for the transition probabilities associated with the four branches of the tree. In order to specify these branch transition probabilities and, hence, calculate the likelihood of tree topology T_1, a model of the evolutionary process at the molecular level together with other auxiliary assumptions is invoked. In detail, these additional auxiliary assumptions are given by a specification of the parameter values of the model (denoted as θ), and the specification of the vector of times during which evolutionary change occurs between the nodes of a tree (denoted as v). The vector specifying the physical times during which evolutionary change happens is also referred to as the vector of 'branch lengths'.[20]

Parameters such as branch lengths, the parameters of a model of molecular evolution and the model of molecular evolution itself are called 'nuisance parameters' in the statistical literature. To the statistician they are a "nuisance" because they have to be invoked to calculate numerical probabilities while they are not of interest in the particular inference. The Bayesian approach to statistical inference treats each nuisance parameter as a random variable and assigns a prior probability distribution. Some statisticians, such as Robert,[21] consider this way of handling nuisance parameters as a key attraction of the Bayesian approach.

In order to simplify the notation, let us assume a particular model of molecular evolution, and consider the vector of branch lengths v and the vector of model parameters θ as random variables with a prior probability distribution (denoted as $f(v)$ and $g(\theta)$). Under these assumptions the likelihood of the tree topology T_i can be calculated as follows:

$$P(D|T_i) = \int_v \int_\theta P(D|T_i, v, \theta) f(v)\, g(\theta)\, dv\, d\theta$$

That is, the likelihood $P(D/T_i)$ results from integrating the conditional probability of the data D given tree topology T_i, branch lengths v and model parameters θ (i.e., $P(D|T_i,v,\theta)$). The likelihood $P(D/T_i)$ is called a 'marginal likelihood' since the nuisance parameters v and θ are integrated out (or 'marginalized') given the prior probability distributions $f(v)$ and $g(\theta)$. Together with prior probabilities of the tree topologies (denoted as $P(T_j)$ for $j = 1, \ldots, m$) the posterior probability $P(T_i/D)$ can then be calculated as follows:

$$P(T_i|D) = \frac{\int_v \int_\theta P(D|T_i, v, \theta) f(v)\, g(\theta)\, dv\, d\theta * P(T_i)}{\sum_{j=1}^{m} \int_v \int_\theta P(D|T_j, v, \theta) f(v)\, g(\theta)\, dv\, d\theta * P(T_j)}$$

20 While here 'branch length' is understood as measuring physical time, this term also has a different meaning in the biological literature. In graphical representations of a phylogenetic tree the length of a branch is seen as a measure of the amount of evolutionary change occurring along the branch. In that case 'branch length' means the expected number of substitutions which is a function of time *and* the rate of substitutions.

21 Christian Robert, "Evidence and Evolution: A Review", in: *Human Genomics*, to appear.

The formula for the posterior probabilities of tree topologies looks very complex. In fact, it is generally too complex to be calculated analytically. Phylogeneticists therefore use the computer algorithm 'MrBayes'[22] to numerically approximate the posterior probabilities of phylogenetic trees. MrBayes samples phylogenetic trees from the posterior probability distribution of phylogenetic trees. Ideally, the proportion of sampled trees adequately reflects the posterior probability of a particular tree. As a result of this approximation method a posterior probability of a tree topology is equal to one if all sampled trees coincide with this topology.

Of course, the Bayesian approach of assigning prior probability distributions to nuisance parameters and integrating them out raises the question of how to assign these priors. This question is particularly pressing since different assignments of prior probabilities to nuisance parameters can lead to conflicting confirmation claims. The problem can be illustrated by means of a recent study on the phylogeny of baleen whales due to Yang.[23] The study demonstrates that a data set of 12 protein-coding genes of the mitochondrial genome from 5 baleen whales (including the Antarctic minke whale, the fin whale, the blue whale, the grey whale and the pygmy right whale) confirms conflicting phylogenetic trees depending on the choice of priors of models of molecular evolution and their parameters.[24]

Consider just two models of molecular evolution invoked in Yang's study: the Jukes-Cantor model and the codon M0 model. The Jukes-Cantor (JC) model[25] is a continuous-time Markov process which contains a single adjustable parameter λ that represents the instantaneous probability of a change from one nucleotide into another at a nucleotide site. While the term 'model' is used in many ways in science and philosophy, it has a fixed meaning in the statistical context. In statistics a model is a family of probabilistic hypotheses described in terms of one or more adjustable parameters. For instance, in the JC model the family of probabilistic hypotheses is characterised in terms of the parameter λ. Each particular numerical value of this parameter denotes a stochastic process.

While in models of nucleotide substitution, such as the JC model, the single nucleotide is the unit of evolution, the codon constitutes the unit of evolution in models of codon substitution. Codons are triplets of nucleotides. The genetic code consists of 64 (= 4^3) triplets of nucleotides which can be divided into 61 'sense'

22 John Huelsenbeck and Fredrik Ronquist, "MrBayes: Bayesian Inference of Phylogenetic Trees", in: *Bioinformatics* 17, 2001, pp. 754–755.

23 Ziheng Yang, "Empirical evaluation of a prior for Bayesian phylogenetic inference", in: *Philosophical Transactions of the Royal Society B* 363, 2008, pp. 4031–4039.

24 Remember that I treat models of molecular evolution itself as nuisance "parameters". This approach corresponds to the Bayesian idea of model averaging (e.g., see John Huelsenbeck, Bret Larget, and Michael Alfaro, "Bayesian Phylogenetic Model Selection Using Reversible Jump Markov Chain Monte Carlo", in: *Molecular Biology and Evolution* 21, 2004, pp. 1123–1133).

25 Thomas Jukes and Charles Cantor, "Evolution of protein molecules", in: Hamish Munro (Ed.), *Mammalian protein metabolism*. New York: Academic Press 1969.

and 3 'non-sense' or 'stop' codons. Every sense codon encodes an amino acid. For instance, the codon 'CTA' encodes the amino acid Leucine. The stop codons are typically not considered in the model since they are not allowed in a functional protein.

The codon M0 model[26] distinguishes between synonymous and nonsynonymous substitutions when describing the substitution rates from one (sense) codon to another. A 'synonymous' (or 'silent') substitution is the substitution of a nucleotide that does not change the encoded amino acid. In contrast, a 'nonsynonymous' substitution is the substitution of a nucleotide that changes the encoded amino acid. For instance, the substitution of the nucleotide 'A' by the nucleotide 'G' at the third codon position of the codon 'CTA' is a synonymous substitution since 'CTG' also encodes the amino acid Leucine. In contrast, substituting the nucleotide 'C' at the first codon position by the nucleotide 'A' represents a nonsynonymous substitution since the resulting codon 'ATA' encodes the amino acid Isoleucine. In Goldman and Yang's codon model the substitution rate from one codon to another is zero when the two codons differ at two or three codon positions. The substitution rate between two codons which differ only at one codon position depends on whether the substitution is synonymous or nonsynonymous as well as whether the substitution constitutes a transition or a transversion.[27]

Returning to the phylogeny of baleen whales, suppose that we are interested in the phylogenetic relationship of the Antarctic minke whale (A), the fin whale (F), the blue whale (B) and the grey whale (G). By assigning a full prior to the JC model (together with the MrBayes default prior to its parameter λ), the posterior probability of tree topology $T = (A, ((F, B), G))$ equals 0.93 based on the data in Yang's study.[28] Put differently, the data set (denoted as D) confirms tree topology T independent of whether the absolute or the relative notion of confirmation is assumed.[29] However, by assigning a full prior to the codon M0 model (as well as the MrBayes default prior to its parameters) data D confirm tree topology $T^* = (G, ((F, B), A))$ since $P(T^*|D) = 0.51$. Again, this result holds independently of whether the absolute or the relative notion of confirmation is assumed. The two inferred trees are in conflict since tree topology T asserts that blue whales are more closely related to grey whales than to Antarctic minke whales while tree topology T^* claims that blue whales are more closely related to Antarctic minke whales than to grey whales.

---

26 Nick Goldman and Ziheng Yang, "A codon-based model of nucleotide substitution for protein-coding DNA sequences", in: *Molecular Biology and Evolution* 11, 1994, pp. 725–736.

27 Transitions are changes between the nucleotides A and G and between C and T, while all other changes are transversions.

28 Yang, *loc. cit.*, Table 3.

29 Here, it is assumed that the 15 possible (binary, rooted) tree topologies for 4 species have equal prior probability (i.e., 1/15) and that $k = 1/2$ in the account of absolute confirmation.

Summing up, Yang's study demonstrates that the sensitivity of Bayesian confirmation claims to the priors of nuisance parameters is not just a possibility occurring only in highly idealized cases discussed by philosophers. Rather, the sensitivity of confirmation claims to priors of nuisance parameters manifests itself in scientific practice. Where does this leave us regarding the 'washing out of priors' argument? Can we reasonably expect that if we had a larger data set at hand, the posterior probabilities of phylogenetic trees would merge?

20.3 Convergence and Phylogenetic Inference

This section discusses the application of convergence to the truth theorems to phylogenetic inference. In particular, I examine the theorems suggested by Savage and Gaifman and Snir. The reason for this choice of theorems is that Savage's theorem has traditionally been invoked in the Bayesian literature.[30] Any discussion of washing out of priors should therefore revisit these more established debates. Gaifman and Snir's theorem relaxes a key assumption of Savage's theorem, that is, the assumption of well-defined (or objective) likelihoods. Due to this generalisation of Savage's result, Gaifman and Snir's theorem is of interest in the phylogenetic context where tree topologies alone do not assign probabilities to observed data.

Before turning to these particular theorems some general remarks are in order. Broadly speaking, one can distinguish between convergence results supplemented with and without an estimate of the rates of convergence. Convergence results coming without an estimate of the rates of convergence might be called 'ultimate convergence results'. An ultimate convergence result asserts that there exists some point in the future at which a certain amount of convergence will, with high probability, have occurred. It remains silent on how far in the future that point lies and nothing is guaranteed until that point is reached. Convergence results of this type seem of very limited use in underwriting the 'washing out of priors' argument for scientific practice. Scientists are always working with finite data sets. Without knowing anything about the rates of convergence, the promise that at *some* point in the future convergence will kick in with high probability seems to be an absolute minimum requirement for the washing out of priors. Ideally, scientists not only would like to know the rate of convergence but also be in the situation that the rate of convergence is relatively fast in order to buy into the washing out of priors reasoning.

---

30 For instance, Edwards, Lindman and Savage, *loc. cit.*, Hesse, *loc. cit.*, and Glymour, *loc. cit.*

20.3.1 Savage's Convergence Theorem

Savage[31] presents an argument to the effect that a person typically becomes almost certain of the truth when the data set grows larger and larger. Savage's convergence to the truth argument is based on the following mathematical theorem. Suppose that $\{H_1, H_2, \ldots, H_m\}$ is a set of mutually exclusive and jointly exhaustive hypotheses, each with prior probability $P(H_i)$. Let $X_1, X_2, X_3, \ldots$ denote a sequence of independent and identically distributed (i.i.d.) random variables with a finite set of values and $X^{(n)}$ the first n of these variables. Further, suppose that no two hypotheses have the same likelihood function, that is, for $i \neq j$ it is not the case that for all realizations $x = (x_1, x_2, x_3, \ldots)$ of the sequence of random variables X we have $P(X = x \mid H_i) = P(X = x \mid H_j)$ Then, the probability that the posterior probability of the true (but unknown) hypothesis given $X^{(n)} = x^{(n)}$ will be greater than α is given by

$$\sum_{i=1}^{m} P(H_i) \times P[P(H_i \mid X^{(n)} = x^{(n)}) > \alpha) \mid H_i],$$

where summation is restricted to those hypotheses with non-zero prior probability (i.e., $P(H_i) > 0$). Savage shows that this probability goes to 1 as n approaches infinity. More informally, the theorem asserts that the opinions of all agents regarding the hypotheses H_i will almost surely merge since each agent almost surely converges to certainty on the true hypothesis.

What about the rate of convergence? Savage's theorem can be supplemented with an estimate of the rate of convergence if we make the additional assumption that the prior probability of any hypothesis with non-zero prior has a lower bound ε.[32] That is, for any hypothesis H_i with $P(H_i) > 0$, we have to assume that $P(H_i) \geq \varepsilon > 0$ for a fixed, positive constant ε. Without a lower limit on the non-zero prior probabilities of hypotheses no statement can be made about the rate of convergence since in any finite time any given hypothesis of the set will have the highest posterior probability for some distribution of prior probabilities.[33]

While the mathematics of Savage's theorem is not in doubt, its relevance for the washing out of priors argument has been questioned by several authors. Hesse[34] particularly criticises the independence assumption and, what she calls, the 'randomness assumption' underlying the theorem. Let me start with the independence assumption. The independence assumption implies that the probability of observing a particular event at stage n is unaffected by having already observed particular events at earlier stages in the sequence. For instance,

31 Savage, *loc. cit.*

32 Hesse, *loc. cit.*, p. 117.

33 One way of measuring the concentration of the posterior probability distribution is to calculate the reciprocal of the variance of this distribution. If we do so, then the concentration of the posterior can be expected to grow as $\sqrt{n}$ in a 'Savage-type' setting. For more details, see Earman, *loc. cit.*, p. 148.

34 Hesse, *loc. cit.*

independence is assumed when balls are sampled with replacement from an urn. Hesse[35] argues that the independence assumption is generally not warranted in scientific practice since scientific experiments are often designed and scientific hypotheses modified in the light of already observed data. In particular, she makes the case that conducting a limited number of experiments based on already observed data and directed towards modified hypotheses is a more efficient procedure than mechanically conducting a long series of independent tests of the same hypothesis.

While the independence assumption is clearly stated in Hesse's writings, matters are more difficult when it comes to her 'randomness assumption'. According to Hesse,[36] randomness asserts that "given a particular hypothesis [H_i], the probability of making the test which yields [observation x_n] is independent of the particular order in which it and other evidence is observed". Hesse's notion of randomness seems to refer to the statistical assumption that the individual observations follow the same probability distribution when she writes that the randomness assumption contemplates "a situation in which we know nothing about the conditions which differentiate one observation from another".[37] If this reading is correct, then the assumption of a sequence of identically distributed random variables might be violated in some scientific applications and, hence, threaten the relevance of Savage's theorem in a particular domain.

Glymour[38] endorses Hesse's criticism and adds some further critical points in his discussion of Savage's theorem. In particular, Glymour argues that the role of second-order probabilities in the theorem weakens its argumentative force. Since for a Bayesian second-order probabilities as well as first-order probabilities represent degrees of belief, the theorem asserts that in the limit as n approaches infinity a Bayesian has degree of belief 1 that a Bayesian, whose degrees of belief are described by the theorem, has degree of belief, given data x, greater than in whatever hypothesis of the partition which actually obtains. Glymour[39] summarizes the implications of Savage's theorem as follows:

> The theorem does not tell us that in the limit any rational Bayesian will assign probability 1 to the true hypothesis and probability 0 to the rest; it only tells us that rational Bayesians are certain that he will.

Glymour[40] concludes that while Savage's theorem "may reassure those who are already Bayesians, but it is hardly grounds for conversion".

While both Glymour's and Hesse's criticism has its merits, the points they raise are not specific to the phylogenetic context. So, can something more subject

35 Hesse, *Ibid.*, pp. 118–119.
36 Hesse, *Ibid.*, p. 118.
37 Hesse, *Ibid.*, p. 118.
38 Glymour, *loc. cit.*
39 Glymour, *Ibid.*, p. 73.
40 Glymour, *Ibid.*, p. 73.

specific be said regarding the application of Savage's theorem? First, one should note that phylogeneticists examine a single data set rather than a growing sequence of data as assumed in Savage's theorem. That is, Bayesian phylogeneticists perform a single step of Bayesian updating rather than engage in continuous updating. An alternative approach would be to divide the single, large data set into a sequence of growing sub-data sets. For instance, one could start with a fixed number of codons and add a single codon (or three nucleotides) in each step. This sequence of sequence data sets could then be used to perform several Bayesian updates including the final one with the complete data set. The latter approach would be more congenial for examining the convergence of posterior tree probabilities understood as a *process*. One could examine whether and if so at what speed posterior tree probabilities approach each other.

That said, the single step analysis performed in the phylogenetic literature could still show the convergence of posterior tree probabilities understood as an *outcome*. That is, the posterior tree probabilities could be very close to each other for different assignments of prior probabilities. Why does this not happen in the study on the phylogeny of baleen whales? A first proposal might be that the amount of data analysed is too small. While this might be the case in some particular studies this does not seem to be the reason in general. To the contrary, phylogeneticists have very large data sets available for their analyses; that is, they routinely analyse sequence data sets of 10 kb or even 10 Mb. For instance, in Yang's study on the phylogeny of baleen whales the alignment e contains 3535 codons (i.e., 3535 * 3 = 10,605 nucleotides). In fact, some phylogeneticists, such as Yang,[41] argue that they have "too much data". If convergence understood as an outcome is supposed to have any impact for scientific practice – which is always operating on finite data sets – then they surely should apply to data sets of the size encountered in the phylogenetic context. Think about an analogy. A data set of 10 kb could be a 0-1 sequence representing the outcome of 10,000 coin tosses. We surely would hope that the posterior probabilities of the parameter of the Bernoulli distribution describing the i.i.d. sequence of coin tosses are very close to each other after analysing 10,000 coin tosses when starting with different assignments of prior probabilities if the 'washing out of priors' argument is supposed to have any bite.

A second and more important difference with the setting of Savage's theorem is to be found in the properties of the likelihoods involved. While Savage assumes that the hypotheses H_i have well-defined likelihoods without invoking any auxiliary assumptions, the likelihood of a tree topology depends on the prior probabilities of auxiliary assumptions regarding the evolutionary process on the molecular level. Put differently, the hypotheses H_i assign probabilities to any element in the event space while a tree topology alone does not assign a probability to a set of sequence data. Hence, the hypotheses in Savage's theorem cannot represent phylogenetic tree hypotheses.

41 Yang, *loc. cit.*, p. 4037.

20.3.2 Gaifman and Snir's Convergence Theorem

In order to overcome the difficulties associated with Savage's theorem it is desirable to establish a convergence theorem which does not require that hypotheses have well-defined, objective likelihoods $P(D/H)$. Gaifman and Snir[42] present a result to this effect. It might come as a surprise that such theorems have been suggested in the first place. If likelihoods are supposed to vary freely (or at least as freely as the probability calculus allows), then it seems astonishing that convergence to the truth and hence, convergence of opinion will occur. Consider the following example.[43] Two theologians might come to different conclusion regarding the conditional probability of the occurrence of human tragedies and natural catastrophes given the assumption that God exists. Suppose that D denotes the occurrence of human tragedies and that H denotes the hypothesis that God exists. For theologian A God might be mean spirited and, hence, the likelihood $P_A(D/H)$ very high, while for theologian B God might be benevolent and, hence, $P_B(D/H)$ very low. It seems unclear how observing what is happening in the world can lead to agreement of opinion between A and B regarding the question of whether or not God exists.

Gaifman and Snir's theorem is formulated in terms of concepts that are more akin to traditional philosophical discussions of confirmation theory where probabilities are assigned to sentences in some formal language. As a result some additional terminology has to be introduced to present their result. Suppose that L denotes a formal language that results from adding finitely many empirical predicates and empirical function symbols to first-order arithmetic containing names for each of the natural numbers. A model for L consists of an interpretation of the empirical symbols (i.e., empirical predicates and empirical function symbols). The set of all models for L is denoted as Mod_L. Gaifman and Snir sometimes refer to the elements of Mod_L as 'worlds'.

Further, for every world and every sentence φ let $\varphi^{(w)}$ be either φ or $-\varphi$ depending on whether or not φ is true in w. If $\varphi_1, \varphi_2, \ldots$ denotes a sequence of sentences, then $\bigcap_{i<n} \varphi_i^{(w)}$ denotes the available data in world w at stage n. Gaifman and Snir[44] suggest that the φ_i are "relatively simple sentences whose truth values can be found by available testing methods". Examples include atomic empirical sentences. In contrast, it is assumed that the truth or falsity of the hypothesis of interest cannot be tested directly in this way.

Finally, Gaifman and Snir introduce the concept of separation, which plays a crucial role in their theorem. A sentence φ is said to 'separate two worlds w_1 and w_2' if and only if it is true in one of them and false in the other. A class of sentences Φ 'separates a set of worlds X' if and only if every two worlds in X are separated by

42 Gaifman and Snir, *loc. cit.*

43 Example adapted from Michael Strevens, "Notes on Bayesian Confirmation Theory", unpublished manuscript, pp. 93–94.

44 Gaifman and Snir, *loc. cit.*, p. 507.

some $\varphi \in \Phi$. And, a class of sentences Φ 'is separating' if and only if it separates Mod_L.

Suppose that ψ is the hypothesis of interest and that $\{\varphi_1, \varphi_2, \dots\}$ is separating. Let $[\psi](w)$ denote the characteristic function corresponding to ψ which is defined as 1 if ψ is true in w and 0 otherwise. Then subject to some additional technical assumptions Gaifman and Snir show that:

$$P(\psi \mid \bigcap_{i<n} \varphi_i^{(w)}) \rightarrow [\psi](w) \quad \text{for } n \rightarrow \infty \text{ almost everywhere.}$$ [45]

In more informal terms, the theorem asserts that the truth or falsity of hypothesis ψ in w is revealed in the long run. It implies that if P^* is a probability measure that assigns probability zero to the same elements of the probability space as P, then

$$\left(P^*(\psi \mid \bigcap_{i<n} \varphi_i^{(w)}) - P(\psi \mid \bigcap_{i<n} \varphi_i^{(w)})\right) \rightarrow 0 \quad \text{for } n \rightarrow \infty \text{ holds almost everywhere.}$$ [46]

Put informally, agents starting with different but equally dogmatic prior probabilities (i.e., they assign zero probability to the same elements of the probability space) converge in their probability assignments in the long run. The observation of more and more data not only washes out differences in priors of hypotheses but also differences in subjective likelihoods based on differences in priors of nuisance parameters.

What about the rate of convergence in Gaifman and Snir's theorem? In contrast to Savage's theorem it does not come with an estimate of the rate of convergence. Even worse, as Earman[47] points out, it does not seem possible to derive informative estimates in the general setting of the theorem. So, its relevance for scientific practice is already very limited. But can something more specific be said about its application to the phylogenetic context? Not surprisingly, one of the assumptions of the theorem is violated. The culprit is to be found in the separation assumption of the convergence theorem. The separation condition requires that for any two worlds, there exist some (possibly quantified) description of the data that holds in one, and is false in the other.

The working of the separation condition can be illustrated in the case of the two theologians. Suppose that the set of possible worlds is divided into subsets: those worlds in which God exist and those in which He (or She) does not. Further suppose that God exists in w_1 and God does not exist in w_2. The separation condition requires that there is an observational sentence $D^*(w_1, w_2)$ that separates these

45 The expression 'almost everywhere' means that the statement under consideration is true for all worlds belonging to some set of probability 1.

46 Note that here 'almost everywhere' means the same with respect to probability measure P as it does for probability measure P^* since the sets of probability 1 are the same for the two probability measures.

47 Earman, *loc.cit.*, p. 148.

two worlds. In order to simplify the example assume that this observational sentence separates any two worlds that come from the two different subsets. Hence, we can drop the reference to the particular worlds w_1 and w_2 in $D^*(w_1, w_2)$, that is, $D^*(w_1, w_2) = D^*$. Say D^* is true in worlds in which God exists, and false in worlds where God does not exist. For instance, D^* might refer to the occurrence of a 'miracle'. Since D^* is implied by hypothesis H, the probability calculus requires that the likelihood $P(D^*/H)$ is equal to one. Further, since $-D^*$ is implied by $-H$, the likelihood $P(-D^*|-H)$ is also equal to one. While the two theologians differ regarding the probability they attribute to the event of observing human tragedies (i.e., event D) under the assumption that God exists (i.e., hypothesis H), the two scholars must agree in their assignments of $P(D^*/H)$ and $P(D^*/-H)$. Roughly the theorem then works as follows: as more data are collected, 'separators', such as D^*, come into effect by ruling out alternatives to the truth by means of these objective likelihoods.

Returning to the phylogenetic context, suppose that w_1 is a world in which fin whales (F) are more closely related to grey whales (G) than to minke whales (A), that is, the tree topology ((F, G), A) is true in w_1. Further, suppose that w_2 is a world in which fin whales (F) are more closely related to minke whales (A) than to grey whales (G), that is, the tree topology ((F, G), M) is true in w_2. Finally, let $\{\varphi_1, \varphi_2, \ldots\}$ denote the sequence of descriptions of DNA sequence data from the three species under consideration. Obviously, the genome of any organism has only a finite number of nucleotides but let us assume for the sake of argument that these organisms have DNA sequences of infinite length. In that case suppose that φ_i denotes the alignment of i nucleotides of each species (i.e., $\varphi_i = (a_i, f_i, g_i)$) where a_i denotes the first i nucleotide of species A etc.). Now, the question is whether there is any sentence in $\{\varphi_1, \varphi_2, \ldots\}$ that separates the two worlds w_1 and w_2. Put differently, is there any alignment of DNA sequence data of the three species that is true in one of the two worlds and false in the other? No, any alignment is compatible with both tree topologies given our current understanding of evolutionary processes on the molecular level. It is possible that in both worlds w_1 and w_2 the same alignments obtain. The difference is just that depending on the model of molecular evolution assumed, the two tree topologies assign different probabilities to an alignment. Gaifman and Snir's separation assumption is not satisfied in the phylogenetic context.

20.4 Conclusion

While philosophical discussions of the washing out of priors strategy typically focus on the priors of the hypotheses of interest, this paper discussed the priors of nuisance parameters in the context of molecular phylogenetics. Based on a study on the phylogeny of baleen whales it was shown that confirmation statements are sensitive to the priors of nuisance parameters. Further, it was argued that two prominent convergence theorems fail to apply in molecular phylogenetics. This, of course, is no proof that *no* convergence theorem applies in the phylogenetic context. Other convergence theorems have been established in the literature and additional theorems might be proven in the future.[48] However, the role of nuisance parameters in a Bayesian phylogenetic analysis typically renders these results inapplicable and creates a serious challenge for any future convergence theorems to be established.[49] Returning to Earman's two strategies of addressing the problem of the priors, this leaves the phylogeneticist with the constraining of priors strategy. This, however, is the topic of another paper.

Department of Philosophy, Logic and Scientific Method
London School of Economics
Houghton Street
WC2A 2AE, London
United Kingdom
B.C.Autzen@lse.ac.uk

48 See, for instance, James Hawthorne, "Confirmation Theory", in: Prasanta Bandyopadhyay and Malcolm Forster (Eds.), *Philosophy of Statistics, Handbook of the Philosophy of Science, Volume 7*. Elsevier (to appear).

49 For instance, Hawthorne's 'Likelihood Ratio Convergence Theorem' (Hawthorne, loc. cit.) invokes the 'Directional Agreement Condition' when dealing with subjective likelihoods. This condition is generally not satisfied in the phylogenetic context.

Team C
Philosophy of the Cultural and Social Sciences

CHAPTER 21

ILKKA NIINILUOTO

QUANTITIES AS REALISTIC IDEALIZATIONS

The quantitative method is a powerful tool in the natural and social sciences. Depending on the research problem, the use of quantities may give significant methodological gains. Therefore, it is an important task of philosophers to clarify two related questions: in what sense do mathematical objects and structures exist? What justifies the assignment of numbers and numerals to real objects? The former question is discussed in the ontology of mathematics, the latter in the theory of measurement. This paper defends constructive realism in mathematics: numbers and other mathematical entities are human constructions which can be applied to natural and social reality by means of representation theorems. The axioms of such representation theorems are typically idealizations in the sense analyzed by critical scientific realists. If the axioms are truthlike, then quantities can be regarded as realistic idealizations.

21.1 THE QUANTITATIVE METHOD

By the *quantitative method* we mean the use of quantities, quantitative concepts and mathematical methods in science. The history of this method started in the antiquity within optics, astronomy, and statics. In the early modern times, it was successfully applied in dynamics and mechanics by Galileo, Kepler, Descartes, Newton, and Leibniz. In the nineteenth century, statistical methods were developed in genetics and the social sciences. In the twentieth century, mathematical models have been employed as a tool in economics and psychology. Today, computers with numerical and simulation methods can be used in all fields of research.[1]

Philosophers of language have traditionally distinguished the extension and intension of a linguistic expression. *Concepts* are intensions of terms. In particular, the intension of a monadic predicate is a property and its extension is a class: the term "red" expresses the property of redness and denotes the class of all red objects. The intension of a two-place predicates is a relational concept, and its extension is a relation. The intension of a function term or functor is a function concept and its extension is a function.

1 Cf. Wenceslao J. Gonzalez, "The Role of Experiments in the Social Sciences: The Case of Economics", in: Theo Kuipers (Ed.), *Handbook of the Philosophy of Sciences: General Philosophy of Science – Focal Issues*, Amsterdam: Elsevier 2007, pp. 275–301.

D. Dieks et al. (eds.), *Probabilities, Laws, and Structures*, The Philosophy of Science in a European Perspective 3, DOI 10.1007/978-94-007-3030-4_21,

A related distinction can be given between classificatory, comparative, and quantitative concepts.[2] The extensions of *classificatory* terms, like "hot", are classes. *Comparative* terms, like "is warmer than", can express comparisons of objects with respect to some property. Quantitative terms, and *quantitative concepts* or *quantities* as their intensions, like temperature in degrees centigrade, attribute single numbers (scalars) or several numbers (vectors) to objects. One usually thinks that concept formation goes from classificatory to quantitative concepts via comparative ones, but movement in the opposite direction is also viable.

Quantitative methods are contrasted with *qualitative research*, including classification by descriptive terms, functional and teleological explanations, and hermeneutic or interpretative methods.[3] The use of quantities has many methodological virtues[4]: explication or the replacement of everyday terms with exact and unambiguous concepts improves precision and accuracy, information and communication. Quantitative vocabulary facilitates the treatment and estimation of empirical uncertainty. In empirical science "concept formation and theory formation go hand in hand"[5]: quantities allow us to formulate causal laws, dynamic laws, theories and models with great systematic power, thereby helping the basic goals of scientific inference: reduction, explanation, and prediction. Further, quantitative hypotheses may have a high degree of testability.

In the positivist tradition, the quantitative approach is regarded as the only acceptable method in science. Some critics of positivism have mistakenly claimed that this approach presupposes that reality can be divided into countable "units" or "atoms" – this claim ignores the rich and flexible framework of measurement scales. In spite of the importance of measuring instruments (e.g., thermometers[6]), already Hempel[7] argued that the operationalist school naively assumes that measuring devices define concepts, when they in fact help to test quantitative statements. The claim that quantitative approaches lead only to trivial results reflects the fact that sometimes its proponents have failed to use imagination and boldness in their formulation of research problems.

In general, the potential gains of the use of the quantitative method depend on the research problem, cognitive interests of the researcher, and matters of fact about the object of inquiry. For a philosopher, the most important critical issue

---

2 Carl Gustav Hempel, *Fundamentals of Concept Formation in Empirical Science*, Chicago: The University of Chicago Press 1952; and Rudolf Carnap, *An Introduction to the Philosophy of Science*, New York: Basic Books 1966.

3 Norman K. Denzin, and Yvonna S. Lincoln (Eds.), *Handbook of Qualitative Research*, London: SAGE 1994.

4 Abraham Kaplan, *The Conduct of Inquiry: Methodology for Behavioral Science*, New York: Chandler 1964.

5 Carl Gustav Hempel, *Ibid.*

6 See Hasok Chang, *Inventing Temperature: Measurement and Scientific Progress*, Oxford: Oxford University Press 2004.

7 Carl Gustav Hempel, *Ibid.*

concerns the scope of the quantitative method: are all important aspects of reality numerically measurable?

21.2 Mathematics

Quantities provide a bridge between empirical science and mathematics. Therefore, it is important to start with the ontological question about the existence of mathematical entities.

For this purpose, it is useful to refer to Popper's ontological distinction: World 1 consists of physical or material things and processes, World 2 subjective mental states and events, and World 3 abstract entities.

Platonists place mathematical objects in World 3, understood as a pre-existing and ultimate domain of ideal entities. Intuitionist place them as mental constructions in World 2, formalists as material signs in World 1.[8] These three positions are essentially the same as the views in the classical debate on universals: realism, conceptualism, and nominalism.

Various kinds of naturalists, physicalists, and empiricists locate mathematical structures in World 1. Here they agree with the mathematical realism of Galileo who argued that the Book of Nature is "written in the mathematical language, and the symbols are triangles, circles, and other geometrical figures". Locke also argued that "primary qualities" (extension, figure, motion or rest, solidity, number) are objective, while "secondary qualities" (color, taste, smell) are subjective.

The logical empiricists made a sharp distinction between pure mathematics, which is a priori and analytic, and interpreted mathematics, which is synthetic and a posteriori. For example, physical geometry is a branch of natural science.[9]

Popper's approach in *Objective Knowledge*[10] suggests that one can be at the same time a realist and constructivist in the philosophy of mathematics.[11] According to Popper, World 3 is human-made: we create abstract objects, among them mathematical structures, cultural objects, artefacts, works of art, language, and social institutions. Such public constructions are real, sustained by their documen-

8 See Paul Benacerraf and Hilary Putnam (Eds.), *Philosophy of Mathematics: Selected Readings*, Oxford: Blackwell 1964.

9 Rudolf Carnap, *Ibid.*

10 Karl Raimund Popper, *Objective Knowledge*, Oxford: Oxford University Press 1972.

11 See Ilkka Niiniluoto, "Reality, Truth, and Confirmation in Mathematics – Reflections on the Quasi-Empiricist Programme", in: Javier Echeverria, Andoni Ibarra, and Thomas Mormann (Eds.), *The Space of Mathematics: Philosophical, Epistemological, and Historical Explorations*, Berlin: Walter de Gruyter 1992, pp. 60–78; Ilkka Niiniluoto, "World 3: A Critical Defence", in: Ian Jarvie, Karl Milford, and David Miller (Eds.), *Karl Popper: A Centenary Assessment, vol. II. Metaphysics and Epistemology*, Aldershot: Ashgate 2006, pp. 59–69; and Donald Gillies, "Informational Realism and World 3", in: *Knowledge Technology and Policy*, 23, 1–2, 2010, pp. 7–24.

tations in World 1 and manifestations in World 2. Yet they transcend their makers by not being entirely transparent to us.

Mathematical structures can be applied to Worlds 1, 2, and 3, even though infinite structures like N (natural numbers) and R (real numbers) cannot be exhausted by their physical or mental interpretations. Such applications can be explained by modern theories of measurement which show how "metrization" leads from comparative to quantitative concepts. The representation theorems do not prove the existence of mathematical entities, as the structures N and R are presupposed, but they justify the assignment of numerals to physical objects and mental states.

21.3 Measurement

Campbell distinguished in 1920 "quantities" or "extensive properties" which are capable of addition and "qualities" or "intensive properties" which are not. He assumed that only quantities can be fundamentally measured.[12] However, the model-theoretical account of measurement allows for the metrization of intensive properties.[13] This study, sometimes called "mathematical psychology", has been important for the development of mathematical methodology in economics.

An empirical system $(E, \triangleright)$ is a set E of objects with a binary relation $\triangleright$. A real-valued measure function $m: E \rightarrow R$ agrees with $\triangleright$ if for all a, b in E,

$$a \triangleright b \text{ iff } m(a) \geq m(b).$$

Scales of measurement are then defined by the uniqueness condition on measure m:

a one-to-one function:	nominal scale
a strictly increasing function:	ordinal scale
an affine transformation:	interval scale
a similarity transformation:	ratio scale
the identity function:	absolute scale.

The scale determines which mathematical operations (e.g., addition and multiplication) may be performed with numbers, which is an important presupposition

12 See Norman Robert Campbell, *Physics: The Elements*, Cambridge: Cambridge University Press 1920. Reprinted as *Foundations of Science: The Philosophy of Theory and Experiment*, New York: Dover 1957.

13 Dana Scott and Patrick Suppes, "Foundational Aspects of Theories of Measurement", in: *Journal of Symbolic Logic*, 23, 1958, pp. 113–128; David Krantz, Duncan Luce, Patrick Suppes, and Amos Tversky, *Foundations of Measurement*, vol I. New York: Academic Press 1971.

in the use of quantities in the formation of mathematical models and systematic theories.

Let us consider the example of extensive measurement. Let E be a set of physical objects, $\triangleright$ a binary relation on E, and o a binary operation on E (concatenation). Define $a \sim b$ iff $a \triangleright b$ and $b \triangleright a$. Let $1a = a$, $(n+1)a = (na \text{ o } a)$. Then $<E, \triangleright, o>$ is an *extensive system* if

(i) $\triangleright$ is reflexive, transitive and connected in E
(ii) $a \text{ o } (b \text{ o } c) \sim (a \text{ o } b) \text{ o } c$
(iii) $a \triangleright b$ iff $(a \text{ o } c) \triangleright (b \text{ o } c)$ iff $(c \text{ o } a) \triangleright (c \text{ o } b)$
(iv) $(a \text{ o } b) \triangleright a$ but not $a \triangleright (a \text{ o } b)$
(v) if $a \triangleright b$, then for all c, d in E there exists a positive n in N such that $(na \text{ o } c) \triangleright (nb \text{ o } d)$

The following representation theorem can be proved for extensive measurement: $<E, \triangleright, o>$ is an extensive system iff there exists a positive function m: $E \rightarrow R$ such that for all a, b in E

$a \triangleright b$ iff $m(a) \geq m(b)$
$m(a \text{ o } b) = m(a) + m(b)$.

Measure function m is unique up to a similarity transformation $m' = \alpha m$ for $\alpha > 0$ (ratio scale).

For example, define $a \triangleright b$ by the condition: a is at least as low as b in an equal arm balance, and let a o b be the combined object a and b. Then conditions (i)–(iv) are factually true, and the measure function m(a) is the mass of object a. When the scale has been fixed, we can make factually true statements of the form "2 kg + 1 kg = 3 kg". Similar constructions can be given for the quantities of length, time duration, resistance, and velocity.

The possibility of applying arithmetic to reality can also be explained by the theory of measurement. Frege argued quite convincingly that Mill, who defended empiricism in mathematics,[14] confused the applications of arithmetical propositions and the pure mathematical proposition itself.[15] For example, in arithmetic the equation "2+1 = 3" can be proved. In the pebble arithmetic it is the case that two pebbles and one pebble equal three pebbles. But this holds only if the operation of heaping up pebbles factually satisfies the conditions of measurement with an absolute scale. A similar rule of addition does not hold for water drops.

Let $A \triangleright B$ mean that agent X regards event A at least as probable as event B. Assume that

14 See John Stuart Mill, *A System of Logic*, 8th ed., London: Longmans, Green, and Co. 1906.

15 See Gottlob Frege, *The Foundations of Arithmetic*, Oxford: Blackwell 1959, p. 73.

$\triangleright$ is transitive and connected,
A $\triangleright$ Ø (impossible event),
not Ø $\triangleright$ E (sure event),
if $A \cap C = B \cap C = \emptyset$, then A $\triangleright$ B iff AUC $\triangleright$ BUC.

Two events A and B are equal if A $\triangleright$ B and B $\triangleright$ A. De Finetti's theorem for qualitative probability states that if E can be divided into n equal parts for each n in N, then there is a unique probability measure P such that A $\triangleright$ B iff $P(A) \geq P(B)$.[16] Necessary and sufficient conditions for the representation of qualitative probability have been found by Scott in the 1960s.

Let A $\triangleright$ B mean that agent X regards option A at least as good as option B. A function u which agrees with this preference relation is ordinal utility: A $\triangleright$ B iff $u(A) \geq u(B)$.

Cardinal utilities have been axiomatized by Ramsey in 1926 and von Neumann and Morgenstern in 1944: if u is a utility function, so is $u' = au + b$ (interval scale). Savage[17] gave conditions for the existence of a subjective probability measure P and a utility function u which satisfy the *Principle of Subjective Expected Utility* for all acts A and B:

(SEU) A $\triangleright$ B iff $Exp^P u(A) \geq Exp^P u(B)$.

Causal decision theory gives up Savage's assumption that the probabilities of states of nature are independent of the performed act, and this alternative to SEU can again be justified by its own representation theorem.[18]

The theory of measurement shows that the presupposition of applying the quantitative method is not Galilean mathematical realism. One form of this realism in contemporary ontology is the doctrine of quantitative tropes: for example, the relation 'Sam is sadder than Hans' is internal or grounded in the sadness tropes of Sam and Hans.[19] Instead, in the representation theorems, comparative relations are treated as primary, and the properties of these relations justify the assignment of degrees or numerical values to compared objects. This approach is thus compatible with the view that reality as such is qualitative, and the quantitative approach is a way of describing and systematizing a research area. The use of this method is tantamount to a choice of language.[20] The choice between quantitative and qualitative approaches is factual and methodological rather than metaphysical: quantities

16 See Leonard Jimmie Savage, *The Foundations of Statistics*, New York: John Wiley 1954.
17 Leonard Jimmie Savage, *Ibid.*
18 See Brad Armendt, "A Foundation for Causal Decision Theory", in: *Topoi*, 5, 1986, pp. 3–19.
19 See Kevin Mulligan, "Internal Relations", in: Jaegwon Kim and Ernest Sosa (Eds.), *A Companion to Metaphysics*, Oxford: Blackwell 1995, pp. 245–246.
20 See Rudolf Carnap, *Ibid.*, p. 59.

are justified to the extent that the assumptions of the relevant representation theorems are true.[21]

One should add that in the social world there is an important example of a real quantity which is exemplified in coins and banknotes. However, based on social conventions and institutions, money is an abstract entity whose existence and value can be explained in the same way as mathematical objects in World 3.[22]

21.4 Realism and Idealizations

It is well known that many assumptions of mathematical models in the social sciences are unrealistic or false about real-life agents. For example, economic theories and models typically assume that human beings and business firms are perfectly rational agents maximizing profit on the basis of complete information.[23]

Axioms for qualitative probability and utility are often called "rationality" conditions for a person or for her beliefs and preferences. Suppes distinguishes rationality axioms and structural axioms.[24] Typically the former are universal (e.g., transitivity: if A $\rhd$B and B $\rhd$C, then A $\rhd$C), the latter existential (e.g., existence of equal n-fold partitions, solvability). Structural assumptions are nonnecessary axioms which limit the domain of the applicability of the representation theorems.[25]

There are many experimental studies which try to show that decision theory is not a descriptively correct account of human behavior.[26] The criticism does not concern only the structural assumptions needed to prove representations theorems for probability and utility. For example, Savage's model neglects the phenomenon of risk aversion and assumes, as necessary conditions, axioms which are often violated by the behavior or intuitions of real-life agents (e.g., the transitivity of preferences, the sure-thing principle).

One reaction to these observations is *instrumentalism*: mathematical models are understood as devices for the systematization of observational statements, adopted for the sake of convenience and simplicity. For example, Machlup treats neoclassical firms as fictions, and Friedman regards economic theories as merely

21 Ilkka Niiniluoto, "Reality, Truth, and Confirmation in Mathematics — Reflections on the Quasi-Empiricist Programme", *op. cit.*

22 See Ilkka Niiniluoto, "World 3: A Critical Defence", *op. cit*; Gillies, *Ibid.*

23 See Bert Hamminga and Neil De Marchi, (Eds.), *Idealization in Economics*, Amsterdam: Rodopi 1994; Uskali Mäki (Ed.), *Fact and Fiction in Economics*, Cambridge: Cambridge University Press 2002.

24 See Patrick Suppes, *Studies in the Methodology and Foundations of Science: Selected Papers from 1951 to 1969*, Dordrecht: D. Reidel 1969, p. 95.

25 David Krantz et al, *Ibid.*, p. 95.

26 E.g., Amos Tversky, "A Critique of Expected Utility Theory: Descriptive and Normative Considerations", in: *Erkenntnis*, 9, 1975, pp. 163–174.

predictive models. The SEU model is often expressed in an instrumentalist spirit: if the assumptions of preferences were true, then human agents would behave *as if* they were maximizing cardinal utilities relative to subjective probabilities.

Another alternative is the *normative* interpretation: human beings should satisfy the pure axioms of rationality, and mathematical models like SEU express a normative constraint for real-life agents.[27] On the normative reading, mathematical models are not falsifiable statements about human behavior. Many human agents are willing to modify their behavior if violation of transitivity is explicitly shown to them.[28] An interesting comment on this was given by Savage:

> I am not familiar with any serious analysis of the notion that a theory is only slightly inexact or is almost true, though philosophers of science have perhaps presented some. Even if valid analyses of the notion have been made, or are made in the future, for the ordinary theories science, it is not to be expected that those analyses will be immediately applicable to the theory of personal probability, normatively interpreted; because that theory is a code of consistency for the person applying it, not a system of predictions about the world around him.[29]

Another relevant remark is by Mill who in his *A System of Logic* in 1843 pointed out that the necessity of conclusions in geometry means that they deductively or "correctly follow from the suppositions", but these suppositions "are so far from being necessary that they are not even true; they purposively depart, more or less, widely from the truth".[30] Mill thus realized that the application of mathematics to reality involves simplifying and idealizing assumptions. This is a point which has been systematically developed by critical scientific realists.[31]

According to *critical scientific realism*, scientific theories are attempts to give true or truthlike descriptions of reality. A theory can represented as a set of possible worlds (a disjunction of constituents), and its truthlikeness depends on the distances of its elements from the actual world (the true constituent). Such distances can be defined between constituents as linguistic entities. Alternatively, a theory is true about a model and the model is similar to the real system. More precisely, a theory is *approximately true* if some of its model is close to the real system, and *truthlike* if all of its models are close to the real system. For true theories truthlikeness covaries with logical strength, but also some false theories may be so close to the truth that they are truthlike. This approach can be developed

---

27 See Peter Gärdenfors and Nils-Eric Sahlin, (Eds.), *Decision, Probability, and Utility: Selected Readings*, Cambridge: Cambridge University Press 1988.

28 David Krantz et al, *Ibid.*, p. 418.

29 Leonard Jimmie Savage, *Ibid.*, p. 59.

30 See John Stuart Mill, *Ibid.*

31 See Ilkka Niiniluoto, *Critical Scientific Realism*, Oxford: Oxford University Press 1999.

to all kinds of singular and general statements with classificatory, relational, and quantitative terms.[32]

For a critical realist, science makes progress so far as its successive theories succeed in gaining increasingly truthlike information about real systems. In this dynamical picture of progress, rival successive theories can refer to same partly unknown entities: by the principle of Charitable Reference, theoretical terms in a theory H refer to those real entities which make the theory H most truthlike.

Idealization is an important feature of the natural sciences as well. Already Galileo knew that theories in physics typically contain idealizations and approximations: some factors are ignored (e.g., resistance of air), some exaggerated (e.g., the velocity of light c is infinite or 1/c is zero). Idealized theories can be formulated as counterfactual conditionals (if idealizing assumptions were true, then …), and in this form they may be true or truthlike. The method of idealization and concretization,[33] i.e. the introduction and removal of idealizational assumptions, is a progressive way of approaching to the truth.[34]

Examples of idealizations can be found in the representation theorems of measurement theories. The axioms of extensive measurement of mass include the idealization that all massive bodies of noninteractive substances can be compared in a frictionless equal arm balance.[35] The transitivity of preferences is a "quasi-idealization" in Nowak's[36] sense: it may hold for some agents in some situations but may also fail in other circumstances. Transitivity excludes temporal considerations or presupposes that the person has a reliable memory and does not change her mind. It also presupposes that the preference scale is one-dimensional: violations of transitivity may occur in multi-dimensional choice situations which resemble the famous voting paradoxes.[37] While the representation theorems are valid mathematical statements, their premises are truthlike idealizations which hold in worlds similar to the actual world. This may be the case for the SEU model and its concretizations in causal decision theory. For a critical realist, this means that cardinal utility functions as mathematical constructions may help to give truthlike descriptions of rational human decision-making, even though such a quantification of preference does not actually exist in the human mind.[38]

---

32 Ilkka Niiniluoto, *Truthlikess*, Dordrecht: D. Reidel 1987.

33 Leszek Nowak, *The Structure of Idealization*, Dordrecht: D. Reidel 1980.

34 Ilkka Niiniluoto, "Idealization, Counterfactuals, and Truthlikeness", in: Jerzy Brzezinski et al. (Eds.), *The Courage of Doing Philosophy: Essays Presented to Leszek Nowak*, Amsterdam: Rodopi 2007, pp. 103–122.

35 David Krantz et al., *Ibid.*, p. 89.

36 Leszek Nowak, *Ibid.*

37 Amos Tversky, *Ibid.*

38 Ilkka Niiniluoto, "Truthlikeness and Economic Theories", in: Uskali Mäki (Ed.), *Fact and Fiction in Economics*, Cambridge: Cambridge University Press 2002, pp. 214–228.

What could be the actual charitable reference of the utility function? One of the many candidates is money as a real quantity in the social world – even though we know that the utility of money for a person may differ from its actual monetary worth. Other alternatives might include personal degrees of satisfaction and degrees of technical usefulness.[39]

21.5 Conclusion

The use of quantities allows the scientists to construe fruitful theories and mathematical models and to apply the powerful method of idealization and concretization. But the introduction of quantities can also be regarded as an idealization: if the axioms of representation theorems are truthlike or approximately true, then quantities exist in possible worlds that are near to the actual world. In this sense, quantities are realistic idealizations.

Department of Philosophy, History, Culture and Art Studies
University of Helsinki
Unioninkatu 40 A
00014, Helsinki
Fiinland
ilkka.niiniluoto@helsinki.fi

39 Bengt Hansson, "Risk Aversion as a Problem of Conjoint Measurement", in: Peter Gärdenfors and Nils-Eric Sahlin, (Eds.), *Decision, Probability, and Utility: Selected Readings*, Cambridge: Cambridge University Press 1988, pp. 136–158.

Chapter 22

Marcel Boumans

Mathematics as Quasi-Matter to Build Models as Instruments

22.1 Introduction

I have argued elsewhere that models should be distinguished from theories.[1] They are not theories about the world but instruments through which we can see the world and so gain some understanding of it. As mathematical representations, models should also be distinguished from pure formal objects. They should be seen as devices that help us to see the phenomena more clearly. Models are instruments of investigation, epistemological equivalent to the microscope and the telescope. In a textbook on optical instruments, we find the following description:

> The primary function of a lens or lens system will usually be that of making a pictorial representation or record of some object or other, and this record will usually be much more suitable for the purpose for which it is required than the original object.[2]

If one replaces "lens or lens system" by "model", one has an adequate description of the way that models are understood and treated in this paper.

One usually associates the word instrument with a physical device, such as a microscope or telescope. Models, however, are not material objects, they are mathematical objects. The absence of materiality makes that the physical methods used to test material instruments, such as control and insulation, cannot be applied to models.[3] This means that we cannot easily borrow from the philosophy of technology, which is geared to physical objects. Models, being "quasi-material" objects belonging to a world in between the immaterial world of theoretical ideas and the material world of physical objects, require an alternative epistemology.

1 Marcel Boumans, *How Economists Model the World into Numbers*, London and New York: Routledge 2005.

2 Ronald John Bracey, *The Technique of Optical Instrument Design*, London: The English University Press 1960, p. 15.

3 This requirement of materiality for controllability (in the usual meaning of this term) has also been discussed in Marcel Boumans, and Mary S. Morgan, "Ceteris Paribus Conditions: Materiality and the Applications of Economic Theories", in: *Journal of Economic Methodology*, 8, 1, 2001, pp. 11–26. This essay also treats the kinds of controllability that are possible in the case of quasi-material or non-material experiments.

D. Dieks et al. (eds.), *Probabilities, Laws, and Structures*, The Philosophy of Science in a European Perspective 3, DOI 10.1007/978-94-007-3030-4_22,

In comparing the epistemological difference between models and experiments, Morgan[4] argues that although both function as "epistemic mediators"[5] and can be understood to work in an experimental mode, experiments offer greater epistemic power than models as a means to investigate the world. This outcome rests on the distinction that whereas experiments are *versions of* the real world captured within an artificial laboratory environment, models are artificial worlds built to *represent* the real world. The model world is artificial because made out of mathematics. This difference in ontology has epistemic consequences: experiments have greater potential to make strong inferences back to the world.[6] This latter power is manifest in the possibility that whereas working with models may lead to "surprise", experimental results may be unexplainable within existing theory and so "confound" the experimenter.

According to Morgan, the reason that working with mathematical models only surprises is that the model-builder knows the resources that went into the model. Using the model may reveal some surprising, and perhaps unexpected, aspects of the model behaviour. But in principle, the constraints on the model's behaviour are set, however opaque they may be, by the model builder so that however unexpected the model outcomes, they can be traced back to, and re-explained in terms of, the model. "That possibility may not be open to us with material experiments where ignorance may prevent us from explaining why a particular set of results occur",[7] because we might have "the wrong account of theory about what will happen or our knowledge of the world might be seriously incomplete".[8]

Another reason for a difference in epistemic power is the "potential for independent action" by nature in laboratory experiments from which new phenomena emerge which confounds the experimenter. This "potential for independent action" is, according to Morgan, an important consideration in the design of experiments: experiments need to be set up with a certain degree of freedom so that the behaviour in the experiment is not totally determined by the theory involved, nor by the

4 Mary S. Morgan, "Experiments versus Models: New Phenomena, Inference and Surprise", in: *Journal of Economic Methodology*, 12, 2, 2005, pp. 317–329. See also Mary S. Morgan, "Experiments without Material Intervention: Model Experiments, Virtual Experiments, and Virtually Experiments", in: Hans Radder (Ed.), *The Philosophy of Scientific Experimentation*, Pittsburgh: University of Pittsburgh Press 2003, pp. 216–235.

5 This term is coined by Lorenzo Magnani, "Epistemic Mediators and Model-Based Discovery in Science", in: Lorenzo Magnani and Nancy J. Nersessian (Eds.), *Model-Based Reasoning. Science, Technology, Values*, New York: Kluwer Academic/Plenum Publishers 2002, pp. 305–329. This article nicely captures the functioning of both models and experiments in empirical research.

6 See for this latter claim: Francesco Guala, "Models, Simulations, and Experiments", in: Lorenzo Magnani and Nancy J. Nersessian (Eds.), *Model-Based Reasoning. Science, Technology, Values*, pp. 59–74.

7 Mary Morgan, "Experiments without Material Intervention", *loc. cit.*, p. 220.

8 *Ibid.*, p. 220.

rules of the experiment. If the experimental behaviour is totally predetermined, there is no "potential for unexpected patterns to emerge".[9] There must be potential to confound the experimenter with noteworthy results which are both surprising and unexplainable within the given realm of theory. This potential for laboratory experiments to surprise and confound contrasts, according to Morgan, with the potential for mathematical modelling only to surprise. The indeterminateness of nature – even domesticated in a laboratory – allows not only for surprises but also will confound the experimenter.

The implicit assumptions Morgan uses to distinguish between material experiments and non-material models are that in the artificial world of the model there is no potential for unexpected patters, no potential for independent action, no potential for discovery of new phenomena. As soon as the model-builder knows the resources that went into the model, knows the constraints, than only surprise is what can be achieved, surprise that can be traced back, that can be explained. Moreover, the modeller is much less ignorant about (the theories of) the mathematical tools, concept and structures being used, it is only that "we do not already know about how those structures behave when the parts of the model are put together or when we vary certain things in the model".[10] Discoveries only appear because we cannot see all the mathematically derivable consequences of a self-constructed mathematical world. The world of the model is much more determined compared to the world of the experiment.

This paper will show that mathematical models are quasi-empirical objects, objects that are made of quasi-material mathematics, and so have potential to confound the modeller because of the indeterminateness of the mathematical matter. Mathematical matter is indeterminate in two different ways: (1) indeterminate with respect to ignorance: the possibility of having false or incomplete theories of the mathematical structures; (2) indeterminate with respect to potential independence: despite the structure may be determined, there is still a certain degree of freedom, of indeterminacy, for the patterns of behaviour of the mathematical objects.

22.2 Lost Materiality

In mathematics and physics, the term "model" originally specifically referred to material objects: "a representation in three dimensions of some projected or existing structure, or of some material object artificial or natural, showing the proportions and arrangement of its component parts", or "an object or figure in clay, wax, or the like, and intended to be reproduced in a more durable material".[11]

9 Mary Morgan, "Experiments versus Models", *loc. cit.*, p. 324.
10 Mary Morgan, "Experiments without Material Intervention", *loc. cit.*, p. 220.
11 Oxford English Dictionary, Oxford: Clarendon Press 1933.

Boltzmann's entry for "Model" in the *Encyclopaedia Britannica*[12] also indicates its material roots: "a tangible representation, whether the size be equal, or greater, or smaller, of an object which is either in actual existence, or has to be constructed in fact or thought".[13] To Boltzmann, models could only be material.

At the beginning of the twentieth century, the term "mathematical model" referred to a physical three-dimensional representation of a mathematical entity. Models lost their materiality halfway the 1930s. Usually the term "scheme" was used to denote a non-material, mathematical representation. This shift in terminology from scheme to (mathematical) model gave name to a new practice of "explicit mathematizing as technique" which matched with an empiric-oriented reaction to the logical view on mathematics.[14]

This non-material concept of a model as a *Darstellung*[15] came from Hertz.[16] While a "model" was still considered by Hertz as something material, it stood in the same relation to the system of inquiry as the images (*Bilder*) we made of this system. Both image and model should satisfy the "first fundamental requirement", also called the requirement of "correctness":

> For if we regard the condition of the model as the representation of the condition of the system, then the consequents of this representation, which according to the laws of this representation must appear, are also the representation of the consequents which must proceed from the original object according to the laws of this original object.[17]

Because various images of the same object are possible, Hertz formulated two additional requirements an image should fulfil. First, the requirement of "logical permissibility": "We should at once denote as inadmissible all images which implicitly contradict the laws of our thought".[18] But two permissible and correct images of the same system may yet differ in respect of "appropriateness".

> Of two images of the same object that is the more appropriate which pictures more of the essential relations of the object, – the one which we may call the more distinct. Of two images of equal distinctness the more appropriate is the one which contains, in addition to the essential characteristics, the smaller number of superfluous or empty relations, – the

12 Ludwig Boltzmann, "Model", in: Brian McGuinness (Ed.), *Theoretical Physics and Philosophical Problems*, Dordrecht: Reidel 1974, pp. 213–220.

13 *Ibid.*, p. 213.

14 Cf. Gerard Alberts, Jaren van Berekening. *Toepassingsgerichte Initiatieven in de Nederlandse Wiskundebeoefening 1945–1960*, Amsterdam: Amsterdam University Press 1998, pp. 134–135.

15 In German philosophy, there is a distinction between "Darstellung" and "Vorstellung". While a "Vorstellung" is a passive mental image of a sense datum. A "Darstellung" is a consciously constructed scheme for knowing.

16 Heinrich Hertz, *The Principles of Mechanics Presented in a New Form*, New York: Dover 1956.

17 *Ibid.*, p. 177.

18 *Ibid.*, p. 2.

simpler of the two. Empty relations cannot be altogether avoided: they enter into the images because they are simply images, – images produced by our mind and necessarily affected by the characteristics of its mode of portrayal.[19]

In short, the three requirements a representation of a phenomenon should fulfil are "conformity" between the relations of the representation and those of the phenomenon, logical correctness, and containing the essential characteristics as simple as possible. Whether an image satisfies the first two requirements can be decided without ambiguity, but

we cannot decide without ambiguity whether an image is appropriate or not; as to this differences of opinion may arise. One image may be more suitable for one purpose, another for another; only by gradually testing many images can we finally succeed in obtaining the most appropriate.[20]

22.3 Rigor

The fulfilment of the second requirement of logical permissibility will appear to be dividing line whether a model is considered to be a formal object or a quasi-material instrument. To see this, I will use Morgan's distinction between two meanings of "formalization".[21] The first meaning is giving form to, shaping or providing an outline of something. This will be discussed below. The second meaning, Morgan attach to "formalization", is based on the contrast of "formal" with "informal", meaning lacking in exactness or in rules whereas "formal" implies something rule bound, following prescribed forms.

The second meaning of formalization is most often taken that the form has to be permissible, in the sense of Hertz's model requirements above; and, permissibility is often taken as the requirement that the formalization should be done rigorously, that is, the rules that one has to follow are considered to be the rules of logic. For example, it is not permissible to have two inconsistent (formal) statements. Moreover, a model of contradictory features is impossible: "If contradictory attributes be assigned to a concept, I say, that mathematically the concept does not exist".[22]

The semantic view of theories, including its model theory, was highly influenced by Hilbert's axiomatization program. According to this view, a model for a theory is considered as an interpretation on which all the axioms of that theory are true. If the axioms are inconsistent, a model does not exist.

19 *Ibid.*, p. 2.

20 *Ibid.*, p. 3.

21 Mary S. Morgan, "Modelling as a Method of Enquiry", in: *The World in the Model*, Cambridge University Press, forthcoming.

22 David Hilbert, "Mathematical Problems", in: *Bulletin of the American Mathematical Society*, 8, 1902, p. 448.

This identification of rigor with axiomatics has, however, not always been made. Israel shows that the interpretation of rigor had changed under the influence of Hilbert's axiomatization program.[23] As a result of that program,

> a rigorous argument was reconceptualized as a logically consistent argument instead of as an argument that connected the problematic phenomenon to a physical phenomenon by use of empirical data. Propositions were henceforth 'true' within the system considered because they were consistent with the assumptions instead of being 'true' because they could be grounded in 'real phenomena'.[24]

Israel discussed the distinction between rigor and axiomatics in relation to the "crisis of present-day mathematics", namely that the axiomatic trend has emptied mathematical research of any external determination and content to such as extent, that the relation to applications has been lost. Although the role of mathematics in applied sciences is growing rapidly, mathematics is still deeply separated from the applied sciences. "What appears to be missing, is a codification of the rules which should define and guide the use of mathematics as an instrument for the description, interpretation and control of phenomena".[25]

Models as instruments ask for a different set of rules than those of logic, so rigor in instrument making will be different from rigor in an axiomatic system. If in an axiomatic system a mathematical object cannot exist when it should fulfil contradictory requirements, it is still possible that it practically can be built and can be used as an instrument for calculations, measurement or other purposes. To understand modelling practices in which models are considered as instruments of investigation, a more general idea of mathematical rigorousness is needed, namely one that demands that a mathematical object only exist when it is constructed according specific rules, not necessarily consistency. Compare this again with the approach in instrument making, like the design of a lens system, see above:

> Sometimes control with a single lens is impossible since some incompatible features are required and a compromise becomes necessary calling for further judgement on the part of the designer as to which error should be reduced and to what degree.[26]

In his design of index numbers (mathematical models that function as measuring instruments to measure price levels), Irving Fisher[27] makes a similar comparison:

---

23 Giorgio Israel, "'Rigor' and 'Axiomatics' in Modern Mathematics", in: *Fundamenta Scientiae*, 2, 2, 1981, pp. 205–219.

24 *Ibid.*, p. 237.

25 *Ibid.*, p. 219.

26 Ronald John Bracey, *ibid.*, p. 18.

27 American economist (1867–1947). Fisher may be considered as one of the first model builder in economics, see Mary S. Morgan, "Learning from Models", in: Mary S. Morgan and Margaret Morrison (Eds.), *Models as Mediators*, Cambridge: Cambridge University Press 1999, pp. 350–351.

[A]lthough in the science of optics we learn that a perfect lens is theoretically impossible, nevertheless, for all practical purposes lenses be constructed so nearly perfect that it is well worth while to study and construct them. So, also, while it seems theoretically impossible to devise an index number, *P*, which shall satisfy all of the tests we should like to impose, it is, nevertheless, possible to construct index numbers which satisfy these tests so well for practical purposes that we may profitably devote serious attention to the study and construction of index numbers.[28]

The practice of model building where models are supposed to function as instruments it seems that formalization should be done in an appropriate way, in the sense of Hertz's model requirements: whether one formalization is more suitable for one, e.g. measuring, purpose than another formalization should be tested on the comparison of both models in how much they are able to attain that purpose.

22.4 The Making of an Instrument

The reason that most model are more "appropriate" for its purpose than "logical permissible" is that they are built by fitting together bits from disparate sources. Model building is comparable to baking a cake without a recipe.[29] It is a trial and error process. You create a new pastry by estimating which ingredients to add and in what order, on the basis of your knowledge and experience in baking a similar, but not identical, cake. A comparable view on model building is expressed by Clive Granger[30]:

I think of a modeler as starting with some disparate pieces – some wood, a few bricks, some nails, and so forth – and attempting to build an object for which he (or she) has only a very inadequate plan, or theory. The modeler can look at related constructs and can use institutional information and will eventually arrive at an approximation of the object that they are trying to represent, perhaps after several attempts.[31]

Others compared model building with "*basteln*" – tinkering – to denote the "art" of model building.[32] The reason that I prefer the analogy of baking is that one of its

28 Irving Fisher, *The Purchasing Power of Money; Its Determination and Relation to Credit, Interest and Crises*, New York: Kelley 1963, p. 200.

29 Marcel Boumans, "Built-in Justification", in: Mary S. Morgan and Margaret Morrison (Eds.), *Models as Mediators*, pp. 66–96.

30 American economist (1934–2009), 2003 Nobel Prize in economics.

31 Clive W. J. Granger, *Empirical Modeling in Economics: Specification and Evaluation*, Cambridge: Cambridge University Press 1999, pp. 6–7.

32 E.g. Frank Stehling, "Wolfgang Eichhorn and the Art of Model Building", in: Walter Erwin Diewert, Klaus Spremann and Frank Stehling (Eds.), *Mathematical Modelling in Economics; Essays in Honor of Wolfgang Eichhorn*, Berlin: Springer-Verlag 1993, pp. vii–xi.

characteristics is that in the end product you can no longer distinguish the separate ingredients.

In a model, the ingredients are theoretical ideas, norms and values, mathematical concepts and techniques, metaphors and analogies, stylized facts and empirical data. Integration takes place by reshaping the ingredients into a mathematical form and merging them into one framework.

Mathematics is the stuff non-material models are made of. The selection of mathematical forms must be such that the disparate ingredients can be harmonized and homogenized into one effective model. Modelling is a process of committing oneself to how aspects of a system should mathematically be represented and at the same time being constrained by the selected mathematical forms. Moreover, not every element in the mathematical model necessarily has an empirical meaning. To make the model workable, sometimes, elements of convenience or fiction have to be introduced.[33]

An important element in the modelling process is mathematical moulding. Mathematical moulding is shaping the ingredients in such a mathematical form that integration is possible. As a result, the choice if the mathematical formalism ingredient is important. It determines the possibilities of the mathematical modelling. However, which formalism should be chosen is not obvious. It is often assumed that mathematics is an efficient and transparent language. One of the most well-known supporters of this view is Paul Samuelson.[34] He considers mathematics to be a transparent mode of communication and that it is this transparency that will stop people making the wrong deductive inferences. As we will see below, mathematics is not always transparent (neither, some would say, is language) and it does not necessarily function as a language.

22.5 Mathematics as Quasi-Matter

Physical instruments are made of matter, like metal, wood, or plastic. Models are made of mathematical matter. To explore what the epistemological implications are of this distinction, Fleischhacker's discussion of substance and quasi-substance will be used.[35] According to Fleischhacker, mathematical objects can be characterised as "quasi-substantial". This means that it is thought of as substantial, whereas

---

33 A similar view, the simulacrum account of models, is developed by Nancy Cartwright, How the Laws of Physics Lie, Oxford: Clarendon Press 1983.

34 American economist (1915–2009), 1970 Nobel Prize in economics. Paul Anthony Samuelson, "Economic Theory and Mathematics – An Appraisal", in: *The American Economic Review, Papers and Proceedings*, 42, 2, 1952, pp. 56–66.

35 Louk Fleischhacker, "Mathematical Abstraction, Idealisation and Intelligibility in Science", in: Craig Dilworth (Ed.), *Idealization IV: Intelligibility in Science*, Amsterdam, Atlanta: Rodopi 1992, pp. 243–263.

in content it is nothing but structure. In other words, a mathematical object is analyzable into "matter" and "form", but its matter-principle is not physical but abstract in the sense that it is understood as the pure principle of structurability. While structurability is a real property of the physical world, in its abstract form it is the intelligible material principle of the world of mathematical objectivity. Structurability is an intelligible aspect of physical indeterminateness, which is the physical matter-principle.

This indeterminateness cannot be completely intelligible, because it is a purely *passive* potency and therefore does not offer any determinate hold to the intellect. It is however known to us by the experience of the senses, just like the physical qualities, the concepts of which remain essentially dependent on experience.[36]

Fleischhacker uses the term "quasi-substance" to indicate that mathematical objects are analyzable into matter and form. To emphasize their matter-aspect, this paper uses the term quasi-matter.

The inderminateness of quasi-matter seems to be the underlying assumption of Lakatos view on the history of mathematics.[37] Lakatos considered mathematical objects as "quasi-empirical objects", and showed that mathematics grows as an "informal, quasi-empirical" discipline. [38] His logic of mathematical discovery was a critique and even an "ultimate rejection" of "formalism", that is, "the school of mathematical philosophy which tends to identify mathematics with its formal axiomatic abstraction".[39]

But what can one *discover* in a formalised theory? Two sorts of things. *First*, one can discover the solution to problems which a suitable programmed Turing machine could solve in a finite time [...]. No mathematician is interested in following out the dreary mechanical 'method' prescribed by such decision procedures. *Secondly*, one can discover the solutions to problems [...], where one can be guided only by the 'method' of 'unregimented insight and good fortune'.[40]

According to Lakatos, the methodology of the "growth" of mathematical knowledge is more similar to the methodology of empirical research than to the methodology of a formal deductive science.

36 *Ibid.*, p. 248.

37 Imre Lakatos, *Proofs and Refutations: The Logic of Mathematical Discovery*, Cambridge: Cambridge University Press 1976.

38 As "quasi-matter", mathematical objects are, in certain respects, parts of Popper's third world, a world that is not yet fully explored. See Karl Raimund Popper, "Epistemology without a Knowing Subject", in: B. van Rootselaar and John F. Staal (Eds.), *Logic, Methodology and Philosophy of Science III*, Amsterdam: North-Holland 1968, pp. 333–373.

39 Imre Lakatos, *ibid.*, p. 1.

40 *Ibid.*, pp. 3–4.

22.6 Two Illustrative Cases

22.6.1 Cycle Model or Not?

An illustrative case of mathematics as quasi-matter is the assumption held by mathematical economists in the 1930s, that mixed difference-differential equations are the most suitable formalism for business-cycle models. In general, it is difficult to solve mixed differential-difference equations. Moreover, in the 1930s, there were hardly any systematic accounts available. Systematic overviews on mixed differential-difference equations did not appear until the early 1950s.[41] As a consequence, they were studied as if they were the same as the more familiar differential equations. The general solution of this latter kind of equation is a finite weighted sum of trigonometric and exponential functions, so that their periodic behaviour can easily be analyzed. In contrast, the general solution of a mixed difference-differential equation is an infinite weighted sum of these functions. This is not necessarily a periodic movement if the weights are not further specified.

In a more recent study, a well-known model of the business cycle, Frisch's 1933 "Rocking Horse Model", a system of three mixed difference-differential equations was analyzed and worked out using computer simulations.[42] It appeared that this system was not a cycle model because when it is subjected to an external shock it evolves back to the equilibrium in a non-cyclical manner.

> Why such a paradoxically result found in [Frisch 1933] went unnoticed for almost sixty years is an intriguing question that, in my opinion, should be of interest to scholars of mathematical economics, business-cycle theory and history of economic thought.[43]

I would like to add here that it also should be of interest to philosophers of science with an interest in mathematical modelling.

This case illustrates that working with a mathematical model, in this case Zambelli's computer simulation, can be confounding. This computer simulation

41 E.g., Richard Bellman and Kenneth L. Cooke, *Differential-Difference Equations*, New York: Academic Press 1963. However, the various mathematical aspects of this kind of equations already attracted attention on the 1930s. In the first place, there is Ragnar Frisch and Harald Holme, "The Characteristic Solutions of a Mixed Difference and Differential Equation Occurring in Economic Dynamics", in: *Econometrica*, 3, 1935, pp. 225–239, but also three papers by R.W. James and Maurice Henry Belz: "On a Mixed Difference and Differential Equation", in: *Econometrica*, 4, 1936, pp. 157–160; "The Influence of Distributed Lags on Kalecki's Theory of the Trade Cycle", in: *Econometrica*, 6, 1938, pp. 159–162; "The Significance of the Characteristic Solutions of Mixed Difference and Differential Equations", in: *Econometrica*, 6, 1938, pp. 326–343.

42 Stefano Zambelli, "The Wooden Horse that Wouldn't Rock: Reconsidering Frisch", in: Kumaraswamy Velupillai (Ed.), *Nonlinearities, Disequilibria and Simulation*, Basingstoke: Macmillan 1992, pp. 27–54.

43 *Ibid.*, p. 53.

did not tell were the former studies went wrong, the surprise could not be traced back, but it showed that the theory on mixed difference-differential equations was not complete.

22.6.2 Spurious Result or Not?

Many textbooks of statistics warn against the use of filters or moving averages because they might produce artificial oscillations due solely to the statistical treatment of the data.[44] This is the so-called (Yule-)Slutzky effect, after two statisticians who studied it in detail. A filter used in current macroeconomics for detrending time series, that is, filtering the trend component out of the time series to extract the business cycle component, is the Hodrick-Prescott filter (HP-filter). In the 1990s however several articles appeared claiming that the HP-filter may extract spurious cycles.[45]

The functioning of such filters is mainly discussed in terms of frequencies extracted by taking the Fourier transform of a linear filter, also called spectral analysis. The problem, however, is that these analyses are not conclusive, because spectral analysis can be only applied to stationary time series, non-stationary time series do not have a periodic decomposition. So, it is only shown for stationary time series that the HP-filter operates like a detrending filter. Hence it is not clear yet what the effect of the HP-filter is when applied to non-stationary time series.

Macroeconomic time series often have an upward trend which makes them non-stationary, and one of the objectives of filtering is transformation to induce stationarity. To analyse the HP-filter effect for these non-stationary cases, the HP-filter is split into two parts. One part is chosen to make the time series stationary so that subsequently the resulting part can be analysed to see its effect on the stationary data. It was shown that this resulting part taken on its own as a filter leads to a Slutzky effect. But one cannot infer from this result that the complete HP-filter has this effect, too. Properties of the split parts of the filter do not necessarily sum to the properties of the complete filter, they may cancel each other out. The Slutzky effect of one part of the filter may be nullified by the other part.[46]

---

44 E.g., Maurice G. Kendall and Alan Stuart, *The Advanced Theory of Statistics*, Volume 3: Design and Analysis, and Time-Series, London: Charles Griffin 1966.

45 E.g., Timothy Cogley and James M. Nason, "Effects of the Hodrick-Prescott Filter on Trend and Difference Stationary Time Series: Implications for Business Cycle Research", in: *Journal of Economic Dynamics and Control*, 19, 1995, pp. 253–278; Andrew C. Harvey and Albert Jaeger, "Detrending, Stylized Facts and the Business Cycle", in: *Journal of Applied Econometrics*, 8, 1993, pp. 231–247; Albert Jaeger, "Mechanical Detrending by Hordick-Prescott Filtering: A Note", in: *Empirical Economic*, 19, 1994, pp. 493–500.

46 For a more detailed discussion of this case see Marcel Boumans, "Calibration of Models in Experiments", in: Lorenzo Magnani and Nancy J. Nersessian (Eds.), *Model-Based Reasoning. Science, Technology, Values*, pp. 75–93.

This case also shows that the theory is not complete, and so different analyses lead to different results. Showing that the HP-filter may lead to spurious results (or not) is confounding because these results can not easily be feed back to a theory.

22.7 Conclusion

The ontological difference between the two epistemic mediators model and experiment has consequences for their individual epistemic powers. Because an experiment is built of the same matter as the phenomenon, it has a greater potential to make strong inferences back to the world than models which are only representations of that world. Morgan uses two different labels to indicate the difference between these epistemic powers: models can surprise, whereas experiments can confound, where surprise is considered to be a weaker epistemic result. Models (can) only surprise because unexpected outcomes can be traced back and re-explained by theory. An experiment (can) confound because of a larger extent of ignorance: we may have a false or incomplete theory. Parts of the world are still not discovered and so new (confounding) phenomena may appear in an experiment.

The underlying assumption for this distinction is that mathematical theories are true and complete, in contrast to empirical theories which are admittedly incomplete and even false. This paper has, however, argued that because mathematical worlds have a similar indeterminatess as the physical world, experimenting on a mathematical world can also lead to confounding results. Mathematical objects have a materiality epistemological similar to that of the physical world of which our knowledge of its structure is dependent on experience. So, our mathematical theories can be wrong, or incomplete. Mathematics is not a transparent language to describe the world, it is quasi-matter out of which we can mould representations of the world.

Department of Economics
University of Amsterdam
Roetersstraat 11
1018 WB, Amsterdam
The Netherlands
m.j.boumans@uva.nl

CHAPTER 23

DAVID F. HENDRY

MATHEMATICAL MODELS AND ECONOMIC FORECASTING: SOME USES AND MIS-USES OF MATHEMATICS IN ECONOMICS

ABSTRACT

We consider three "cases studies" of the uses and mis-uses of mathematics in economics and econometrics. The first concerns economic forecasting, where a mathematical analysis is essential, and is independent of the specific forecasting model and how the process being forecast behaves. The second concerns model selection with more candidate variables than the number of observations. Again, an understanding of the properties of extended general-to-specific procedures is impossible without advanced mathematical analysis. The third concerns inter-temporal optimization and the formation of "rational expectations", where misleading results follow from present mathematical approaches for realistic economies. The appropriate mathematics remains to be developed, and may end "problem specific" rather than generic.

23.1 INTRODUCTION

Mathematics is ubiquitous in modern economics and especially in econometrics. We draw on two examples from the latter where mathematics is essential in order to understand the properties of economic forecasts and the outcomes of empirical model selection exercises respectively. We then use the findings from the first of these to demonstrate important flaws in present approaches to the mathematics of inter-temporal optimization and the formation of expectations, in particular, so-called "rational expectations" as applied to realistic economic time series. The three examples are drawn from work by the author jointly with a number of co-authors. What prompts the need to discuss the obvious, namely the use of mathematics in economics, since that discipline intrinsically includes econometrics? First, it is a long-standing debate, with historical roots in the 19th century.[1] A particularly amusing complaint about the "excessive use of advanced mathematics" is

1 Discussed by John N. Keynes, *The Scope and Method of Political Economy*, New York: Kelley and Millman 1891, and Joseph Schumpeter, "The Common Sense of Econometrics" in: *Econometrica*, 1, 1933, pp. 5–12, among others (see David F. Hendry and

D. Dieks et al. (eds.), *Probabilities, Laws, and Structures*, The Philosophy of Science in a European Perspective 3, DOI 10.1007/978-94-007-3030-4_23,

the discussant of the brilliant analysis of nonsense regressions when first read to the *Royal Statistical Society*,[2] although most economics undergraduates today would find the mathematics straightforward. Second, recent criticisms have elicited a torrent of supportive responses.[3] Third, many non-professional economists seem to suspect that formalization in economics was a partial cause of not foreseeing the financial crisis of 2007–2011. For example, HM Queen Elizabeth II questioned why UK economists had not done so, and the ensuing debate revealed that viewpoint (more precisely, "a failure of the collective imagination of many bright people").[4] Fourth, there are formal attacks on our "excessive ambitions".[5] Finally, the ESF-PSE Workshop on *The Debate on Mathematical Modeling in the Social Sciences* reflects a widespread desire to reconsider our tools. As ever, there is something to be said on both sides of the debate.

First, my forecasting case study is adapted from research which developed a theory of economic forecasting for settings where the model is mis-specified in unknown ways for an economic process that unexpectedly shifts at unknown times by unknown magnitudes.[6] That work shows that a general mathematical analysis is both feasible and insightful, radically changing the interpretation of the outcomes of forecasting competitions,[7] and what can be learned from forecast failures.[8] Building on such results, later research led to explanations as to why

---

Mary S. Morgan, *The Foundations of Econometric Analysis*, Cambridge: Cambridge University Press 1995).

2 By G. Udny Yule, "Why Do We Sometimes Get Nonsense-correlations between Time-series? A Study in Sampling and the Nature of Time Series (With Discussion)" in: *Journal of the Royal Statistical Society*, 89, 1926, pp. 1–64.

3 See e.g., John Llewellyn, "It's Possible To Subtract Mathematics From Economics", in: *The Observer*, 16 August, 2009 http://www.guardian.co.uk/business/2009/aug/16/economics-economics.

4 See e.g., http://www.guardian.co.uk/uk/2009/jul/26/monarchy-credit-crunch.

5 Like that to which I responded in "Comment on 'Excessive ambitions' (by Jon Elster)", in: *Capitalism and Society*, 4, 2009, DOI: 10.2202/1932-0213.1056.

6 Taken from David F. Hendry and Bent Nielsen, *Econometric Modeling: A Likelihood Approach*, Princeton: Princeton University Press 2007, building on Michael P. Clements and David F. Hendry, *Forecasting Non-stationary Economic Time Series*, Cambridge, Mass.: MIT Press 1999.

7 See Spyros Makridakis and Michelle Hibon, "The M3-competition: Results, Conclusions and Implications", in: *International Journal of Forecasting*, 16, pp. 451–476, and Robert Fildes and Keith Ord , "Forecasting Competitions–Their Role In Improving Forecasting Practice and Research", in: Michael P. Clements and David F. Hendry (Eds.), *A Companion to Economic Forecasting*, Oxford: Blackwells 2002, and compare Michael P. Clements and David F. Hendry, 2001, "Explaining the Results of the M3 Forecasting Competition" in: *International Journal of Forecasting*, 17, pp. 550–554.

8 See David F. Hendry and Jurgen A. Doornik, "The Implications for Econometric Modelling of Forecast Failure" in: *Scottish Journal of Political Economy*, 44, 1997, pp. 437–461, and Michael P. Clements and David F. Hendry, "Explaining Forecast Fail-

some forecasting methods are "robust" to location shifts after they have occurred,[9] as well as suggesting possible approaches to forecasting breaks and during breaks.[10]

The second case study concerns model selection.[11] Since the forms, magnitudes and timings of breaks are usually unknown, a "portmanteau" approach to their detection is required that allows for potential location shifts at every possible point in the sample. Impulse-indicator saturation (IIS) includes an impulse indicator for every observation in the set of candidate regressors, so adds T variables for T observations, then selects significant indicators from that saturating set.[12] Its ability to detect multiple breaks is established,[13] and IIS allows an automatic test for super exogeneity.[14] The properties of model selection in general, especially when there are more candidate variables, N, for inclusion in the analysis than the

---

ure in Macroeconomics", in: Clements and Hendry (Eds.), *A Companion to Economic Forecasting*, *op. cit.*

9 See David F. Hendry, "Robustifying Forecasts From Equilibrium-correction Models" in: *Journal of Econometrics*, 135, 2006, pp. 399–426.

10 See Jennifer L. Castle, Nicholas W. P. Fawcett, and David F. Hendry, "Forecasting Breaks and During Breaks", in: Michael P. Clements and David F. Hendry, (Eds.), *Oxford Handbook of Economic Forecasting*, Oxford: Oxford University Press, 2011.

11 This draws on research by David F. Hendry and Hans-Martin Krolzig, "The Properties of Automatic Gets Modelling", in: *Economic Journal*, 115, 2005, pp. C32–C61; Jennifer L. Castle, Jurgen A. Doornik and David F. Hendry, "Evaluating Automatic Model Selection", in: *Journal of Time Series Econometrics*, 3 (1), DOI: 10.2202/1941-1928.1097; Jennifer L. Castle and David F. Hendry, "Automatic Selection of Nonlinear Models", in: Liuping Wang, Hugues Garnier and T. Jackman (Eds.), *System Identification, Environmental Modelling and Control*, New York: Springer forthcoming; Jurgen A. Doornik, "Autometrics", in: Jennifer L. Castle and Neil Shephard (Eds.), *The Methodology and Practice of Econometrics*, Oxford: Oxford University Press 2009, pp. 88–121; David F. Hendry and Grayham E. Mizon, "Econometric Modelling of Time Series with Outlying Observations", in: *Journal of Time Series Econometrics*, 3 (1), DOI: 10.2202/1941-1928.1100.

12 Its properties in a simple setting are analyzed in David F. Hendry, Søren Johansen and Carlos Santos, "Automatic Selection of Indicators in a Fully Saturated Regression", *Computational Statistics*, 33, 2008, pp. 317–335, Erratum, pp. 337–339, and extended by Søren Johansen and Bent Nielsen, "An Analysis of the Indicator Saturation Estimator as a Robust Regression Estimator" in: Castle and Shephard (Eds.), *The Methodology and Practice of Econometrics*, *op.cit.*, pp. 1–36, to both stationary and unit-root autoregressive models.

13 See Jennifer L. Castle, Jurgen A. Doornik and David F. Hendry, 2011, "Model Selection when There Are Multiple Breaks", in: *Journal of Econometrics*, forthcoming.

14 See David F. Hendry and Carlos Santos, "An Automatic Test of Super Exogeneity", in: Mark W. Watson, Tim Bollerslev, and Jeff Russell (Eds.), *Volatility and Time Series Econometrics*, Oxford: Oxford University Press 2010, pp. 164–193, extending earlier research by Robert F. Engle, David F. Hendry and Jean-Francois Richard, "Exogeneity", in: *Econometrica*, 51, 1983, pp. 277–304, and Robert F. Engle and David F. Hendry, "Testing Super Exogeneity and Invariance in Regression Models", in: *Journal of Econometrics*, 56, 1993, pp. 119–139.

number of observations, T–as must occur with IIS–can only be resolved by mathematical analysis and its numerical sister of Monte Carlo simulations. Again, an understanding of the astonishingly good properties of extended general-to-specific procedures would be impossible without advanced mathematical analysis.

The third example concerns the mathematics of inter-temporal optimization and the formation of expectations, in particular, so-called "rational expectations" (RE), where misleading results follow from present approaches applied to realistic economies. When unanticipated location shifts occur, estimated econometric models experience forecast failure, as noted above. However, that finding also entails that conditional expectations formed today of a future period after such a shift will be biased and potentially far from the minimum mean square error predictor "proved" in most textbooks under the unstated assumption that the distributions involved are unchanged. Unfortunately, in economics, location shifts and other forms of structural break are all too common.[15] Conclusions drawn on the "as if" basis that breaks do not occur are inapplicable when they do: on a much grander scale, Euclidean geometry was long believed to be "true", and many theorems, such as "the sum of the angles of a triangle add to 180°", were proved on that basis—until Riemann established the existence of non-Euclidean geometries in which the sum can exceed or fall short of 180° depending on the curvature of the space. Thus, the additional assumption was needed for Euclidean geometry that space was flat–an assumption that holds approximately locally, but is violated on the surface of a globe. Similarly, theorems about conditional expectations and the law of iterated expectations require the additional assumption that distributions do not shift, and are inapplicable otherwise.[16] The appropriate mathematics for settings where distributions shift remains to be developed, and may end being "problem specific" rather than generic.

The structure of the chapter is as follows. We first explain how mathematics was crucial in developing a theory of economic forecasting relevant to the practical setting where models are mis-specified and the world experiences intermittent unanticipated location shifts, and illustrates some surprising implications that could not have been deduced without a mathematical analysis. Then we consider the formalization of model selection when there are more candidate regressors, $N > T$, than observations, T, although fewer variables, $n < T$, actually matter. Finally we draw the implications of forecast failure for inter-temporal optimization theory, and conclude.

15 See e.g., James H. Stock and Mark W. Watson, "Evidence on Structural Instability in Macroeconomic Time Series Relations", in: *Journal of Business and Economic Statistics*, 14, 1996, pp. 11–30, and Ray Barrell, "Forecasting the world economy", in: David F. Hendry and Neil R. Ericsson (Eds.), *Understanding Economic Forecasts*, Cambridge, Mass.: MIT Press, 2001, pp. 149–169.

16 See David F. Hendry and Grayham E. Mizon, "On the mathematical basis of inter-temporal optimization", Working paper 497, Economics Department, Oxford, 2010.

23.2 Formalizing Forecasting Theory

There is a well-developed theory of economic forecasting based on the assumption that the econometric model coincides with a stationary economic data generation process (DGP).[17] Consider an $n \times 1$ vector of variables to be forecast denoted $\mathbf{x}_t \sim \mathsf{D}_{\mathbf{x}_t}(\mathbf{x}_t|\mathbf{X}_{t-1}, \theta)$ for $\theta \in \mathbf{\Theta} \subseteq \mathsf{R}^k$, where $\mathbf{X}_{t-1} = (\ldots \mathbf{x}_1 \ldots \mathbf{x}_{t-1})$ and $\mathsf{D}_{\mathbf{x}_t}(\mathbf{x}_t|\mathbf{X}_{t-1}, \theta)$ is its distribution. A statistical forecast $\widetilde{\mathbf{x}}_{T+h|T} = \mathbf{f}_h\left(\mathbf{X}_T^1\right)$ is made at time T (the forecast origin) for a future date $T+h$ (the forecast horizon). The key question in this setting is how to select $\mathbf{f}_h$.

The answer was "well known": the conditional expectation $\widehat{\mathbf{x}}_{T+h|T} = \mathsf{E}[\mathbf{x}_{T+h}|\mathbf{X}_T^1]$ is unbiased, with $\mathsf{E}[(\mathbf{x}_{T+h} - \widehat{\mathbf{x}}_{T+h|T})|\mathbf{X}_T^1] = 0$. Further, $\widehat{\mathbf{x}}_{T+h|T}$ has the smallest mean-square forecast-error matrix:

$$\mathsf{M}\left[\widehat{\mathbf{x}}_{T+h|T} \mid \mathbf{X}_T^1\right] = \mathsf{E}\left[\left(\mathbf{x}_{T+h} - \widehat{\mathbf{x}}_{T+h|T}\right)\left(\mathbf{x}_{T+h} - \widehat{\mathbf{x}}_{T+h|T}\right)' \mid \mathbf{X}_T^1\right].$$

However, that analysis finesses ten distinct problems. The first six concern problems learning about $\mathsf{D}_{\mathbf{X}_T^1}(\cdot)$ and θ from the available sample information, and the last four relate to the forecast period:[18]

(1) *Specification* of the set of relevant variables $\{\mathbf{x}_t\}$;
(2) *Measurement* of the xs;
(3) *Formulation* of $\mathsf{D}_{\mathbf{X}_T^1}(\cdot)$;
(4) *Modelling* of the relationships;
(5) *Estimation* of θ, and;
(6) *Properties* of $\mathsf{D}_{\mathbf{X}_T^1}(\cdot)$, which determine the "intrinsic" uncertainty.

All of these introduce in-sample uncertainties. Next, over the forecast horizon:

(7) *Properties* of $\mathsf{D}_{\mathbf{X}_{T+H}^{T+1}}(\cdot)$ determine forecast uncertainty;
(8) *Which grows* as H increases;
(9) Especially for *integrated* data;
(10) Increased by *changes* in $\mathsf{D}_{\mathbf{X}_{T+H}^{T+1}}(\cdot)$ or θ.

These ten issues structure the analysis of forecasting. We now illustrate with a simple example, although its implications are generic, and hold for all forecasting models and DGPs, irrespective of the correctness (or otherwise) of the specification of the model, and the properties of the DGP (stationary or integrated, with or without breaks of unknown timing, magnitude and form), and the data accuracy.[19]

---

17 See the famous treatise of Trygve Haavelmo, "The probability approach in econometrics", in: *Econometrica*, 12, 1944, pp. 1–118, extended by Lawrence R. Klein, *An Essay on the Theory of Economic Prediction*, Chicago: Markham Publishing Company 1971.

18 See David F. Hendry and Bent Nielsen, *Econometric Modeling: A Likelihood Approach., op. cit.*

19 Cf. Michael P. Clements and David F. Hendry, "Forecasting with Breaks in Data Processes", in: Graham Elliott, Clive W. J. Granger and Allan Timmermann (Eds.), *Handbook of Econometrics on Forecasting*, Amsterdam: Elsevier 2006, pp. 605–657, who provide a general "non-parametric" analysis.

23.2.1 Stationary Scalar Example

Consider a simple first-order autoregressive DGP with a known exogenous variable $\{z_t\}$:

$$y_t = \rho y_{t-1} + \gamma z_t + \epsilon_t \text{ where } \epsilon_t \sim \mathsf{IN}\left[0, \sigma_\epsilon^2\right] \text{ with } |\rho| < 1, \tag{23.1}$$

and $\mathsf{IN}[0, \sigma_\epsilon^2]$ denotes an independent normal distribution with mean, $\mathsf{E}[\epsilon_t] = 0$, and variance $\mathsf{V}[\epsilon_t] = \sigma_\epsilon^2$. When ρ and γ are known and constant, the optimal forecast for $T+1$ from y_T for known z_{T+1} is:

$$\overline{y}_{T+1|T} = \rho y_T + \gamma z_{T+1} \tag{23.2}$$

In terms of the general analysis above, $\mathsf{D}_{\mathbf{X}_T^1}(\cdot)$ implies $\mathsf{D}_{\mathbf{X}_{T+1}^{T+1}}(\cdot)$, producing an unbiased forecast:

$$\mathsf{E}\left[\left(y_{T+1} - \overline{y}_{T+1|T}\right) \mid y_T, z_{T+1}\right] = (\rho - \rho)\, y_T + (\gamma - \gamma)\, z_{T+1} + \mathsf{E}\left[\epsilon_{T+1}\right] = 0,$$

with the smallest possible variance determined by $\mathsf{D}_{\mathbf{X}_T^1}(\cdot)$:

$$\mathsf{V}\left[\left(y_{T+1} - \overline{y}_{T+1|T}\right)\right] = \sigma_\epsilon^2.$$

Thus, in this specific case, $\mathsf{D}_{\mathbf{X}_{T+1}^{T+1}}(\cdot)$ implies $y_{T+1} \sim \mathsf{IN}[\rho y_T + \gamma z_{T+1}, \sigma_\epsilon^2]$. There will indeed be no forecasting problems, as issues (1)–(10) are "assumed away". However, the ten potential problems return when omniscience is unavailable, even if z_{T+1} is known:

[1] *Specification is incomplete* if (e.g.) $\mathbf{x}_t$ is a vector not a scalar.
[2] *Measurement is incorrect* if (e.g.) observe $\widetilde{\mathbf{x}}_t$ not $\mathbf{x}_t$.
[3] *Formulation is inadequate* if (e.g.) an intercept is needed.
[4] *Modelling is wrong* if (e.g.) the wrong variables or lags are selected.
[5] *Estimating* ρ and γ may add biases $((\rho - \mathsf{E}[\widehat{\rho}])$, and $(\gamma - \mathsf{E}[\widehat{\gamma}]))$, and variances $\mathsf{V}[\widehat{\rho}, \widehat{\gamma}]$.
[6] *Properties of* $\mathsf{D}(\epsilon_t) = \mathsf{IN}\left[0, \sigma_\epsilon^2\right]$ determine $\mathsf{V}[y_t]$.
[7] *Assumed* $\epsilon_{T+1} \sim \mathsf{IN}\left[0, \sigma_\epsilon^2\right]$ but $\mathsf{V}[\epsilon_{T+1}]$ *could differ.*
[8] *Multi-step forecast errors cumulate* $\sum_{h=1}^{H} \rho^{h-1} \epsilon_{T+h}$ with $\mathsf{V} = \frac{1-\rho^{2H}}{1-\rho^2}\sigma_\epsilon^2$.
[9] $\rho = 1$ induces a *trending forecast variance*, $H\sigma_\epsilon^2$.
[10] *If* ρ *changes*, forecast failure could occur.

A forecaster must be prepared for risks from all of [1]–[10], but some matter more.

To illustrate, we will first "undo" problem (5), so the specification is correct, but (ρ, γ) have to be estimated from sample data, $t = 1, \ldots, T$. Next, we will also violate [1] by omitting z_t, then [10] by changing ρ. Figure 1 illustrates for Monte Carlo simulated data from (23.1) when $z_t \sim \mathsf{IN}[0, 1]$ with $\rho = 0.8$, $\gamma = 1$ and $\sigma_\epsilon^2 = 1$ when $T = 40$ and $H = 5$. We consider the six panels in turn.

Panel a records forecasts from a single draw of the process in (23.1), both when (ρ, γ) are known ($\overline{y}_{T+i|T+i-1}$ from (23.2) with error bands of $\pm 2\widehat{\sigma}$) and

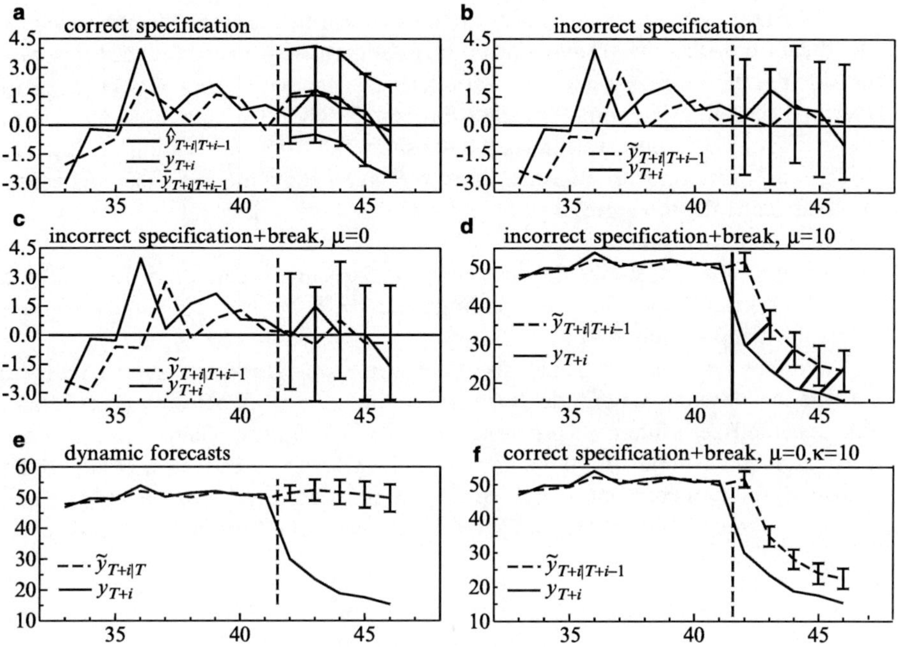

Figure 23.1: Forecasts under different scenarios

when estimating them ($\widehat{y}_{T+i|T+i-1}$ with bars). The forecasts are almost identical, and there is only a small increase in uncertainty from estimation relative to knowing true parameter values. So not problem [5].

Panel b reports forecasts when z_t is omitted both in estimation and forecasting: the forecasts are poorer, but remain well within their *ex ante* forecast intervals. So not problem [1].

Panel c adds a shift in ρ at $T = 41$ to 0.4, so all of [1], [5] and [10] are violated, yet there is little noticeable impact from *halving* ρ: the forecasts are close to those in panel b, and well within the forecast intervals. In fact, a parameter constancy test barely rejects the false null more often for a halved ρ than for a constant one. Such changes hardly seem disastrous: moreover, similar results will be found if white noise measurement errors are added; or model selection is undertaken when the precise specification is not known. Is forecasting really that resilient in the face of estimation, mis-specification, selection and breaks?

Consider a slight change to the DGP in (23.1), namely introducing a non-zero intercept $\mu = 10$:

$$y_t = \mu + \rho y_{t-1} + \gamma z_t + \epsilon_t \text{ where } \epsilon_t \sim \mathsf{IN}\left[0, \sigma_\epsilon^2\right] \text{ and } |\rho| < 1. \qquad (23.3)$$

when everything else remains the same, including the change in ρ of the same magnitude, sign and timing. Since economic data are often indices or have arbitrary units (millions versus billions), μ is relatively arbitrary.

As *Panel d* shows, the forecasts are now catastrophically bad, emphasized by the dreadful 5-step ahead (dynamic) forecasts in *Panel e*: every forecast lies well outside the 95% forecast interval. Parameter constancy tests now reject 100% of the time. Such an outcome is called forecast failure. The data are identical to those in *Panel a* till observation 40, but the sharp fall from observation 41 onwards is obviously different. The dashed lines in *Panel d* show that the 1-step forecasts are systematically too high: at every point, the data are falling yet the forecasts are above the previous outcome.

Finally, *Panel f* shows the forecasts for the same break when $\mu = 0$, the model is correctly specified by including z_t, but $\mathsf{E}[z_t] = \kappa = 10$: forecast failure is manifest and similar to *Panel d*.

Without a mathematical analysis of the properties of forecasts, such a dramatic change for the same magnitude, form and timing of a break between no failure in mis-specified zero-intercept processes and massive failure when there are non-zero intercepts, would simply be an unexplained surprise. In fact, it is due to the impact of the non-constant ρ on the pre-existing mean, $\mathsf{E}[y_t] = \theta$. In (23.1) when $\mu = \kappa = 0$, then $\mathsf{E}[y_t] = 0$ before and after the shift in ρ. But in the second case:

$$\mathsf{E}[y_t] = \theta = \frac{\mu + \gamma\kappa}{(1 - \rho)}$$

shifts markedly from $\theta = 50$ before the break in ρ to $\theta^* = 17$ after. Writing the model in (23.1) as:

$$\Delta y_t = (\rho - 1)(y_{t-1} - \theta) + \gamma(z_t - \kappa) + \epsilon_t \qquad (23.4)$$

reveals it is an equilibrium-correction model (EqCM), where the equilibrium built-in to the model is θ, so the forecasts will converge back to θ *irrespective of what the data do*. Thus, if $\theta > \theta^*$, the data will fall, but the forecasts will continually return towards θ. This *location shift* is clearly pernicious for forecasting, and explains *Panel f* as θ shifts when $\kappa \neq 0$. Perhaps more surprising, location shifts are the main problem likely to induce forecast failure, as we now describe, another result that cannot be established without mathematical analysis.

23.2.2 *Forecast-Error Taxonomy*

We now change the DGP to involve lagged rather than current z:

$$y_t = \theta + \rho(y_{t-1} - \theta) + \gamma(z_{t-1} - \kappa) + \epsilon_t \text{ for } t = 1, \ldots, T \qquad (23.5)$$

where $\epsilon_t \sim \mathsf{IN}[0, \sigma_\epsilon^2]$, $\mathsf{E}[y_t] = \theta$ and $\mathsf{E}[z_t] = \kappa$ with $\gamma \neq 0$, but z_{t-1} is omitted from the model:

$$y_t = \mu + \rho y_{t-1} + v_t$$

The break occurs at T, which leads to the post-break DGP:

$$y_t = \theta^* + \rho^*(y_{t-1} - \theta^*) + \gamma^*(z_{t-1} - \kappa^*) + \epsilon_t \text{ for } t = T + 1, \ldots \qquad (23.6)$$

The forecasting model:

$$\widehat{y}_{T+1|T} = \widehat{\theta} + \widehat{\rho}\left(\widehat{y}_T - \widehat{\theta}\right) \tag{23.7}$$

is estimated over $t = 1, \ldots, T$ delivering parameter estimates $(\widehat{\theta}, \widehat{\rho})$. The omitted variable and the dynamics induce biases, so $\mathsf{E}[\widehat{\theta}] = \theta_e$ and $\mathsf{E}[\widehat{\rho}] = \rho_e$. The forecast from an estimated $\widehat{y}_T$ at the forecast origin yields a forecast error of $\widehat{\epsilon}_{T+1|T} = y_{T+1} - \widehat{y}_{T+1|T}$. Ignoring interaction terms (corresponding to estimation covariances of $\mathsf{O}_p(T^{-1})$), the forecast error can be decomposed into the following taxonomy.

$\widehat{\epsilon}_{T+1\|T} \simeq$	Component	Expectation	Variance
$(1-\rho^*)(\theta^* - \theta)$	*(ia)*	$(1-\rho^*)(\theta^* - \theta)$	0
$+ (\rho^* - \rho)(y_T - \theta)$	*(ib)*	0	$(\rho^* - \rho)^2 \mathsf{V}[y_T]$
$+ (1-\rho)(\theta - \theta_e)$	*(iia)*	$(1-\rho)(\theta - \theta_e)$	0
$+ (\rho - \rho_e)(y_T - \theta)$	*(iib)*	0	$(\rho - \rho_e)^2 \mathsf{V}[y_T]$
$-\rho(\widehat{y}_T - y_T)$	*(iii)*	$-\rho(\mathsf{E}[\widehat{y}_T] - y_T)$	$\rho^2 \mathsf{V}[\widehat{y}_T - y_T]$
$-(1-\rho)\left(\widehat{\theta} - \theta_e\right)$	*(iva)*	0	$\mathsf{O}_p(T^{-1})$
$-(\widehat{\rho} - \rho_e)(y_T - \theta)$	*(ivb)*	$\simeq 0$	$\mathsf{O}_p(T^{-1})$
$+\gamma^*(z_T - \kappa^*)$	*(v)*	0	$(\gamma^*)^2 \mathsf{V}[z_T]$
$+\epsilon_{T+1}$	*(vi)*	0	σ_ϵ^2

(23.8)

The third and fourth columns give the game away, but starting at the foot of the table:
(*vi*): the **innovation error** has $\mathsf{E}[\epsilon_{T+1}] = 0$ and $\mathsf{V}[\epsilon_{T+1}] = \sigma_\epsilon^2$ so there is no bias, but an $\mathsf{O}_p(1)$ variance component that is irreducible when $\{\epsilon_t\}$ is, and remains, an innovation error;
(*v*): the **omitted variable** again has $\mathsf{E}[\gamma^*(z_T - \kappa^*)] = 0$ and $\mathsf{V}[\gamma^*(z_T - \kappa^*)] = (\gamma^*)^2\sigma_z^2$, so there is also no bias despite the omission and the change in parameters, but an $\mathsf{O}_p(1)$ variance component (reducible if $\{z_{t-1}\}$ is included as a regressor with an offsetting estimation variance effect of $\mathsf{O}_p(T^{-1})$);
(*ivb*): **slope estimation** has $\mathsf{E}[(\widehat{\rho} - \rho_e)(y_T - \theta)] \simeq 0$ as $\mathsf{E}[\widehat{\rho} - \rho_e] = 0$ and $\mathsf{E}[y_T - \theta] = 0$, with a variance from estimation of $\mathsf{O}_p(T^{-1})$;
(*iva*): **equilibrium-mean estimation** has $\mathsf{E}[(1-\rho)(\widehat{\theta} - \theta_e)] = 0$ with an estimation variance of $\mathsf{O}_p(T^{-1})$;
(*iii*): **forecast-origin uncertainty** only has $\mathsf{E}[\rho(\widehat{y}_T - y_T)] = 0$ if the forecast origin is unbiasedly estimated, but that can be achieved using modern methods of model selection applied to "Nowcasting"[20] and has a variance component, probably of $\mathsf{O}_p(1)$;

20 See e.g., Jennifer L. Castle, Nicholas W. P. Fawcett, and David F. Hendry, "Nowcasting Is Not Just Contemporaneous Forecasting", in: *National Institute Economic Review*, 210, 2009, pp. 71–89, and Jennifer L. Castle and David F. Hendry, "Nowcasting From Disaggregates in the Face of Location Shifts", in: *Journal of Forecasting*, 29, 2010, pp. 200–214.

(*iib*) **slope mis-specification** again has $\mathsf{E}[(\rho - \rho_e)(y_T - \theta)] = 0$ and an $\mathsf{O}_p(1)$ variance component unconditionally;
(*iia*) **equilibrium-mean mis-specification** is the first potentially serious component as $\theta \neq \theta_e$ is possible if there have been earlier in-sample location shifts that were not modelled, but IIS could resolve that difficulty;
(*ib*) **slope change** surprisingly has $\mathsf{E}[(\rho^* - \rho)(y_T - \theta)] = 0$ as $\mathsf{E}[y_T - \theta] = 0$ irrespective of $\rho^* \neq \rho$, a point illustrated above;
(*ia*) **equilibrium-mean change** is the fundamental problem: $\theta^* \neq \theta$ induces forecast failure.

In summary, once in-sample breaks are removed, from good forecast origin estimates:

$$\mathsf{E}[\widehat{\epsilon}_{T+1|T}] \simeq (1 - \rho^*)(\theta^* - \theta) \tag{23.9}$$

and that bias persists at $\widehat{\epsilon}_{T+2|T+1}$ etc., so long as (23.7) is used, even though no further breaks ensue. Keeping μ constant while shifting ρ to ρ^* induces a shift in θ to θ^*. The power of that insight is exemplified by (a) changing *both* μ and ρ by large magnitudes, such that $\theta = \theta^*$, then demonstrating that the outcome is isomorphic to $\mu = \mu^* = 0$ (and hence $\theta = \theta^*$) as above, so no break is detected;[21] and (b) when $\mu = \mu^* = 0$ and z_{t-1} is correctly included, then $\kappa \neq \kappa^*$ induces forecast failure by shifting θ when ρ changes.[22]

The specificity of the example is irrelevant to the entailed result, which applies to all models in the equilibrium-correction class: they fail systematically when $\mathsf{E}[y]$ changes as the models' forecasts are forced to converge back to θ irrespective of the value of θ^*. The class of EqCMs is huge and comprises all regression models; autoregressions; dynamic systems; vector autoregressions (VARs); dynamic-stochastic general equilibrium systems (DSGEs); autoregressive conditional heteroscedastic (ARCH) models; and generalized ARCH (GARCH) among others. Shifts in means are a pervasive and pernicious problem affecting forecasts from all such models.

23.2.3 Empirically-Relevant Theory

Such a theory needs to allow for the model being mis-specified for the DGP, with parameters estimated from inaccurate observations, on an integrated-cointegrated system, intermittently altering unexpectedly from structural breaks. That theory has achieved some success as it explains the prevalence of forecast failure, accounts for the results of forecasting competitions, and explains much of the good performance of "consensus" forecasts. Of equal importance, it corrects some "folklore" of forecasting, namely that forecast failure is *not* due to "poor econo-

---

21 Figure 1, *Panel a*: see e.g., David F. Hendry, "On Detectable and Non-detectable Structural Change", in: *Structural Change and Economic Dynamics*, 11, 2000, pp. 45–65.

22 See e.g., David F. Hendry and Grayham E. Mizon, "On the mathematical basis of intertemporal optimization", *op.cit.*

metric methods", "inaccurate data", "incorrect estimation", or "data-based model selection".[23]

Location shifts are the key to break detection: if there were no such shifts, forecast failure at 1% would be a 1 in 100 event. A crucial feature of (23.8) is that forecast errors persist unless the model is revised or abandoned. The former is difficult, as the cause of the forecast failure needs to be rapidly diagnosed and treated, and unfortunately, previous findings on forecasting breaks and during breaks show the large uncertainty attached to such attempts.[24] The latter requires a new model, which is even harder after a large unanticipated location shift. Fortunately, there is another approach–transform the initial model to avoid systematic forecast failure after location shifts.[25]

To illustrate that result, reconsider the forecasting model in (23.7), but instead of using the level, which depends on θ, difference the model, retaining the original estimated parameter values:

$$\Delta\widetilde{y}_{T+1|T} = \widehat{\rho}\Delta\left(\widehat{y}_T - \widehat{\theta}\right)$$

written as:

$$\widetilde{y}_{T+1|T} = \widehat{y}_T + \widehat{\rho}\Delta\widehat{y}_T \tag{23.10}$$

At the break point at time T, (23.10) makes the same magnitude forecast error as (23.7) precisely because the break is unpredicted. But one period later:

$$\widetilde{y}_{T+2|T+1} = \widehat{y}_{T+1} + \widehat{\rho}\Delta\widehat{y}_{T+1} = y_{T+1} + \widehat{\rho}\Delta\widehat{y}_{T+1} + \left(\widehat{y}_{T+1} - y_{T+1}\right) \tag{23.11}$$

where (23.11) distinguishes an incorrect estimate of the forecast origin from the consequences of a break. When unbiased forecast origin estimates are available, so $\mathsf{E}\left[\widehat{y}_{T+1}\right] = y_{T+1}$ and the "noise term" $\widehat{\rho}\Delta\widehat{y}_{T+1}$ is omitted to highlight the key point, then:

$$\widetilde{y}_{T+2|T+1} = y_{T+1} = \theta^* + \rho^*\left(y_T - \theta^*\right) + \gamma^*\left(z_T - \kappa^*\right) + \epsilon_{T+1}.$$

23 See, e.g., the unsubstantiated assertions on what went wrong in economic forecasting at the Bank of Canada by Don Coletti, Ben Hunt, David Rose and Robert J. Tetlow, "The Bank of Canada's New Quarterly Projection Model, Part 3. The Dynamic Model: QPM", in: *Technical report 75*, 1996, Bank of Canada, Ottawa: "the inability of relatively unstructured, estimated models to predict well for any length of time outside their estimation period seemed to indicate that small-sample econometric problems were perhaps more fundamental than had been appreciated and that too much attention had been paid to capturing the idiosyncrasies of particular samples."

24 See e.g., Jennifer L. Castle, Nicholas W. P. Fawcett and David F. Hendry, "Forecasting with Equilibrium-correction Models during Structural Breaks", in: *Journal of Econometrics*, 158, 2010, pp. 25–36.

25 See David F. Hendry, "Robustifying Forecasts from Equilibrium-correction Models", *op. cit.*, and for a non-technical account, see Michael P. Clements and David F. Hendry, "Economic Forecasting in a Changing World", in: *Capitalism and Society*, 3, 2008, pp. 1–18

Consequently, the forecast error is:

$$\begin{aligned} y_{T+2} - \widetilde{y}_{T+2|T+1} &= \theta^* + \rho^* \left(y_{T+1} - \theta^*\right) + \gamma^* \left(z_{T+1} - \kappa^*\right) + \epsilon_{T+2} \\ &\quad - \left(\theta^* + \rho^* \left(y_T - \theta^*\right) + \gamma^* \left(z_T - \kappa^*\right) + \epsilon_{T+1}\right) \\ &= \rho^* \Delta y_{T+1} + \gamma^* \Delta z_{T+1} + \Delta \epsilon_{T+2} \end{aligned} \tag{23.12}$$

which is noisy, but not systematic, and delivers near unbiased forecasts because:

$$y_{T+1} = \theta^* + \rho^* \left(y_T - \theta^*\right) + \gamma^* \left(z_T - \kappa^*\right) + \epsilon_{T+1}$$

"contains" z_T despite the omission of z_T from the forecasting model.

Whereas the estimated in-sample DGP suffers from all the main sources of forecast error, namely stochastic and deterministic breaks, omitted variables, inconsistent parameters, estimation uncertainty and innovation errors, the "differenced" transform reflects all the effects needed–parameter changes, differences of omitted variables, with no estimation components. There are two drawbacks, namely the unwanted presence of ϵ_{T+1} in (23.12), which doubles the innovation error variance; and all variables are lagged one extra period, which adds the "noise" of $\mathsf{I}(-1)$ effects. Nevertheless, there is a clear trade-off between avoiding systematic forecast failure and adding somewhat to the forecast-error variance when no location shifts occur. After the unanticipated occurrence of a location shift, as with the recent financial crisis, forecast failure is ubiquitous in EqCMs, but not in differenced variants thereof, so there need be no connection between the in-sample "quality" (or verisimilitude) of a model and that of its later forecasts.

23.2.4 Designing Monte Carlo Simulations

Simulation evidence is complementary to mathematical analysis in that, while mathematics is fundamental to understanding the analytically-tractable cases, simulation analysis helps examine empirically-relevant cases that may be intractable analytically.[26] The insights from mathematical analysis remain essential when designing Monte Carlo studies to focus on "canonical" cases, isolating aspects that are invariant across the simulations. For example, in a mean-zero autoregressive process, the units of the error standard deviation are irrelevant, but cease to be so if there is a non-zero intercept in the data generating process. When all parameters shift but leave the equilibrium mean constant is isomorphic to a zero mean, so allows a specific simulation to entail general results.

---

26 A more detailed analysis of efficient Monte Carlo simulation is provided in David F. Hendry, "Monte Carlo Experimentation in Econometrics", in: Zvi Griliches and Michael D. Intriligator (Eds.), *Handbook of Econometrics*, 2, Amsterdam: North-Holland 1984, pp. 937–976.

23.3 Selecting Econometric Models from a Mass of Candidate Variables

There are many critical analyses of model selection, almost all of which assume "correct" models with constant parameters where simply fitting the given specification dominates selection. This is not a realistic characterization of the situation confronting empirical investigators of economic time series. Data processes are complicated and evolving, so models derived from economic theory provide only a guide to some of the main variables, and rarely address breaks or outliers which vitiate any *ceteris paribus* assumptions. Thus, model selection is inevitable in practice, where only some substantively relevant aspects are correctly included, some are omitted, and some irrelevant aspects are also included, usually correlated with omitted variables.

Selection is essential when there are large numbers of potential explanatory variables. But can model selection work well in that setting? The canonical case of more variables than observations, $N > T$, is including an impulse indicator for every observation in the candidate regressor set. In the simplest analysis (the "split-half" case), one regression only includes the first $T/2$ of these indicators initially. By dummying out that first subset of observations, estimates are based on the remaining data, and any observations in the first half that are discrepant will result in significant indicators.[27] The location of the significant indicators is recorded, then the first $T/2$ are replaced by the second half and the procedure repeated. The two sets of significant indicators are then added to the general model for selection of those that remain significant together with selecting over the non-dummy variables. This is the approach called impulse-indicator saturation (IIS) above.[28] IIS is an efficient method: under the null of no breaks, outliers or data contamination, the cost of applying IIS at a significance level α is the loss of αT of the sample, so at $\alpha = 0.01$ and $T = 100$, IIS is 99% efficient. This follows because under the null, αT indicators will be retained by chance sampling, and each merely "dummies out" an observation. Thus, despite adding as many indicator variables as observations to the set of candidate variables to be selected from, when IIS is not needed the costs are almost negligible; and if IIS is required, the most pernicious effects of induced location shifts on non-constant intercepts, slopes and equation standard errors can be corrected.

---

27 In essence, that lies behind the approach for testing parameter constancy using indicators in David S. Salkever, "The Use of Dummy Variables to Compute Predictions, Prediction Errors and Confidence Intervals", in: *Journal of Econometrics*, 4, 1976, pp. 393–397.

28 See David F. Hendry, Søren Johansen and Carlos Santos, "Automatic Selection of Indicators in a Fully Saturated Regression", *op. cit.*, and Søren Johansen and Bent Nielsen, "An Analysis of the Indicator Saturation Estimator as a Robust Regression Estimator", *op. cit.*, who derive the distributions of estimators after IIS when there are no outliers or breaks, and relate IIS to robust estimation.

The mathematical analyses supporting these claims are in the references, and are consistent with a wide range of Monte Carlo simulations. No amount of non-mathematical thinking could have delivered such an astonishing insight: indeed most reactions are that adding $N > T$ candidate variables to the model search cannot be done, and if it could, it would produce garbage. But in fact it is easy to do, and almost costless.

23.3.1 *As Many Candidate Variables as Observations*

The analytic approach to understanding IIS can be applied when there are $N = T$ IID mutually orthogonal candidate regressors $z_{i,t}$, where none matters under the null. Formally, the DGP is:

$$y_t = \epsilon_t \tag{23.13}$$

and the general, but inestimable, model can be expressed as:

$$y_t = \sum_{j=1}^{N} \delta_j z_{j,t} + \epsilon_t \tag{23.14}$$

where $\delta_j = 0 \ \forall j = 1, \ldots, N$. We consider the analogue of the "split-half" analysis from IIS. Thus, add the first $N/2$, and select those with $|\mathsf{t}_{\delta_j=0}| > c_\alpha$ at significance level $\alpha = 1/T = 1/N$. Record which were significant, and drop them all. Now add the second block of $N/2$, again select those with $|\mathsf{t}_{\delta_j=0}| > c_\alpha$ at significance level $\alpha = 1/N$, and record which are significant. Finally, combine the recorded variables from the two stages (if any), and select again at significance level $\alpha = 1/N$. At both sub-steps, on average $\alpha N/2 = 1/2$ of a variable will be retained by chance under the null, so on average $\alpha N = 1$ will be retained from the combined stage. Again, despite examining the relevance of $N = T$ additional irrelevant variables, almost none is retained and the statistical analysis is 99% efficient under the null at eliminating irrelevant variables, merely costing one degree of freedom on average.

23.3.2 *More Candidate Variables Than Observations*

These results can be extended to having $N > T$ general candidate variables in the search, where $n < N$ are relevant.[29] The k theory-determined variables are not selected over, so are forced to be retained by the search. When the theory is correctly specified, the costs of searching over the remaining $N - k$ candidates is trivial for small α, as now $\alpha(N - k)$ irrelevant variables will be retained by chance and each merely costs a "degree of freedom". The real surprise is that the distribution of the estimates of the coefficients of the relevant variables are exactly

29 See David F. Hendry and Søren Johansen, "Model Selection when Forcing Retention of Theory Variables", Unpublished paper, Economics Department, University of Oxford 2010.

the same as if no search was undertaken at all. This is because the relevant and irrelevant variables can be orthogonalized without loss of generality, and as the latter are irrelevant, orthogonalizing does not alter the parameters of the relevant variables–and it is well known that estimator distributions are unaffected by the omission or inclusion of orthogonal variables. Without an advanced mathematical analysis, such a result is unimaginable.

Most economists and econometricians believe model selection is a pernicious but necessary activity–as shown above, it is in fact almost costless despite $N > T$, and invaluable when needed. Their beliefs were not based on sound mathematics, and that signals the dangers of not using powerful analytical tools. The practical difficulty is to be sure the tool is correctly based, and relevant to the target situation, a problem to which we now turn.

23.4 Models of Expectations

The very notation used for the mathematics of expectations in economics is inadvertently designed to mislead. Instead of $\mathsf{E}[\mathbf{x}_{T+h}|\mathbf{X}_T^1]$ as above, one must write conditional expectations as:

$$\mathsf{E}_{T+h}[\mathbf{x}_{T+h}|\mathbf{X}_T^1].$$

Thus *three* time subscripts are needed: that for the date of the conditioning information (here $\mathbf{X}_T^1$); that for the date of the variable being expected (here $\mathbf{x}_{T+h}$); and that for the distribution over which the expectation is formed (here E_{T+h}). If the distribution is stationary, then $\mathsf{E}_{T+h} = \mathsf{E}_T$, where the latter is the only feasible distribution at the time the expectation is formed. Otherwise, we have a paradox if $\mathsf{D}_{\mathbf{x}_t}(\cdot)$ is not constant as one needs to know the whole future distribution to derive the forecast. Worse, one cannot prove that $\widetilde{\mathbf{x}}_{T+h|T} = \mathsf{E}_T[\mathbf{x}_{T+h}|\mathbf{X}_T^1]$ is a useful forecast if $\mathsf{D}_{\mathbf{x}_{T+h}}(\cdot) \neq \mathsf{D}_{\mathbf{x}_T}(\cdot)$.

Theories of expectations must face the realities of forecasting discussed above. "Rational" expectations (RE) correspond to the conditional expectation given available information (denoted $\mathcal{I}_t$):

$$y_{t+1}^{re} = \mathsf{E}\left[y_{t+1} \mid \mathcal{I}_t\right]. \tag{23.15}$$

RE assumes free information, unlimited computing power, and the discovery of the form of $\mathsf{E}\left[y_{t+1}|\mathcal{I}_t\right]$ by economic agents. If (23.15) is to be useful, it should be written as (for a density $f_{t+1}(.)$):

$$y_{t+1}^{e} = \mathsf{E}_{t+1}\left[y_{t+1} \mid \mathcal{I}_t\right] = \int y_{t+1} f_{t+1}\left(y_{t+1} \mid \mathcal{I}_t\right) \mathrm{d}y_{t+1}. \tag{23.16}$$

Only then is y_{t+1}^e even unbiased for y_{t+1}. But (23.16) requires a crystal ball for *future* $f_{t+1}(y_{t+1}|\mathcal{I}_t)$. The best an agent can do is to form a "sensible expectation", y_{t+1}^{se}, forecasting $f_{t+1}(\cdot)$ by $\widehat{f}_{t+1}(\cdot)$:

$$y_{t+1}^{se} = \int y_{t+1} \widehat{f}_{t+1}\left(y_{t+1} \mid \mathcal{I}_t\right) \mathrm{d}y_{t+1}. \tag{23.17}$$

If the moments of $f_{t+1}(y_{t+1}|\mathcal{I}_t)$ alter, there are no good rules for $\widehat{f}_{t+1}(\cdot)$, but $\widehat{f}_{t+1}(y_{t+1}|\mathcal{I}_t) = f_t(\cdot)$ is not a good choice. Agents cannot know how $\mathcal{I}_t$ will enter $f_{t+1}(\cdot)$ if there is no time invariance.

When $f_{t+1}(\cdot) \neq f_t(\cdot)$, forecasting devices robust to location shifts avoid systematic mis-forecasting after breaks, as illustrated above. But if agents use robust predictors, and are not endowed with prescience that sustains an unbiased RE, then one needs to re-specify expectations in economic-theory models. But the problem is unfortunately even worse. Consider a very simple example–if $x_t \sim \mathsf{IN}[\mu_t, \sigma_x^2]$, then:

$$\begin{aligned} \mathsf{E}_t\left[x_t \mid \mathbf{X}_{t-1}\right] &= \mu_t \\ \mathsf{E}_{t-1}\left[x_t \mid \mathbf{X}_{t-1}\right] &= \mu_{t-1} \end{aligned}$$

when the mean changes, so letting $\epsilon_t = x_t - \mathsf{E}_{t-1}[x_t|\mathbf{X}_{t-1}]$:

$$\mathsf{E}_t\left[\epsilon_t\right] = \mu_t - \mu_{t-1} \neq 0 \tag{23.18}$$

shows that the conditional expectation is biased. But such a result also entails that the law of iterated expectations does not hold inter-temporally without the additional assumption that the distribution does not shift, and is inapplicable otherwise.[30] If the distribution shifts, many of the "mathematical derivations of inter-temporal optimization" are invalid in the same way that Euclidean calculations are invalid on a sphere. And as with non-Euclidean geometry, a different mathematics is needed depending on the shape of the relevant space, so here, different calculations will be required depending on the unanticipated breaks experienced by economies, the abilities of economic agents to learn what those breaks entail, and the speeds with which they reform their plans and expectations about the future. Thus, more powerful mathematical tools are urgently required to enable such analyses.

23.5 Conclusion

The paper has considered three possible situations of the use or mis-use of mathematics in economics and econometrics. The first concerned the properties of economic forecasts and forecast failure in particular, where a mathematical analysis was both essential and highly revealing. While only a specific example was given, the analysis holds independently of how well or badly specified the forecasting model is, and how the process being forecast actually behaves. Location shifts were isolated as the primary cause of forecast failure, with the myriad of other possible model mis-specifications and data mis-measurements playing a secondary role, despite prior intuitions to the contrary.

30 As shown in David F. Hendry and Grayham E. Mizon, "On the mathematical basis of inter-temporal optimization", *op.cit.*; a non-technical discussion is provided in David F. Hendry and Grayham E. Mizon, 2011, "What Needs Rethinking in Macroeconomics?", in: *Global Policy*, 2, pp. 176–183.

The second situation concerned model selection when there are more candidate variables N than the number of observations T. Again, an understanding of the astonishingly good properties of extended general-to-specific based procedures would be impossible without advanced mathematical analysis. That is particularly true of the finding that the distributions of the estimated coefficients of a correct theory model's forced variables are not affected by selecting over any number of irrelevant candidate variables. Yet there are innumerable assertions in the econometrics literature (and beyond) that selection is pernicious "data mining", leads to "over-fitting", etc., all without substantive mathematical proofs.

The third concerned the mathematics of inter-temporal optimization and the formation of expectations, in particular, so-called "rational expectations", where misleading results followed from present approaches applied to realistic economies. Conventional notation fails to address the three different times relevant to expectations formation, namely that of the available conditioning information, of the target variable to be forecast and of the time the expectation is formed. Consequently, the effects of shifts in distributions over which expectations are calculated have been hidden. Conditional expectations formed today for an outcome tomorrow need not be unbiased nor minimum variance. The appropriate mathematics remains to be developed, and may end being "problem specific" rather than generic. Nevertheless, the conclusion is inexorable: the solution is more powerful and more general mathematical techniques, with assumptions that more closely match "economic reality".

Acknowledgements: This research was supported in part by grants from the Open Society Institute and the Oxford Martin School. I am grateful to Jennifer L. Castle and Grayham E. Mizon for helpful comments.

Economics Department and Institute for New Economic Thinking
Oxford Martin School
University of Oxford
United Kingdom
david.hendry@nuffield.ox.ac.uk

Chapter 24

Javier Echeverria

Technomathematical Models in the Social Sciences[1]

24.1 Sciences and Technosciences

The sciences that universities and scientific societies developed during the modern era underwent a radical transformation over the twentieth century. They experienced a structural mutation that affects, above all, the organization of scientific practice, as well as the ways of producing, distributing, teaching, and using scientific knowledge. As a result, the technosciences, a hybrid between science and technology, have appeared. Because science has changed, the philosophy of science must also change. These are the basic hypotheses that I will use as a starting point for this contribution.

Different conceptual proposals for analyzing this change have been made. Ziman distinguished between *academic and post-academic science*, for the purpose of characterizing the "radical, irreversible, and worldwide transformation of the way science is organized and carried out".[2] Latour proposed the term *technoscience* to underline the close ties between twentieth-century science and technology and to "avoid the interminable expression *science and technology*".[3] In his actor-network theory, he also pointed out the existence of a non-human agency, that is, technological agency, in research activity. Since 1992, Silvio Funtowicz and Jerome Ravetz have been talking about a *post-normal science* that deals with problems that exceed Kuhn's disciplinary matrixes.[4] They have also insisted that, in these cases, scientists act in conditions of uncertainty, so scientific research is not subject to any kind of determinism. Ilkka Niiniluoto[5] referred to the design sciences, a proposal that Wenceslao J. Gonzalez[6] and other philosophers of science

1 This paper has been written in the framework of the Research Project FFI 2008- 03599/ FISO financed by the Spanish Ministry of Science and Innovation.

2 John Ziman, *Real Science: What It is and What It means*. Cambridge UK: Cambridge University Press 2000, p. 7.

3 Bruno Latour, *Science in Action*. Baltimore: John Hopkins University 1983.

4 Silvio Funtowicz and Jerome Ravetz, "Science for the Post-Normal Age", in: *Futures*, 25, 7, 1993, pp. 739–755.

5 Ilkka Niiniluoto, "The aim and structure of applied research", in: *Erkenntnis*, 38, 1, 1993, pp. 1–21.

6 Wenceslao J. Gonzalez (Ed.), *Las Ciencias de Diseño: Racionalidad limitada, predicción y prescripción*, A Coruña: Netbiblo 2007.

D. Dieks et al. (eds.), *Probabilities, Laws, and Structures*, The Philosophy of Science in a European Perspective 3, DOI 10.1007/978-94-007-3030-4_24,

have studied. Previously, Herbert Simon used the expression "sciences of the artificial" since 1969. Nowotny, Scott, Gibbons, and others[7] declared that a new way to produce scientific knowledge has appeared, Mode 2, which is transdisciplinary, heterogeneous, and non-hierarchical, in contrast to the academic mode, which has traditionally been disciplinary, homogeneous, and hierarchical. In 1997, Etzkowitz proposed the triple helix model (Academy, Industry, and Government), which he and Leydesdorff have successfully developed over the last decade.

All of these authors, and many others devoted to studying science and technology, coincide in stating that, since the emergence of *Big Science*, science has changed radically, particularly due to the eruption of the information and communications technologies (ICT). The expression *e-science* is another name for this transformation of contemporary science that is related to the emergence of the information society and the economy of knowledge. The conceptual proposals of different authors differ, because some emphasize one characteristic or property and others, another. What no one denies is the fact that science has changed, becoming strongly tied to technology, in particular, to the ICTs.

I think that the expression *technoscience* is the most adequate term for this new form of science. However, I do not use it as a container term that covers everything. In a previous book,[8] I tried to specify the concept of technoscience, distinguishing among techniques, technologies, sciences, and technosciences. They all exist at present, but we must not confuse them. The conceptual framework that I support can be summarized as follows:

1. From a philosophical perspective, the main difference between science and technoscience refers to knowledge: for science, the search for knowledge is an end in itself, for technoscience it is a means. Technoscientific companies and agencies are interested in the search for knowledge, as well as its validity, but their objectives go beyond knowledge. In I+D+i systems, the lines of scientific research that are given priority are the ones that generate technological developments and, in the end, innovations. It is science when the search for knowledge continues to be the main objective. It is technoscience when scientific knowledge becomes a means to generate new technologies and innovations.

2. Science aspires to explain the world (phenomena) and, when relevant, to predict them. Technoscience, on the other hand, aspires to transform the world, not only the natural world, but also the social world. Beyond the debate about explanation and comprehension in the social sciences, the philosophy of technoscience must deal with the types of social science that aspire, above all, *to transform social phenomena*, either

7 Nowotny, Scott, Gibbons, and others. Helga Nowotny, Peter Scott and Michael Gibbons, *Re-thinking Science: Knowledge and the Public in the Age of Uncertainty*, London: Polity Press & Blackwell Publishers 2001; and Michael Gibbons, Camille Limoges, Helga Nowotny et al., *The New Production of Knowledge. The Dynamics of Science and Research in Contemporary Societies*, London: Sage Publications 1994.

8 Javier Echeverría, *La revolución tecnocientífica*, Madrid: Fondo de Cultura Económica 2003.

on a large or a small scale. *Marketing* is a very clear example of social technoscience, but so are designing logos, corporative images, and electoral campaigns.

3. The sciences continue to exist, *not everything is technoscience*. Latour, Hottois and other authors tend to state that technoscience has absorbed science. I think that this is not true and that we philosophers must distinguish carefully between sciences and technosciences, including social sciences and social technosciences. It is important to analyze the moment when a technoscientific discipline emerges, and how this happens. We will see below that the social technosciences use complex, non-deterministic computing models that are quite different from traditional mathematical models.

4. The transformation mentioned affects not only knowledge but, above all, *scientific practice*. On this point, I disagree with Gibbons, Nowotny, and those who propose Mode 2. Not only has the way of producing scientific knowledge changed, but the way of presenting it, evaluating it, distributing it (publishing it), storing it, and using it have also changed. There are abundant examples. The evaluations of impact indexes (Thomson Reuters) are another example of social technoscience, which modifies scientific practice and its valuation criteria. The same thing can be said about teaching science online (online campuses), about online publications (Scopus, ArXiv, electronic journals), about the evaluation procedures for research projects, and about digital repositories (Berlin Declaration 2004 in favor of the Open Access). The ICTs are penetrating all areas of technoscientific activity, not only the production of knowledge. Therefore, my hypothesis is that a *technoscientific revolution* took place during the second half of the twentieth Century, and that the national I+D+i systems, whose ultimate objective is innovation, with research subordinated to it, have arisen as a result of this revolution.

5. From a sociological perspective, academic science is done by scientific communities, as Merton showed. Thomas Kuhn even correlated the notions of scientific community and paradigm quite closely. Technoscience, in contrast, is done by technoscientific companies and agencies, who are the ones who define the *technoscientific agendas*, that is, what should be done in science, technology, and innovation (STI). In the economy of knowledge, scientists become *knowledge workers*, losing a large part of their traditional autonomy. Technoscientific practice necessarily includes internal conflicts, owing to the structure of its agency, which is very different from the agency of science.

6. Therefore, in the tewnty-first Century, doing philosophy of science, or the history of science, is not enough. It is also necessary to do philosophy and history of technoscience, including a *philosophy of innovation*, which is not the same thing as the philosophy of scientific research. Neither epistemology nor methodology are sufficient for analyzing technoscience. Above all, a *philosophy of technoscientific practice* that includes a theory of scientific and technoscientific action is needed, not just a theory of scientific knowledge. I feel that the notion of a technoscientific agenda, specifically, is a key notion, as important as the notion of scientific theory is in the philosophy of science. In the case of the social technosciences, it

is necessary to analyze the practices of technosocial agents and companies, as well as their agendas, not only their theories and mathematical models, although these are also important.

Independently of the divergences between different authors, hardly anyone doubts that science has undergone a very profound change during the twentieth century. Scientific research on its own is not enough for technoscience. We can express this second change by saying that scientific communities are no longer the agent-subject of science; public or private *technoscientific companies*, whose strategies are guided by the imperative of innovating, take their place. Managing scientific research in a business-like manner and stating that innovation is the final objective are two of the distinctive features of present-day technoscience.

24.2 The Emergence of Technomathematics

Technoscience arose in the USA in the 1940s in the sphere of the experimental sciences (*Radiation Laboratories*, *Manhattan Project*), but technomathematics emerged almost simultaneously (ENIAC Project of the Moore School of Pennsylvania, 1944). In the first period, it focused on numerical calculation (von Neumann and non-lineal problems), but right from the start, it also paid attention to symbolic calculus (Turing and cryptology). At MIT in 1930, Vannevar Bush had built a *differential analyzer* that solved non-lineal equations that were very important in electric circuit theory. The German Konrad Zuse invented a *universal calculator*, the Z3, finished in 1941. Zuse's Z4 was used in 1943 for operations against Allied ships in the Mediterranean.[9] Another similar technomathematics project was the MARK I, started by Howard H. Aiken at Harvard in 1937. Aiken introduced a data register in that machine that later became the memory in computers. Financed by IBM, the MARK I was presented in 1944 and was immediately offered to the military because of its calculating power. All of these machines were electromechanical. The introduction of vacuum tube technology (Atanasoff and Berry, with their ABC in 1939) made it possible to create the first electronic calculators, as well as a digital representation of numbers as opposed to a decimal representation.

The construction of the ENIAC at the Moore School of Pennsylvania meant the integration of the abovementioned advances in technomathematics. Construction was begun in 1943, although von Neumann introduced important improvements to its design in 1945. Eckert, an engineer, Mauchly, a consultant, and Goldstine, the military manager of the project, which was a classified project (project PX in the Office of Ballistic Materials) collaborated with him. The ENIAC had 17,648 vacuum tubes, 70,000 resistances, 10,000 capacitances, 1,500 relays, and 6,000 manual switches. It was a large, complex machine and a lot of technical abilities

9 See Philippe Breton, *Historia y crítica de la informática*, Madrid: Cátedra 1989, pp. 69–81.

were necessary for working it. If just one single tube broke, the calculation was interrupted and had to be started all over. It cost a fortune, $500,000 at the time, but it worked at great speed, it was programmable, and it did different kinds of calculations. Its consumption of electricity and heat emission were enormous, so it had to be continuously cooled. After von Neumann joined the team, its automation improved significantly; the EDVAC was produced, and after that a saga of technomathematical artifacts designed according to "von Neumann architecture". Financed by the United States Army, the EDVAC can be considered the first computer in the present-day sense of the term.

In later decades, technomathematics continued to develop, while at the same time, in parallel, mathematical research followed its own course. Branches of mathematics such as Algebra and Differential and Integral Calculus were expanded by technomathematics, through the creation of different packages for mathematical and symbolic computation (*Macsyma*, *Reduce*, *Mathematica*, *SPSS*, etc.). The same could be said about Geometry, as computers made it possible to draw and solve geometric figures much more easily and rapidly than the classical techniques. All of this mathematical *software* is based on mathematical knowledge, of course, but in order to work, it also requires technological and computer knowledge and, in particular, programming languages. The mathematical *hardware* and *software* not only increased the capacity for operating considerably; they also generated new mathematical objects, such as fractals, data bases, and coding, decoding, compression, and ciphering systems. The protocols for interconnecting computers online can, in turn, be considered canonical technomathematical artifacts. The novelty is that the computers carry out numerous mathematical actions better than people, and they can even do operations that the human brain cannot. Technomathematics is a kind of mathematics that is determined by the ICTs and, in particular, by digitalization and by the algorithms and machine languages that make it possible. Obviously, this does not mean that mathematics disappears. What happens is that in the mid-1940s, a new way of *doing mathematics* was shaped, a way that later was disseminated throughout all the scientific disciplines, determining them all, even some social sciences. This hybridization between mathematics and ICTs is the foundation of what I call technomathematics.

There were important consequences to this. In Number Theory, *Computational Number Theory* appeared, fundamental for cryptography and for dealing with some classic problems such as Goldbach's and Riemann's conjectures. The majority of the problems of Lineal Algebra can be approached using computer programs from *Computer Algebra*. The same can be said of Mathematical Analysis, an area where there has been great progress in its computerization. One of the most outstanding examples of technomathematics was the proof of the four color theorem in topology, especially because it introduced radical changes in one of the most typical mathematical actions: the *action of proving*, whose result is the proof. An important part of this proof can only be carried out by computer, so that technological determination also reached proofs.

A third technomathematical canon was the creation in the 1980s of a new mathematical language, *TEX*, designed by Knuth and widely disseminated throughout the world. Today, mathematicians write in one of the several variants of *TEX*, all sharing one computer language. This mathematical info-writing technique has not eliminated the different systems of signs mathematicians use, but rather joined them. Another example is infography, and many more examples could be mentioned. In short: the history of technomathematics is waiting to be written.

24.3 Mathematical Models and Technomodels

Mathematical models have had a very important function in modern science: on one hand, for representing phenomena and data and, on the other, for expressing scientific laws. Mathematics has contributed formal languages that have later had different semantic interpretations. Interpreting a formal language consists of assigning meanings to its symbols, so that the formulas that express axioms, laws, and properties will be true. In this way, mathematical formulas became the canonical expression of scientific laws, thanks to the universality, the precision, and the rigor of their formulations. The mathematical formulas that express the scientific laws of a theory, for example, Newton's laws, do not stop at describing the trajectory of falling objects. These trajectories also *verify* the corresponding equations, which provide *truth value* for the empirical statements. Scientific laws explain and predict the movements of falling objects precisely because the mathematical formulas produce true statements when they are interpreted in empirical terms.

The notion of the model became one of the main analytical tools in the philosophy of science. Carnap, Braithwaite, Nagel, and Suppes, among others, defined models as interpretations of a language or formal system, based on the logical theory of models and, more specifically, on the definition of Tarski, according to which "a possible realization in which all valid sentences of a theory T are satisfied is called a model of T".[10] Other philosophers of science (Achinstein, Hesse, etc.), in contrast, drew up a notion of mathematical model that was very close to the way scientists use these models in their research. We should also remember that the philosophers of science who defended the structural conception (Sneed, Stegmüller, Balzer, Moulines, etc.) analyzed the structure of theories by distinguishing classes of models that verified the laws of the theory, as well as the observational statements. The same thing happened in the case of van Fraassen and of Giere, who gave greater importance to the notion of model than to the notion

10 Alfred Tarski, "A General Method Proofs of Undecidability", in: Alfred Tarski, Andrzej Mostowski and Raphael M. Robinson (Eds.), *Undecidable Theories*, Amsterdam: North Holland 1953, p. 11.

of theory, identifying theories with families of models.[11] In short: the theory of models has been important not only for the logic and the philosophy of mathematics, but for the philosophy of science in general. Obviously, mathematical models have been much used by scientists; as Herfel emphasized: "by and large, scientists spend their time building models, testing models, comparing models, discussing models and revising models".[12]

Analytical mathematical models have been surpassed by computing simulations that are based not on formulas but on algorithms. When sufficiently complex systems are studied, their behavior cannot be reduced to mathematical formulas, not even to systems of equations and formulas, precisely because they are phenomena with degrees of complexity that are not polynomial or even exponential. Instead of formulas, it is necessary to use different algorithms to represent and interpret these systems, for example, in the dynamics of complex systems. Bush, Zuse, von Neumann, and many others designed their computing machines to be able to deal with physical problems (missile trajectories, explosions, critical masses) which could not be analyzed using algebraic or differential equations, sometimes not even using statistical models, due to the large amount of data to be managed and the difficulty of computing these problems with traditional means. One of the first successes of emerging technomathematics was in the Manhattan Project, where the Montecarlo algorithm was used to calculate the interactions among 12 hard, mutually impenetrable spheres, an issue that could not be dealt with using analytical methods. Since then, computational methods have been applied to multiple issues (cosmology, elementary particles, fluid dynamics, wind tunnels, flight simulation, population dynamics, ecosystems, weather forecasting, climate change, macroeconomy, market studies, cognitive sciences, sociological surveys, migratory flows, etc.), producing the distinction between *analytical methods* and *computational methods* which is the basis of my distinction between mathematical models and technomodels. Computational models (technonumerical models, in my terminology) represent phenomena and data on computer screens, like mathematical models, but they do not fulfill the second function that was mentioned, that is, they do not formulate scientific laws because that is not their methodological objective. Technomodels are not built to explain phenomena, but to represent possible situations, modifying the corresponding parameters. Their function is, above all, heuristic and

11 See Bas van Fraassen *The Scientific Image*, Oxford: Oxford University Press 1980, and Ronald Giere, *Explaining Science, A Cognitive Approach*. Chicago: University of Chicago Press 1988, pp. 79–86.

12 William E. Herfel et al. (Eds.), *Theories and models in scientific processes*, *Pozna Studies in the Philosophy of the Sciences and the Humanities*, 44. Dordrecht: Rodopi 1995, p. 70.

experimental,[13] although it can also be predictive without, however, resorting to a nomological-deductive model. Changes in the state of physical, biological, social, and economic systems are represented by sequences of bits, making it possible to control later changes and simulate what would happen in different initial conditions and contexts. Technomodels do not attempt to explain what might happen, because they work in conditions of uncertainty, but they can represent what might happen in one set of circumstances or another. This is more than enough in the case of the social sciences, because they provide an empirical basis for making decisions based on data, even though these data are simulations of facts, more than scientific facts in the traditional sense of the term.

Technomodels have a different epistemic function because the final objective of the social technosciences is not to obtain knowledge with a truth value, but to transform the data, phenomena, and systems studied in an efficient manner, whether these systems are economic, social, semiotic, perceptual, or other ones, in order to work with them and modify them, having a kind of forecast of what can happen if one action or another is taken. As for the objects studied, the social technosciences deal with very complex systems and operate in non-deterministic conditions, and even in conditions of uncertainty. Both because of the transformational tendency of technoscience and because of the conditions of uncertainty of the objects and processes studied, computer simulations have shown themselves to be a very powerful tool to study social phenomena. Thanks to computers and computer simulations, the technosciences have been able to deal with very complex scientific, technological, economic, and social problems which were beyond the scope of traditional mathematical models. This would not have been possible without the ICTs and this is why it is better to call this new kind of mathematical model the technomodel, because technologies are indispensable for working with these models. Scientific research has always resorted to scientific instruments, but in the case of the technosciences, we are not talking about simple instruments but authentic *determinations of scientific knowledge*.

Several philosophers of science have paid attention to computing simulations, as a new conception of mathematical models. Stephan Hartmann stated that "although simulations are therefore of considerable importance in science, philosophers of science have almost entirely ignored them".[14] Frigg and Reiss, criticizing some authors, as Galison, Humphreys, Rohrlich and Wingsberg, said that computer simulations pose no significant philosophical question: "Simulations, far from

---

13 Uskali Mäki, "Models are Experiments, Experiments are Models", in: *Journal of Economic Methodology*, 12, 2, 2005, pp. 303–315; Anouk Barberousse, Simon Franceschelli and Claude Imbert, "Computer simulations as experiments", in: *Synthese*, 169, 3, 2009, pp. 557–574.

14 Stephan Hartmann, "The World as a Process: Simulations in the Natural and Social Sciences", in: Rainer Hegselmann et al. (Eds.), *Modelling and Simulation in the Social Sciences from the Philosophy of Science Point of View*, Dordrecht: Reidel 1996, p. 78.

demanding a new metaphysics, epistemology, semantics and methodology, raise few if any new philosophical problem".[15]

This kind of statement shows, in my view, an insufficient understanding of the philosophy of science. Reiss and Frigg reduce it to epistemology, semantics and methodology (more metaphysics), excluding however the praxiology (Kotarbinski) and the axiology (Rescher, Agazzi, etc.), together with the pragmatics of science (theory of automated action, for example). For his part, Paul Humphreys has disagreed with Frigg and Reiss in a recent issue of the journal *Synthese*, saying that their thesis is false and implies a profound ignorance of the peculiarities of computer simulation methods. Their arguments are as follows:

> Computational science introduces new issues into the philosophy of science because it uses methods that push humans away from the centre of the epistemological enterprise. Until recently, the philosophy of science has always treated science as an activity that humans carry out and analyze.[16]

To my way of thinking, this kind of discussion is produced because the distinction between science and technoscience has not yet been sufficiently accepted by the philosophy of science, despite the recent turn toward the philosophy of scientific practice. In any case, I fully support Hartmann's and Humphreys's proposals, although I prefer to use the expression *technomathematical models* (*or technomodels*) because it is more general than the expression *computing simulations*. Going back to Simon's proposals about the new *design sciences*, I believe that the use of mathematical models characterizes science, while the predominant use of technomodels (including computer simulations) is one of the distinctive features of the technosciences.

It is not easy to define the notion of model, as several authors have argued,[17] because of the great diversity of models that scientists use. The same is true for the case of technomodels. However, Hartmann[18] provided a very interesting definition of computing simulations: "I maintain that the most significant feature of a simulation is that it allows scientists to *imitate one process by another process; process* here refers solely to a temporal sequence of states of a system".[19] According to

15 Roman Frigg and Julian Reiss, "The Philosophy of Simulation: Hot New Issues or Same Old Stew?", in: *Synthese*, 169, 3, 2009, p. 593.

16 Paul W. Humphrey, "The Philosophical Novelty of Computer Simulation Methods", in: *Synthese*, 169, 3, 2009, p. 616.

17 See, for example, Joseph Agassi, "Why there Is no theory of models?", in: William E. Herfel et al. (Eds.), *Theories and models in scientific processes*, Poznań Studies in the Philosophy of the Sciences and the Humanities, 44, Dordrecht: Rodopi 1995, pp. 17–26.

18 Stephan Hartmann, *ibid*.

19 Robert Axelrod, "Advancing the Art of Simulation in the Social Sciences", in: Rosario Conte, Rainer Hegselmann and Pietro Terna (Eds.), *Simulating Social Phenomena*, Berlin: Springer 1997, p. 27.

this definition, a physical, biological, economic, social, cognitive or other process can be simulated using computers, because digitalization and programming languages cause the numerical simulations of a specific object or state to change their state, in turn, which is shown, perceptually, as a modification of the images on the screen. Therefore, technomodels are not mathematical formulas or formal languages, but sequences of bits that are represented by artificial images and sounds that resemble the objects and processes studied, in physics or in chemistry or in the social sciences. Once they are constructed, technomodels can be controlled and operated, so that we can represent natural and social processes using their corresponding digital representations, which change as time goes by. This makes it possible to represent different possible situations, not only the ones that actually occur. Without being predictive in the nomological-deductive sense, technomodels make it possible to represent the evolution of different types of systems, as well as the changes in their states, all independently of the complexity of the phenomena studied. Technomodels are interesting because they make it possible to fully control the computer processes that take place inside the computers, at least as far as the results go. By modifying the parameters and the algorithms, we cannot manage to know what is true, nor can we formulate scientific laws, but we can forecast, to a certain extent, what will happen in the digitally-represented systems, in one set of circumstances or another. Not much more is needed to make decisions and, therefore, to prefer one set of representations over another, in the social sciences. The different alternatives can be evaluated, not according to the truth values of the model theory, but according to other values, such as precision, the efficiency of the algorithms, complexity, fruitfulness, empirical fit, etc. In short, technomodels and technosciences refer to other kinds of values which are not only epistemic but also technological, economic, political, and social.

I conclude that the philosophy of technoscience should introduce new tools for conceptual analysis, for example, the notions of *technoscience* and *technomodels*. A good number of the models generated by the ICTs cannot be considered to be mathematical models, because they are not interpretations of formal languages that express true or even likely laws and properties. As Axelrod says, "exploratory models should be judged by their fruitfulness, not by their accuracy".[20] The social technosciences intend to be effective and fruitful, not true. This is because they work with assemblies and sequences of signs, not with statements and propositions. The traditional categories of the philosophy of science are insufficient for technoscience.

20 *Ibid.*, p. 22.

References

Peter Achinstein, "Models, Analogies and Theories", in: *Philosophy of Science*, 31, 1964, pp. 328–350.

Joseph Agassi, "Why there Is no Theory of Models?", in: William E. Herfel et al. (Eds.), *Theories and Models in Scientific Processes*, Poznań Studies in the Philosophy of the Sciences and the Humanities, 44, Dordrecht: Rodopi 1995, pp. 17–26.

Evandro Agazzi, Javier Echeverria and Amparo Gómez (Eds.), *Epistemology and the Social*, Poznań Studies in the Philosophy of Science 96. Amsterdam & New York: Rodopi 2008.

Robert Axelrod, "Advancing the Art of Simulation in the Social Sciences", in: Rosario Conte, Rainer Hegselmann and Pietro Terna (Eds.), *Simulating Social Phenomena*, Berlin: Springer 1997, pp. 21–40.

Wolfgang Balzer, Ulises Moulines and Joseph D. Sneed, *An Architectonic for Science*, Dordrecht: Reidel/Kluwer 1987.

Anouk Barberousse, Simon Francaschelli and Claude Imbert, "Computer Simulations as Experiments", in: *Synthese*, 169, 3, 2009, pp. 557–574.

Richard B. Braithwaite, *Scientific Explanation*, Cambridge: Cambridge University Press 1963.

Philippe Breton, *Historia y crítica de la informática*, Madrid: Cátedra 1989.

Vannevar Bush, *Science, the Endless Frontier*, Washington D.C.: US Government Printing Office 1945.

Rudolf Carnap, *Logical Foundations of Probability*, 2nd edition, Chicago, Il.: University of Chicago Press, 1962 (1st edition 1950).

Javier Echeverría, *La revolución tecnocientífica*, Madrid: Fondo de Cultura Económica 2003.

Javier Echeverría, "Towards a Philosophy of Scientific Practice: From Scientific Theories to Scientific Agendas", in: Fabio Minazzi (Ed.), *Filosofia, Scienza e Bioetica nel dibattito contemporaneo*, Roma: Istituto Poligrafico e Zecca dello Stato 2007, pp. 511–524.

Javier Echeverría and José Francisco Alvarez, "Bounded rationality in Social Sciences", in: Evandro Agazzi, Javier Echeverria and Amparo Gómez (Eds.) *Epistemology and the Social*, Poznań Studies in the Philosophy of Science 96, Amsterdam & New York: Rodopi 2008, pp. 173–190.

Henry Etzkowitz, "The Triple Helix Academy-Industry-Government Relations and the Growth of Neo-Corporatist Industrial Policy in the US", in: Sergio Campodall'Orto (Ed.), *Managing Technological Knowledge Transfer*, Bruxelles: EC Social Sciences COST A3, vol. 4, EC Directorate General, Science, Research and Development 2004.

Henry Etzkowitz and Loet Leydesdorff, "The Dynamics of Innovation: From National Systems and 'Mode 2' to a Triple Helix of University-Industry-Government Relations", in: *Research Policy*, 29, 2, 2001, pp. 109–123.

Roman Frigg, Stephan Hartmann and Claude Imbert (Eds.), *Models and Simulations*, in: Special Issue, *Synthese*, 169, 3, 2009, pp. 593–613.

Silvio Funtowicz and Jerome Ravetz, "Science for the Post-Normal Age", in: *Futures*, 25, 7, 1993, pp. 739–755.

Peter Galison, "Computer Simulation and the Trading Zone", in: Peter Galison and David Stump (Eds.), *Disunity of Science: Boundaries, Contexts and Power*, California: Stanford University Press 1996, pp. 118–157.

Michael Gibbons, Camille Limoges, Helga Nowotny et al., *The New Production of Knowledge. The Dynamics of Science and Research in Contemporary Societies*, London: Sage Publications 1994.

Ronald Giere, *Explaining Science: A Cognitive Approach*, Chicago: University of Chicago Press 1988.

Wenceslao J. Gonzalez (Ed.), *Las Ciencias de Diseño: Racionalidad limitada, predicción y prescripción*, A Coruña: Netbiblo 2007.

Wenceslao J. Gonzalez, "Trends and Problems in Philosophy of Social and Cultural Sciences: A European Perspective", in: Friedrich Stadler (Ed.), *The Present Situation in the Philosophy of Science*, Dordrecht: Springer 2010, pp. 221–242.

Stephan Hartmann, "The World as a Process: Simulations in the Natural and Social Sciences", in: Rainer Hegselmann et al. (Eds.), *Modelling and Simulation in the Social Sciences from the Philosophy of Science Point of View*, Dordrecht: Reidel 1996, pp. 77–100.

William E. Herfel et al. (Eds.), *Theories and Models in Scientific Processes, Poznań Studies in the Philosophy of the Sciences and the Humanities*, 44, Dordrecht: Rodopi 1995.

Mary Hesse, *Models and Analogies in Science*, London: Thomas Nelson and Sons 1963.

Paul W. Humphrey, "Computational Models", in: *Philosophy of Science*, 69, 2002, pp. 1–11.

Paul W. Humphrey, "The Philosophical Novelty of Computer Simulation Methods", in: *Synthese*, 169, 3, 2009, pp. 615–626.

Donald Knuth, *TEX and METAFONT: New Directions in Typesetting*, Providence, RI: American Mathematical Society 1979.

Donald Knuth, *The TEX Book*, Reading, Mass.: Addison Wesley 1984.

Thomas S. Kuhn, *The Structure of Scientific Revolutions*, 2nd edition, Chicago, Il.: University of Chicago Press 1970 (1st edition 1962).

Bruno Latour, *Science in Action*, Baltimore: John Hopkins University 1983.

Robert K. Merton, *The Sociology of Science: Theoretical and Empirical Investigations*, Chicago, Il.: Chicago University Press 1979.

Uskali Mäki, "Models are Experiments, Experiments are Models", in: *Journal of Economic Methodology*, 12, 2, 2005, pp. 303–315.

Ulises Moulines, *Exploraciones metacientíficas*, Madrid: Alianza 1982.

Ernst Nagel, *The Structure of Science*, New York: Harcourt 1961.

Ilkka Niiniluoto, "The Aim and Structure of Applied Research", in: *Erkenntnis*, 38, 1, 1993, pp. 1–21.

Helga Nowotny, Peter Scott and Michael Gibbons, *Re-thinking Science: Knowledge and the Public in the Age of Uncertainty*, London: Polity Press & Blackwell Publishers 2001.

Fritz Rohrlich, "Computer Simulation in the Physical Sciences", in *PSA 1990*, II, 1990, pp. 507–518.

Herbert Simon, *The Sciences of the Artificial*, 3rd edition, Cambridge: The MIT Press, 1996 (1st edition 1969).

Joseph D. Sneed, *The Logical Structure of Mathematical Physics*, Dordrecht: Reidel 1971.

Wolfang Stegmüller, *The Structuralist View of Theories*, Berlin: Springer 1979.

Patrick Suppes, *A Comparison of the Meaning and Uses of Models in Mathematics and the Empirical Sciences*, Technical Report No. 33 (August 25, 1960), Stanford, Cal.: Stanford University.

Alfred Tarski, "A General Method in Proofs of Undecidability", in: Alfred Tarski, Andrezj Mostowski and Raphael M. Robinson (Eds.), *Undecidable Theories*, Amsterdam: North Holland 1953, pp. 1–36.

Bas van Fraassen, *The Scientific Image*, Oxford: Oxford University Press 1980.

Eric Wingsberg, "Simulated Experiments: Methodology for a Virtual World", in: *Philosophy of Science*, 70, 2003, pp. 105–125.

John Ziman, *An Introduction to Science Studies*, Cambridge, UK: Cambridge University Press 1984.

John Ziman, *Real Science: What It is and What It means*, Cambridge UK: Cambridge University Press 2000.

Centro Carlos Santamaria
University of the Basque Country / Iberbasque
Campus de Ibaeta
20018 San Sebastián
Spain
javier_echeverria@ehu.es

Chapter 25

Donald Gillies

The Use of Mathematics in Physics and Economics: A Comparison

25.1 The Use of Mathematics in Physics

The aim of this paper is to compare the use of mathematics in physics and in economics. So I will begin in the first section by considering the case of physics. I will claim that mathematics has been used in physics to obtain (i) precise explanations and (ii) successful predictions.

A "precise explanation" can be characterised as follows. Suppose physicists are studying a particular phenomenon, and connected with this phenomenon there is a parameter, Θ say, which can be measured very precisely. If there is a mathematical theory, T say, of the phenomenon in question from which a theoretical value for Θ can be derived, and, if this theoretical value agrees with the observed value within the limits of experimental error, then T gives a precise explanation of Θ.

A famous example of a precise explanation concerned the motion of the perihelion of the planet Mercury. The perihelion of a planet is the point at which it is closest to the Sun. The motion of the perihelion of Mercury was calculated using Newtonian theory in the 19th century, but the theoretical value differed from the observed value by a small amount. Newcomb in 1898 gave the value of this discrepancy as $41.24'' \pm 2.09''$ per century; that is, less than an eightieth part of a degree per century. This is a tiny anomaly, and yet even this anomaly was successfully explained by the general theory of relativity which Einstein introduced in 1915. Einstein's calculations using his new mathematics gave a value for the anomalous advance of the perihelion of Mercury as $42.89''$ per century – a figure well within the bounds set by Newcomb.

Let us now turn to successful predictions. A very nice example here is Maxwell's prediction of the existence of radio waves. James Clerk Maxwell carried out research into electricity and magnetism in the period from 1855. He published his results in definitive and rigorous form in his famous *Treatise on Electricity and Magnetism* in 1873. In this work he formulated a number of equations, now known as Maxwell's equations, which apply to electrical and magnetic phenomena. One consequence of these equations was that there should exist electromagnetic waves travelling at the velocity of light. This led Maxwell to postulate that light was an electromagnetic radiation. However, his equations also indicated that there should

D. Dieks et al. (eds.), *Probabilities, Laws, and Structures*, The Philosophy of Science in a European Perspective 3, DOI 10.1007/978-94-007-3030-4_25,

be electromagnetic waves having a much longer wavelength than light. These electromagnetic waves, now known as radio waves, were generated by Heinrich Hertz in 1887. It is worth noting that mathematics played an essential role in the work of both Maxwell and Hertz. Maxwell's prediction of radio waves was only possible using his complicated mathematical equations, and Hertz also used Maxwell's equations to devise a method for generating radio waves.

Another example of the use of mathematics in physics to obtain a successful prediction is provided by Pauli's prediction of the existence of the neutrino. Pauli postulated the existence of a new particle in 1930 as a result of his mathematical study of the radioactive phenomenon of β decay. Pauli's mathematical calculations showed that the laws of conversation of energy, momentum, and angular momentum were not satisfied in β decay, if account was taken only of the particles which had so far been observed. Pauli therefore postulated that there must be a hitherto unobserved particle, whose characteristics would preserve the conservation laws. This particle was named "the neutrino" by Fermi in 1934 when he developed Pauli's theory of β decay. Neutrinos were detected for the first time in 1956.

We see from these examples, and of course many more could be given, that mathematics has been used in physics to obtain precise explanations and successful predictions. Let us now turn to the case of economics.

25.2 A Complication Caused by the Many Different Schools of Economics

Economics presents a complication which does not occur in physics. There is a consensus within the physics community about the fundamental principles of their subject. Virtually all contemporary physicists accept relativity theory and quantum mechanics. In Kuhnian terms they share a paradigm. The situation is very different in economics. The economics community is divided into different schools. The members of each of these schools may indeed share a paradigm, but the paradigm of one school can be very different from that of another. Moreover the members of one school are often extremely critical of the views of members of another school. The school of economics which has the most adherents at present is neoclassical economics. The majority of economists are neoclassical, and this approach can justly be referred to as the mainstream. Indeed Weintraub says:

> When it comes to broad economic theory, most economists agree. … 'We're all neoclassicals now, even the Keynesians', because what is taught to students, what is mainstream economics today, is neoclassical economics.[1]

---

1 E. Roy Weintraub, "Neoclassical", in: David R. Henderson (Ed.) *The Fortune Encyclopedia of Economics*, New York: Warner Books 1992. Quotations are from the online version in *The Concise Encyclopedia of Economics*, Library of Economics and Liberty, http://www.econlib.org, 2002, p. 1. (The page numbers are from an A4 print-out of the article.)

There is some truth in what Weintraub says here, and yet he also exaggerates in some respects. While most economists are indeed neoclassicals, there is a small, but very vocal, minority who reject the neoclassical approach completely. They are known as heterodox economists. Weintraub is also correct to say that some Keynesians do accept neoclassical economics. Versions of Keynes' original theory have been produced which fit in with the neoclassical framework. This is known as the neoclassical synthesis. However, Keynes himself did not accept neoclassical economics, and many Keynesians both in the past and today have been sharply critical of neoclassical economics.

Of course Weintraub is aware of this, and he goes on to say:

> Some have argued that there are several schools of thought in present-day economics. They identify (neo)-Marxian economics, neo-Austrian economics, post-Keynesian economics, or (neo)-institutional economics as alternative metatheoretical frameworks for constructing economic theories. To be sure, societies and journals promulgate the ideas associated with these perspectives. … But to the extent these schools reject the core building blocks of neoclassical economics … they are regarded by mainstream neoclassical economists as defenders of lost causes or as kooks, misguided critics, and antiscientific oddballs. The status of non-neoclassical economists in the economics departments in English-speaking universities is similar to that of flat-earthers in geography departments: it is safer to voice such opinions after one has tenure, if at all.[2]

One can certainly agree with Weintraub that it is difficult for heterodox economists to obtain permanent posts in universities, and that, even if they do obtain such a post, they may well be treated badly by their neoclassical colleagues. However, despite these handicaps, there still remain a significant number of heterodox economists who are active in the academic world. They are divided into a number of schools. Leaving out some of the "neos", Weintraub mentions: Marxist, Austrian, Post-Keynesian, and Institutionalist economists, and one could add some more. There are Sraffian, or neo-Ricardian economists, who are followers of the Italian economist Sraffa who worked at Cambridge with Keynes, but developed his own system. There are also evolutionary economists and economists who use complexity theory.

Weintraub states correctly that neoclassical economists have a low opinion of heterodox economists, but equally most heterodox economists have a low opinion of neoclassical economics. Every few years a book appears by one or more heterodox economists denouncing neoclassical economics as intellectual rubbish. A well-known example of this genre is Steve Keen, *Debunking Economics. The Naked Emperor of the Social Science*.[3] Steve Keen is a Sraffian economist. The economics which he debunks is neoclassical economics. According to him it is like the naked emperor of Hans Christian Andersen's fairy tale. Another more

2 *Ibid.*, pp. 2–3.

3 Steve Keen, *Debunking Economics. The Naked Emperor of the Social Science*, London & New York: Zed Books 2001.

recent example is Edward Fullbrook, *A Guide to What's Wrong with Economics*.[4] This is a collection of papers by contributors most of whom criticize neoclassical economics very sharply. The general scene in economics then, with its various schools which criticize each other harshly, is quite different from that in physics. There just is no group of heterodox physicists who spend their time denouncing relativity theory and quantum mechanics as valueless theories.

The situation in economics which I have just sketched makes it difficulty to give a simple answer to the question of how mathematics is used in economics because different schools of economics have very different attitudes to mathematics. There is one point on which nearly all economists agree, namely that mathematical statistics is useful for analysing economic data. However, when it comes to the use of mathematics in constructing economic theories to explain the data, opinions differ. Let us begin with the neoclassical school. The neoclassicals have from the start been very favourable to the use of mathematics in economics. Indeed the first neoclassical economist, Jevons, declared that economics should be a mathematical science. This was at a time when mathematics was not much used in economics. So the neoclassicals were responsible for the mathematization of economics.[5] Most neoclassical economists today use a great deal of mathematics in their work.

The mature Keynes, however, had a very different attitude to the use of mathematics in economics from the neoclassicals. Keynes started his career as a neoclassical economist, but he abandoned this approach because he thought that neoclassical economics was unable to explain the Wall Street crash of 1929, and the great depression of the 1930s. However, Keynes abandoned not only neoclassical theory, but also the neoclassical use of mathematics, though, as a wrangler in mathematics at Cambridge, he was himself well-trained in mathematics. Keynes reached the conclusion that mathematics was not an appropriate tool for economics, and indeed that neoclassical economists were led astray by their use of mathematics. In his most famous book, perhaps the most famous work of economics in the twentieth century, the *General Theory* of 1936, Keynes uses very little mathematics.

Most members of the Post-Keynesian school follow Keynes in rejecting the use of mathematics in economic theory, though, the Keynesians who accept the neoclassical synthesis mentioned earlier, are prepared to used mathematics.

The Sraffian school are happy to use mathematics in economic theorising. Sraffa himself wrote a sophisticated work of mathematical economics (*Production of Commodities by means of Commodities*).[6] Keen, who is a Sraffian, strongly

4 Edward Fullbrook (Ed.), *A Guide to What's Wrong with Economics*, London: Anthem Press 2004.

5 For further details see Margaret Schabas, *A World Ruled by Number. William Stanley Jevons and the Rise of Mathematical Economics*. Princeton: Princeton University Press 1990.

6 Piero Sraffa, *Production of Commodities by means of Commodities*. Cambridge: Cambridge University Press 1960.

rejects neo-classical economics in his 2001 book,[7] but in Chap. 12 of the book, wittily entitled: "Don't shoot me I'm only the piano", he argues that the use of mathematics by the neoclassical school is not the problem, and that mathematics has a legitimate use in economic theory. The Austrian economists, while sharing with neoclassical economists a love and admiration for the market, hold that there is little or no role for mathematics in economic analysis.

In a short paper like this, I cannot analyse in detail the attitudes towards mathematics of all the various schools in economics. So for the rest of the paper, I will focus exclusively on the use of mathematics in mainstream, that is, neoclassical, economics. As the present section has indicated, this is a considerable simplification, but it is not altogether valueless, since, the majority of economists are neoclassical economists, and the overwhelming majority of mathematical economics has been carried out within the neoclassical paradigm.

25.3 The Use of Mathematics in Mainstream (Neoclassical) Economics

Neoclassical economics began in the nineteenth century, but, in this paper, I will confine myself to neoclassical economics since the Second World War. In Sect. 25. 1 I argued that mathematics has been used in physics to obtain precise explanations and successful predictions. This naturally raises the questions of whether the use of mathematics in mainstream economics since 1945 has produced any precise explanations or successful predictions. My own reading of the literature of neoclassical economics has suggested the following conjecture. The use of mathematics in neoclassical economics since 1945 has produced no precise explanations or successful predictions. This seems to me the main difference between the use of mathematics in physics and the use of mathematics in neoclassical economics.

I say that this is a conjecture, and it is indeed a conjecture which is difficult to establish. To do so in a strict sense, a researcher would need to read every single paper and book on mathematical neoclassical economics written since 1945, and to check whether this work contains a precise explanation or a successful prediction. This is something which I have not done, and which would be a virtually impossible task. The best that can be done is to propose a conjecture, and to invite others to produce counter-examples, and I would certainly be very happy if anyone could send me an alleged counter-example to the conjecture for consideration.
However, although it is very difficult to establish this conjecture, it is nonetheless possible to produce some evidence supporting it, and this is what I propose to do in the rest of the paper. My idea is to examine the most well-known works of a selection of the most famous neoclassical economists in the period from 1945

7 Cf. Steve Keen, *ibid.*

to the present. Surely if precise explanations and successful predictions have indeed been produced by mathematical neoclassical economists, they are likely to be found in a sample of this kind rather than in the papers of less well-known economists, perhaps published in obscure journals. So if our sample yields no examples of precise explanations or successful predictions, this provides support for my conjecture.

It is easy to select a sample of well-known neoclassical economists since Nobel Prizes for economics have been awarded since 1969, and a good number of them have gone to practitioners of mathematical neoclassical economics. Actually Nobel himself did not create a prize in economics, but in 1968 Sweden's central bank established a prize in economics in memory of Alfred Nobel.

I have chosen a sample of four mainstream mathematical economists who won the Nobel Prize in economics. They are the following, arranged by the date at which they won the Nobel Prize.

Name	**Year of Nobel Prize in Economics**
Paul A. Samuelson	1970
Kenneth J. Arrow	1972
Gerard Debreu	1983
Edward C. Prescott	2004

I will now examine some of their best-known works (books or papers) to see whether they contain any precise explanations or successful predictions.[8]

Let us start with Paul Samuelson. Perhaps his most famous work is his book *Foundations of Economic Analysis*.[9] This is one of the classics of mathematical economics and has been widely used for teaching purposes in elite universities. Let us ask how it compares with classics of mathematical physics. As we have seen, one of the great successes of mathematical physicists consisted in their being able to use mathematics to calculate from their theories results which could be compared to observational data and which were found to agree with observational data to an often amazingly high degree of accuracy. Now if mathematical economists are even to begin to emulate this success, the first step must be to use mathematics to calculate from their theories results which could be compared to observational data. The extraordinary thing is that Samuelson in his classic book *does not even take this first step*. The book consists, in the 1963 edition, of 439

8 In preparing this part of the paper I was greatly helped by my wife, Grazia Letto-Gillies, who is an economist. I am also grateful to Guy Lipman who introduced to the equity premium puzzle, and drew my attention to the work of Mehra and Prescott cited below.

9 Paul Samuelson, *Foundations of Economic Analysis*, Cambridge: Harvard University Press 1947.

pages almost all of them filled with mathematical formulas, but not even one result is derived which could be compared with observational data. Indeed there is no mention of observational data in the entire book. One has to conclude that this book, far from emulating the successes of mathematical physics, seems more like a work of pure mathematics which lacks any empirical content whatever.

Let us now go on to consider Kenneth Arrow and Gerard Debreu. A joint paper, entitled: "Existence of an Equilibrium in a Competitive Economy", which they published in 1954 in *Econometrica* is regarded as one of the seminal papers in contemporary mathematical neoclassical economics. To explain why this paper has such a central importance for contemporary neoclassical economists, we must introduce the concept of general equilibrium theory.

Once again it is worth quoting Weintraub's clear and concise account. He says:

> Neoclassical economists conceptualized the agents, households and firms, as rational actors. Agents are modelled as optimizers who were led to 'better' outcomes. The resulting equilibrium was 'best' in the sense that any other allocation of goods and services would leave someone worse off.[10]

In a neoclassical general equilibrium model, we have firms which arrange their production in order to maximize their profits, given the existing technology; and households which arrange their consumption in order to maximize their utility, given their income. It is then shown that, if there is a market with free competition, this behaviour leads to an equilibrium which is Pareto-optimal. The conclusion which is drawn from this result is that any interference with a freely competitive market will produce a sub-optimal outcome.

The general equilibrium approach was introduced by Walras, and it has become the core of the neoclassical paradigm. This explains incidentally why the neoclassicals had to introduce mathematics, because it is impossible to consider the maximisation of sums of quantities under constraints without using calculus. Walras did indeed represent his economy as a system of simultaneous equations, but he was unable to show that these equations have a solution. The search for a solution was a problem which he bequeathed to his successors.

Let us now examine what contribution Arrow and Debreu made to this problem in their famous 1954 paper. They begin by saying that it has been established that a competitive equilibrium, if it exists, is Pareto-optimal. In their words:

> It is well known that, under suitable assumptions on the preferences of consumers and the production possibilities of producers, the allocation of resources in a competitive equilibrium is optimal in the sense of Pareto (no redistribution of goods or productive resources can improve the position of one individual without making at least one other individual

10 E. Roy Weintraub, *Ibid.*, p. 3.

worse off), and conversely every Pareto-optimal allocation of resources can be realised by a competitive equilibrium …[11]

However, it still needs to be established that a competitive equilibrium exists. They then go on to prove two theorems concerning the existence of a competitive equilibrium.

One initial point which could be made is that an equilibrium, if it exists, might be unstable. That is to say, if the economy moved for a moment into the equilibrium state, a slight disturbance would move it immediately away from equilibrium. If policy conclusions are to be drawn, it needs to be shown that the economy moves into a stable equilibrium, since an unstable equilibrium would hardly be a satisfactory outcome. Yet Arrow and Debreu do not show that their equilibrium is a stable one. They say:

Neither the uniqueness nor the stability of the competitive solution is investigated in this paper. The latter study would require specification of the dynamics of a competitive market as well as the definition of equilibrium.[12]

The next thing we have to examine is the realism of the assumptions under which the two theorems are proved, since if the assumptions are quite unrealistic, there is no reason to suppose that the theorems will hold for any actual competitive market. Theorem I is the following:

"*For any economic system satisfying Assumptions I-IV, there is a competitive equilibrium*".[13]

But what are these assumptions? Curiously enough Arrow and Debreu themselves state that one of them (assumption IV) is clearly unrealistic. They say:

The second half of IV.a. asserts in effect that every individual could consume out of his initial stock in some feasible way and still have a positive amount of *each* commodity available for trading in the market. This assumption is clearly unrealistic.[14]

The assumption in effect is that every individual in the economy possesses a positive amount of every commodity produced in that economy. One could hardly imagine an assumption so obviously false and so outrageously unrealistic. The need for such an assumption casts very great doubt on whether theorem I could be successfully applied to any competitive market.

Arrow and Debreu admit that this is the case, and they try to correct the situation by proving their Theorem II which states:

11 Kenneth Arrow and Gerard Debreu, "Existence of an Equilibrium for a Competitive Economy", in: *Econometrica*, 22, 3, 1954, p. 265.

12 *Ibid.*, p. 266.

13 *Ibid.*, p. 272.

14 *Ibid.*, p. 270.

"*For an economic system satisfying Assumptions I-III, IV′, and V-VII, there is a competitive equilibrium*".[15]

Here assumption IV is replaced by IV′, and three additional assumptions V-VII are added. But what is the new assumption IV′? Arrow and Debreu explain it as follows:

As noted ... Assumption IVa, which states in effect that a consumption unit has initially a positive amount of every commodity available for trading, is clearly unrealistic, and a weakening is very desirable. ... IV′.a. is a weakening of IV.a.; it is now only supposed that the individual is capable of supplying at least one type of productive labor.[16]

However, Arrow and Debreu immediately go on to describe a case in which the new assumption IV′ is not satisfied. This is what they say:

It is easy to see ... how an equilibrium may be impossible. Given the amount of complementary resources initially available, there will be a maximum to the quantity of labor that can be employed in the sense that no further increase in the labor force will increase the output of any commodity. ... as real wages tend to zero, the supply will not necessarily become zero; on the contrary, as real incomes decrease, the necessity of satisfying more and more pressing needs may even work in the direction of increasing the willingness to work despite the increasingly less favorable terms offered. It is, therefore, quite possible that for any positive level of real wages, the supply of labor will exceed the maximum employable and hence *a fortiori* the demand by firms. Thus, there can be no equilibrium at positive levels of real wages. At zero real wages, on the contrary, demand will indeed by positive but of course supply of labor will be zero, so that again there will be no equilibrium.[17]

This counter-example to one of their own assumptions, concerns an economy, suffering from chronic unemployment, in which the firms cannot increase their output of any commodity by employing more workers. For such an economy, many of the unemployed workers will be unable to supply productive labor, and so assumption IV′ will not be satisfied. Actually this case could easily occur for many real competitive economies. For example, many developing economies may often be in this situation.

We see from this that it is very doubtful whether the general equilibrium models presented by Arrow and Debreu apply to any real world competitive markets. One might therefore expect that the two authors would go on to compare their models with data supplied by actual competitive markets to see if agreement can be found. However, they do not do this. On the contrary, they behave exactly like Samuelson. They do not derive even one result which could be compared with observational data, and indeed do not mention observational data in their paper.

---

15 *Ibid.*, p. 281.

16 *Ibid.*, pp. 279-80.

17 *Ibid.*, p. 281.

Our next Nobel Prize laureate is Edward C. Prescott who won the Nobel Prize for economics in 2004. His most famous paper, entitled: "The Equity Premium. A Puzzle", was written jointly with Rajinish Mehra and published in 1985. It is what is called a seminal paper and gave rise to a very considerable literature. In this paper an attempt is made to compare an Arrow-Debreu general equilibrium model of an economy with data obtained from a real economy, namely the US economy in the period 1889–1978. I will now give a brief account of the contents of this paper.[18]

Let me first explain what is meant by the equity premium. Investors can put their money into short-term virtually default-free debt, such as, to give Mehra and Prescott's own example,[19] ninety-day U.S. Treasury Bills; or they can buy equities, i.e. shares in companies. Now, as Mehra and Prescott say: "Historically the average return on equity has far exceeded the average return on short-term virtually default-free debt".[20]

The difference in average returns is known as the equity premium, or sometimes as the equity risk premium. The latter expression arises because it is thought that it is a greater risk to buy equities than virtually riskless government securities. This greater risk can, however, be rewarded by the equity risk premium.

Mehra and Prescott begin by estimating the equity premium for the U.S. economy in the period 1889–1978. Their results are as follows:

> The average real returns on relatively riskless, short-term securities over the 1889–1978 period was 0.80 percent. … The average real return on the Standard and Poor's 500 Composite Stock Index over the ninety years considered was 6.98 percent per annum. This leads to an average equity premium of 6.18 percent (standard error 1.76 percent).[21]

They then build a model of the Arrow-Debreu General Equilibrium type to try to explain this observed value. Their model has five parameters: α, β, μ, δ, and φ. β is by definition in the range $0 < \beta < 1$. μ, δ, and φ were estimated from: "the sample values for the U.S. economy between 1889–1978. … The resulting parameter values were $\mu = 0.018$, $\delta = 0.036$ and $\varphi = 0.43$".[22] This leaves the parameter α which theoretically could take any value in the range $0 < \alpha < \infty$. The meaning of this parameter is explained as follows: "The parameter α … measures peoples' willingness to substitute consumption between successive yearly time periods …"[23]

---

18 Cf. Rajish Mehra and Edward Prescott, "The Equity Premium – A Puzzle", in: *Journal of Monetary Economics*, 15, 1985, pp. 145–161.

19 Cf. *Ibid.*, p. 147.

20 *Ibid.*, p. 145.

21 *Ibid.*, pp. 155–6.

22 *Ibid.*, p. 154.

23 *Ibid.*, p. 154.

Mehra and Prescott go on to quote a series of estimates of a by a number of different authors.[24] These estimates are as follows:

Arrow (1971)	$\alpha \approx 1$
Tobin and Dolde (1971)	$\alpha \approx 1.5$
Friend and Blume (1975)	$\alpha \approx 2$
Kydland and Prescott (1982)	$1 < \alpha < 2$
Altug (1983)	$\alpha \approx 0$
Hildreth and Knowles (1982)	$1 < \alpha < 2$
Kehoe (1984)	$\alpha \approx 1$

These seven estimates all agree on putting α in the range $0 \leq \alpha \leq 2$. In the light of this, Mehra and Prescott put the following restriction on α:

> Any of the above cited studies can be challenged on a number of grounds but together they constitute an *a priori* justification for restricting the value of α to be a maximum of ten, as we do in this study. This is an important restriction, for with large α virtually any pair of average equity and risk-free returns can be obtained by making small changes in the process on consumption. With α less than ten, we found the results were essentially the same for very different consumption processes, provided that the mean and variances of growth rates equalled the historically observed values.[25]

So, after all this work of setting up the model and estimating the parameters, what result was obtained? Mehra and Prescott state it as follows: "The largest premium obtainable with the model is 0.35 percent, which is not close to the observed value".[26]

The value obtained from the model is certainly very far from the observed value which was 6.18 % (standard error 1.76 %). However, the situation is really worse even than this statement suggests. The maximum value of 0.35 percent was only obtained with an average risk free rate of 4 %. If we set the average risk free rate to its empirical value of 0.8 %, the average equity premium drops to zero. Mehra and Prescott attempted to alter this result by varying the other parameters (μ, δ, and φ), but without success.

This is clearly not a precise explanation of an observed parameter. It is a result which is completely wrong. Mehra and Prescott's model gives an equity premium of zero, even though they say at the beginning of their paper that historically it has been large.

My survey of well-known works by four famous mathematical neoclassical economists who all won the Nobel Prize for economics, has not revealed any

24 *Ibid.*, p. 154.

25 *Ibid.*, pp. 154–5.

26 *Ibid.*, p. 156.

precise explanations or successful predictions. This supports my conjecture that the use of mathematics in mainstream (or neoclassical) economics has not produced any precise explanations or successful predictions. This, I would claim, is the main difference between neoclassical economics and physics, where both precise explanations and successful predictions have often been obtained by the use of mathematics.

Department of Science and Technology Studies
University College London
Gower Street
WC1E 6BT, London
United Kingdom
donald.gillies@ucl.ac.uk

Chapter 26

Daniel Andler

Mathematics in Cognitive Science

Abstract

What role does mathematics play in cognitive science today, what role should mathematics play in cognitive science tomorrow? The cautious short answers are: to the factual question, a rather modest role, except in peripheral areas; to the normative question, a far greater role, as the periphery's place is reevaluated and as both cognitive science and mathematics grow. This paper aims at providing more detailed, perhaps more contentious answers.

26.1 Clearing the Ground: Mathematics, Models, and Cognitive Science

Cognitive science and mathematics do not relate to one another as two well-defined, stable entities: they evolve and in fact co-evolve. This of course happens whenever a new science starts looking for help from mathematics. Take physics, or economics: in both of these cases, mathematics has profoundly shaped the emerging science, and reciprocally the science has impacted mathematics by making it develop some specific tools (which then become part of a new branch which can be used elsewhere).[1] But cognitive science resembles biology more than these other disciplines: at least up until recently, mathematics was not seen by a majority of cognitive scientists as having an important role to play in their field. Unlike biology however, cognitive science is hardly a mature discipline, in fact it is more of a loose federation of research programs, still searching for unifying principles.

The very fact that mathematics has historically been peripheral to cognitive science, and cognitive science to mathematics, makes it imperative not to assume that the interaction must involve 'core' areas of both field. Logic was never, and arguably still is not a core area of mathematics, yet it was for a long time, and it remains to a large extent, suitably extended, the main representative of mathematics within

1 This converse influence is of course much greater in the case of physics than in the case of economics: a large part of mathematics owes its existence to the requirements of physics, while the branches of mathematics which were developed in response to the specific needs of economics are few.

D. Dieks et al. (eds.), *Probabilities, Laws, and Structures*, The Philosophy of Science in a European Perspective 3, DOI 10.1007/978-94-007-3030-4_26,

cognitive science. Symmetrically, vision and motor control are not core areas of cognitive science, nor is the physiology of the single neuron or of cortical columns, yet they are the main recipients of knowledge stemming from such core areas of mathematics as functional analysis, topology, dynamical systems, group theory or probability. We should therefore keep an open mind as to what belongs to mathematics or cognitive science. Regarding the former, we should not rule out of bounds areas that at present lay at the periphery (say, logic, graph theory, computational geometry or theoretical computer science). Regarding the latter, we should refrain from imposing upon it some preconceived structure, with (cognitive) neuroscience, or artificial intelligence, or developmental psychology, or generative linguistics at its center, and, for example basic neuroscience, computational linguistics, artificial intelligence or motor coordination in subordinate positions. Cognitive science is forever reconfiguring and does not seem any closer to unification than when it emerged some 60 years ago. Only its nominal object, loosely defined as the conjunction of the mind and the brain, has remained fixed, with an increased emphasis, in the last couple of decades, on the context provided by the body.

Now that I have somewhat narrowed down the relata, I should say something about the relation(s) to be examined.

First, among the applications of mathematics to cognitive science, we need to distinguish those that merely (though perhaps importantly) impact one of the component disciplines or sub-disciplines from those that directly impact, or claim to impact, or may impact the enterprise as a whole, in its general methodology.

A second useful distinction we may wish to make is the following. Among the mathematical tools and techniques deployed in the various areas of cognitive science, some are of such general scope as to be equally applicable to areas unconnected to cognitive science: for example, statistical methods for the aggregation and assessment of experimental data that are extensively used in developmental psychology, in linguistics, in neuroscience, in neuropsychology, etc., but have no relevance to these areas *qua* members of the cognitive science federation: they serve the same purpose as they do in any one of the so-called special sciences. On the other hand, certain mathematical tools seem to have a significant impact on the content, or the conceptual structure, of the discipline that deploys them. The distinction is not necessarily sharp: the mathematics of neuroimaging, for example, although quite general – it works for medical imagery and many other kinds of imagery – significantly impacts cognitive neuroscience and in particular raises specific methodological problems. It has also been argued by Gigerenzer that the tools we use sometimes evolve into a structural principle or a general heuristic for the field.[2] Nonetheless, as a first approximation, it is both useful and feasible to concentrate on the second sort of mathematical application.

Third, we can ask whether mathematical modeling – the production of mathematical models of cognitive phenomena – exhausts the topic at hand, or

2 Gerd Gigerenzer, "From Tools to Theories: A Heuristic of Discovery in Cognitive Psychology", in: *Psychological Review* 98, 2, 1991, pp. 254–267.

whether mathematics can relate to cognitive science in a different way. The thought would be that mathematics provides, could or should provide, a *framework* for cognitive science; which would then be, or become, a fully mathematized science, in the way of physics for example. If we take our lead from the "queen of science", taken at the most elementary level of sophistication, we can for example distinguish between, on the one hand, calculus as a mathematical method (whose centrality need not be stressed), and on the other, the differential equations of the Earth-Moon system, or of the tides, or of the propagation of heat in a metallic bar: these are mathematical models of physical phenomena. The two are obviously related, and no less obviously distinct. Perhaps a helpful metaphor might be that calculus is, or is part of the language of physics, while models are descriptions or representations couched in that language. Alternatively, we could perhaps say that mathematics is constitutive of physics, as we know it today, with the consequence that a model in physics is almost by definition a mathematical object; while mathematics is not constitutive, e.g., of biology, whose models (moreover) are infrequently mathematical objects. Yet another way in which the difference is made manifest is that in physics, models are theories; in cognitive science (as in biology), models (in most approaches explored today) never rise to the status of theories – the one major exception being the proposal by classical AI to regard a computer program as a theory.[3]

Finally, we might want to set up a continuum between two polar situations. On one end, we would find "deep" mathematics (mathematical theories with conceptual depth, wide scope, powerful techniques) imparting intelligibility on some deep questions in cognitive science. On the other end, we would find simple mathematics used to describe or systematize fairly limited domains.

These four distinctions, though in large part conceptually independent, rather naturally give rise to two clusters of properties, characteristic of two opposing stances. The mathematically modest perspective is content with viewing mathematics as a toolbox providing methods, and material, some quite general, some more domain-specific, for opportunistically constructing models, piecemeal, of various cognitive phenomena at various levels of description. The mathematically ambitious perspective aims at couching cognition, so to speak, in the language of mathematics, and thereby revealing the deep structure of the mental realm, in which the piecemeal models of specific functions obtained from detailed empirical work would be seen to find their natural place.

3 See, e.g., Herbert Simon, "Artificial Intelligence: An Empirical Science", in: *Artificial Intelligence* 77, 1995, p. 97: "The theory is no more separable from the program than classical mechanics is from the mathematics of the laws of motion".

26.2 From Prehistorical to Postmodern Cognitive Science: Five Stages

As is well known, cognitive science has undergone a number of stages, since its inception, which can be placed in the 1940s. It is important to have this history in mind, in schematic form, for to each stage corresponds a specific framework for the mathematics of cognitive science. The following thumbnail descriptions are provided as no more than an aide-mémoire.

The prehistorical phase (1942–1956) was centered on the recently reborn logic and the just emerging cybernetics.[4] Logic was developed as a branch of mathematics and as a language for representing certain essential mental operations. It was mechanicized in the hands of Turing[5] and others, and biologized by McCulloch and Pitts and others.[6] The broad ambition of cybernetics was to provide an overarching theory of mind, brain and machines, couched in the appropriate language of information and control. It included a branch concerned with higher functions, with logic as its main tool, and a branch concerned with perception and motricity, with some classical and new mathematics distinct from logic.

The first phase of the historical period (roughly 1956–1980) centered on artificial intelligence (AI), broadly understood as the science of "intelligent" information processing, leading up to the so-called classical, or symbolic paradigm in cognitive science.[7] The formal systems of logic provided the language, and theories (at least notionally) took the form of (computer) programs; we would be more comfortable today calling them *models*, but at the time it was important not to let the theoretical ambition of AI be watered down: AI was to be the scientific theory of human intelligence (of cognition), not a mere methodology for producing intelligence-like effects. The needed mathematics was logic, automata theory, and the nascent computer science or informatics. However, a large part of the work was carried out with no visible help from the theoretical parts of these formal disciplines. In fact the deepest contributions concerned the development of programming

4 Cybernetics may in fact be regarded as an early form of cognitive science. See Jean-Pierre Dupuy, *On the Origins of Cognitive Science: The Mechanization of the Mind*. Cambridge, MA: MIT Press 2009; Steve Joshua Heims, *The Cybernetics Group*. Cambridge MA: MIT Press 1991.

5 Alan M. Turing, "On Computable Numbers, with an Application to the *Entscheidungsproblem*", in: *Proceedings of the London Mathematical Society*, 42, 2, 1937, pp. 230–265; reprinted in: Martin Davis (Ed.), *The Undecidable*. Hewlett, NY: Raven Press 1965 and many other collections.

6 Warren S. McCulloch, Walter A. Pitts, "A Logical Calculus of Ideas Immanent in Nervous Activity", in: *Bulletin of Mathematical Biophysics* 5, 1943, pp. 115–133; reprinted in: Warren S. McCulloch, *Embodiments of Mind*. Cambridge, MA: MIT Press 1965; also in: James A. Anderson and Edward Rosenfeld (Eds.), *Neurocomputing. Foundations of Research*. Cambridge, MA: MIT Press 1988.

7 See e.g. Max Lungarella, Fumiya Iida, Josh Bongard and Rolf Pfeifer (Eds.), *50 Years of Artificial Intelligence*. Berlin-Heidelberg: Springer 2007.

languages, first and foremost Lisp, then Prolog and more recently object-oriented languages such as Java, which made writing code for cognitive functions feasible. Actually producing a computer program for, say, chess or checkers playing, or scene recognition, or parsing, or writing a large corporation's paychecks, consisted in armchair construction of information-flow diagrams, an activity that can hardly be taken as part of mathematics. Exception must be made for the study of perceptual and motor functions, which recruited several high-powered mathematical areas, ranging from differential geometry to Fourier analysis and probability theory, and in fact extending them to meet specific requirements.

Next came (*ca.* 1980–1995) connectionism or the neural nets approach, which took up the perceptual strand of cybernetics and extended it into a full-fledged framework for cognitive science (and AI), competing with the classical, symbolic approach.[8] Connectionism, which comprises several rather distinct currents, can be applied at the functional or mental level, at the neuronal level, or again at an intermediate level, abstracted from the neuronal level and reflecting the "micro-structure" of cognition, understood in informational terms. The mathematics is here much more visible than in the symbolic approach, and also much richer and more varied, comprising fragments of linear algebra, of probability and signal theory, of analysis, and of dynamical systems, although seldom reaching great heights of sophistication.

The modern phase, to which the present still belongs, but is morphing into what I venture to call post-modern, is characterized, first and foremost, by the appearance of a new contender for the status of admiral discipline: cognitive neuroscience, supported by functional neuro-imaging technology but also by the strengthening of theoretical neuroscience, which consists in applying the methods of physical modeling to phenomena arising at various levels of organization of the nervous tissue.[9] Mathematical tools have become considerably more sophisticated. Functional imagery calls upon highly complex statistical methods aiming at providing a pictorial representation of the distributed activity in neuronal population, taking a gigantic mass of indirect signals as the basis of an inference to their sources. Theoretical neuroscience helps itself to a vast repertory of mathematical theories. The second most important feature of the modern phase is the return of the body, which appears not only under the guise of the brain, material "siege" of cognition, but also as organism and genuine bearer of cognition. With the body come perception and motricity, which, as we just saw, were never totally

8 See e.g. James A. Anderson, Andras Pellionisz and Edward Rosenfeld (Eds.), *Neurocomputing II*. Cambridge, MA: MIT Press 1990.

9 See e.g. Peter Dayan and Laurence F. Abott, *Theoretical Neuroscience: Computational and Mathematical Modeling of Neural Systems*. Cambridge, MA: MIT Press 2001; Michael A. Arbib, *The Handbook of Brain Theory and Neural Networks*, 2nd ed. Cambridge, MA: MIT Press 2003; Michael Gazzaniga, *The Cognitive Neurosciences*, 4th ed., Cambridge, MA: MIT Press 2009.

neglected, but now take on an entirely novel dimension and call for pretty deep mathematics.

Post-modernism (a notion which I venture to propose here, but which to my knowledge has not been proposed under this or any other name by observers of contemporary cognitive science) is characterized by a breakdown of pragmatic unity and doctrinal consensus. Cognitive science is at a tipping point. Is it on the verge of disintegration, with a majority of programs recategorized inside neuroscience (and more broadly biology), and the rest reintegrating other main disciplines, or is it headed towards a fully integrated field, awaiting a new framework in which mathematics is likely to play a fundamental part? While this "crisis" is playing out, though, the working scientists are going into high gear, bringing to bear, in some areas, extremely powerful and wide-scope principles with strong mathematical content. The field is increasingly divided between areas in which a hard-science culture is required, and those in which it isn't, reconstructing, perhaps, the boundary between the natural and the human sciences.

26.3 Mathematics as a Local Player: A Sample

Not only do the stages overlap, so that the mathematical methods characteristic of the various phases actually have co-existed and were sometimes combined, but they abstract away from the divisions, at all times, within cognitive science. All along, besides the general framework to which some programs explicitly refer, specialties have existed and evolved quite independently, developing a proprietary methodology with owed little to the general framework. Typically, vision science, as has been already mentioned, though in a sense central to the project, due in particular to its uncharacteristic success, and yet in another peripheral, due to the its de-emphasis on *human* as opposed to *machine* (robotic) or *animal* vision, was off to an early start and not only exploited existing theories from contemporary mathematics, but developed its own mathematical tools. Meanwhile many other branches of cognitive science developed without any mathematics at all. Still, the overall trend has been a growing role of mathematics in the field.

As a second pass then, I offer a quick tour of a number of research programs, some of which were mentioned in passing, and which call on a variety of mathematical techniques or styles, as both language and modeling methodology (as per the distinction sketched in Sect. 26.1).

What follows is a mere sample, by no means an exhaustive list. I have divided it in three parts, which are not clear-cut but rather denote different attitudes and practices in the deployment of mathematics combined with differing approaches to cognition.

(a) Abstract or pure information-processing theories (also known as computational theories)

(i) Logic(s). The study of reasoning is probably the best-known subprogram of early cognitive science, a natural extension of the tradition of logic, taking on board two crucial dimensions, control and computational (neo-mechanical) feasibility, and straddling cognitive psychology and AI. In the widest sense, it can be argued that the guiding assumption of the early period of the field was that cognition is, at base, reasoning. Though this assumption is no longer in favor, reasoning, widely construed, remains a central topic. It encompasses not only deduction from firm premises in eternal propositional format (such things as "2+2 = 4" or "force equals mass times acceleration"), but also a large variety of inferential regimes (inductive, abductive ...) deployed on different materials (non-purely propositional, non-eternal, non-firm, etc.), including coherence maintenance and belief revision, as well as problem-solving and even scientific inquiry. The development of non-standard systems of logic, including defeasible or non-monotonic logics, of algorithmic control systems, and of algorithmic complexity theory, clearly demonstrates the co-evolutionary process affecting cognitive science and mathematics. However, the part of cognitive science directly affected by the more sophisticated mathematical logic involved belongs to AI and computer science, rather than cognitive or developmental psychology or even formal theories of rationality, core areas which recruit no more than primitive mathematical techniques.

(ii) Signal detection theory [SDT]. How to discriminate noise from signal, say in a visual or auditory scene, can be seen as a decision process. Probability theory is thus brought to bear on psychophysics, the study of perceptual systems as physical measurement devices. But SDT extends to a wider set of phenomena, including some that are more clearly cognitive, such as individual or collective decision-making under uncertainty.

(iii) Control theory. Classical robotics relies heavily on control theory, a part of dynamical systems theory that also applies to the study of various biological processes. Here again, the initial target of the (fairly deep) mathematics involved leans towards the machine dimension of cognitive science, or the motoric dimension, long deemed somewhat peripheral. However, on the one hand this motoric dimension has recently been recognized as more important and more closely linked to cognition than was previously thought, and on the other, dynamical systems are propounded as an alternative to classical computational models for cognitive science at large.

(iv) Machine learning. The first attempt at developing an information-theoretic approach to learning was initiated in the 1960s with the aim of formalizing induction in general, and more particularly, the inductive identification of the ambient language by the non-linguistic infant. The acquisition, by a child, of the grammar (syntax) of her mother tongue can be seen, and formalized, as a problem of induction: the innate language faculty provides a set of constraints which limit the set of possible grammars. The child's job is to identify with which of the possible languages she is in fact confronted, on the limited basis of what she hears. This thought has led to the development of formal learning theory, which draws on fairly simple notions from discrete mathematics and recursive functions.[10] It has been extended to the study of scientific inquiry, seen as a process of induction from basic empirical data.[11]

A very different approach to machine learning, now generally preferred, is the PAC paradigm (probably approximate learning) developed in the early 1980s[12]: from a sample of the set to be "learned", PAC learning produces, with high probability, a generalization function which suitably approximates the given set. PAC involves sophisticated tools drawn from or developed within computational complexity theory.

(v) Probability theory. Probability lies of course at the foundation of decision theory, an area that is traditionally claimed by economics as its core theory but is increasingly taken over by "neuro-economics", a joint venture of economics and cognitive science. Less known perhaps outside the field, but quite important, is the attempt to attack a very broad collection of cognitive processes[13] by postulating an optimizing principle operating on non-conscious sensations or data. The so-called Bayesian approach is now pre-eminent in vision science; it is also applied to the study of memory, and mobilizes fairly sophisticated mathematical tools.

(vi) Game theory. Decision theory has gone collective with the help of game theory, specifically invented for that purpose. But again it is not widely known that game theory has become an instrument of choice for the evolutionary approaches of collective behavior, and is thus relevant for the study of social cognition, e.g. the natural basis of other-oriented behavior and norms.[14] The mathematical results required for the latter topic are however quite rudimentary.

(vii) Category theory. Classical first-order logic has been pressed into service in the quest for formal models of natural language–this is the well-known program of Montague semantics. But just as category theory has claimed to provide mathematics with a better foundation than set theory, it has also been promoted as the best framework for the semantics of natural language and the associated field of categorization. Indeed, some authors have argued that category theory should

---

10 Sanjay Jain, Daniel N. Osherson, James S. Royer and Arun Sharma, *Systems That Learn: An Introduction to Learning Theory (Learning, Development, and Conceptual Change)*, 2nd ed., Cambridge, MA: MIT Press 1999.

11 Eric Martin and Daniel N. Osherson, *Elements of Scientific Inquiry.* Cambridge, MA: MIT Press 1998.

12 Leslie Valiant, "A Theory of the Learnable", in: *Communications of the ACM* 27, 11, 1984, pp.1134–1142.

13 And even cognition as a whole; see Nick Chater and Mike Oaksford (Eds.), *The Probabilistic Mind: Prospects for Bayesian Cognitive Science.* New York: Oxford University Press 2008.

14 Robert Axelrod, *The Evolution of Cooperation*, Revised Edition. New York: Basic Books 2006.

replace logic (which in some sense it generalizes) as the *organon* of cognition.[15] Not surprisingly, this last example leads us back to the first as far as conceptual ambition goes: like logic in the early days, category theory is poised not (only) as a (modeling) tool for this or that cognitive process, but as the true language of cognition. It must be remarked that this is a minority view, ignored by a vast majority of scientists and philosophers of cognitive science.

(b) Neural dynamics. In this group belong some applications of core mathematical theories to systems that take their inspiration from a general view of the basic structure of the brain.

(i) Linear algebra and statistical physics are the indispensable tools to study the dynamics and learning capabilities of feed-forward layered networks of threshold automata, which constitute a large and well-studied family of neural nets. In the "parallel distributed processing" (PDP) view,[16] such systems are capable of supporting a wide variety of cognitive functions and are regarded as a basic architecture competing with the von Neumann computer. Mathematical analysis is essential in order to determine the conditions under which a system will stabilize, hence provide a definite output in response to a given input and even more importantly in order to define learning algorithms that work, for example retropropagation. Learning is of the essence for PDP as it allows a network to implement a given input-output function by being exposed to a set of examples.

(ii) Statistical physics is also used, but at a deeper level, in the study of another family of neural nets, those that are fully interconnected (as opposed to feed-forward) and can thus be regarded as autonomous dynamical systems (they are sometimes called "attractor neural networks" or ANN).[17] The first example of this approach was proposed by physicist John Hopfield who exported a modeling technique perfected by solid-state physicists, the Ising model, to the study of a neural net that he could interpret as a device with a content-adressable memory.[18]

(iii) Tools from advanced analysis (ordinary non-linear differential equations, partial differential equations, Fourier analysis and wavelets ...) and from dynamical

15 François Magnan and Gonzalo E. Reyes, "Category Theory as a Conceptual Tool in the Study of Cognition", in: John Macnamara and Gonzalo E. Reyes (Eds.), *The Logical Foundations of Cognition*. New York Oxford: Oxford University Press 1994; Jaime Gómez and Ricardo Sanz, "Modeling Cognitive Systems with Category Theory. Towards Rigor in Cognitive Sciences", Tech. Report Universidad Politécnica de Madrid 2009.

16 David E. Rumelhart and James L. McClelland (Eds.), *Parallel Distributed Processing: Explorations in the Microstructure of Cognition*. Cambridge, MA: MIT Press 1986.

17 Daniel J. Amit, *Modeling Brain Function: The World of Attractor Neural Networks*. Cambridge: Cambridge University Press 1992.

18 John Hopfield, "Neural Networks and Physical Systems with Emergent Collective Computational Abilities", in: *Proceedings of the National Academy of Sciences USA* 79, 1982, pp. 2554–2558.

systems theory, often combined with stochastic methods, are deployed in several areas[19]:

– in basic (i.e. neuron-level) neuroscience, the propagation of the nerve impulse; the receptor field of the ganglia cells of the retina, which play a crucial role in early visual processing,[20] or of the simple cells in V1, which help the visual system deal with noisy stimuli; the functional architecture of V1, etc.[21]

– at a higher level of integration, the theory of networks of weakly coupled oscillators provides models for complex representations in the brain (i.e. representations that include independently identifiable components);

– at a functional level, abstracting away from neural implementation, image processing by biologically plausible or by artificial systems calls on variational methods.

(iv) In the theory of motor control (balance, gait, reach, grasp, navigation ...), geometry is increasingly regarded as encoded at a deep level in the relevant brain areas, which literally solve complex geometrical problems. Differential geometry and kinematics are thus seen not only as descriptive tools, but as naturally realized faculties of the brain.[22] On the side of artificial systems, robotics has directly or indirectly attracted the interest of top mathematicians working in such fields as algebraic and differential geometry, Lie theory, optimization theory as far back as the 19th century. Nowadays, robotics calls on a large spectrum of powerful theories, ranging from dynamical systems to Bayesian statistics, discrete and computational geometry or topology.[23]

26.4 Deep vs. Shallow Engagement

I have up until now more or less explicitly indexed the depth of mathematization of a research program on the depth of the mathematics deployed. On that count, the mathematization of classical or symbolic AI or cognitive science is considerably shallower, than that of connectionism, and species of connectionism range from relatively less deep to quite deep; or again, formal learning theory is mathematically shallower than PAC learning. Similarly, theories of specific functions, such as language acquisition, phonology, pattern recognition, vision, motor control

19 Alain Berthoz, "Rapport sur les liens entre mathématiques et neurosciences", in: *Rapports sur la science et la technologie* 20, 2005, pp. 175–211.

20 David Marr, *Vision*. San Francisco: W.H.Freeman & Co Ltd 1982.

21 Jean Petitot, *Cognitive Morphodynamics: Dynamical Morphological Models of Constituency in Perception and Syntax*. New York: Peter Lang Pub Inc 2011.

22 Nikolai Bernstein, *The Coordination and Regulation of Movement*. New York: Pergamon Press 1967; Alain Berthoz, *The Brain's Sense of Movement*. Harvard: Harvard University Press 2002.

23 "The Interplay between Mathematics and Robotics", Summary of a Workshop, National Science Foundation, Arlington, VA, 15–17 May, 2000.

etc. can be roughly ordered according to the sophistication of the mathematics they use. This criterion also can serve to identify a general pattern: mathematical sophistication tends to increase with time, but quite unevenly, leaving some non-marginal areas in their essentially non-mathematized initial state, while others have undergone radical mathematization.

However sophisticated the mathematics involved, they do not necessarily have a profound effect on the field. First, there may be a "lamppost" effect, when a mathematical technique is developed on its own impetus, perhaps to the point of creating an entire academic field, but without actually furthering the original problematic (as seen, at least, from a limited time perspective). But second, and more importantly, even a successful mathematization may concern a strictly limited area, without bringing consequential changes to the overall landscape. Besides the depth of mathematical methods, it is therefore important to distinguish between research programs that aim at engaging the entire field of cognitive science with mathematics and programs which result in minimal engagement, either due to the shallowness of the mathematics employed, or to the limited scope of the program.

Scientific temperaments vary. To some, grand schemes are suspect and their formulation and examination are basically a waste of time. Scholars of that bend will therefore be inclined to turn their attention to well-defined problem areas where mathematics has a serious potential. Others are loth to abandon the initial ambitions of cognitive science, viz. to produce in the fullness of time an integrated account of mind and brain, with a density of conceptual connections at least comparable to that of biology, if not that of physics. And so, in the face of the increasing fragmentation of cognitive science, they turn to mathematics. Some specific proposals of mathematically-induced unification of the field are on offer. To some of these I now turn, by evoking a few representative theorists.

In his landmark monograph,[24] the late physicist Daniel Amit proposed the most elaborated view of cognitive science as the study of the cooperative properties of the brain tissue, in the tradition initiated by Hopfield, but with a novel concern with neurobiological realism. By applying the know-how of the physicist in deploying statistical mechanics and dynamical systems theory to nature's most complex system, the brain, examined with the utmost care at every level, one can hope, according to Amit, to develop a unified theory that would stand to the brain in roughly the same relation as state-of-the-art mathematical physics to (non-biological) natural systems, by establishing systematic links between the various levels of organization. Cognitive science would thus be unified, though not reduced, under the banner of a highly sophisticated mathematical physics.

Paul Smolensky, also trained as a physicist, went on to become the most articulate and powerful theorist of the PDP school,[25] to which he contributed early on a

24 Daniel J. Amit, *op. cit.*

25 Paul Smolensky, "On the Proper Treatment of Connectionism", in: *The Behavioral and Brain Sciences* 11, 1988, pp. 1–23.

unifying framework which he called "harmony theory".[26] Cognitive processes, in a wide (and ever widening) spectrum of cases, consist in attempting to honor a (usually large) number of "soft constraints". It is seldom possible to honor them all, so that the desired outcome is a state where the sum total of violations is minimal: a system that reaches such a state has achieved the highest possible 'harmony'. Now what turns this idea from metaphor to theoretical principle is the mathematics (linear algebra, dynamical systems, probability theory) that shows that, under suitable conditions, a feed-forward multi-layered network can actually achieve a harmony maximum. Characteristically, Smolensky did not rest content with this perspective, which failed to connect with the classical principles and concepts of "classical" or "symbolic" cognitive science. He now regards the central challenge to be to precisely characterize the kind of abstraction that bridges the biophysical properties of the brain to the computational properties of mental representations and knowledge – in short, to the *mind*, and he has taken up the challenge in the area of linguistics.[27] His mental representations are to be understood as abstract theoretical constructs that must be characterized precisely through formal systems developed using the methods of mathematics. Thus, he writes, "The ultimate goal of my work is to help usher cognitive science through a fundamental transition into a truly mathematical discipline."[28]

Methodology, according to Smolensky, has been the great weakness of cognitive science, causing a sterile battle of "isms", speculative theses regarding the true nature of cognition. Between the "ism" level and the "model" level (highly specialized accounts of lab-generated data on very specific behaviors) there lacks what he calls the level of *general theory*, which according to him is "largely missing because sophisticated use of mathematics is required" much of which remains to be created *by adequately trained cognitive scientists*: co-evolution again, as the mathematics that cognitive science requires to come of age is itself yet in limbo. Now why exactly, one may ask, would mathematics be the means to reach the prescribed end? Smolensky's answer can be broken down in two components. First, formalization is indispensable to regiment and justify the use of abstractions (so as not to smuggle in occult properties in the guise of theoretic entities), and convincing formalizations must yield accounts of complex phenomena from small number of principles governing a small number of variables. Mathematics is the only known discipline that can achieve this. Second, cognitive science presents a special challenge, which is to bridge the gap between the essentially continuous physical substratum and the discrete manifestations at the mental level: again, only the mathematics of emergence deployed in nonlinear physics are known to achieve

26 Paul Smolensky, "Information Processing in Dynamical Systems: Foundations of Harmony Theory", in: David E. Rumelhart and James L. McClelland (Eds.), *op. cit.*

27 Paul Smolensky and Géraldine Legendre, *The Harmonic Mind: From Neural Computation to Optimality-Theoretic Grammar Volume I: Cognitive Architecture; Volume II: Linguistic and Philosophical Implications*. Cambridge, MA: MIT Press 2006.

28 Personal communication.

this, for the fairly simple systems studied thus far – and a similar bridge awaits to be constructed between the physical brain and the mind.

Other strong programs are advocated by authors such as Jean Petitot, Scott Kelso, Chris Eliasmith, or Mark Bickhard.[29] Due to space constraints, no attempt will be made to give the reader so much as a flavor of their respective proposals. Petitot's main theoretical sources are dynamical systems theory, nonlinear physics (emergence in disordered systems, phase transitions), differential geometry, on the one hand, and on the other, strikingly, Husserlian phenomenology turned, so to speak, on its head, and thus "naturalized" by virtue of the new mathematical physics which Husserl could not fathom. Kelso takes his inspiration also from dynamical systems and more particularly from Hermann Haken's theory of self-organized nonequilibrium phase transitions. Eliasmith sees control theory, a branch of theoretical computer science, as providing a unifying framework for the necessarily pluralistic theories of the mind and brain. Bickhard develops an approach of his own, "interactivism", which aims at reconfiguring cognitive science by way of rethinking the naturally emerging high-level properties of living organisms that give rise to representations.

Most of these programs, next to several others, are often grouped under the general label of "dynamicism", and blanket arguments are proffered in favor of what is presented as a shared approach. For example, R. D. Beer claims that "By supplying a common language for cognition, for the neurophysiological processes that support it, for non-cognitive human behavior, and for the adaptive behavior of simpler animals, a dynamical approach holds the promise of providing a unified theoretical framework for cognitive science, as well as an understanding of the emergence of cognition in development and evolution."[30] The trouble with such

29 J. A. Scott Kelso, *Dynamic Patterns.* Cambridge, MA: MIT Press 1997; Chris Eliasmith and Charles H. Anderson, *Neural Engineering: Computation, Representation and Dynamics in Neurobiological Systems.* Cambridge, MA: MIT Press 2003; Jean Petitot, Francisco Varela, Bernard Pachoud and Jean-Michel Roy (Eds.), *Naturalizing Phenomenology.* Stanford: Stanford University Press 1999; M.H. Bickhard, "The Biological Foundations of Cognitive Science", in: *New Ideas in Psychology* 27, 1, 2009, pp. 75–84. Other programs originate in AI (the new wave of so-called "Artificial General Intelligence", see Ben Goertzel and Cassio Pennachin, *Artificial General Intelligence*, 1st ed. New York: Springer 2007 ; Ben Goertzel, *The Hidden Pattern. A Patternist Philosophy of Mind.* Florida: BrownWalker Press 2006, or robotics (Rodney Brooks, *Cambrian Intelligence: The Early History of the New AI.* Cambridge, MA: MIT Press 1999; Patti Maes, *Designing Autonomous Agents.* Cambridge, MA: MIT Press 1990.

30 Randall D. Beer, "Dynamical Approaches in Cognitive Science", in: *Trends in Cognitive Sciences* 4, 3, 2000, pp. 91–99. It is generally accepted that dynamicism came into existence as a self-aware and visible orientation within cognitive science with the publication of two collections: Robert Port and Tim van Gelder (Eds.), *Mind as Motion: Explorations in the Dynamics of Cognition.* Cambridge, MA : MIT Press 1995; Esther Thelen and Linda B. Smith (Eds.), *A Dynamic System Approach to the Development of Cognition and Action.* Cambridge, MA: MIT Press 1996; see also Lawrence M. Ward,

global views is that they lead to a battle of "isms", as Smolensky has argued, a battle that no side can win, while true theoretical progress lies in unearthing the scientific substance of the initial slogans. Better then, perhaps, to examine the various programs and judge them on their own merits rather than on their adherence to overly general principles.

26.5 Strong Programs, Conceptual Reform and the Co-evolution of Cognitive Science and Mathematics vs. Pluralism and the Toolbox Philosophy

Still, it is a striking and crucial fact that all of these strong programs aim at putting order into chaos by virtue of bringing cognition under the jurisdiction of mathematics. And here lie three seemingly major difficulties.

First, by their very plurality, they add to the buzzing, booming confusion of a field that they claim to be desperately in need of regimenting. Second, what their proponents are counting on to make this happen, viz. mathematics, cannot yet deliver: the requisite mathematical tools do not exist at the present pioneering stage. Third, of all the specialized languages and disciplines of science, mathematics is the most impenetrable not only to the practitioners of the human and social sciences, philosophers included, but to many biologists and even computer scientists, who are, on the side of the natural sciences, the ones with the strongest ties to cognitive science. It would seem then that most scientists have at best hands-on knowledge of cognition, or of mathematics, but not both, while both are claimed to be necessary if cognitive science is ever going to attain maturity.

The first problem can be counted on being overcome by the mere passage of time. Cognitive science is at a stage where it suffers from an acute case of the "toothbrush problem" – every major figure in the field with a general theory wants to use his own theory, and nobody else's, but the reasonable hope is that this won't last forever, and more particularly that mathematical models will gain wider acceptance and accelerate convergence.

The second problem is more interesting. One lesson to be gleaned from the most casual inspection of research programs such as those just mentioned is that no program for a fully mathematized cognitive science can succeed without a worked-out program for conceptual reform: just like the founders of the field, tomorrow's architects must provide a structural hypothesis, or, in Newell and Simon's terms,[31] a "law of qualitative structure" regarding the ontology of cognition, together with a unifying methodology. This is of course in line with more mature disciplines such as physics and (molecular) biology. Mathematization invariably goes hand in hand

Dynamical Cognitive Science. Cambridge, MA: MIT Press 2001.

31 Allen Newell and Herbert A. Simon, "Computer Science as Empirical Inquiry: Symbols and Search", in: *Communications of the ACM* 19, 3, 1976, pp. 113–126.

with a set of principles, which are part ontological and part epistemic, the latter regulating the necessary abstractions. The principles, in turn, provide traction only insofar as they make the phenomena accessible to mathematics. And, with rare exceptions, the specific mathematical tools must be developed in tandem with the principles (a point forcefully made, in particular, by Petitot and Smolensky, but also included or implied in just about all the detailed proclamations of new paradigms in cognitive science). This situation therefore calls for conceptual reform driven by co-evolution of cognitive science and mathematics.

The third problem is more vexing, and it is not restricted to cognitive science. I can think of two optimistic and one pessimistic responses. First, we can hope (like Smolensky) that a new generation of mathematically savvy cognitive scientists is now emerging from a few pioneering graduate programs, who will be the moving force of cognitive science in the coming years and decades. Second, we can imagine a situation of distributed scientific competence, where the sophisticated mathematics lies in one group of brains, the advanced cognitive science in another group, without there being many brains, or any for that matter, in both groups. Third, and this is the less sanguine view, we can imagine a future where the two orientations remain at an increasing, rather than diminishing, distance from one another. Mathematical cognitive science would evolve into a separate field, with or without occupying center stage: economics and biology are perhaps examples of each scenario.

Yet we should not forget that overarching methodologies stand at one end of a continuum, whose other pole reflects a pure "hands on" philosophy, one which recommends context-sensitive, case-by-case model construction, and sometimes evokes evolutionary theory and the modularity thesis to bolster the case of tinkering as the proper method in cognitive science. This stance countenances a thoroughgoing pluralism, with at least three dimensions along which models can vary: the level of aggregation, or level of reality dimension, from (say) the synaptic cleft to consciousness or culture; the genus of models (what counts as a model), as determined by the basic science and the methodology; and finally the task domain, from (say) navigation to chess playing, from face recognition to economic behavior, and so forth. The mathematics provides a toolbox to the working cognitive scientist, who constructs models of systems whose function is known or hypothesized, and whose neural realization is sought. The extent to which this plurality of models can be brought under a unifying scheme, and the importance of mathematics in that scheme, remain to be determined.

UFR de Philosphie et Sociologie
University of Paris-Sorbonne (Paris IV)
1, rue Victor Cousin
75005, Paris
France
daniel.andler@ens.fr

Chapter 27

Ladislav Kvasz

What Can the Social Sciences Learn from the Process of Mathematization in the Natural Sciences

Abstract

The paper tries to put the conflict of the natural and the human sciences into its historical context. It describes the changes in classification of scientific disciplines that accompany a scientific revolution, and offers an alternative to Kuhn's theory. Instead of a conflict between the proponents and opponents of the new paradigm it interprets the revolution as a conflict between the mixed disciplines and the metaphorical realm of the old paradigm.

27.1 Introduction

For almost two centuries there has been a tension between the natural and the social sciences. As Thomas S. Kuhn writes in *The Structure of Scientific Revolutions*,[1] it was this tension that led him to the creation of the notion of a paradigm. According to Kuhn the difference between natural and social sciences consists in the fact that while in natural sciences we have to do with research in the framework of *normal science* based on a widely *accepted paradigm*, in social sciences there is nothing comparable to paradigms and so scholars again and again question the foundations of their disciplines. Kuhn thus drew attention to an important difference between these two areas. Nevertheless, according to Kuhn this difference does not create a gap between them:

> I'm aware of no principle that bars the possibility that one or another part of some human science might find a paradigm capable of supporting normal, puzzle-solving research. … Very probably the transition I'm suggesting is already under way in some current specialties within the human sciences. My impression is that in parts of economics and psychology, the case might already be made.[2]

1 Thomas S. Kuhn, *The Structure of Scientific Revolutions*, Chicago: University of Chicago Press 1962.

2 Thomas S. Kuhn, "The Natural and the Human Sciences", in: Thomas S. Kuhn, *The Road since Structure*, Chicago: University of Chicago Press 2000, pp. 222–223.

D. Dieks et al. (eds.), *Probabilities, Laws, and Structures*, The Philosophy of Science in a European Perspective 3, DOI 10.1007/978-94-007-3030-4_27,

If we want to understand this problem it is expedient to look at the tension between the natural and social sciences in a broader historical perspective.

The first thing which we probably notice after turning to a broader historical perspective is that the conflict between natural and social sciences is not as old as it might seem. In the Classical era there was no conflict between the way how people understood human and social phenomena on the one hand, and how they approached nature on the other. This, of course, does not mean that in the Classical era the whole knowledge would form a harmonic whole. Also in Greek science there was a conflict that in many respects resembles the tension between the natural and the social sciences that we encounter in modern times. The border, along which the tension manifested itself, nevertheless, ran elsewhere. It did not separate knowledge of nature from the knowledge of human and social phenomena but rather it separated the mathematical knowledge (based on the deductive method and using categories such as number, proportion, and shape) from the "organic" realm (based on causal explanation and using categories such as purpose, goal, and action). In this second realm we could find biological as well as social disciplines, i.e. disciplines which according our classification lie on the opposite sides of the barricade that separates the natural from the social sciences. Ancient Greeks approached in a similar way the study of the "*generation of animals*" and the study of "the psyche" or politics. Starting from the seventeenth century onwards the study of the "*generation of animals*" was gradually incorporated into the realm of the newly constituted natural science, while the study of "the psyche" became one of the crystallization cores of the emerging social sciences. Therefore, one of the first aims of the present paper is to propose a framework for the reconstruction of the shifts in the classification of scientific disciplines.

27.2 Classification of Scientific Disciplines According to Their Relation to the Paradigm

In order to be able to understand the transitions of scientific disciplines between the categories of "hard" and "soft" sciences it is useful to form a more differentiated image of the "topography of the scientific landscape" that lies between these two poles. As a first move we suggest to abandon the terminology of dividing the scientific disciplines into "hard" and "soft". Instead let us call the "hard" disciplines *paradigmatic disciplines*. In contemporary science the paradigm is formed by physics and so the paradigmatic disciplines are all those disciplines in which the methods of quantification and measurement lead to success. For a more precise characterization of a particular area of "soft" disciplines I suggest to introduce the term *elusive region of the paradigm*. It comprises those disciplines where the methods and approaches of the particular paradigm cannot be employed. Besides these two kinds of scientific disciplines we introduce two other kinds which lie somewhere between the paradigmatic region and the elusive region of the paradigm.[3]

The first sort of scientific disciplines that lie between the paradigmatic and the elusive region are the *mixed disciplines*. This term is used by historians to describe a remarkable set of disciplines from late Antiquity, such as Euclidean optics, Archimedean theory of the lever, the theory of simple machines, or Ptolemaic astronomy.[4] These disciplines have in common the use of exact mathematical language in the description of situations which according to the ancient understanding of science should not be described using mathematics because matter plays a substantial role in them. These disciplines cannot be fully deductive and, therefore, they do not fulfill the standards of mathematics. On the other hand, these disciplines do not use explanations based on the notions of aim and purpose (final cause), that Aristotle considered being the explanation of phenomena that belong to the elusive region of the ancient paradigm. Thus, in a whole range of cases the practice of ancient science did not follow the standards laid down by Aristotle and it formed disciplines the methodological status of which was rather unclear. The fact that a lever, a mirror, or a pulley are material objects, but in spite of this, in their description scholars use mathematics, is from the ancient point of view inconsistent. The mixed disciplines played an important role during the scientific revolution of the seventeenth century. Galileo made important discoveries in the theory of simple machines, while Fermat and Descartes created theories of refraction of light. We may say that it were the mixed disciplines where the fundamental notions of the paradigm of modern science were born.

The second category of disciplines lying between the paradigmatic and the elusive region can be called the *metaphorical region of the paradigm*. It forms a counterpart to the mixed disciplines. While in the case of the mixed disciplines

---

3 Kuhn's notion of paradigm had many meanings. Later Kuhn restricted the scope of this notion (see Thomas S. Kuhn, "Second Thoughts on Paradigms", in: Thomas S. Kuhn, *The Essential Tension: Selected Studies in Scientific Tradition and Change*, Chicago: University of Chicago Press 1977, pp. 293–319). But it still remained rather broad. In Ladislav Kvasz, "On Classification of Scientific Revolutions", in: *Journal for General Philosophy of Science*, 30, 1999, pp. 201–232, I suggested to distinguish three kinds of scientific revolutions and three kinds of paradigms: the paradigm of idealization, of representation and of objectification. For the present paper the paradigm of idealization is the most relevant one and in the text that follows, by paradigm I will understand the paradigm of idealization.

4 I suggest (in contrast to Kuhn) to consider Euclid's *Elements* as the paradigm of Ancient science. It may sound unusual to call *Elements* a paradigmatic theory. We understand paradigms as a part of science while for us mathematics does not belong to science. Nevertheless, it is problematic to use our contemporary classification of disciplines in interpreting antiquity. If we look at Ancient science not from our but from its own viewpoint, it is rather the *Elements* than the *Almagest* that had a paradigmatic status. Therefore, the Ptolemaic astronomy that Kuhn characterized as paradigmatic I prefer to include among the mixed disciplines.

the notions and methods of the paradigm are used in a precise and unambiguous way, and the problem is only that they are being used outside the area where their use can be justified by the paradigm's methodology, in the *metaphorical region* the fundamental notions of the paradigm are used with a *transferred, distorted* and *stretched* meaning. As a representative of the metaphorical region of the ancient paradigm we can consider *Aristotle's theory of local motions*, according to which heavy bodies fall downwards while light bodies float upwards. In a paper on Cartesian physics I argued that the Aristotelian theory of local motions is a geometrical theory.[5] It is based on the image of a geometrically ordered universe and it understands motion as a transition between different places of this geometrical order. Nevertheless, geometry (the paradigmatic discipline of ancient science) is used here in a different manner from that in the mixed disciplines. Geometry does not enter the Aristotelian view of the order of the cosmos in an explicit way as a set of exact notions and methods for making constructions and proving theorems (as it enters the Archimedean theory of the lever), but only implicitly, as a set of metaphors, by means of which we can discern order and meaning in the phenomena. Thus, even though Aristotle's understanding of motion is biological (or organic) and, therefore, belongs into the elusive region of the ancient paradigm, a fraction of it – the theory of local motions – is based on geometrical metaphors.

We see that besides the *paradigmatic region*, i.e. the realm of disciplines that use the notions and methods of the paradigm in accordance with the methodological standards of the paradigm, and the *elusive region*, i.e. the realm which defies the use of the notions and methods of the paradigm, there are at least two other areas of scientific disciplines that are constructed using the means of the paradigm. On the one hand, the paradigm offers the technical tools for the formation of the *region of the mixed disciplines*, i.e. disciplines that use the notions and methods of the paradigm in a precise and correct manner but apply them to phenomena which were not foreseen by the creators of the paradigm and where it is not possible to fully comply with the methodological standards dictated by the paradigm. Further, the paradigm leads to the formation of the *metaphorical realm of the paradigm*, which comprises those phenomena that are too complex, and so a precise technical use of the notions and methods of the paradigm is not possible. Nevertheless, the paradigm offers a whole range of metaphors that make it possible to understand these phenomena at least in a qualitative manner and so to incorporate them into the rational discourse created by the paradigm. If we wish to understand the relation of natural sciences (forming the paradigmatic region of contemporary science) and humanities (lying to a great extent in the elusive region of that paradigm), it seems reasonable to replace the opposition of the "hard" and "soft" sciences by the following scheme[6]:

5 Ladislav Kvasz, "The Mathematisation of Nature and Cartesian Physics", in: *Philosophia Naturalis*, 40, 2003, pp. 157–182.

6 The scheme represents the topography of the scientific landscape. The *horizontal arrows* separate the strict use from the metaphorical use of the basic notions (in the paradigmatic region and in the mixed disciplines the notions of the paradigm are used

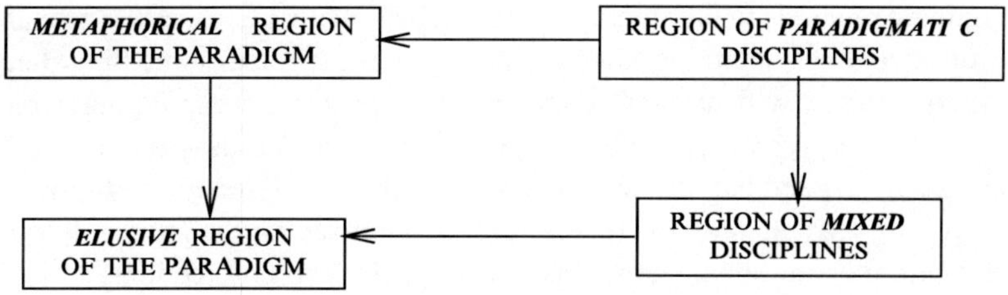

27.3 An Outline of a Reconstruction of the Scientific Revolution

The above scheme makes it possible to soften the contrast between the paradigmatic region that is formed by the "hard" disciplines and the elusive region of the paradigm that is formed by the "soft" disciplines and so opens a new possibility for a rational reconstruction of the scientific revolution of the seventeenth century. It turns out that it were the mixed disciplines and their conflict with the metaphorical realm of the paradigm which were the driving force of that revolution. Newtonian physics was created not inside the paradigmatic region of the old paradigm. The paradigmatic region of ancient science was mathematics. The birth of Newtonian physics stimulated the creation of several new mathematical disciplines, but despite of this, we cannot say that inside of mathematics there occurred some massive refutation of the previous research (which would be a case if a revolution occurred in this region). Also the elusive region of the old paradigm (the realm of the organic) did not undergo radical changes. Biology was during the scientific revolution of the seventeenth century on the fringe of the scientific interest. It came into the center of interest towards the end of the eighteenth century when the scientific revolution already reached its consummation. It is fair to say that the scientific revolution of the seventeenth century took place on the contact of the mixed disciplines of the ancient paradigm (astronomy, optics, the theory of simple machines) and the metaphorical region of that paradigm (the geocentric view of the cosmos). And this is rather natural.

In the *paradigmatic region* of ancient science, i.e. mathematics, the methodological standards are so strict and well founded that a refutation of the overall picture is improbable. On the other hand, the *elusive region of the paradigm* (i.e. the realm of biology) is not sufficiently stable and, therefore, changes happen there too often to be able to cause some deeper considerations. It is precisely the *mixed sciences* where the methods of the paradigm offer sufficiently effective means of research so that their progress is intensive. It is so because the application of the

in the strict sense, while in the metaphorical and in the elusive regions they are used in a distorted sense). The *vertical arrows* separate the intended area from the unintended one (in the paradigmatic region the methods, in the metaphoric region the metaphors are applied to those situations, for which they were introduced, while in the region of the mixed disciplines and in the elusive region the methods or the metaphors are applied to situations, for which they were originally not intended).

paradigmatic methods to unintended areas of phenomena increases the probability of the discovery of something radically new and unexpected, something that will be in sharp contrast with all that we are used to expect in the paradigmatic region. The *metaphorical region of the paradigm* is important for another reason. There the research is carried out on the fringe of what the paradigm allows to thematize and, therefore, the metaphorical region is often the place for the basic cultural projections with the emotional charge that accompanies such projections. The mixed disciplines alone would probably never have led to a revolution. Had Galileo accepted the suggestions of the Church and discussed the Copernican system only as a hypothesis, i.e. if he had restricted himself to the technical realm of the mixed disciplines and had not confronted this system with the geocentric world-view, it is probable that he Church would have succeeded in keeping the new astronomical discoveries on the periphery of the interest of the public as an incomprehensible, innocuous technical hypotheses. The dynamic of the scientific revolution of the 17th century was driven precisely by the conflict of the mixed disciplines with the metaphorical region when not absolutely sure results of scientific inquiry got into conflict with metaphors by means of which we articulate our place in the universe.

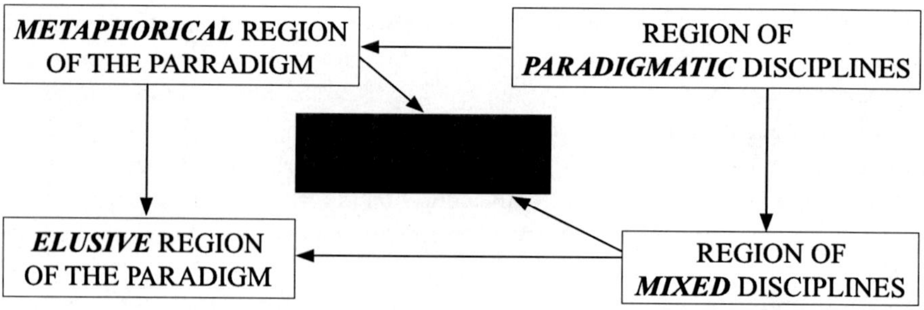

In this scheme paradigmatic disciplines are the paradigmatic disciplines of the *old* paradigm, and the same holds for the mixed disciplines as well as for the metaphorical and the elusive region. If we restrict ourselves to the scientific revolution of the seventeenth century, the above scheme expresses the fact that *the paradigm of modern physics originated neither in the paradigmatic nor in the elusive region of the ancient paradigm, but in the area between them*. The paradigmatic region of the ancient science was mathematics, which during the seventeenth century underwent a dramatic development (creation of the analytic geometry and of the calculus), but this development occurred in the framework of normal science. In mathematics of the seventeenth century nobody seriously questioned the results of the past. The elusive region of the ancient science was the realm of the organic, and the founders of modern physics almost completely avoided the discussion of questions of biology. Galileo marginally discussed the question of size of organisms and Descartes made occasional dissections.

It is important to realize that the new paradigm rises from a conflict between the mixed disciplines and the metaphorical region of the old paradigm. This indicates where to look for the source of revolutionary changes in the contemporary social sciences. The paradigm of the modern science is physics while its elusive region is the realm of the subjective (the Cartesian *res cogitans*), i.e. the area of social sciences. The above scheme shows that all those who were waiting for "*Newton of the social sciences*", waited at the wrong door. Social sciences form the elusive region of the physical paradigm. The elusive region is inaccessible to scientific methods and, therefore, it will not play any important role in the contemporary revolutionary changes. The next fundamental change in science will take place *not in the elusive region of the physical paradigm, i.e. in the social sciences, but on the border of the mixed disciplines and the metaphorical region of physics.* So let us have a closer look at this border.

27.4 The Revolution in Biology

If we want to form a clearer idea about the scientific revolution that is happening in contemporary science, it is useful to turn to a scheme, which would contain the paradigmatic, mixed, metaphorical, and elusive regions not only of the ancient science (that we analyzed in the previous section), but also of the paradigm of the contemporary science.

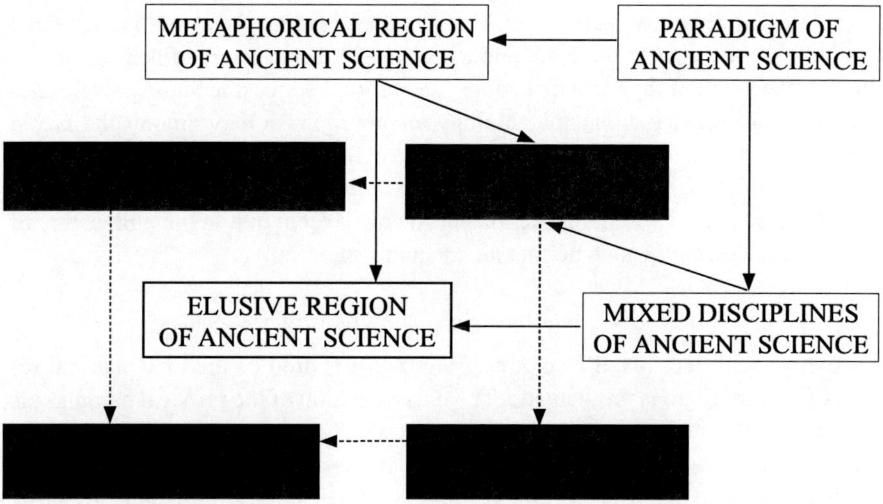

As we already mentioned, the elusive region of the ancient paradigm, representing the realm of the organic phenomena, did not play any important role in the formation of the Newtonian paradigm. Nevertheless, the elusive region underwent a radical change. *The elusive region of the ancient science became the center of the*

area between the metaphorical region and the region of the mixed disciplines of the Newtonian paradigm.

Because scientific revolutions happen in the area between the region of the mixed sciences and the metaphorical region, it is reasonable to conclude that the contemporary scientific revolution is taking place in biology, in the science of the organic. The new biological paradigm will emerge from the conflict between the mixed disciplines and the metaphorical region of the physical paradigm. So let us analyze these regions more thoroughly.

27.4.1 The Mixed Disciplines of Modern Science

The mixed disciplines of modern science use the technical and theoretical tools of physics (its experimental methods and laboratory equipments, its theoretical notions and mathematical formalism) in the study of nonphysical systems. For the mixed disciplines it is characteristic that they use these tools in an exact and methodologically correct way; the only problem is that they use these tools in the study of biological systems, i.e. systems where strict repeatability of experiments is impossible. Despite these difficulties we witness a spectacular progress of experimental techniques leading from the discovery of the microscope through the Roentgen apparatus to the computer tomography. Among the recent developments are magnetic resonance imaging and positron emission tomography which make it possible to visualize the brain activity during cognitive processes.[7] The discovery of a new physical method of registration of data leads to a new breakthrough in biology and medicine. A similarly spectacular development has also occurred in the area of chemical analysis of living matter, leading from the first artificial synthesis of uric acid through the understanding of the structure of hemoglobin to the decipherment of the human genome. Therefore, I suggest including disciplines such as biochemistry, molecular biology, or neurophysiology among the *mixed disciplines, the methodological status of which is analogous to Euclidean optics or Archimedean theory of the lever in the antiquity.* This region of the landscape of science does not present any serious problems, except that in the philosophy of science these disciplines do not get an adequate attention.

27.4.2 The Metaphorical Region of Modern Science

In contrast with the mixed disciplines, the interpretation of the metaphorical region of the paradigm is problematic. The elusive region of the physical paradigm is the realm of the subjective. It is not important whether we define it metaphysically as Descartes did at the dawn of the physical paradigm, or epistemologically as did Dilthey, who witnessed its climax at the end of the nineteenth century. What is

7 See Thomas Koenig and Dietrich Lehmann, "Microstates in Language-Related Brain Potential Maps Show Noun-Verb Differences", in: *Brain and Language*, 53, 1996, pp. 169–182, or Naho Ikuta et al., "Brain Activation During the Course of Sentence Comprehension", in: *Brain and Language*, 97, 2006, pp. 154–161.

important is to realize the elusive nature of the subjective, i.e. the fact that it cannot be dealt with by means of the physical paradigm. From this elusive region of the physical paradigm gradually a small part separated itself in a similar way as from the Aristotelian organic theory of motion the theory of local motions was separated. It is the part that makes use of the metaphors of the paradigmatic disciplines. As an example I would like to mention the *association psychology*, developing the ideas of David Hume, the *economic theory of the circulation of capital* initiated by Francois Quesnay, or *classical sociology* initiated by August Comte.[8] All these disciplines use notions like process, dynamics, speed, intensity, increase or force. Nevertheless, *the use of notions as "mental process", "intensity of emotional experience", "speed of associations" has very the same epistemological status as the use of notions "upwards" and "downwards" in the Aristotelian theory of local motions*. The point is that these notions are used not in their strict meaning, determined by the physical paradigm. Physical processes take place in space and the metric structure of this space enables us to speak about their velocity. Associations do not happen in any space that would have a straightforward metric structure and so the term "process" is being used here only in a metaphorical way. Similarly, Comte used the terms "social statics" and "social dynamics" in a metaphorical way. In the strict sense, i.e. the sense fixed by the physical paradigm, the term statics refers to the science studying the equilibria of forces. Force is a physical quantity that is measured in unequivocally defined units ($kg.m.s^{-2}$). On the other hand "forces" acting in society have no units in which we could measure them and so we can speak about equilibria in the social context only in a metaphorical sense. Similarly metaphorical is the notion of labor force in economics. Labor force is not a real force in the physical sense of the term force; it cannot be measured by means of the physical units by means of which we measure gravitational or electric forces. Similarly the use of the notion of work in economics is a metaphorical use of the physical notion of work, which is defined as a path integral of force. In economics it is not clear what forces we have to integrate along what path.

The metaphors used in these disciplines cannot be conceptually clarified: from the metaphor of social forces or labor forces it is impossible to create a notion that would be at least approximately as clear and unambiguous as the notion of force in classical physics. It is precisely due to this vagueness and ambiguity of the basic notions of sociology and economics why I suggest including them into the metaphorical region of the physical paradigm. I would like to suggest that disciplines such as association psychology, political economy, or classical sociology are trying to understand their particular subject matter using metaphors coming

8 See David Lewisohn, "Mill and Comte on the Methods of Social Science", in: *Journal of the History of Ideas*, 33, 1972, pp. 315–324. I could mention also historicism the view that in human history there are laws similar to physical laws. The discussion of historical explanation falls outside the scope of the present paper (see Eugen Zelenak, "On Explanatory Relata in Singular Causal Explanation", in: *Theoria*, 75, 2009, pp. 179–195).

from physics. At the same time the phenomena to which these disciplines apply their metaphors come from the elusive region of the paradigm, i.e. from the region to which the concepts and methods of the paradigm cannot be unambiguously applied. That is the reason why these disciplines have a problematic status (when compared with the paradigmatic disciplines), but on the other hand, just like the Aristotelian theory of local motion, these disciplines are the place for basic cultural projections and have a great potential for a radical transformation in the course of the next scientific revolution.

27.5 The Biological Revolution and Social Sciences

The main weakness of all discussions about the differences between natural and social sciences is the dominance of physics and ignorance of biology. All ruminations on the alleged different character of the social sciences can be seen as an articulation of the fact that the *social sciences have their origin in the elusive region of the physical paradigm*. This may be correct but the example of biology shows that the paradigmatic disciplines of physics do not exhaust the entire region of natural sciences. Between the physical disciplines, which are usually taken as paradigmatic examples of science, and the biological sciences there are many deep differences in the nature of their empirical basis, epistemological status of fundamental categories as well as logical structure of the whole theory.[9] Biological data (say in ecology or in the theory of evolution) are often qualitative; the theory often contains notions of different levels of complexity. If we extrapolate the scheme presented in the previous chapter one step further, it seems probable that a breakthrough in the area of social sciences will occur only when the biological paradigm matures, so that it will develop its own mixed disciplines and own metaphorical region. The social sciences forming at present the elusive region of the physical paradigm will then be clinched between the mixed disciplines and the metaphorical region of the biological paradigm. Biology will thus lead to a fundamental change of the social sciences, similar to that which physics brought about in the sphere of the organic.[10]

---

9 See Allan Franklin, "The Role of Experiments in the Natural Sciences: Examples from Physics and Biology", in: Theo Kuipers (Ed.), *Handbook of the Philosophy of Science: General Philosophy of Science – Focal Issues*, Amsterdam: Elsevier 2007, pp. 219–274, and William Bechtel and Andrew Hamilton, "Reduction, Integration, and the Unity of Science: Natural, Behavioral, and Social Sciences and the Humanities", in: Theo Kuipers (Ed.), *Handbook of the Philosophy of Science: General Philosophy of Science – Focal Issues*, pp. 377–430.

10 This change is already on the way under the heading of the "Naturalist Turn" (see e.g. Wenceslao J. Gonzalez, "Trends and Problems in Philosophy of Social and Cultural Sciences: A European Perspective", in: Friedrich Stadler et al. (Eds.), *The Present Situation in the Philosophy of Science*, Vienna: Springer 2010, pp. 227–232.)

Aristotelian theory of local motion, which was initially close to the elusive region of the ancient paradigm, was in the course of the scientific revolution of the seventeenth century shifted into the very center of the newly emerging mathematical physics. It is probable that during the biological revolution a similar shift awaits also the metaphorical region of the physical paradigm, namely psychology, economics, and sociology. These disciplines will be shifted from the elusive region of the physical paradigm (from the realm of social sciences) to the very center of the new paradigm of biology. Nevertheless, this will at the same time transform biology as well. Similarly as Newtonian physics was no longer physics in the Aristotelian sense of this word. It was not based on the four Aristotelian causes. It is probable that biology, after it absorbs psychology, economics and sociology, will be not the same science as we know it now. It will be not the theory of living systems (i.e. a discipline defined in the contraposition to the theory of non-living systems which are the subject matter of physics) but rather it will be the theory of systems with biological information (i.e. information understood as a code – in contrast to theories of information understood as symbol). From an informational point of view a cognitive scheme, the price of a commodity or a social hierarchy is similar to the genetic code. The information content can be interpreted as a code that represents the degree of adaptation of the system to its environment (the cognitive task, the market, the social, or the natural environment). The affinity of these disciplines is visible also from the increasing role which game theory plays in them.[11] If we define biology by the means of description it uses rather than by the subject matter to which these means are applied, then psychology, economics and sociology will become biological disciplines, however strange this might sound.

Acknowledgements: I would like to thank Donald Gillies and Marek Tomecek for valuable comments. The paper is a part of the project VEGA 1/3621/06 *Historical and philosophical aspects of exact disciplines* granted by the Slovak Grant Agency.

Faculty of Mathematics and Physics
Comenius University
Mlynská dolina
842 15, Bratislava
Slovak Republic
kvasz@fmph.uniba.sk

11 See Wenceslao J. Gonzalez, "The Role of Experiments in the Social Sciences: The Case of Economics", in: Theo Kuipers (Ed.), *Handbook of the Philosophy of Science: General Philosophy of Science – Focal Issues*, pp. 292–294.

CHAPTER 28

MARIA CARLA GALAVOTTI

PROBABILITY, STATISTICS, AND LAW

28.1 FOREWORD

In recent decades probability and statistics have gradually made their way into the realm of law. This has been favoured by the proliferation of forensic techniques including identification by means of fingerprints, DNA evidence, marks on bullets, etc., and by the ever-increasing amount of epidemiological and medical data, and the refinement of risk analysis. These developments have forced those involved in forensic matters, particularly if acting in court, to pay increasing attention to science. Important steps in that direction were taken by a number of court decisions including the 1993 U.S. Supreme Court's *Daubert decision* (Daubert v. Merrell Dow Pharmaceuticals) which ruled that the standard for admitting expert testimony in federal courts should meet the demands of scientific method. A few years later, the 2000 revised version of the Federal Rules of Evidence fixed the requirements of *reliability* and *relevance* for the admissibility of testimony in court, adding that fulfilment of such requirements depends on compliance with scientific methodology.[1] This obviously brings probability and statistics to the foreground.

28.2 TWO CONTROVERSIAL CASES

Let us start by discussing two controversial cases, namely those of Lucia de Berk and Sally Clark, which exemplify the problems raised by the use, or better the misuse, of statistics in court. Both of these cases provoked ample debate and media coverage, also attracting the attention of lawyers, forensic experts, statisticians, scientists operating in various fields, and epistemologists.[2]

In 1996 Sally Clark's first son was found dead in his cot at 11 weeks, and in 1997 her second son died at 8 weeks in the same way. Sally was accused of having murdered both babies, and in 1999 was convicted of murder and sentenced to life imprisonment. The jury's verdict was based on paediatrician Roy Meadow's

1 A critical survey of U.S. decisions and rules regarding testimony is to be found in Susan Haack, "Entangled in the Bramble-Bush", in: Susan Haack, *Defending Science – Within Reason*, Amherst: Prometheus Books 2003, pp. 233–264.

2 See the Wikipedia articles on both of these cases, with bibliography.

D. Dieks et al. (eds.), *Probabilities, Laws, and Structures*, The Philosophy of Science in a European Perspective 3, DOI 10.1007/978-94-007-3030-4_28,

testimony. As an expert witness he estimated that the probability of one death from natural causes (SIDS: sudden infant death syndrome) in one family was about 1 in 8,543, and the probability of two such deaths 1 in 73 million (8,543 × 8,543). The low probability of 1 in 73 million was presumably taken by the jury as ruling out the possibility that both Sally's children had died from natural causes. In 2003, after it turned out that the prosecutor's pathologist had failed to disclose microbiological reports suggesting that one of her sons had died from natural causes, the trial was reopened and she was released from prison, but a few years later she died from acute alcohol poisoning. The case has been discussed in some detail by the statistician Philip Dawid in a number of writings.

In 2003, the Dutch nurse Lucia de Berk was convicted for seven murders and three attempted murders, and sentenced to life imprisonment. There was no evidence supporting the charge against her apart from the fact that a number of resuscitations had occurred during Lucia's shifts. As in Sally Clark's case, the verdict was presumably influenced by statistical calculations brought to court by expert witnesses. The probability that so many incidents could have happened by accident during one nurse's shifts was calculated by the expert witness Henk Elffers to be 1 in 7 billion. Later on, other calculations gave a probability of 1 in 342 million. These numbers were interpreted as ruling out the possibility that the deaths which occurred during Lucia's shifts were due to natural causes. The Dutch philosopher of science Ton Derksen called attention to the case, arguing that Lucia's conviction has been the result of misused statistics.[3] Lucia's case was reopened in 2008 and in April 2010 she was acquitted.

The most striking analogy between these two cases was that neither of them produced evidence that a criminal offence had taken place. The deaths of Sally Clark's two children and of seven patients during Lucia's shifts were classified as murders simply because they were judged *too improbable to be accidental*, or due to natural causes. In Lucia's case there was no evidence apart from the frequency of the emergency calls during Lucia's shifts, and for that reason the case is also referred to as the "nurse/roster problem".

The data brought to court to incriminate Lucia were biased in various ways. The frequency of incidents that caused Lucia's incrimination was calculated on the basis of a disputable concept of "incident". No clear definition of what counted as incident was given: the prosecutor simply took the number of times Lucia called a doctor to a patient's bed, and compared it with the number of times doctors were called during other nurses' shifts. As observed by Gill and Groeneboom, "incidents were never formally defined. However, if doctors were expressly called to the bed of the patient by nursing staff, then that soon qualified as an accident, especially if Lucia was somehow involved".[4] This is obviously not an objective criterion,

3 The English speaking reader is addressed to Ton Derksen and Monica Meijsing, "The Fabrication of Facts: the Lure of the Incredible Coincidence", in: Hendrik Haptein, Henry Prakken and Bart Verheij (Eds.), *Legal Evidence and Proof*, Farnham: Ashgate 2009, pp. 39–70.

4 Richard Gill and Piet Groeneboom, "Elementary Statistics on Trial (the case of Lucia de B.)", 2009, p. 5 (online publication available from the Wikipedia article on Lucia de B.).

for the inclination to call for a doctor when faced with a critical situation could be influenced by myriad factors, including experience, ability/inability to detect critical situations, psychological elements, and so on. In addition, as observed by Derksen and Meijsing, the data used by the prosecutor concerned only three of the five wards Lucia had been working on, and only 1 year and 3.5 months out of the 11.75 years that she worked in those hospitals. Also puzzling is the fact that no search was made for incidents outside Lucia's shifts.

Various commentators deem it disputable that the data on the frequency of incidents, taken as evidence against Lucia, were used twice in the incriminating calculations: once in order to state that a criminal offence had taken place, and once more to infer that Lucia was responsible.[5] In the debate that followed Lucia's sentence, doubts were raised against the statistical calculations that led to the figure of 1 in 342 million, taken as the probability that the incidents happened during Lucia's shifts could have occurred by chance (the null hypothesis). A number of authors, including Derksen, Gill and Groeneboom, objected to the use of Fisher's exact test made by the expert of the prosecution Henk Elffers. Other calculations obtained by the same method, but taking into account a more complete body of evidential data and avoiding certain assumptions (like independence) resulted in totally different figures.[6]

The statistical calculations that led to the prosecution of both Sally Clark and Lucia de Berk made use of *independence assumptions*, with no sound justification. In Sally Clark's case the figure of 1 in 73 million was calculated by squaring the probability that one infant dies of SIDS within a population having the characteristics as the parents of the dead children, therefore assuming the independence of the two deaths even though commonsense suggests that the deaths of two brothers can hardly be judged as independent events.[7] Also in Lucia's case a number of independence assumptions were made. Meester, Collins, Gill and van Lambalgen list the following (among others): (1) the probability of an incident during a night shift is the same as during a day shift (although more people die during the night); (2) the probability of an incident during a shift does not depend on the prevailing atmospheric conditions (although these have an effect on respiratory problems); (3) all nurses have an equal probability of witnessing incidents (whereas on the

5 See Ton Derksen and Monica Meijsing, "The Fabrication of Facts: the Lure of the Incredible Coincidence", *op. cit.*, and David Lucy, "Commentary on Meester et. al. 'On the (Ab)use of Statistics in the Legal Case against Lucia de B.'", in: *Law, Probability, and Risk*, 5, 2006, pp. 251–254.

6 See Ton Derksen and Monica Meijsing, "The Fabrication of Facts: the Lure of the Incredible Coincidence", *op. cit.*

7 See Philip Dawid, "Bayes' Theorem and Weighing of Evidence by Juries", in: Richard Swinburne (Ed.), *Bayes's Theorem* (*Proceedings of the British Academy* 113), Oxford: Oxford University Press 2002, pp. 71–90.

contrary terminally ill patients often die in the presence of the nurse they feel most comfortable with).[8]

Last but not least, at both trials use was made of the argument known as the *prosecutor's fallacy*. In Sally Clark's case, the figure of 1 in 73 million, calculated as the measure of the initial rarity of the event "two SIDS deaths", was interpreted as the probability that that particular event had in fact happened and was then taken as the probability that Sally Clark was innocent, giving a very high probability of her being guilty. Also in Lucia's case the same kind of fallacy was committed. As put by Derksen and Meijsing, the wrong question was addressed: instead of asking "Assuming Lucia's innocence, what is the probability that she meets with such a coincidence (the number of incidents) by chance?" the court should have asked "Given the coincidence, is there reason to convict Lucia?".[9]

The prosecutor's fallacy deserves closer inspection. It typically occurs as follows. Take the so-called "match probability" p, namely the probability that a given piece of evidence – such as a trace left at the murder scene, like blood, hair or other organic material – is to be ascribed to an individual taken at random from a reference population. The fallacy occurs when such probability is interpreted as the probability that the defendant is not guilty, and the conclusion is drawn that the probability of his guilt is calculated as $1-p$. For instance: a match probability $p\,(M\,/\,-G) = 1/10{,}000{,}000$ [M = a trace that was found on the murder scene; $-G$ = the defendant is not responsible for it, namely the trace was left by an individual chosen randomly from the reference population] is confused with $p\,(-G \mid M)$ [the probability that the defendant is not guilty, given the piece of evidence found at the murder scene]. The probability of the defendant being guilty is then obtained as $1 - 1/10{,}000{,}000$. This is clearly a fallacious way of obtaining a very high probability of guilt of the defendant, based on the confusion between the probability that a certain trace was left by an unknown individual randomly chosen from the population and not by the defendant, and the probability that the defendant is not guilty, given the piece of evidence found at the murder scene.

28.3 The Strength of Comparison

The cases of Sally Clark and Lucia de Berk – like many others not mentioned here – exemplify misuse of statistics in court. How can one make good use of statistics for forensic purposes? In the first place, good use of statistics requires data to be collected carefully to avoid bias, and all assumptions to be spelled out and justified. In addition, a number of authors stress the need to use statistics as

8 Cf. Ronald Meester, Marieke Collins, Richard Gill and Michiel van Lambalgen, "On the (Ab)use of Statistics in the Legal Case Against the Nurse Lucia de B.", in: *Law, Probability and Risk* 5 (2006), pp. 233–250.

9 See Ton Derksen and Monica Meijsing, "The Fabrication of Facts: the Lure of the Incredible Coincidence", *op. cit.*

a *means for comparison*. Using relative in place of absolute values not only conveys more accurate information, it also avoids to incur in bad arguments like the prosecutor's fallacy. In Sally Clark's case, this line of reasoning led to comparing the probability that two infants die of SIDS with the probability that two brothers are murdered by their mother (under similar circumstances). As a result of such a comparison, Dawid calculates the odds as follows:

$$(1/2 \text{ billion}) / (1/73 \text{ million}) = 0.0365.$$

If this value were to be interpreted as a hint of Sally's guilt or innocence, it would obviously speak against conviction beyond any reasonable doubt.[10]

The appropriate tool for statistical comparison is the *likelihood ratio* (LR). Not itself a probability, the LR results from comparing two probabilities. Typically, it allows the weight of a given body of evidence to be compared to alternative hypotheses:

$$\text{LR} = p(E/H)/p\,(E/G)$$

or

$$\text{LR} = p(E/H)/p\,(E/-H)$$

if one wanted to weigh some body of evidence with respect to a given hypothesis and its negation.

The likelihood ratio relates naturally to the notion of relevance: when its value equals 1 the given body of evidence is irrelevant to the hypothesis, when its value differs from 1 the given body of evidence is relevant. More particularly, a likelihood ratio greater than 1 indicates how much a given body of evidence favours the truth of a certain hypothesis against the alternative which is being considered, and conversely if the likelihood ratio is less than 1. The LR is extensively used in court to convey information on how evidence can affect the probability of competing hypotheses (typical case: the use of DNA evidence in a paternity dispute). Following Evett, Robertson and Vignaux define a likelihood ratio for adoption in court as "weak" in the range 1–33, "fair" in the range 33–100, "good", in the range 100–330, "strong" in the range 330–1,000, and "very strong" a ratio greater than 1,000.[11]

Although the likelihood ratio can be used alone, as it yields useful information of the kind described, supporters of the Bayesian method favour its adoption within the Bayesian framework, stressing its crucial role in connection with the shift from prior to posterior probabilities. This appears evident if Bayes' rule is expressed in terms of odds:

10 See Philip Dawid, "Statistics and the Law", in: Andrew Bell, John Swenson-Wright, Karin Tybjerg (Eds.), *Evidence*, Cambridge: Cambridge University Press 2008, pp. 119–148.

11 Cf. Bernard Robertson and G. A. Vignaux, *Interpreting Evidence*, Chichester: Wiley 1995, p. 12. I. W. Evett, "Interpretation: A Personal Odissey", in: C. G. G. Aitken and David Stoney (Eds.), *The Use of Statistics in Forensic Science*, New York: Ellis Horwood 1991, pp. 9–22.

$$[p(H/E)/p(-H/E)] = [p(H)/p(-H)] \times [p(E|H)/p(E|-H)].$$

By considering the shift from priors to posteriors one can evaluate how a given body of evidence is apt to influence the comparison between alternative hypotheses, such as hypotheses regarding the cause of a certain event (for instance, someone's death). Forensic literature abounds in examples of the application of the likelihood ratio to identification problems by means of evidence related to DNA, the glass refraction index, cloth fibres or other materials. In all such cases the likelihood ratio can sometimes favour one of two given hypotheses so that even in the presence of very little information concerning prior probabilities it is often possible to assign it a very high probability.

The following table gives an idea of the effect of the likelihood ratio in the shift from prior probability $p\ (H)$ to posterior probability $p\ (H \mid E)$.[12] For a likelihood ratio $p\ (E \mid H) = 100$:

$p(H)$	0.001	0.01	0.1	0.3	0.5	0.7	0.9
$p(H\|E)$	0.09	0.50	0.92	0.98	0.99	0.996	0.999

Assuming that one wanted to apply Bayes' reasoning to the two hypotheses of guilt (G) and innocence ($-G$) of a defendant:

$$[p(G/E)/p(-G/E)] = [p(G)/p(-G)] \times [p(E|G)/p(E|-G)]$$

the table shows how much a piece of evidence which according to the likelihood ratio is 100 times more likely to be conditional on the guilt than on the innocence hypothesis, affects various priors. In order to obtain a posterior probability of at least 99% – that is to say a value that would satisfy the standard of proof of beyond any reasonable doubt (BARD)[13] – the prior probability, namely the probability of guilt before the piece of evidence E is taken into account, needs to be at least 50%.

Obviously, the value of priors has to be established on solid grounds, on the basis of myriad elements not susceptible of quantitative analysis. For this reason, a number of authors recommend application of the Bayesian method at an advanced stage of the trial.

12 See Philip Dawid, "Probability and Statistics in Court", a section of the appendix online ("Probability and Proof") to the second edition of Terence Anderson, David Schum and William Twining, *Analysis of Evidence*, Cambridge: Cambridge University Press 2005.

13 See Dennis Lindley, "Probabilities and the Law", in: Dirk Wendt and Charles Vlek (Eds.), *Utility, Probability, and Human Decision Making*, Dordrecht–Boston: Reidel 1975, pp. 223–232.

28.4 Criticism and Replies

An influential criticism against the use of probability and statistics in court has been raised by Laurence Tribe, who objects to the use of probability and statistics in court on account of (1) the presumption of innocence (PI) and (2) the BARD standard, which he takes to be moral principles that impinge respectively upon the beginning and the end of the trial.[14] According to Tribe, the moral nature of PI and BARD should discourage the idea that the hypothesis of guilt be expressed by means of a probability value, calculated by means of Bayes' rule. He is especially concerned that posterior probability of guilt is not conflated with the BARD standard.

A possible rejoinder to Tribe's caveat can be found in Lindley's writings, where the author maintains that when probability is applied to the hypothesis of guilt it refers "to the event that the defendant committed the crime with which he has been charged [...] not to the judgment of guilt".[15] In other words, the *hypothesis* of guilt is taken by Bayesians as a useful working hypothesis, not to be confused with the *judgment* of guilt, which falls within the competence of the judge, or juror, who will formulate it on the basis of a complex body of elements usually not reducible to the mere quantitative evidence. The same holds for the BARD standard, whose nature is too complex to be expressed by means of a probability value. This is emphasized by Robertson and Vignaux, who claim that BARD "is a matter for the court", not for statisticians.[16]

Obviously, Bayesian inference crucially depends on the value assigned to priors, and the process of fixing priors is a most delicate matter. Larry Laudan objects to the use of Bayes' method in court on the grounds that prior probabilities are the expression of the "subjective hunches" of those who fix them.[17] However, Laudan's conviction clashes with a vast literature testifying to a totally different attitude on the part of supporters of the use of Bayes' method in court. Among them David Kaye, who made serious attempts to rebut the idea that prior probabilities are merely the expression of personal feelings. In his words: "there appears to be no reason in principle why a juror could not generate a prior probability that could be described in terms of objective, relative-frequency sort of probability".[18] For instance, in identification cases the information on the frequency with which

14 See Laurence Tribe, "Trial by Mathematics: Precision and Ritual in the Legal Process", in: *Harvard Law Review*, 84, 1971, pp. 1329–1393.

15 Dennis Lindley, "Probability", in: C. G. G. Aitken and David Stoney (Eds.), *The Use of Statistics in Forensic Science*, *op. cit.*, p. 27.

16 Cf. Bernard Robertson and G. A. Vignaux, *Interpreting Evidence*, *op. cit.*, p. 78.

17 See Larry Laudan, "Is Reasonable Doubt Reasonable?", in: *Legal Theory*, 9, 2003, pp. 295–331, and Larry Laudan, *Truth, Error, and Criminal Law*, Cambridge: Cambridge University Press, 2006.

18 David Kaye, "The Laws of Probability and the Laws of the Land", in: *The University of Chicago Law Review*, 47, 1979, p. 55.

relevant characters occur in the reference population is used to determine prior probabilities.[19] The literature on the topic is constantly growing.

As a further objection against the adoption of Bayes' method in court, Laudan maintains that the presumption of innocence should impose that the probability of guilt is assigned the value 0 or at best a value so low as to make it impossible to obtain a significant posterior probability, no matter how much evidence is brought to court. This argument is mistaken for at least two reasons. First, as pointed out by Lindley, prior probability of guilt should never be 0, on the ground of "Cromwell's rule" (after Cromwell's advice to the Church of Scotland: "I beseech you, … think it possible you may be mistaken").[20] Secondly, a very strong likelihood ratio can raise the smallest prior to a significantly high posterior.

Supporters of the Bayesian method, like Lindley, Dawid, Kaye, Robertson and Vignaux, and many others recommend its adoption in court as a heuristic device to help the parties involved in a trial in *interpreting evidence*. In this spirit, Richard Lempert holds that the Bayesian method can promote understanding of a cluster of issues related to relevance, such as "the meaning of logical relevance" and "the principle that only relevant evidence is admissible".[21] Obviously, there is no unique way of applying Bayes' method to a given problem, therefore it is crucial to state explicitly what assumptions are made in each particular context. To be sure, utilizing the Bayesian method is no easy task. For one thing, the calculation of likelihoods can be problematic, and equally tricky is the choice of priors. Also important is to make the calculations formulated by experts understood to those whose responsibility it is to formulate the final judgment of guilt. These issues are the focus of a vast literature.[22]

Behind the application of Bayesian methods, and indeed of any kind of statistical inferences (in court as well as anywhere else) lurks the problem of using the appropriate reference class. Identifying a suitable reference class for base rates requires that no relevant variables are omitted (in order to avoid confounding), and that data are plausibly collected. This challenging problem admits of no simple and general solution, but is attracting increasing attention in the realm of law.[23]

---

19 See for instance Ira Mark Ellmann and David Kaye, "Probability and Proof: Can HLA and Blood Group Testing Prove Paternity?", in: *New York University Law Review*, 54, 1979, pp. 1131–1162.

20 Dennis Lindley, "Probability", *op. cit.*, p. 43.

21 Richard Lempert, "Modeling Relevance", in: *Michigan Law Review*, 75, 1977, p. 1031.

22 See for instance the discussion in Stephen Fienberg and Michael Finkelstein, "Bayesian Statistics and the Law", in: José Bernardo, James Berger, Philip Dawid, and Adrian Smith (Eds.), *Bayesian Statistics*, Oxford: Oxford University Press 1996, pp. 129–146. For an interesting comparison between the Bayesian and frequentist approaches to a DNA identification problem see David Kaye, "Case Comment – *People v. Nelson*: a Tale of Two Statistics", in: *Law, Probability and Risk*, 7, 2008, pp. 249–257.

23 See for instance Aaron Taggart and Wayne Blackmon, "Statistical Base and Background Rates: The Silent Issue not Addressed in Massachusetts v. EPA", in: *Law, Prob-*

It should not go without mention that Jonathan Cohen strongly objected to the use of standard (or Pascalian) probabilities in court. His arguments, which cannot be recollected here, have been analysed in some detail by a number of authors who put forward convincing rejoinders.[24]

Instead, the adoption of standard probability in court has been vigorously defended by Dennis Lindley, who claims that

> a simple and effective reason for using probability is that it works. I know of no situation in which probability is inadequate or fails to yield a reasonable answer. Sometimes the calculations are horrendous and cannot at the moment be done: but that is a technical difficulty that adequate research will, one hopes, overcome. Sometimes it is hard to relate the probability model to the actual situation: but again, when it can be done the result is illuminating.[25]

Kaye also patronizes the use of standard probability in the context of criminal trial, observing that

> the equations of the axiomatized theory of probability – like the rules of logic and arithmetic – work admirably in other contexts. [...] Surely the probability axioms work sufficiently well for objectively estimated probabilities. Why should they not serve as well when applied to thoughtful, subjective estimates?[26]

28.5 A Plea for Subjective Probability

As suggested by the above passage, Kaye argues that subjective probability can find useful applications in court. A similar conclusion is reached by Philip Dawid, who maintains:

> The subjectivist philosophy holds that complete objectivity is an illusion, and thus that there is no such thing as 'the' probability of any uncertain event – rather, each individual is entitled to his or her own subjective probability. This is not, however, to say that anything goes: in the light of whatever relevant evidence may be available, certain opinions will be more

---

ability and Risk, 7, 2008, pp. 275–304; and David Kaye, "Logical Relevance: Problems with the Reference Population and DNA Mixtures in People v. Pizarro", in: *Law, Probability, and Risk*, 3, 2004, pp. 211–220.

24 See Jonathan Cohen, *The Probable and the Provable*, Oxford: Clarendon Press 1977. For comments on Cohen's arguments see issue n. 4 (1981) of the journal *The Behavioural and Brain Sciences*; Philip Dawid, "The Difficulty about Conjunction", in: *The Statistician*, 36, 1987, pp. 91–97; Stephen Fienberg, "Misunderstanding, beyond a Reasonable Doubt", in: *Boston Law Review*, 66, 1986, pp. 651–656; and David Kaye, "Do we Need a Calculus of Weight to Understand Proof beyond a Reasonable Doubt?", in: *Boston Law Review*, 66, 1986, pp. 657–672.

25 Dennis Lindley, "Probability", *op. cit.*, p. 37.

26 David Kaye, "The Laws of Probability and the Laws of the Land", *op. cit.*, p. 55.

reasonable than others. This seems to correspond to the legal conceptions of the 'reasonable man' and 'reasonable doubt'.[27]

The opinion of these authors is in tune with the position of one of the "fathers" of the subjective interpretation of probability, namely Bruno de Finetti, who used to distinguish between the *definition* of probability as *degree of belief* and its *evaluation*, regarded as a complex procedure that depends on objective as well as subjective elements. The evaluation of probability should take into account all available evidence including, whenever available, frequencies and symmetries, but for de Finetti it would be a mistake to put these elements, which are useful ingredients of the evaluation of probability, at the core of its definition. While opposing *objectivism*, namely the idea that probability is a property of objects, having a true value that is usually unknown, de Finetti takes very seriously the issue of *objectivity*, namely the problem of using good probability appraisers.[28] He recommends that whenever empirical information is available it should be taken into account, because ignoring such information would conflict with the Bayesian ideal of rationality, which requires that probability evaluations be guided by evidence. Once it is acknowledged that probability is subjective and that there is no unique "rational" way of assessing probability, room can be made for a whole array of elements to influence and improve probability evaluations. This problem was extensively addressed by de Finetti, especially from the Sixties onwards. The approach adopted is based on scoring rules such as "Brier's rule", named after the meteorologist Brier who applied it to weather forecasts. Scoring rules can be used to improve probability forecasts made both by a single person and by several people, providing a tool for obtaining robust evaluation methods.[29]

Regrettably, the wrong idea that subjectivism is some sort of "anything goes" approach to probability is still widespread, and there is little awareness that the subjective interpretation of probability is fully compatible with objective assessments. In this vein, Laudan claims that adopting subjective probabilities in court would mean admitting arbitrary and discretional probability judgments, and Redmayne maintains that the assessment of evidence in court needs objective probability and discards subjectivism because "when the only constraint on rational belief is coherence among a belief set, it can seem that anything goes".[30] Not surprisingly,

27 Philip Dawid, "Probability and Statistics in Court", *op. cit.*

28 More on de Finetti's attitude towards the problem of objectivity in Maria Carla Galavotti, "Subjectivism, Objectivism and Objectivity in Bruno de Finetti's Bayesianism", in: David Corfield and Jon Williamson (Eds.), *Foundations of Bayesianism*, Dordrecht–Boston–London: Kluwer 2001, pp. 173–186, and Maria Carla Galavotti, *Philosophical Introduction to Probability*, Stanford: CSLI 2005.

29 For more on scoring rules see Philip Dawid and Maria Carla Galavotti, "De Finetti's Subjectivism, Objective Probability, and the Empirical Validation of Probability Assessments", in: Maria Carla Galavotti (Ed.), Bruno de Finetti, *Radical Probabilist*, London: College Publications 2009, pp. 97–114.

30 Mike Redmayne, "Objective Probability and the Assessment of Evidence", in: *Law,*

though, Redmayne is unable to identify an objective theory of probability that could adequately represent the uses of probability in court, and turns to "epistemic justification" as a criterion for the acceptability of evidentiary arguments to be presented in a trial. This is not much of a solution, because it remains to be clarified what an epistemic justification should consist of. Incidentally, subjectivists have a lot to say in that connection; suffice it to think of Frank Ramsey's conception of knowledge as "obtained by a reliable process",[31] together with his idea that the criterion for the reliability of inductive inferences is given by their success. The same idea is to be found in de Finetti, and in the vast literature on calibration methods developed by statisticians of subjectivist inspiration.[32] Whether these ideas can find useful applications in court is an open question.

It should be added that in *Expert Evidence and Criminal Justice* Redmayne takes a milder attitude towards subjective probability, claiming that

> various aspects of subjective probability are controversial [...] Nevertheless, the notion of subjective probability captures, if only crudely, something important about our doxastic attitudes: the existence of degrees of belief. Even if the idea of attaching figures to these degrees is a simplistic way of conceptualizing them, this simplicity at least buys a rigorous way of thinking through various problems.[33]

In the same book Redmayne puts forward the conviction that a natural way to think about evidence in criminal trials is by framing it within explanatory accounts. In this connection he refers to the "story model for juror decision making" which has been proposed by Nancy Pennington and Reid Hastie as a model of "the cognitive strategies that individual jurors use to process trial information in order to make a decision prior to deliberation".[34] Redmayne regards the story model and Bayesian theory as "the two views of a Necker cube", claiming that they are different ways

Probability and Risk, 2 , 2003, p. 276.

31 See Frank Plumpton Ramsey, *Philosophical Papers*, ed. by Hugh Mellor, Cambridge: Cambridge University Press 1990, p. 110; Nils-Eric Sahlin, "Obtained by a Reliable Process and always Leading to Success", in: *Theoria*, 17, 1991, pp. 132–149, and Maria Carla Galavotti, "F. P. Ramsey and the Notion of 'Chance'", in: Jaakko Hintikka and Klaus Puhl (Eds.), *The British Tradition in the 20th Century Philosophy. Proceedings of the 17th International Wittgenstein Symposium*, Wien: Hölder-Pichler-Tempsky 1995, pp. 330–340.

32 See for instance Leonard J. Savage, "Elicitation of Personal Probabilities and Expectations", in: *Journal of the American Statistical Association*, 66, 1971, pp. 783–801, and Philip Dawid, "Probability Forecasting", in: Samuel Kotz, Norman Johnson and C. B. Read (Eds.), *Encyclopedia of Statistical Sciences*, 7, New York: Wiley 1986, pp. 210–218.

33 Mike Redmayne, *Expert Evidence and Criminal Justice*, Oxford: Oxford University Press 2001, pp. 54–55.

34 Nancy Pennington and Reid Hastie, "The Story Model for Juror Decision Making", in: Reid Hastie (Ed.), *Inside the Juror*, Cambridge: Cambridge University Press 1993, p. 192.

of representing evidence that "it is not easy to mix".[35] In point of fact, Pennington and Hastie account for these approaches as two contrasting decision models, standing in opposition to each other. By contrast, we have seen that most authors do not recommend the Bayesian method as a means for taking decisions in court, but rather as a tool for helping decision-makers by conveying information on the weigh of available evidence. From this standpoint it seems perfectly feasible, albeit by no means easy, to combine Bayesian method with an explanatory account of facts.

Department of Philosophy
University of Bologna
Via Zamboni 38
40126 Bologna
Italy
mariacarla.galavotti@unibo.it

35 Mike Redmayne, *Expert Evidence and Criminal Justice*, *op. cit.*, p. 70.

Chapter 29

Adrian Miroiu

Experiments in Political Science: The Case of the Voting Rules

Nearly two centuries ago, in his essay *On the Definition of Political Economy; and on the Method of Investigation Proper to It*, John Stuart Mill developed the view that in moral sciences the only certain or scientific mode of investigation is the a priori method, or that of "abstract speculation". The following quotation concentrates his main argument:

> There is a property common to almost all the moral sciences, and by which they are distinguished from many of the physical; this is, that it is seldom in our power to make experiments in them. ... We cannot try forms of government and systems of national policy on a diminutive scale in our laboratories, shaping our experiments as we think they may most conduce to the advancement of knowledge. We therefore study nature under circumstances of great disadvantage in these sciences; being confined to the limited number of experiments which take place (if we may so speak) of their own accord, without any preparation or management of ours; in circumstances, moreover, of great complexity, and never perfectly known to us; and with the far greater part of the processes concealed from our observation.[1]

For Mill, experiments in political science are not an appropriate means of arriving at truth. However, Mill attaches them another important role: experiments help verify truth, and reducing as much as possible the "uncertainty before alluded to as arising from the complexity of every particular case, and from the difficulty (not to say impossibility) of our being assured *a priori* that we have taken into account all the material circumstances".[2]

Mill's view is still critical for understanding the role of experiments in political science.[3] I shall start by discussing some of the views expressed by economists concerning the role of experiments in moral sciences. The view developed by

---

1 John Stuart Mill (1874). *Essays on Some Unsettled Questions of Political Economy*, Second Edition, Batoche Books, Kitchener, 2000, p. 103.

2 John Stuart Mill (1874). *Essays on Some Unsettled Questions of Political Economy*, p. 107.

3 For a general discussion on the role of experiments in social sciences, see Wenceslao J. Gonzalez, "The Role of Experiments in the Social Sciences: The Case of Economics", in: Theo Kuipers (Ed.), *Handbook of the Philosophy of Science: General Philosophy of Science – Focal Issues*, Amsterdam: Elsevier 2007, pp. 275–301.

D. Dieks et al. (eds.), *Probabilities, Laws, and Structures*, The Philosophy of Science in a European Perspective 3, DOI 10.1007/978-94-007-3030-4_29,

Vernon Smith is specifically relevant is this context. A main reason is that Smith gives institutions a core role in theory construction as well as in experimental settings. Voting rules, i.e. rules to transform the electorate's votes into a group decision, are clear examples of institutions. Experiments performed with these rules will be discussed. The main argument of this paper is that there is much to gain in the experimental approaches by taking into account the study of the voting rules by means of the social choice techniques. Social choice theorists showed that voting rules can be characterized by appealing to sets of properties they uniquely satisfy. Therefore, it is tempting to study not only how voters behave when confronted with situations in which a certain voting rule works, but also their attitudes towards such properties. For example, one such property some voting rules have is that of anonymity. Roughly, it states that all voters should be treated as equals. Then a large collection of experiments concerning the topic of voters' attitudes toward equality and fairness becomes relevant for the experimental study of voting rules.

29.1

Vernon Smith received the Nobel Prize in 2002 for his contribution in experimental economics. According to him, there are at least seven reasons for a researcher to devise and conduct experiments.[4] She may want to: (i) test a theory, or discriminate between theories; (ii) explore the causes of a theory's failure; (iii) establish empirical regularities as a basis for new theory (in the laboratory, especially with computerization, institutions with complex trading rules are as easier to study); (iv) compare environments; (v) compare institutions (using identical environments, but varying the market rules of exchange, has been the means by which the comparative properties of institutions has been established); (vi) evaluate policy proposals; (vii) treat the laboratory as a testing ground for institutional design, for examining the performance properties of new institutions.

Smith acknowledges that to accept that experiments have such roles is at odds with the standard, received view on the way economics is commonly researched, taught, and practiced.[5] On this view economics is conceived as an a priori science consisting in logically correct, internally consistent theories and models, while experiments can only be used to "test" alternative model specifications. It is then counterintuitive for people trained in this tradition to understand key features of the experimentalist economists' methodology. When confronted with economists working in this paradigm, the experimental researcher essentially sees himself as a kind of an anthropologist on Mars: he and the traditional economist live in different ways of thinking, have different two world views.[6]

4 Cf. Vernon L. Smith, "Economics in the Laboratory", in: *The Journal of Economic Perspectives*, 8, 1, 1994, pp. 113–131.

5 Cf. Vernon L. Smith, *Rationality in Economics. Constructivist and Ecological Forms*, Cambridge: Cambridge University Press, 2008.

6 From the point of view of a deductivist economist, allocation mechanisms require agents to have complete information, but not mechanism designers. But the experi-

As an institutionalist theorist, V. Smith is aware of the fact that experimentalist economists have been largely influenced by institution-specific theory that began to develop about 1960. The lesson they learned is that institutions matter: agent incentives in the choice of messages (like bids) are affected by the institutional rules that convert messages into outcomes. Institutions are a core element of a theory and, as we shall immediately see, of an experimental setting.[7] Let us take as an example a special class of economic theories: microeconomic theories. Smith distinguishes three ingredients of these theories: the *environment*, the *institution* and the *behaviour* of the actors.[8] The first two ingredients help define the micro-economic system to be studied. The third concern the way in which agents choose messages. All three components allow for an assessment of the system performance.[9]

The environment can be specified by describing the agents' characteristics: first, the number of the economic agents; secondly, the list of the commodities or goods among which they are to choose; third, relevant characteristics of the economic agents, such as the agent's utility or preference function, the endowment of agents with resources (technology and knowledge), and the production or cost functions. Hence, a microeconomic environment is specified by a set of initial circumstances that cannot be altered by the agents or the institutions within which they interact. This final aspect is especially important. In an experimental setting, the environment should include some circumstances that cannot be altered by the agents because they are control variables fixed by the experiment.

Institutions, in D. North's famous phrase, define the rules of the game under which agents may communicate and exchange or transform commodities or goods for the purpose of modifying initial endowments in accordance with their private tastes and knowledge. The institution specifies first, a language: the set of mes-

---

mentalist thinks in a quite different manner: "The whole idea of laboratory experiments was to evaluate mechanisms in an environment where the Pareto optimal outcome was known by the experimental designer but not by the agents so that performance comparisons could be made", Vernon L. Smith, *Rationality in Economics. Constructivist and Ecological Forms*, Cambridge: Cambridge University Press, 2008, p. 294.

7 As Bottom et al. write, "Experiments are uniquely suited for examining institutional effects". William P. Bottom, Ronald A. King and Larry Handlin, "Miller, G. J., Institutional Modifications of Majority Rule", in: Vernon L. Smith, Charles R. Plott (Eds.), *Handbook of Experimental Economics Results*, Amsterdam: North-Holland, 2008, p. 857. The experimental strategy is to hold preferences constant and randomly assign subjects to treatments distinguished only by variations in institutional rules. The obvious interpretation is that the resulting differences in behavior are to be ascribed to the institutional differences. Significantly, the degree of confidence reached would be impossible in natural political settings.

8 Cf. Vernon L. Smith, "Theory, Experiment and Economics", in: *The Journal of Economic Perspectives*, 3, 1, 1989, pp. 151–169.

9 Cf. Vernon L. Smith, "Microeconomic Systems as an Experimental Science", in: *American Economic Review*, 72, 1982, pp. 923–55.

sages that can be sent by each of the agents. A message might be a bid, an offer, or an acceptance. Secondly, it specifies the rules: (a) allocation rules – which is the resulting commodity or goods allocation to each agent as a function of the messages sent by all agents; a subclass of these rules include the imputation rules, which specify the payment to be made by each agent as a function of the messages sent by all agents; (b) adjustment process rules. In general, these rules consist of a starting rule specifying the time or conditions under which the exchange of messages shall begin, a transition rule (or rules) governing the sequencing and exchange of messages, and a stopping rule under which the exchange of messages is terminated.

The third ingredient of the theory is the behaviour of the actors. First, theories introduce assumptions about agent behaviour, e.g. that agents maximize utility, or expected utility, that common information yields common expectations, that agents make choices as if they are risk averse, that expectations adjust using Bayes rule, that transactions costs (the cost of thinking, deciding, acting) are negligible, etc. The theoretical scheme is this: agents choose messages, and institutions determine the outcomes – the allocations – via the rules that carry messages into allocations. The scheme can be used to explain or to make predictions: for example the bid(s) that an agent will submit at a sealed bid auction, the price that will be posted by an oligopolist, the reservation price below which a price searching agent will buy, and so on.

Now let us move to experiments. The crucial point is that Smith regards the structure of the experiment as a replica of the theory.[10] Experiments also have three ingredients: an environment, an institution, and the observed behaviour of the agents. The characteristic of the experiments is control. "Control is the essence of experimental methodology, and in experimental exchange studies it is important that one be able to state that, as between two experiments, individual values (e.g., demand or supply) either do or do not differ in a specified way".[11] Control infuses the first two ingredients of the experiment. The environment is controlled using monetary rewards to induce the desired specific value/cost configuration.[12] The institution is defined by the experimental instructions which describe the messages and procedures of the market, which are most often computer controlled.[13]

10 Cf. Vernon L. Smith, "Economics in the Laboratory", in: *The Journal of Economic Perspectives*, 8, 1, 1994, pp. 113–131.

11 Vernon L. Smith, "Experimental Economics: Induced Value Theory", in: *The American Economic Review*, 66, 2, 1976, p. 275.

12 A "nonsatiation" condition is here assumed (cf. Vernon L. Smith, "Microeconomic Systems as an Experimental Science", *op. cit.*): given a costless choice between two alternatives, identical (i.e., equivalent) except that the first yields more of a reward medium than the second, individuals will always chose the first over the second.

13 Smith acknowledges, however, that full control is an illusion. "I want simply to note here that there are similar illusions that control is a panacea for ensuring the quality of the information we gather in experiments", Vernon L. Smith, *Rationality in Economics. Constructivist and Ecological Forms*, *op. cit.*, p. 295.

29.2

I shall use the framework developed by V. Smith to sketch a picture of the way in which voting rules can be studied under laboratory conditions. For our purposes, the environment can be defined by a set of players, called the voters, and sets of policies offered by competing parties. The voters are endowed with votes. Usually, each voter is supposed to have exactly one vote. The voters can offer they vote in a mass election to one of the competing parties. Since the number of the parties as well as they position concerning an electoral agenda are not variables that depend upon the behaviour of the agents, they are also taken as circumstances that cannot be altered by the agents. Finally, the agents are supposed to have preferences over the competing sets of policies, which translate into preferences over competing parties.

The institution is the voting rule. Given the messages (votes) received from the voters, the voting rule allocates seats to the parties in the Parliament. Of course, indirectly the rule determines if the policies preferred by an actor will be among those promoted by the winning party. Various assumptions concerning the behaviour of the voters have been proposed. Most general are those that voters are rational – they are endowed with a complete and transitive preference relation – and that they have common knowledge of the voting situation. Others are more specific; the voters are supposed: to have single picked preferences (Black); to vote for the most preferred party most likely to win (Duverger); to vote for the party closest to their ideal point (Downs), etc.

Quite often the role of the voting rules is presented by reference to the so-called fundamental equation of politics: as Plott phrased it, the outcomes are function of the preferences and the voting rule.[14] We can keep the institution constant and let preferences change; or we can keep preferences constant and see which outcomes are reached under different voting rules. For experimental research, it is provoking to see what happens when players are presented with different rules of the game, how their behaviour is affected.

One of the most celebrated pieces of work in political science is due to Maurice Duverger. By comparing electoral systems he concluded that the plurality system, or the simple majority single ballot system, tends to favour a two-party pattern, while proportional representation creates conditions favourable to foster multiparty development.[15] To account for these differences, Duverger relied on a distinction between mechanical and psychological effects. The mechanical effect corresponds to the transformation of votes into seats. So it expresses the working of the institution. The psychological effect can be viewed as the anticipation of the mechanical system: voters are aware that there is a threshold of representation

14 Cf. Charles R. Plott, "Will Economics Become an Experimental Science?", in: *Southern Economic Journal*, 57, 1991, pp. 901–919.

15 Cf. Maurice Duverger, *Les partis politiques*, Paris: Armand Colin 1951.

and they decide not to support parties that are likely to be excluded because of the mechanical effect. Suppose that there are three parties. Under the plurality rule the voters realize that their votes are wasted if they give them to the third party. So they decide to transfer their votes to the party which in their order of preference is on a higher position. Their "natural tendency" is to choose the less evil and to prevent the greater evil. When the simple majority single ballot system is in place, the result is then that a polarization effect works: the institution is detrimental to the new party or the less favoured of the existing parties. So, the theory predicts that under an institutional setting, actors curb their messages, i.e. the way they vote, in a specific way. Duverger's psychological effects are paradigmatic instances of such changes in the agents' behaviour induced by institutions like voting rules.

Since the time of Duverger, the psychological effect is generally explained as an instance of strategic voting.[16] Theorists developed sophisticated, but appealing models of individual voting based on the idea that individuals are rational and vote strategically. In the past decades the view, earlier associated with political scientists like W. H. Riker, that strategic voting has a high explanatory capacity, got a large support.[17]

However, the methodology of formal analysis is subject to at least two types of critics.[18] First, one may wonder about the validity of its assumptions. The (more or less) rational voter hypothesis was subject to numerous criticisms. Some of them focused on limitations of the individuals' capacities to behave rationally: are ordinary people able to produce complete and/or coherent preference relations, or utility functions? Are they able to devise strategic voting procedures? Are they able to acquire and process the information required for a rational choice among the alternatives? In sum, does strategic voting occurs in real world elections in a relevant proportion? Others questioned the whole methodology behind the rational voter hypothesis.[19]

Secondly, there is an epistemological problem of the empirical testing. On the one hand, we need to clearly define the consequences of the actors' behaviour. But in many situations this cannot be well-defined. Usually the approaches associated with game theory look for the existence of Nash equilibria. The trouble is that

16 Cf. Gary W. Cox, *Making Votes Count: Strategic Coordination in the World's Electoral Systems*, Cambridge: Cambridge University Press 1997.

17 "The evidence renders it undeniable that a large amount of sophisticated voting occurs – mostly to the disadvantage of the third parties nationwide – so the force of Duverger's psychological factor must be considerable", William H. Riker, "The Two-Party System and Duverger's Law: An Essay on the History of Political Science", in: *American Political Science Review*, 76, 1982, p. 764.

18 Cf. Jean-François Laslier and M. Remzi Sanver (Eds.), *Handbook on Approval Voting*, Springer-Verlag: Berlin, Heidelberg 2010; Cf. André Blais, Jean-François Laslier, Annie Laurent, Nicolas Sauger, and Karine Van der Straeten, "One Round versus Two Round Elections: An Experimental Study", in: *French Politics*, 5, 2007, pp. 278–286.

19 Cf. Donald P. Green and Ian Shapiro, *Pathologies of Rational Choice Theory: A Critique of Applications in Political Science*, New Haven: Yale University Press 1994.

many games have more than one Nash equilibrium, and there seems to be no way to predict which equilibrium will be reached (and also how the individuals behave at a particular equilibrium).[20] Laslier observes that this difficulty goes to the heart of our conception of democracy: for in the case of elections it comes to the idea that the outcome of voting cannot be predicted from individual opinions.[21] On the other hand, to test the existence of rational strategic behaviour of the individuals we need to measure voters' preferences among the various candidates as well as their beliefs on how other voters will behave in the election and also on how their own vote will affect the outcome of the election. Beliefs cannot be directly observed, so we need to use instead proxies for the relevant beliefs.

A similar difficulty is faced when we try to determine the voters' preferences. Preferences are not observable; only choices are revealed. When the institution is the plurality rule, the voters are asked to express only their top preference. But if a psychological effect is appealed to, then we are also required to consider at least which alternative ranks second and third in the individuals' preferences. Duverger's argument is that under the plurality rule the voter does not vote for her first preference; rather she votes for the second one, in order that her third option would have smaller chances to be elected. But empirically we are again presented with (at most) one chosen alternative for each individual voter. We have no way to find out the entire preference order of the individuals.[22] So when studying the real world behaviour of the individual voters, how can we conclude that their vote was the expression of a psychological effect or not?

One way to overcome these difficulties is to radically change the strategy of research, and adopt an experimental setting. The basic principle of the experiments[23] "is to observe individual behaviour in situations where the experimenter can control individual preferences. The classical way to induce and control preferences is to use money, that is to pay the subjects more or less, depending on what they do and, in group experiments, what the other subjects do".[24] Under an experimental setting, beliefs are also controlled, by letting subjects know relevant information about the others' situation (and also, if applicable, about the way the other subjects behaved in previous rounds). Since the experimental situation is simple, it is reasonable to assume that subjects will behave in a rational way.

---

20 Cf. Thomas Schelling, *The Strategy of Conflict*, Cambridge, MA: Harvard University Press, 1960.

21 Cf. Jean-François Laslier and M. Remzi Sanver (Eds.), *Handbook on Approval Voting*, *op. cit.*

22 The Borda rule requires that the voters reveal more than their top alternative, but not necessarily all the preferences.

23 See also Vernon L. Smith, *Rationality in Economics. Constructivist and Ecological Forms*, *op. cit.*, pp. 293–294, on public goods experiments.

24 Jean-François Laslier, "Laboratory Experiments on Approval Voting", in: Jean-François Laslier and M. Remzi Sanver (Eds.), *Handbook on Approval Voting*, *op. cit.*, p. 339.

A voting rule can be described simply by pointing to the move the voter is allowed to take in a given situation. There are extremely many voting rules discussed in the literature. Three examples are the plurality rule, the Borda rule and the approval rule. Under the plurality rule, individuals are required to pick up exactly one candidate. Under the approval rule, they may cast one vote for as many candidates as they wish. In its simplest form, the Borda rules requires that individuals give two votes to one candidate and one vote to one of the other candidates. Most laboratory experiments use such simple statements of the voting rules. As Laslier observes, "these rules are so simple that, in the laboratory, one does not have to explain how ballots are counted: people naturally understand that votes are added".[25] So the fact that people can take into account the possibility to vote strategically is quite straightforward.

Experiments in political science concerning voting rules have a long history.[26] However, it is only in the past two decades that their use in political research has boomed. One best known field researcher is Elinor Ostrom, a political scientist who recently (in 2009) received a Nobel Prize for economics.

> Given my reputation as an avid field researcher, colleagues often ask why I "bother" with conducting experiments. They ask questions such as "Why would you pay any attention to outcomes in an experiment?" and "What more can you possibly learn about institutions and resource governance from laboratory experiments that you have not already learned in the field?".[27]

She advances two reasons. The first is very general: we should learn more from multiple research methods applied to the same question than from a single method. For the scientific community, confidence is higher when the results of more methods are corroborated. Secondly, in a field research "one of the frustrating aspects is that so many variables are involved that one is never certain that one has isolated the specific variable (or limited set of variables) that causes an outcome". Therefore, the possibility to control is a main rationale for the use of lab experiments.[28] However, control in the lab is often criticized for factoring out the wider political

25 Jean-François Laslier, "Laboratory Experiments on Approval Voting", *op. cit.*, p. 346.

26 Cf. David A. Bositis and Douglas Steinel, "A Synoptic History and Typology of Experimental Research in Political Science", in: *Political Behavior*, 9, 1987, pp. 263–284.

27 Elinor Ostrom, "The Value-Added of Laboratory Experiments for the Study of Institutions and Common-Pool Resources", in: *Journal of Economic Behavior & Organization*, 61, 2006, p. 149.

28 One of the main conclusions Ostrom derives from studying lab experiments on the actors' behavior in commons-dilemma situations is that individuals initially rely on a battery of heuristics in response to complexity; while without communication and agreements on joint strategies, these heuristics lead to overuse, individuals are still willing to discuss ways to increase their own and others' payoffs over a sequence of rounds, cf. Elinor Ostrom, "Coping with the Tragedy of the Commons", in: *Annual Review of Political Science*, 2, 1999, p. 507.

context: the real behaviour of the voters in a real election, as well as their strategic information and beliefs are largely distorted in the lab.

But control in the lab can be criticized from the opposite side, for being too loose: since they leave too much for individual freedom in choosing, lab experiments remain too complex. This complexity is not subject to mathematical models, but "open", in the sense that it not within the control of the researcher. Moreover, if the experimental setting is expanded to include more constraints and variables, then the experiments itself become hard to manage; on the other hand, conducting a theoretical analysis of a more complicated mathematical model would be very difficult. The alternative approach that has been proposed is to implement a computer simulation. The principal advantage of a computer simulation is that it can be arbitrarily complex. Since the famous tournament experiments of R. Axelrod, nearly thirty years ago, this approach was extensively used to observe comparative advantages of voting rules.

For example, McCabe-Dansted and Slinko studied comparatively 26 rules.[29] Since most of these rules have never been applied in real world group choices, it is infeasible to compare them empirically. Therefore, the authors had to artificially generate the data. They fixed three parameters: the size of the group, the number of alternatives, and a parameter of group homogeneity. The group was formed of 85 agents who could choose among five alternatives (this number is sufficiently large to discriminate among the rules). Out of the immense number of possible profiles of this group, a subclass is chosen. The authors used in simulations sets of about one million profiles. For example, if profiles are randomly chosen, and no dependency between agents is assumed, their collection is called impartial. Given the set of profiles, it is possible to construct a matrix of dissimilarities between the rules based on frequency data. Computer simulations show that departing from the impartiality assumption brings about considerable changes in the results obtained under different rules, and thus offers a new means of comparing voting rules, and see similarities between them.[30]

29.3

In this final section I first argue that the voting rules are much more complex than it is usually assumed. In this sense, arguments from social choice theory will be briefly discussed. Then I suggest that experimental research on voting rule may largely benefit from connections with some quite different experimental research.

29 Cf. John C. McCabe-Dansted and Arkadii Slinko, "Exploratory Analysis of Similarities between Social Choice Rules", in: *Group Decision and Negotiation*, 15, 2006, pp. 77–107.

30 For a randomly generated set of profiles using the same parameter of homogeneity the estimated dissimilarity between rules can be defined by appeal to the frequency that rules fail to pick the same winning alternative.

In most experiments on voting rules, they are assumed to be stated in a simple and easy to understand way, as we saw with the plurality rule, the approval or the Borda rule. There are of course some more complicated rules. Consider for example the Hare rule (also known as Single Transferable Vote or Alternative Vote). By this rule, if one alternative's plurality score is greater than *n*/2 (*n* is the number of voters), then that alternative is the Hare's winner; otherwise, eliminate the alternative with the lowest plurality score; continue until one alternative remains. (The plurality score of an alternative is the number of votes for it.) The Hare rule is only a bit more complicated than the first three rules, but there are ones much more difficult to understand and to compute. However, all the rules are defined by reference to the aggregation mechanism they use. The votes are aggregated in different ways, and sometimes the results are different (while sometimes they are not). So it looks that voting rules are very simple institutions, especially as we compare them with other political institutions, like the presidential system or federalism. It is precisely this characteristic that accounts for the prominent role they played in experimental research.

However, some of the most interesting results on voting rules consist in the proof of so-called characterization results. The proof goes as follows. First, properties a voting rule may or may not satisfy are defined. For example, a voting rule may treat all the members of the electorate as equal; others do not. Majority rule paradigmatically treats all the voters on the same par. But consider the Chairperson tie rule. According to it, if the votes of the members of a group go for an alternative, then it is chosen; but if there is tie, then the vote of the chairperson is decisive. Obviously, the chairperson is attached a special position by this rule. Secondly, we can then form different collections of such properties of the voting rules. The properties included in such a collection can be satisfied by more rules, by no rule, of by exactly one rule. The second and the third case gained a special interest in the social choice literature. K. Arrow's celebrated impossibility theorem states that reasonable such properties cannot be simultaneously satisfied by any rule.[31] May proved that the simple majority rule is the only aggregation procedure that jointly satisfies four such properties:[32] universal domain, anonymity, neutrality, and positive responsiveness.[33] Fishburn and Young gave similar characterizations

31 Cf. Keneth Joseph Arrow, *Social Choice and Individual Values*, New York: Wiley 1951.

32 Cf. Kenneth O. May, "A Set of Independent, Necessary and Sufficient Conditions for Simple Majoritary Decision", in: *Econometrica*, 20, 1952, pp. 680–684.

33 The properties referred to in the above mentioned theorems can be defined rigorously in the frame of social choice theory. A rule satisfies universal domain if it accepts all logically possible profiles of votes as admissible input. Neutrality basically says that the names of candidates should not play any role in determining winning candidates. Analogously, anonymity requires that the identity of individual voters does not affect the outcome. By positive responsiveness, if one or more voters change their votes in favour of an option that is winning or tied and no other voters change theirs, then that option is uniquely winning after the change.

of the approval voting, respectively of the Borda rule.[34] Goodin and List generalized the classic result of May to the plurality rule.[35]

So, a voting rule can be identified with a collection of more abstract rules or properties that define the voting situation. In this sense, voting rules are complex institutions, including different clusters of rules. Some of them are agenda rules: who are the candidates for choice, how are they nominated, etc. Others are allocation rules[36]: who are the members of the electorate, how many votes they have, what is their relative position, etc.; still others are domain rules: which are the allowed preference profiles, how are they related, etc.

For example, simple majority rule and absolute majority rule differ in respect to the agenda rules that constrain the voters who act under each of them. Indeed, the simple majority voting requires the individuals to behave by treating all the candidates in an election as equal. But in an absolute majority voting the electorate is allowed to weight higher the incumbent president, if he is among the candidates, or to favour the present law and make it harder the adoption of an alternative regulation. Voting rules differ very much with respect to the allocation rules they contain. Under the plurality voting, each voter is attached exactly one vote, while under the approval rule each voter can give one vote to as many candidates as she wants. But under both voting rules voters are treated in a fair way: no one is assumed to have a privileged position. However, some voting rules, weighted majority rule among them, require that voters be treated unequally. This means that they include rules that define the ways in which individuals are not equal in the voting procedure. Domain rules help characterize voting procedures as complex institutions. They specify the way in which a collection of profiles is generated. As already mentioned, computer simulations have been used to investigate different "cultures", i.e. generations of collections of profiles. Different rules behave differently on such domains.[37]

Now, the idea is that to experimentally investigate a voting rule turns to be quite complicated. It does not simply consist in a simple statement one can easily agree or disagree with. The experimenter may try to see how subjects behave when faced with different agenda, position or domain rules, etc. Given a domain, which agenda rule is preferred by the subjects? How do people react to cases in which the neutrality condition is questioned? For example, how do actors behave in situations in which candidates are treated asymmetrically? A large collection of experiments

34 Cf. Peter C. Fishburn, "Axioms for Approval Voting: Direct Proof", in: *Journal of Economic Theory*, 19, 1978, pp. 180–185; and H. Peyton Young, "An Axiomatization of Borda's Rule", in: *Journal of Economic Theory*, 9, 1974, pp. 43–52.

35 Cf. Robert E. Goodin and Christian List, "A Conditional Defense of Plurality Rule: Generalizing May's Theorem in a Restricted Informational Environment", in: *American Journal of Political Science*, 50, 4, 2006, pp. 940–949.

36 Cf. Vernon L. Smith, "Microeconomic Systems as an Experimental Science", *op. cit.*

37 Cf. Jean-François Laslier, "*In Silico* Voting Experiments", in: Jean-François Laslier and M. Remzi Sanver (Eds.), *Handbook on Approval Voting*, *op. cit.*, pp. 311–335.

concerning the topic of fairness becomes relevant when allocation rules are taken into account.[38] How favourable are the subjects to fairness properties like anonymity or weaker alternatives to it? Or, when domain rules are investigated, how much do subjects agree with an impartial culture or with a distributive one?[39]

So, the theoretical results on the axiomatizations of the voting rules may open the experimental research to a new class of approaches.

References

Kenneth Joseph Arrow, *Social Choice and Individual Values*, New York: Wiley 1951.

André Blais, Jean-François Laslier, Annie Laurent, Nicolas Sauger, and Karine Van der Straeten, "One Round versus Two Round Elections: An Experimental Study", in: *French Politics*, 5, 2007, pp. 278–286.

David A. Bositis and Douglas Steinel, "A Synoptic History and Typology of Experimental Research in Political Science", in: *Political Behavior*, 9, 1987, pp. 263–284.

William P. Bottom, Ronald A. King and Larry Handlin, "Miller, G. J., Institutional Modifications of Majority Rule", in: Vernon L. Smith, Charles R. Plott (Eds.), *Handbook of Experimental Economics Results*, Amsterdam: North-Holland 2008, pp. 857–871.

Gary W. Cox, *Making Votes Count: Strategic Coordination in the World's Electoral Systems*, Cambridge: Cambridge University Press 1997.

Maurice Duverger, *Les partis politiques*, Paris: Armand Colin 1951.

Peter C. Fishburn, "Axioms for Approval Voting: Direct Proof", in: *Journal of Economic Theory*, 19, 1978, pp. 180–185.

Wenceslao J. Gonzalez, "The Role of Experiments in the Social Sciences: the Case of Economics", in: Theo Kuipers (Ed.), *Handbook of the Philosophy of Science: General Philosophy of Science - Focal Issues*, Amsterdam: Elsevier 2007, pp. 275–301.

---

38 Cf. James Konow, "Which Is the Fairest One of All? A Positive Analysis of Justice Theories", in: *Journal of Economic Literature*, 41, 4, 2003, pp. 1188–1239.

39 An impartial culture allows of profiles in which individuals are free to choose their preferences as they wish; in a distributive culture individuals are in a complete antagonism: given a divisible good, they wish to get a share as much as possible of it, and do not care about the others' shares.

Robert E. Goodin and Christian List, "A Conditional Defense of Plurality Rule: Generalizing May's Theorem in a Restricted Informational Environment", in: *American Journal of Political Science*, 50, 4, 2006, pp. 940–949.

Donald P. Green and Ian Shapiro, *Pathologies of Rational Choice Theory: A Critique of Applications in Political Science*, New Haven: Yale University Press 1994.

James Konow, "Which Is the Fairest One of All? A Positive Analysis of Justice Theories", in: *Journal of Economic Literature*, 41, 4, 2003, pp. 1188–1239.

Jean-François Laslier and M. Remzi Sanver (Eds.), *Handbook on Approval Voting*, Berlin, Heidelberg: Springer-Verlag 2010.

Jean-François Laslier, "*In Silico* Voting Experiments", in: Jean-François Laslier and M. Remzi Sanver (Eds.), *Handbook on Approval Voting*, Berlin, Heidelberg: Springer-Verlag 2010, pp. 311–335.

Jean-François Laslier, "Laboratory Experiments on Approval Voting", in: Jean-François Laslier and M. Remzi Sanver (Eds.), *Handbook on Approval Voting*, Berlin, Heidelberg: Springer-Verlag 2010, pp. 339–356.

Kenneth O. May, "A Set of Independent, Necessary and Sufficient Conditions for Simple Majoritary Decision", in: *Econometrica*, 20, 1952, pp. 680–684.

John C. McCabe-Dansted and Arkadii Slinko, "Exploratory Analysis of Similarities between Social Choice Rules", in: *Group Decision and Negotiation*, 15, 2006, pp. 77–107.

John Stuart Mill, (1874). *Essays on Some Unsettled Questions of Political Economy*, Second Edition, Batoche Books, Kitchener, 2000. (Second Edition London, Longmans, Green, Reader, And Dyer.)

Elinor Ostrom, "Coping with the Tragedy of the Commons", in: *Annual Review of Political Science*, 2, 1999, pp. 493–535.

Elinor Ostrom, "The Value-Added of Laboratory Experiments for the Study of Institutions and Common-Pool Resources", in: *Journal of Economic Behavior & Organization*, 61, 2006, pp. 149–163.

Charles R. Plott, "Will Economics Become an Experimental Science?", in: *Southern Economic Journal*, 57, 1991, pp. 901–919.

William H. Riker, "The Two-Party System and Duverger's Law: An Essay on the History of Political Science", in: *American Political Science Review*, 76, 1982, pp. 753–766.

Thomas Schelling, *The Strategy of Conflict*, Cambridge, MA: Harvard University Press 1960.

Vernon L. Smith, “Experimental Economics: Induced Value Theory”, in: *The American Economic Review*, 66, 2, 1976, pp. 274–279.

Vernon L. Smith, “Microeconomic Systems as an Experimental Science”, in: *American Economic Review*, 72, 1982, pp. 923–955.

Vernon L. Smith, “Experimental Economics: Reply”, in: *The American Economic Review*, 75, 1, 1985, pp. 265–272.

Vernon L. Smith, “Theory, Experiment and Economics”, in: *The Journal of Economic Perspectives*, 3, 1, 1989, pp. 151–169.

Vernon L. Smith, “Economics in the Laboratory”, in: *The Journal of Economic Perspectives*, 8, 1, 1994, pp. 113–131.

Vernon L. Smith, *Rationality in Economics. Constructivist and Ecological Forms*, Cambridge: Cambridge University Press 2008.

H. Peyton Young, “An Axiomatization of Borda’s Rule”, in: *Journal of Economic Theory*, 9, 1974, pp. 43–52.

Political Science Department
National School of Political Studies and Public Administration
Povernei St. 6
10643, Bucharest
Romania
admiroiu@snspa.ro

Team E
History of the Philosophy of Science

CHAPTER 30

VOLKER PECKHAUS

THE BEGINNING OF MODEL THEORY IN THE ALGEBRA OF LOGIC

30.1 INTRODUCTION

Model Theory is commonly closely connected to the work of Alfred Tarski. The main thesis developed in the present paper is that basic ideas of Model Theory, in particular as regards its structural and semantical respects, were anticipated in the work of Ernst Schröder (1844–1902), the main German representative of nineteenth century Algebra of Logic.[1] The second section is devoted to the clarification of the notion "Algebra of Logic". In the third section Schröder's programme of an Absolute Algebra is described in order to show that this programme has some features in common with modern Model Theory. According to Wilfried Hodges "[...] in a broader sense, model theory is the study of the interpretation of any language, formal or natural, by means of set-theoretic structures, with Alfred Tarski's truth definition as a paradigm."[2] The relation between notation, interpretation, and modelling will be discussed in the fourth section. The paper is concluded with some observations about Schröder's notion of modality. An algebraic notation which is claimed to be suitable for all of logic should be able to express modalities as well. Schröder's work contains only some preliminary epistemological remarks for such algebraic theory of modalities.

1 This paper draws upon results published earlier, e.g., in Volker Peckhaus, *Logik, Mathesis universalis und allgemeine Wissenschaft. Leibniz und die Wiederentdeckung der formalen Logik im 19. Jahrhundert*. Berlin: Akademie-Verlag 1997, ch. 6, pp. 233–296; Volker Peckhaus, "Calculus Ratiocinator vs. Characteristica Universalis? Two Traditions in Logic, Revisited", in: *History and Philosophy of Logic* 25, 1, 2004, pp. 3–14; Volker Peckhaus, "Schröder's Logic", in: Dov M. Gabbay, John Woods (Eds.), *Handbook of the History of Logic*. Vol. 3: *The Rise of Modern Logic: From Leibnz to Frege*. Amsterdam et. al.: Elsevier North Holland 2004, pp. 557–609. In the present paper the role of semantics in the Algebra of Logic and its relationship to Model Theory is stressed. I would like to thank Anna-Sophie Heinemann, Paderborn, for valuable comments on an earlier version of the present paper.

2 Wilfried Hodges, "Model Theory", in: *The Stanford Encyclopedia of Philosophy (Fall 2009 Edition)*, Edward N. Zalta (Ed.), URL = <http://plato.stanford.edu/archives/fall2009/entries/model-theory/>.

D. Dieks et al. (eds.), *Probabilities, Laws, and Structures*, The Philosophy of Science in a European Perspective 3, DOI 10.1007/978-94-007-3030-4_30,

30.2 What is Algebra of Logic?

Standard answers to the question "What is Algebra of Logic?" were given by Jan van Heijenoort and Jaakko Hintikka. Jan van Heijenoort characterized the Algebra of Logic as "logic as calculus", distinguishing it from the Fregean kind of logic, called "logic as language".[3] This distinction was later modified by Jaakko Hintikka who opposed Algebra of Logic as "language as calculus" to Fregean style logic, i.e., "language as a universal medium".[4] Both authors take up Leibniz's distinction between *calculus ratiocinator* and *lingua characterica*,[5] obviously inspired by the dispute between Gottlob Frege and Ernst Schröder about the question whose logical system represented best the alleged Leibnizian idea of a *lingua characterica*.[6]

---

3 Cf. Jan van Heijenoort, "Logic as Calculus and Logic as Language", in: *Synthese* 17, 1967, pp. 324–330. For a critical discussion cf. Peckhaus, "Calculus Ratiocinator vs. Characteristica Universalis?", *loc. cit.*

4 Cf. Jaakko Hintikka, "On the Development of the Model-Theoretic Viewpoint in Logical Theory", in: *Synthese*, 77, 1988, pp. 1–36; reprinted in Jaakko Hintikka, *Lingua Universalis vs. Calculus Ratiocinator: An Ultimate Presupposition of Twentieth-Century Philosophy*. Dordrecht/Boston/London: Kluwer 1997, pp. 104–139.

5 The expression "lingua characterica" does not appear in Leibniz's works. Leibniz spoke of "lingua generalis", "lingua universalis", "lingua rationalis", "lingua philosophica", all meaning basically the same. He also introduced the terms "characteristica" viz. "characteristica universalis" representing his general theory of signs. Frege obviously took the term "lingua characterica" from Adolf Trendelenburg who used the expression "lingua characterica universalis" in Friedrich Adolf Trendelenburg, "Über Leibnizens Entwurf einer allgemeinen Charakteristik", in: *Philosophische Abhandlungen der Königlichen Akademie der Wissenschaften zu Berlin. Aus dem Jahr 1856*. Berlin: Commission Dümmler 1857, pp. 36–69; reprinted Friedrich Adolf Trendelenburg, *Historische Beiträge zur Philosophie*. Vol. 3: *Vermischte Abhandlungen*. Berlin: Bethge 1867, pp. 1–47. Cf. Günther Patzig "Einleitung", in: Gottlob Frege, *Logische Untersuchungen*, edited by Günther Patzig, 2nd rev. ed., Göttingen: Vandenhoeck & Ruprecht 1976 (1st ed. 1966), p. 10, n. 8; also Peckhaus, *Logik, Mathesis universalis und allgemeine Wissenschaft*, *op. cit.*, pp. 178–181; on Trendelenburg's influence in the history of logic cf. *ibid.*, ch. 4, Risto Vilkko, *A Hundred Years of Logical Investigations: Reform Efforts of Logic in Germany 1781–1879*. Paderborn: Mentis 2002, ch. 4.

6 This dispute was prompted by Ernst Schröder's review of Frege's *Begriffsschrift* (1879) in: *Zeitschrift für Mathematik und Physik, historisch-literarische Abt.*, 25, 1880, pp. 81–94. Gottlob Frege responded in the unpublished paper "Booles rechnende Logik und die Begriffsschrift" (1880/81), in: Gottlob Frege, *Nachgelassene Schriften*. Edited by Hans Hermes, Friedrich Kambartel and Friedrich Kaulbach, 2nd rev. ed., Hamburg: Felix Meiner 1983, pp. 9–52; and in the paper Gottlob Frege, "Ueber den Zweck der Begriffsschrift", in: *Sitzungsberiche der Jenaischen Gesellschaft für Medizin und Naturwissenschaft für das Jahr 1882*, supplement to *Jenaische Zeitschrift für Naturwissenschaft*, 16, 1882/1883, reprinted in Gottlob Frege, *Begriffsschrift und andere Aufsätze. Dritte Auflage. Mit E. Husserls und H. Scholz' Anmerkungen*. Edited by Ignacio Angelelli, Darmstadt: Wissenschaftliche Buchgesellschaft 1977, pp. 97–106.

Dissenting from the standard view on Algebra of Logic represented by van Heijenoort's and Hintikka's distinctions, the present paper characterizes it as such:

- Algebra of Logic is no logic, but the *algebra* of logic.
- It is an algebraic structure, based on algebraic connecting operations, interpreted logically.
- The objects connected are logical objects in the traditional sense: (extensions of) concepts, judgements (propositions) and inferences.
- These objects allow further interpretations, leading to models of the algebraic structures in other fields.

30.3 Absolute Algebra

The key for this understanding of the Algebra of Logic is the conception of an Absolute Algebra proposed by the German mathematician Ernst Schröder. Schröder's first publications on logic were published in the 1870s when he was a grammar school teacher in Baden-Baden. In 1874 he became Professor of Mathematics at the Polytechnic in Darmstadt, changing to the Polytechnic in Karlsruhe in 1876.[7] He stood in the tradition of Combinatorical Analysis (founded by Gottfried Wilhelm Leibniz, Carl Friedrich Hindenburg) and Algebraic Analysis (founded by Heinrich August Rothe, Martin Ohm). According to his own words, the development of an Absolute Algebra constituted "his most original field of research".[8] Absolute Algebra is defined as a "general theory of connections, transcending even the law of associativity".[9] Absolute Algebra is the final step in the development of Formal Algebra. Schröder suggested the following programme[10]:

1. Formal Algebra compiles all assumptions that can serve for defining connectives for numbers of a domain of numbers.
2. Formal Algebra compiles, for every premise or combination of premises, the complete set of inferences, a task that Schröder calls "separation."

7 For a biographical sketch of Schröder, cf. Peckhaus, "Schröder's Logic", *loc. cit.*, pp. 559–564.

8 Ernst Schröder [unsigned], "Grossherzoglich Badischer Hofrat Dr. phil. Ernst Schröder[,] ord. Professor der Mathematik an der Technischen Hochschule in Karlsruhe i. Baden", in: *Geistiges Deutschland. Deutsche Zeitgenossen auf dem Gebiete der Literatur, Wissenschaften und Musik.* Berlin-Charlottenburg: Adolf Eckstein, no year [1901].

9 *Ibid.*

10 Ernst Schröder, *Lehrbuch der Arithmetik und Algebra für Lehrer und Studirende.* Vol. 1: *Die sieben algebraischen Operationen.* Leipzig: B. G. Teubner 1873, pp. 293–294. Only the first volume was published.

3. Formal Algebra investigates in which particular domains of numbers the defined operations hold.

4. Formal Algebra has finally to decide "what geometrical, physical, or generally reasonable meaning these numbers and operations can have, what real substratum they can be given".

Only after having finished the semantical steps (3) and (4), formal algebra becomes an "Absolute Algebra." Absolute Algebra is therefore a Formal Algebra including all the possible models, and logic is only one of them.

At that time, Schröder was not aware of George Boole's Algebra of Logic and its context in the British discussions about Symbolical Algebra. Schröder was influenced by Hermann Günther Graßmann's "General Theory of Forms" and by Robert Graßmann's (H. G. Graßmann's brother) architecture of mathematics.[11]

Hermann Günther Graßmann opened his *Lineale Ausdehnungslehre*[12] with a "survey of a general theory of forms", understood as a series of truths being equally related to all branches of mathematics. The general theory of forms only contains the general concepts of equality and inequality, of connection and separation.[13] The notion of connection is not defined. It is an operation applied to two elements. Brackets indicate the order of these connecting operations in forming complexes. They indicate, e.g., commutativity or associativity. Every synthetic connection of two elements a and b is accompanied by two analytic or "separating" operations, which lead back to a or b respectively if applied to the result of connecting a and b. Graßmann then introduces a second connecting operation (with its inverses), which he regards as a connection of higher level. Both connecting operations are distributive. The resulting algebraic structure is dependent on the features given to these operations.

The general theory of forms was not only applied to the *Lineale Ausdehnungslehre*, but in Graßmann's *Lehrbuch der Arithmetik*[14] to arithmetic as well. The interdependence between the features of the connecting operations and the resulting

11 On this influence cf. Volker Peckhaus, "The Influence of Hermann Günther Grassmann and Robert Grassmann on Ernst Schröder's Algebra of Logic", in: Gerd Schubring (Ed.), *Hermann Günther Graßmann (1809–1877): Visionary Mathematician, Scientist and Neohumanist Scholar. Papers from a Sesquicentennial Conference*. Dordrecht/Boston/London: Kluwer 1996 (*Boston Studies in the Philosophy of Science*; vol. 187), pp. 217–227; Volker Peckhaus, "Robert and Hermann Graßmann's Influence on the History of Formal Logic", in: Hans-Joachim Petsche et al. (Eds.), *Hermann Graßmann. From Past to Future: Graßmann's Work in Context. Graßmann Bicentennial Conference, September 2009*. Basel: Birkhäuser 2011, pp. 221–228.

12 Hermann Günther Graßmann, *Die lineale Ausdehnungslehre ein neuer Zweig der Mathematik dargestellt und durch Anwendungen auf die übrigen Zweige der Mathematik, wie auch auf die Statik, Mechanik, die Lehre vom Magnetismus und die Krystallonomie erläutert*. Leipzig: Otto Wigand 1844; [2]1878.

13 *Ibid.*, p. 1.

14 Hermann Günther Graßmann, *Lehrbuch der Arithmetik für höhere Lehranstalten*. Berlin: Enslin 1861.

structure was further elaborated by the two brothers to an architecture of mathematics, published by Robert Graßmann in six short pamphlets in his *Formenlehre oder Mathematik.*[15]

In the theory ofquantities (*Grösenlehre*, Graßmann introduces the letters a, b, c, ... as syntactical symbols for arbitrary quantities. The letter e stands for special quantities: elements or, in the Graßmann's idiosyncratic German terminology, "Stifte" ("pins"), quantities which do not emerge from other quantities by applying connecting operations. Besides brackets for indicating the order of connections, he introduces the equality sign $=$, the inequality sign $\neq$ (Graßmann himself used a stylized z) and the general sign $\circ$ for designating connecting operations. As special connecting operations he treats "joining" ("fügen") or adding (symbol $+$), and "weaving" ("weben") or multiplying (juxtaposition or symbols $\cdot$, $\times$). These connections can either occur as inner connections (if $e \circ e = e$), or as outer connections (if $e \circ e \neq e$).[16]

The different results of connecting pins with themselves give the criteria for distinguishing between the special parts of the theory of quantities. The "theory of concepts or logic" ("Begriffslehre oder Logik") is the first part, "the most simple and, at the same time, the most inward part", as Graßmann calls it.[17] Inner joining $e + e = e$ and inner weaving $ee = e$ are valid. In the "theory of binding or theory of combinations" ("Bindelehre oder Combinationslehre") as the second part of the theory of forms, inner joining $e + e = e$ and outer weaving $ee \neq e$ are valid; in the "theory of number or arithmetic" ("Zahlenlehre oder Arithmetik") it is outer joining $e+e \neq e$ and inner weaving $ee = e$, respectively $1\times1 = 1$ and $1\times e = e$. In the "theory of the exterior or theory of extensions" finally, the "most complicated and most exterior" part of the theory of forms, outer joining $e + e \neq e$ and outer weaving $ee \neq e$ are valid.[18]

Schröder took up these ideas in his own *Lehrbuch der Arithmetik und Algebra.*[19] There he treated, as he stressed already in the subtitle of this book, "the seven algebraical operations", i.e., the three "direct" operations of adding, multiplying, and raising to a higher power, and their inverses subtracting, dividing, extracting the roots and forming the logarithms (in the beginning only applied to natural numbers). Schröder defines (pure) mathematics as the "theory of numbers".[20] By this definition he deviates from the traditional view of mathematics as the theory of quantities. Although Schröder calls the objects algebraically connected "numbers", he leaves open what kind of objects they are. Hence, the

15 Robert Graßmann, *Die Formenlehre oder Mathematik.* Stettin: R. Graßmann 1872. Reprinted Hildesheim: Georg Olms 1966. Vol. 1: *Die Grösenlehre*; Vol 2: *Die Begriffslehre oder Logik*; Vol. 3: *Die Bindelehre oder Combinationslehre*; Vol. 4: *Die Zahlenlehre oder Arithmetik*; Vol. 5: *Die Ausenlehre oder Ausdehnungslehre.*

16 Robert Graßmann, *Formenlehre* I, pp. 8, 26.

17 *Ibid.*, p. 13.

18 *Ibid.*, pp. 12–13.

19 Ernst Schröder, *Lehrbuch, op. cit.*

20 *Ibid.*, p. 2.

structure erected needs an interpretation. In a pamphlet entitled *Über die formalen Elemente der absoluten Algebra*, which was enclosed to the 1873/1874 issue of the annual report of his school, he became more specific.[21] Schröder proceeds from the existence of an "unlimited manifold [*Mannigfaltigkeit*] of objects (of any kind)" called "domains of numbers" (*Zahlengebiete*). Examples for such "objects constituting a manifold" called "general numbers" are "proper names, concepts, judgements, algorithms, numbers [of pure mathematics], symbols for dimensions or operations, points, systems of points, or any geometrical object, quantities of substances, etc." Schröder's theory of arithmetic was imbedded into a universal algebraic programme, his "Absolute Algebra". And it was also by the Graßmann's influence that logic came into his focus. In a voluminous footnote, running over three pages, he reported about his discoveries in Robert Graßmann's works[22]:

> The author of the respective work uses in the part devoted to logic the + sign for the collective combination, and regards it downright as an *addition* – one could say as a "logical" addition – which has beyond the properties of the usual (numerical) addition the basic property: $a + a = a$ additionally.
>
> [...]
>
> Especially interesting and new for me was [...] the role the author assigns to *multiplication* in the domain of logic. While the *sum* of two concepts is interpreted as the whole of the individuals belonging to the one or the other of these concepts, the *product* of these concepts is a concept which comprises the marks of both. Thus, the real extensional addition is opposed to an intensional addition or *addition* of marks as multiplication. This procedure can indeed not surprise if one takes into account that the basic features of addition and of multiplication are essentially the same, that both operations have an already fixed relation to one another only in usual arithmetic, and that one has therefore in new fields from the beginning the choice between the two conceptions. –

Only in the subsequent years would Schröder learn about the British precursors of his Algebra of Logic.

30.4 Modelling

The further development of Schröder's activities in logic was consistent with his early algebraic programme. For the present he devoted his work to the analysis of a first interpretation of the formal algebraic structure and, with this, to one model of Absolute Algebra: logic. In the *Operationskreis des Logikkalkuls* Schröder concentrated on the duality between logical addition and logical multiplication and, stressing the identity of the algebraic structures of these operations. In the

21 Ernst Schröder, *Über die formalen Elemente der absoluten Algebra. Beilage zum Programm des Pro- und Real-Gymnasiums in Baden-Baden für 1873/74*. Stuttgart: Schweizerbart'sche Buchhandlung 1874, p. 3.

22 Ernst Schröder, *Lehrbuch*, pp. 145–147, footnote. All translations from Schröder's works are mine.

volumes of his monumental, though unfinished *Vorlesungen über die Algebra der Logik*,[23] Schröder distinguished between the object of logic and its structure. He called the calculus an "auxiliary discipline" ("Hülfsdisziplin") that precedes, or goes along with proper logic. The third volume with the subtitle *Algebra und Logik der Relative* provided a further generalization. He always stressed the significance of Charles S. Peirce's influence, especially of Peirce's papers "On the Algebra of Logic".[24] Schröder emphasized the twofold character of the theory of relatives consisting of an algebra and a logic of relatives. In the published first part he presented the algebraic section. The logic of relatives that would have linked his theory to Absolute Algebra was planned for the second part that he could not finish before his death.

Already in his "Note über die Algebra der binären Relative",[25] Schröder illustrated the power of his method by applying it to an example from the mathematics discussed at his time. He symbolized those propositions from Richard Dedekind's theory of chains (*Kettentheorie*)[26] that laid the foundation of complete induction (Theorem 59). Schröder sees the advantage of his presentation in extending the scope of Dedekind's theorems going beyond the validity for definite mappings and "systems", now covering all binary relatives. In addition, Schröder shows that the theory of chains can be simplified at some places when using his symbolism.

The aim of this example is evident. The possibilities of the new symbolism as a tool for an alternative presentation of (here: mathematical) connecting operations should be shown, thereby demonstrating its advantages in respect to brevity, clarity and simplicity of proofs. This is also exactly the aim Schröder pursued in his two papers "Ueber zwei Definitionen der Endlichkeit und G. Cantor'sche Sätze"[27] and "Die selbständige Definition der Mächtigkeiten 0, 1, 2, 3 und die explizite Gleichzahligkeitsbedingung",[28] where he applied the logic of relatives to Cantorian set theory. In the first paragraph of the first paper, Schröder compares Dedekind's definition of infinity from *Was sind und was sollen die Zahlen?* with

23 Ernst Schröder, *Vorlesungen über die Algebra der Logik (exakte Logik)*. Leipzig: B.G. Teubner, Vol. 1, 1890; Vol. 2, Pt. 1, 1891; Vol. 2, Pt. 2, 1905; Vol. 3, Pt. 1, 1895.

24 Charles S. Peirce, "On the Algebra of Logic", in: *American Journal of Mathematics* 3, 1880, pp. 15–57; Charles S. Peirce, "On the Algebra of Logic. A Contribution to the Philosophy of Notation", in: *American Journal of Mathematics* 7, 1885, pp. 180–202.

25 Ernst Schröder, "Note über die Algebra der binären Relative", in: *Mathematische Annalen* 46, 1895, pp. 144–158.

26 This theory is formulated in Richard Dedekind, *Was sind und was sollen die Zahlen?* Vieweg: Braunschweig 1888, [2]1893, [3]1911, Braunschweig: Vieweg & Sohn [8]1960, Schröder used the second edition of 1893.

27 Ernst Schröder, "Ueber zwei Definitionen der Endlichkeit und G. Cantor'sche Sätze", in: *Nova Acta Leopoldina. Abhandlungen der Kaiserlich Leop.-Carol. Deutschen Akademie der Naturforscher* 71, 6, 1898, pp. 301–362.

28 Ernst Schröder, "Die selbständige Definition der Mächtigkeiten 0, 1, 2, 3 und die explizite Gleichzahligkeitsbedingung", in: *Nova Acta Leopoldina. Abhandlungen der Kaiserlich Leop.-Carol. Deutschen Akademie der Naturforscher* 71, 7, 1898, pp. 364–376.

the one given by Peirce 3 years before.[29] After reformulating these definitions in the language of the logic of relatives, it becomes evident for the reader that these definitions do not formally coincide, but that they can nevertheless be translated into the respective other form. Peirce's definition proved to be shorter and simpler. Schröder obviously tried to show with this application that his symbolism could serve as a criterion for simplicity and economy of mathematical definitions and theorems.

Schröder's considerations in the following sections are of greater systematic significance. There he discusses Cantor's propositions A–E from the first paper of his "Beiträge zur Begründung der transfiniten Mengenlehre".[30] Schröder's proof of Cantor's equivalence theorem B (concerning the equivalence of sets) caused a sensation.[31] Almost at the same time, in winter 1896/1897, Felix Bernstein also found a proof of the equivalence theorem, which was first published by Émile Borel.[32] The theorem remained connected to the names of Schröder and Bernstein, until Alwin Reinhold Korselt published evidence, found already in 1902, that Schröder's proof was based on an implicit and incorrect presupposition.[33] Already in May 1902, Schröder had admitted this fault, and stated in a letter to Korselt that he "leaves the honor of having proved G. Cantor's theorem to Mr F. Bernstein alone."[34]

In the last paragraph (Sect. 30.5) of his paper on Cantorian propositions Schröder discusses further results from Cantor's theory of ordered sets. In his résumé he points out that the Algebra of Logic proves to be able – here Schröder takes up an idea from a correspondence of Aurel Voss – ,

> to provide far more insights which are accessible for verbal thinking and for gaining them the hitherto common mathematical forms of expression seem to be not sufficient.[35]

"With this, the new *Peirceian* discipline", Schröder writes, "has had [...] its opportunity to stand a little acid test. G. *Cantor*'s theory, as well." Schröder was sure that Cantorian set theory could be completely "presented pasigraphically[36] with the designation capital of our algebraic logic."[37]

---

29 In Charles S. Peirce, "On the Algebra of Logic. A Contribution to the Philosophy of Notation", *op. cit.*

30 Georg Cantor, "Beiträge zur Begründung der transfiniten Mengenlehre (Erster Artikel)", in: *Mathematische Annalen* 46, 1895, pp. 481–512, esp. p. 484.

31 Ernst Schröder, "Ueber zwei Definitionen der Endlichkeit", *op. cit.*, § 4, pp. 336–344.

32 Émile Félix Édouard Justin Borel, *Leçons sur la théorie des fonctions*, Paris: Gauthier-Villars 1898 (*Collection de monographies sur la théorie des fonctions*), pp. 103–107.

33 Alwin Reinhold Korselt, "Über einen Beweis des Äquivalenzsatzes", in: *Mathematische Annalen* 70, 1911, pp. 294–296.

34 Schröder to Korselt, dated 25 May 1902; quoted according to Korselt, *op. cit.*, 295.

35 Ernst Schröder, "Ueber zwei Definitionen der Endlichkeit", *op. cit.*, p. 361.

36 Pasigraphy is a general script.

37 *Ibid.*

Also in letters to Felix Klein, to whom he had offered several papers with applications of the algebra of relatives to set theory for publication in the *Mathematische Annalen*, Schröder campaigned for his tools stressing especially the short time period he needed to develop a notational system for set theory standing on the same level as Cantor's own, if not superior to it. He wrote:

Mr G. Cantor – I'm far from comparing my modest talents with his genius – was occupied with the topic of his research for 20 years; although I always thought that it is a desideratum to go further into it, I found the time to do so only after the publication of his last paper in the *Annalen* which was published in November last year. When I now, in a certain sense, caught up with him in the shortest period of time, it might justified to compare my instrument with a "bicycle", with which the most sprightly pedestrian can be caught up quickly (whether it also applies for clearing the way is another question which can only be decided by the future).[38]

In his paper on pasigraphy,[39] Schröder illustrates the "implications of our new logic of relatives", by presenting pasigraphically some of the most important basic concepts of arithmetic: the concept of set, the numbers 0, 1 and 2, the relations of equinumerosity and power equality, the finiteness, the actual infinite, the concepts of function and substitution, the concept of order as well as the relation greater than, the successor relation, factor relations, and the notion of a prime. In the papers mentioned Schröder does not aim at a systematic construction of arithmetic or set theory. He is interested in a clear demonstration of the possibility to represent the basic concepts of arithmetic and set theory with the help of the algebra of relatives. Other examples serve the same goal, e.g. from geometry ("z is a point") and from the domain of human relationship, "which form a not unimportant chapter in the Corpus juris for our students of jurisprudence."[40]

The *Algebra und Logik der Relative* represents the attempt to extend the programme of Absolute Algebra to a foundational programme for all scientific disciplines that can be formalized or that work with formal means. This extended programme was twofold. It consisted of Absolute Algebra as general theory of connecting operations and the Logic of Relatives as general logical theory. The Logic of Relatives provided the notational system, i.e., the formal language which

38 Schröder's letter to Felix Klein, dated Karlsruhe, 16 March 1896, Klein papers, *Staats- und Universitätsbibliothek Göttingen*, Cod. Ms. F. Klein 11. Schröder's correspondences with Felix Klein, editor of the *Mathematische Annalen*, and Paul Carus, editor of *The Monist*, are published in Volker Peckhaus, "Ernst Schröder und die 'pasigraphischen Systeme' von Peano und Peirce", in: *Modern Logic* 1, 1990/91, pp. 174–205.

39 English version Ernst Schröder, "On Pasigraphy. Its Present State and the Pasigraphic Movement in Italy", in: *The Monist* 9, 1, 1899, published 1898, pp. 44–62; Corrigenda, p. 320.

40 Ernst Schröder, "Über Pasigraphie, ihren gegenwärtigen Stand und die pasigraphische Bewegung in Italien", in: Ferdinand Rudio (Ed.), *Verhandlungen des Ersten Internationalen Mathematiker-Kongresses in Zürich vom 9. bis 11. August 1897*. Leipzig: Teubner 1898, pp. 147–162, quote p. 159.

could be applied to various fields given a suitable interpretation of its schematic letters and relative operations.

30.5 MODALITIES

Schröder presented his logic as an interpreted algebraic structure which could be applied (by further interpretation) to mathematics and other domains. Proponents of new systems of logic usually claim to exceed the power of older systems. This implies that the new systems are able to handle traditional problems at least as well as the older systems do. It is therefore not astonishing that Schröder's *Vorlesungen* contain a paragraph "On the Modality of Judgements",[41] where Schröder prepared an application of his calculus to modalities. These considerations show some ideas concerning epistemological links between structure and interpretation.

His starting point is Kant's table of judgements. Under the heading "Modality" Kant distinguishes the following kinds of judgements (in Schröder's presentation)[42]:

- *Apodictic judgement*: "A has to be B", or "A is necessarily B";
- *Assertoric judgement*: "A is B", or "A is really, by chance, B";
- *Problematic judgement*: "A can be B, or "A is probably (perhaps) B".

Schröder claims that in a judgement of the type "A is necessarily/really/ probably B" the adverbs do not belong to the predicate. They also do not inform us about the subject. They inform us about the state of our knowledge concerning the judgment "A is B". The difference between apodictic and assertoric judgments is a difference of cognitive psychology. The apodictic judgment contains an indication of evidence, that cannot be found in assertoric judgments.[43]

Assertoric judgments of traditional logic are in essence always apodictic or problematic. "Logic knows only one absolute certainty – the 1 of the calculus of probabilities."[44] In respect to problematic judgements a new (very important) task arises: to determine the degree of credibility of problematic judgments in the case where degrees of credibility of premises are given in form of mathematical probabilities. This is, according to Schröder, a task of exact (deductive) logic.[45]

Schröder announces to deal with the problem in Vol. 3 of his *Lehrbuch der Arithmetik und Algebra* at the latest.[46] Of this textbook, only Vol. 1 has ever been published.

---

41 Ernst Schröder, *Vorlesungen*, *op. cit.*, Vol. II, § 56, pp. 506–511.

42 *Ibid.*, p. 506.

43 *Ibid.*, p. 508.

44 *Ibid.*, p. 510.

45 *Ibid.*

46 *Ibid.*, p. 511.

30.6 Conclusion

It should have become clear that Schröder's Absolute Algebra and his Algebra of Logic anticipated basic ideas of Model Theory, according to Hodges' determination of Model Theory "in a broader sense" given in the first section as "the study of the interpretation of any language." Set theoretic studies played, however, no significant role in Schröder's programme. Cantor developed his Set Theory at the same time when Schröder introduced his Formal Algebra. In the early time Cantor and Schröder ignored each other, 20 years later Schröder regarded his Algebra of Relatives as the better alternative to Set Theory.

Contrary to the opinion of van Heijenoort and Hintikka, Schröder aimed at a scientific universal language consisting of an hierarchy of sub-languages being interpretations of one another and forming what Schröder called "Absolute Algebra". All sub-languages can be traced back to a formal algebraic structure and its first interpretation as a logic, with logical connecting operations, concepts, judgments (propositions), inferences and relatives as objects. Mathematics is gained only in the next step by interpreting logical objects as mathematical objects. This method is certainly against Fregean standards of the unambiguousness of the meaning of symbols, but it opened the way to Model Theory.

Department of Human Sciences
University of Paderborn
Warburger Str. 100
33098, Paderborn
Germany
volker.peckhaus@upb.de

CHAPTER 31

GRAHAM STEVENS

INCOMPLETE SYMBOLS AND THE THEORY OF LOGICAL TYPES

31.1 LOGICISM AND THE THEORY OF TYPES

Central to Russell's original logicist project as set out in his 1903 *Principles of Mathematics*[1] was the view that there is only one logical type of entity. This metaphysical doctrine had its formal counterpart in what has come to be called the "doctrine of the unrestricted variable", according to which all entities are within the range of the variables of Russell's formal language. Under pressure from the paradoxes afflicting the foundations of set-theory, however, Russell is widely assumed to have abandoned this doctrine by the time of his final statement and demonstration of the logicist thesis in *Principia Mathematica* (1910–1913).[2]

The formal language PM contained in *Principia Mathematica*, is stratified with respect to two distinct (or, at least, distinguishable) hierarchies: the hierarchy of types and the hierarchy of orders. The first hierarchy when applied to a higher-order logic by itself gives us what is usually referred to as 'simple type-theory'; when combined with the second hierarchy, it is yields 'ramified type-theory'. The type hierarchy restricts the arguments a given function may have. Understood in terms of the semantics traditionally attributed to PM (though I will challenge this tradition in what follows), *individuals* are of the lowest type (type 0). Only *functions* of type 1 may take individuals as arguments. Only functions of type 2 can take arguments of type 1. In general, a function of type n can only be argument to a function of type $n + 1$. This hierarchy, incidentally, does not have to be stated in semantic terms (i.e. in terms which take type restrictions to apply to a domain of *values* of variables), we could just as easily phrase the theory by attaching type indices to symbols of PM and, indeed, this would be the usual route, as it allows us to incorporate type-distinctions in setting out formation rules for the well-formed formulas of the system. The order part of the hierarchy restricts quantification. Assuming, inaccurately but for the sake of convenience, that the formulas of PM express propositions, every proposition and propositional function has an order,

1 Bertrand Russell, *The Principles of Mathematics*, Cambridge: Cambridge University Press, 1903.

2 Alfred North Whitehead and Bertrand Russell, *Principia Mathematica*, Cambridge: Cambridge University Press, 1910–1913 (3. Vols).

D. Dieks et al. (eds.), *Probabilities, Laws, and Structures*, The Philosophy of Science in a European Perspective 3, DOI 10.1007/978-94-007-3030-4_31,

such that no proposition can be of equal or lower order than the highest order proposition or propositional function quantified over in that proposition.

As mentioned above, it is well known that logicism ran into problems by assuming that there was one single domain that individual variables ranged over (though Frege and Russell differ slightly on this, both originally counted classes in the same domain as all other individuals without restriction, which led to the problem). Michael Dummett sums up what has become the standard view in reference to Frege, though it could equally have been written about Russell:

> We neither need nor can follow Frege in supposing that one single all-embracing domain will serve for all uses of individual variables: for the most direct lesson of the set-theoretic paradoxes is that, at least when we are concerned with abstract objects, there is no one domain which includes as a subset every domain over which we can legitimately quantify: we cannot give a coherent interpretation of a language, such that every sentence of the language can be taken as having a determinate truth-value, by taking the individual variables to range over everything that answers to the intuitive notion of a set, or that of a cardinal number or that of an ordinal. We must therefore separate Frege's basic intuition, the use of quantification understood as relative to a determinate domain as a fundamental tool in the analysis of language, from his incorrect further assumption, that this domain, by being stipulated to be all-inclusive, can be taken to be the same in all contexts.[3]

This final assumption is one shared by Russell in 1903s *Principles of Mathematics*. Do the paradoxes really show that it is incorrect?

The standard view, of course, is that Russell came to see that unrestricted variation was impossible. Accordingly, a hierarchy of types is found in PM. This hierarchy ensures that the range of a variable is adequately restricted so as to outlaw, for instance, self-(or non-self-) predicating predicates. Were classes to be ultimately present (rather than, as they are, ultimately eliminable by contextual definition in terms of propositional functions), the theory of types would also ensure that classes cannot be members (or non-members) of themselves either. Thus we have no Russell class $\{x : x \notin x\}$, nor a 'Russell function' φ of non-self-predicability leading to the contradictory $\varphi\,\varphi \equiv {\sim}\varphi\varphi$.

Ironically, however, this retraction of unrestricted variation is also the main target of complaints against Russell's version of logicism. There is something decidedly *ad hoc* about the claim that something is neither a member nor a non-member of a class. The class complement of $\{x : Fx\}$ ought to include everything lacking the property F. But, in the case of a class r, the fact that r cannot be a member of r does not imply that r is not a member of r, for this latter statement is just a further violation of type restrictions. Indeed, even more simply, it just seems implausible to think that a class cannot share the same property as its members – the class of all abstract objects, for example, is surely an abstract object.

This description of type-theory and its failings is, admittedly, a little simplistic, but it demonstrates quite clearly the general complaint: types lack a philosophical justification. We infer that there are logical types not from any philosophical analysis

3 Michael Dummett, *Frege: Philosophy of Language*, London: Duckworth 1973, p. 476.

of logical objects, but simply because the acceptance of types gives us a means of avoiding contradiction. The conclusion usually drawn is that logicism failed.

31.2 Incomplete Symbols as the Source of Logical Types

The story just sketched presupposes that type-theory and unrestricted variation are irreconcilable. This is unsurprising: one is tempted to say that they are mutually exclusive doctrines *by definition*. For the supposition of a theory of types, surely, is the supposition that variables are restricted in range to a specific logical type. Surprisingly, however, this presupposition is wrong. To see how these two apparently opposed doctrines can be reconciled, one has to understand the relation between the theories of descriptions and types. This relation was obscured for a long time but has recently been revealed by close studies of Russell's "substitutional theory of classes and relations" developed between 1905 and 1907.

The substitutional theory avoids all mention of classes as entities, replacing talk of classes with reference to *matrices* expressed by symbols of the form 'p/a'. Here both 'p' and 'a' are understood to be names of entities. A matrix, however, is *not* an entity. The nature of a matrix is best explained by the fact that a matrix features in propositions such as that expressed by '$p/a;x!q$' which can be read as 'q results from the substitution of x for a in all those places if any where a occurs in p'. Admitting, as Russell did at this time, that propositions are entities, either p or a may be propositions (though neither has to be). This allows Russell to treat a matrix as if it were a class and thus ensure that all talk of classes is eliminable in favour of talk of matrices. The condition under which an entity x is a member of the "class" p/a is just that there is a true proposition resulting from the substitution of x for a in p.

The connection forged by the substitutional theory between the theories of descriptions and types is evident when we reflect on the form that the Russell paradox would need to take in the substitutional calculus. About the closest we could get would be to write something like '$p/a;p/a$'. But this is mere nonsense, amounting to something like 'the result of replacing a in p by the result of replacing a in p by'. Self-membership becomes impossible according to the substitutional analysis of classes. As Russell says: 'now "x is an x" becomes meaningless, because "x is an a" requires that a should be of the form p/a, and thus not an entity at all. In this way membership of a class can be defined, and at the same time the contradiction is avoided'.[4]

The grammar of substitution yields the kinds of distinctions placed on a standard theory of classes by simple type-theory and, indeed, it is best understood as the original foundation for Russell's mature theory of types. The complaint commonly directed at Russellian type-theory considered above, namely that the theory outlaws certain apparent propositions as nonsense on purely *ad hoc* grounds, is obviously unfair once one sees that the theory of types has its origins in the substitutional theory. From the perspective of the logic of substitution, violations of

4 Bertrand Russell, 'On the Substitutional Theory of Classes and Relations', in: *Essays in Analysis*, ed. D. Lackey, London: Allen and Unwin 1973, p. 172.

type-distinctions are clearly nonsense in a quite unquestionable way. Furthermore, it is because "classes" are incomplete symbols that violations of type-distinctions are violations of sense; thus the substitutional theory provides the link between the theory of descriptions and the theory of types.

The substitutional theory provides a justification for type theory without departing from the doctrine of the unrestricted variable. In Russell's words, 'it adheres with drastic pedantry to the old maxim that, "whatever is, is one"'.[5] Such pedantry is not without costs, however. Types have been brought into consistency with unrestricted variation only at the expense of casting the *things typed* out of Russell's ontology. Classes are no longer mind-independent subsisting things, according to this analysis. Numbers, being classes of similar classes, are similarly bereft of ontological status, as is evident from Russell's own statements:

> The theory which I wish to advocate is that classes, relations, numbers, and indeed almost all the things that mathematics deals with, are 'false abstractions', in the sense in which 'the present King of England', or 'the present King of France' is a false abstraction. Thus e.g. the question 'what is the number *one*?' will have no answer; the question which has an answer is 'what is the meaning of a statement in which the word *one* occurs?' And even this question only has an answer when the word occurs in a proper context.'[6]

This may look like a radical departure from logicism but, I believe, it should be viewed as neither harmful to logicism nor surprising. It should not be surprising because it is a fairly predictable consequence of leaning heavily on the theory of descriptions. The theory invariably achieves great things by slimming ontologies to the point of emaciation. This has always been taken for granted by defenders and critics of the theory. Furthermore, the ontological costs here have little philosophical significance. Russell simply calls on the theory of descriptions (as applied to the substitutional theory) in order to put the metaphysical weight of logicism onto an ontology of propositions rather than classes.[7] Bearing in mind the centrality of the proposition to Russell's early metaphysics, this has no negative consequences (beyond the inevitable complications resulting from the rejection of classes) for Russellian logicism.

As I have said, the ontological cutbacks that tend to follow any use of the theory of descriptions should come as no surprise. What may come as more of a surprise about this particular application of the theory, however, is that it shows how the theory can give as well as take. The treatment of classes as incomplete symbols robs mathematical objects of reality but, simultaneously, provides the philosophical justification for dividing those (now fictional) objects into a hierarchy

---

5 *Ibid.*, p. 189.

6 *Ibid.*, p. 166.

7 For a full explanation of how this is carried out, see Graham Stevens, 'Antirealism and the Theory of Descriptions' in: D. Jacquette and N. Griffin (Eds.), *Russell versus Meinong: 100 Years After 'On Denoting'*, London: Routledge 2009.

of types. The contradictions are avoided and we have an explanation of what led us into contradiction in the first place: our mistake was to think that classes were genuine objects. Once we see that they are in fact incomplete symbols, all we need to do is to treat them correctly (that is, in accordance with theory of descriptions) in order to avoid (and solve) the contradictions.

It is important to be perfectly clear about what the solution to the contradictions being offered here amounts to. Although the solution does entail the restriction over quantification, and over the range of variation, urged in the passage quoted from Dummett above, it shouldn't be conflated with it. It is *because* classes are incomplete symbols that unrestricted quantification or variation over them leads us into contradiction. Recognition of this point is crucial to understanding the role played by the so-called 'Vicious-Circle Principle' (VCP) in Russell's mathematical philosophy. The VCP states, in one form, that no proposition may be a value of a bound variable contained within itself. Hence, e.g., Epimenides' proposition

(1) All propositions asserted by Cretans are false,

cannot be a value of the quantifier contained in (1). In this case, the restriction placed on quantification by the VCP serves to generate the hierarchy of orders in ramified type-theory. If the VCP is understood just as a prescription placed on quantifiers in order to evade contradictions, however, it will be prone to just the same objections that were first levelled at the type-hierarchy at the start of this paper. Like the simple part of type-theory, however, the VCP is intended to be a consequence of a correct philosophical diagnosis and solution of the contradictions. In the case of the VCP, the philosophical diagnosis is also one that isolates the cause of the problem in a false ontological assumption – namely the assumption that propositions are entities. Russell abandoned propositions in favour of the multiple-relation theory of judgement in PM. This decision turned out to be a bad one – the multiple-relation theory was a failure. That is not our concern here, however, as I don't want to be sidetracked into talking about the ramified theory or the multiple-relation theory.[8] Returning to the hierarchy of types, we have already seen how this is intended to be a consequence of the correct (substitutional) analysis of mathematical objects. The theory of types is just the consequence for mathematical logic of properly digesting the content of the claim that classes are not entities.

To summarise so far: we have answered two serious objections to type-theory. Firstly, we have shown that the theory is not *ad hoc* – it can be provided with solid philosophical foundations. Secondly, it does not need to contradict the doctrine of unrestricted variation. Types sit quite comfortably alongside unrestricted variation in the substitutional theory because the theory of types is a logical theory, not an ontological one.

8 See Graham Stevens, *The Russellian Origins of Analytical Philosophy*, London: Routledge 2005 for details on each.

At this point, however, there is a concern that needs addressing. Substitution, quite clearly, doesn't feature in PM. Indeed, it cannot. Substitution only works if an ontology of propositions is assumed. With propositions abandoned in favour of the multiple-relation theory in PM, substitution cannot provide type-theory with its justification. The concern that needs addressing, then, is over the ontological status of type-theory in PM. Without substitution to provide foundations for type-theory, does the type hierarchy just collapse into an ontological theory in PM?

31.3 *Principia Mathematica*

As is well known, PM's symbols for classes are contextually eliminable in a way very similar to the method for contextually eliminating denoting phrases. Just as the denoting phrase '$ix(Fx)$' is to be eliminated in contexts such as '$G(ix(Fx))$' as follows:

$$G(ix(Fx)) =_{\mathrm{Df}} \exists x((Fx \,\&\, \forall y(Fy \supset x{=}y)) \,\&\, Gx)$$

So the class symbol '$\{x\text{: } Fx\}$' can be eliminated from contexts such as '$G\{x\text{: } Fx\}$' as follows[9]:

$$G\{x\text{: } Fx\} =_{\mathrm{Df}} \exists H((\forall x(H!x \equiv Fx)) \,\&\, G(H!\hat{u})).$$

As is equally well known, this contextual definition is more problematic than those given for definite descriptions in section *14 of *Principia*. Most concerning is Russell's quantification over predicate variables. This led Quine to dismiss the 'no-classes' theory of types in PM as little more than a use-mention confusion on Russell's part: Russell thought he had reduced classes to symbolic conveniences; what he actually did, according to Quine, was reduce respectably extensional, though unpalatably abstract, entities to attributes in intension.

One who works to Quine's ontological criterion that reads ontological commitment off of ineliminable existential quantifications will share Quine's conclusion that Russell has bought ontological freedom from classes only at the cost of ontological commitment to whatever the values of PM's predicate variables are supposed to be. Russell and Whitehead are notoriously unclear on how PM should be interpreted. No formal interpretation is offered in the way that is now customary for any presentation of a formal language. Hence we are left to decipher the three dense chapters of philosophical introduction to try and shed some light on the mathematical logic that follows. When we do so, we quickly find that Russell and Whitehead speak as if predicate variables stand for things

9 See Whitehead and Russell, *Principia Vol. 1*, *op. cit.* *20.01. I have modernised the notation.

called 'propositional functions'. If propositional functions are understood along the same lines as they feature in the *Principles of Mathematics*, Quine seems to be right. There, propositional functions are as much a part of Russell's ontology as propositions are. Taking propositional functions to be of this sort in PM instantly turns the theory of types back into an ontological theory. This is evident from the justification offered for the type part of the ramified hierarchy of PM. With classes reduced to functions it is functions that must be divided into types. Russell and Whitehead offer the 'direct inspection' argument as justification for the imposition of this hierarchy: direct inspection of the nature of a propositional function is supposed to reveal, they confidently assert, that:

> not only is it impossible for a function $\varphi\hat{u}$ to have itself or anything derived from it as argument, but that, if $\psi\hat{u}$ is another function such that there are arguments a with which both 'φa' and 'φa' are significant, then $\psi\hat{u}$ and anything derived from it cannot significantly be argument to $\varphi\hat{u}$.[10]

If propositional functions are just universals under a different name, then what we have here is a claim that we can recognise through our acquaintance with those universals that they are essentially incomplete entities akin to Fregean concepts and, as such, must be typed in order to explain how they can come together to form unified complexes.

There are very good reasons for rejecting this realist (as I will henceforth call it) interpretation of propositional functions in PM. For one thing, it is in the unfortunate position of being openly contradicted by one of the authors of the work it is intended to interpret. Russell, in *My Philosophical Development*, explicitly states that propositional functions are linguistic entities only: 'A propositional function is nothing but an expression. It does not, by itself, represent anything. But it can form part of a sentence which does say something, true or false'.[11] Secondly, and perhaps more importantly, *Principia* lacks the ontological resources to support a realist interpretation. We have already seen that propositions are rejected in PM in line with the adoption of the multiple-relation theory of judgement. Bearing in mind that the multiple-relation theory is intended to generate the hierarchy of orders and thus ultimately, play a central role in guaranteeing the consistency of PM, this rejection of propositions is one that must be taken seriously when seeking the correct interpretation for PM. But this now poses a surely insurmountable problem for the realist interpretation, as we can see if we remind ourselves of how Russell and Whitehead describe functions in *Principia*:

---

10 Whitehead and Russell, *Principia Mathematica Vol. 1*, Cambridge: Cambridge University Press, 1910, 2nd edition 1926, p. 47.

11 Bertrand Russell, *My Philosophical Development*, London: Allen and Unwin 1959, p. 69.

Let φx be a statement containing a variable x and such that it becomes a proposition when x is given any fixed determined meaning. Then φx is called a 'propositional function'; it is not a proposition, since owing to the ambiguity of x it really makes no assertion at all.[12]

By a 'propositional function' we mean something which contains a variable x, and expresses a proposition as soon as a value is assigned to x. That is to say, it differs from a proposition solely by the fact that it is ambiguous: it contains a variable of which the value is unassigned.[13]

The problem with this talk of propositional functions 'becoming' or 'expressing' a proposition upon the assignment of a value to x is that this means, according to the official ontology of *Principia*, there is nothing for a propositional function *to* become under such circumstances. Propositions just don't exist.

It is perhaps unsurprising then that determined adherents of the realist interpretation have resorted to simply ignoring what Russell and Whitehead have to say about their ontological commitments. So, for example, Church thinks the only way to make sense of PM is to abandon any attempt to make the interpretation comport with that suggested by Russell and Whitehead:

[W]e take propositions as values of the propositional variables, on the ground that this is what is clearly demanded by the background and purpose of Russell's logic, and in spite of what seems to be an explicit denial by Whitehead and Russell.[14]

The usual excuse for such wilful disregard of what is actually said in *Principia*, is that Russell and Whitehead happily quantify over propositions and propositional functions. Thus, the claim goes, the philosophical parts of the work are out of tune with the formal parts, the latter being best interpreted as concerned with a domain of individuals, functions, and propositions, interpreted in realist terms.

As excuses go, this one is particularly poor. The only valid route to this conclusion relies on the assumption that quantification in PM is objectual. But this is clearly question-begging, for (at least in this context) to interpret the quantifiers objectually just *is* to give a realist interpretation of them. A better alternative (if one wishes to reconcile the formal parts of *Principia* with its philosophical preamble) is to interpret the quantifiers substitutionally. A few attempts have been made to interpret PM in this way (e.g. Gödel 1944; Sainsbury 1979; Landini 1996, 1998).[15] Difficulties quickly arise, however. An early casualty of an over-zealous

12 Whitehead and Russell, *Principia Vol. 1*, *op. cit.* p. 14.

13 *Ibid.*, p. 38.

14 Alonzo Church, 'A Comparison of Russell's Resolution of the Semantic Antinomies with that of Tarski's', *Journal of Symbolic Logic* 41 1978, p. 748, ft. nt. 4.

15 E.g. Kurt Gödel, 'Russell's Mathematical Logic', in: P. A. Schilpp (Ed.), *The Philosophy of Bertrand Russell*, London: Harper and Row; Mark Sainsbury, *Russell*, London: Routledge 1979; Gregory Landini, *Russell's Hidden Substitutional Theory*, Oxford: Oxford University Press 1998.

nominalist interpretation of quantification in PM will be the axiom of infinity. Unless this axiom is taken as applying to *objects* (not their *names*) it stands no chance of being true. However, holding that PM is intended to have a nominalistic semantics is not the same as holding that PM is intended to reflect any commitment to nominalism as a philosophical doctrine. Indeed, this would be absurd. Russell never retracted his ontological commitment to universals. Furthermore, that same ontology is suggested in *Principia* by passages like the following: 'the universe consists of objects having various qualities and standing in various relations'.[16] Other comments made by Russell around the same time make it perfectly clear that the ontology gestured at here is very distant from nominalism: 'A complete description of the existing world would require not only a catalogue of the things, but also a mention of all their qualities and relations'.[17] Evidently, there is no mileage in an interpretation that denies Russell's ontological commitment to universals. But this is not an insurmountable problem for the interpreter of PM. Our original objection to Quine, Church, etc., was that they deviated from the officially stated doctrines of *Principia*. There is no need to commit the same exegetical error by foisting a commitment to nominalism in general on to Russell. The point was, rather, that *propositional functions* should be interpreted nominalistically in order to maintain consistency with Russell and Whitehead's rejection of propositions as entities. An ontological commitment to universals is quite consistent with a rejection of any ontological commitment to propositional functions so long as we resist the temptation to *identify* functions with universals. But this is just what is demanded by a nominalist interpretation of propositional functions anyway – propositonal functions can hardly be universals if we are going to withhold ontological commitment to them.

As only quantification over predicate variables involves quantification over propositional functions, it is only the higher-order quantifiers that demand a nominalist treatment if we are to provide a nominalistic semantics for propositional functions. An interpretation of PM like that offered by Landini captures this idea. Two sorts of quantifier are utilised in the system. The first-order quantifiers are interpreted objectually and are unrestricted. Higher-order quantifiers, however, are interpreted substitutionally. Thus a nominalistic interpretation of predicate variables (propositional functions) is allowed to coexist with a realistic interpretation of e.g. the axiom of infinity.

With higher-order quantification interpreted substitutionally, the values of predicate variables will simply be formulas. A propositional function is nothing more than an open sentence. It is somewhat surprising that this interpretation is at odds with the accepted wisdom on PM, when one considers just how well it squares with what Whitehead and Russell have to say:

16 Whitehead and Russell, *Principia Vol. 1*, *op. cit.* p. 43

17 Bertrand Russell, *Our Knowledge of the External World*, London: Allen and Unwin 1914, p. 60.

Let φx be a statement containing a variable x and such that it becomes a proposition when x is given any fixed determinate meaning. Then φx is called a 'propositional function'; it is not a proposition, since owing to the ambiguity of x it really makes no assertion at all.[18]

The most natural interpretation of this passage, I submit, is the one that just takes 'statement' to be synonymous with 'sentence' and holds variables to be symbols contained in sentences. Consequently, a propositional function is just an open sentence. This interpretation is in tune with the official ontological doctrines of *Principia* and, as we have seen, it also allows us to view the formal aspects of the work as subservient to (and consistent with) the same philosophical ambitions.

I mentioned earlier that I would be mainly interested in the type part of the ramified hierarchy in this paper. But we do need to address an issue that touches on the issue of order. In the absence of propositions, Whitehead and Russell are in need of new truthbearers. These are provided by the multiple-relation theory of judgement (MRTJ). At the heart of the MRTJ is a recursive definition of truth (construed as correspondence between judgements and facts). This definition of truth is intended to provide the philosophical foundations for the order part of the ramified hierarchy. Russell and Whitehead set it out as follows[19]:

That the words "true" and "false" have many different meanings, according to the kind of proposition to which they are applied, is not difficult to see. Let us take any function $f\hat{u}$ and let fa be one of its values. Let us call the sort of truth which is applicable to fa "*first* truth." ... Consider now the proposition $(x).\varphi x$. If this has truth of the sort appropriate to it, that will mean that every value φx has "first truth." Thus if we call the sort of truth which is appropriate to $(x).\varphi x$ "*second* truth," we may define "$\{(x).\varphi x\}$ has second truth" as meaning "every value for $\varphi\hat{u}$ has first truth," *i.e.* "$(x).(\varphi x$ has first truth)." Similarly, if we denote by "$(\exists x).\varphi x$" the proposition "φx sometimes," *i.e.* as we may less accurately express it, "φx with some value of x," we find that $(\exists x).\varphi x$ has second truth if there is an x with which φx has first truth; thus we may define "$\{(\exists x).\varphi x\}$ has second truth" as meaning "some value for $f\hat{u}$ has first truth," i.e. "$(\exists x).(\varphi x$ has first truth)." Similar remarks apply to falsehood.[20]

The MRTJ, like the substitutional theory, is construed by its authors as being, at least in *some* sense, another application of the theory of descriptions. This is clear enough from what Whitehead and Russell say when introducing propositions as 'incomplete symbols' in *Principia*:

Thus "the proposition 'Socrates is human'" uses "Socrates is human" in a way which requires a supplement of some kind before it acquires a complete meaning; but when I

18 Whitehead and Russell, *Principia Vol. 1, op. cit.* p. 14.

19 The circumflex device in the formula '$\phi\hat{u}$ 'is a term-forming operator converting an open sentence into a term designating a propositional function. It is perhaps best understood as a forerunner of the modern lambda abstract '$\lambda x(\phi x)$'.

20 *Ibid.*, p. 42.

judge "Socrates is human," the meaning is completed by the act of judging, and we no longer have an incomplete symbol.[21]

We have to be careful how we read this passage. Although it marks a clear application of the doctrine of incomplete symbols, it does not amount to a claim that propositions are disguised definite descriptions (in the way that Russell thought e.g. names are). Definite descriptions are incomplete symbols whose meaning, while lacking in isolation from a wider sentential context, can be contextually defined by familiar means. The symbols for classes, similarly, can be contextually defined by e.g. PM *20.01. No such claim is being made about propositions, however. Rather, the context we are interested in here is the act of judgement. The context is not a wider sentential context; it is the context of a mental act.

Calling a proposition an 'incomplete symbol' cannot be understood as directly parallel to talk of denoting phrases, or class-expressions, as incomplete symbols. In the case of descriptions and class-expressions we have explicit contextual definitions which license the introduction and elimination of the expressions without importing them into the fundamental vocabulary of our language. Thus descriptions and class abstracts are incomplete symbols in a purely syntactic sense – they are defined expressions which can be eliminated at any point. The account of propositions as incomplete symbols, however, is a semantic doctrine – a claim about the proper interpretation of certain elements of the lexicon. There is no method for eliminating these symbols akin to those given in *14 and *20 for descriptions and class abstracts respectively. Russell's talk of propositions as incomplete symbols is somewhat metaphorical as a result. This notion of an incomplete symbol is a wider, and less precise, one than is licensed by the use of explicit contextual definitions, but we can nonetheless appreciate what Russell has in mind – just as descriptions and class abstracts appear to refer to things but turn out, on analysis, not to; so also sentences, that-clauses, propositional variables, etc., seem to refer to entities (propositions) but turn out on analysis no to do so.

In each case that we have examined, we have seen how the treatment of an expression as an incomplete or contextually eliminable symbol enabled Russell to add to his philosophy while subtracting from his ontology. Classes, propositional functions, and propositions, were eliminated by a process that simultaneously generated and justified a hierarchy of logical types and orders. The elimination of definite descriptions was the model applied and re-applied in each case. In this way, type-theory is provided with a justification and simultaneously answers the objections considered at the beginning of this paper. The main objections to type-theory are objections to (1) the idea that objects should be typed and (2) the apparently ad hoc nature of the theory. But, understood as consequence of applying the theory of descriptions to the symbols for classes, functions, and propositions, the theory of types is immune to both criticisms, for types are generated by the

21 *Ibid.*, p. 46.

elimination of the objects in question. Nor is there anything ad hoc about this process. This does not demonstrate that Russell's logicist thesis succeeded, but it does answer the objection that it relied on an unjustified theory of logical types.

31.4 Conclusion

As far as *Principia* goes, the project is not, ultimately, successful. The success of the project as I have outlined it here is dependent on the success of the MRTJ. The MRTJ, however, was a logical fiction too far for Russell. As I have argued in detail elsewhere,[22] Wittgenstein's criticisms showed that the theory couldn't do the work Russell needed it to do. It is notable that the project of logical construction broke down for Russell at this particular point, for the MRTJ differs in interesting ways, as we have seen, from the other eliminativist applications of the theory of descriptions. Firstly, it seeks to apply the eliminativist strategy at the semantic rather than syntactic level; secondly, and perhaps most importantly, it eliminates the very things that had oiled the wheels of Russell's past successes with the elimination of classes and functions – namely, propositions.

School of Social Sciences
University of Manchester
Oxford Road
M13 9PL, Manchester
United Kingdom
Graham.P.Stevens@manchester.ac.uk

22 Graham Stevens, *Russellian Origins, op. cit.*, Ch. 4.

CHAPTER 32

DONATA ROMIZI

STATISTICAL THINKING BETWEEN NATURAL AND SOCIAL SCIENCES AND THE ISSUE OF THE UNITY OF SCIENCE: FROM QUETELET TO THE VIENNA CIRCLE

32.1 INTRODUCTION

The application of statistical methods and models both in the natural and social sciences is nowadays a trivial fact which nobody would deny. Bold analogies even suggest the application of the same statistical models to fields as different as statistical mechanics and economics, among them the case of the young and controversial discipline of Econophysics.[1] Less trivial, however, is the answer to the philosophical question, which has been raised ever since the possibility of "commuting" statistical thinking and models between natural and social sciences emerged: whether such a methodological kinship would imply some kind of more profound unity of the natural and the social domain.

Starting with Adolphe Quetelet (1796–1874) and ending with the Vienna Circle (from the late 1920s until the 1940s), this paper offers a brief historical and philosophical reconstruction of some important stages in the development of statistics as "commuting" between the natural and the social sciences. This reconstruction is meant to highlight (with respect to the authors under consideration):

1. The existence of a significant correlation between the readiness to "transfer" statistical thinking from natural to social sciences and vice versa, on the one hand, and the standpoints on the issue of the unity/disunity of science, on the other;
2. The historical roots and the fortunes of the analogy between statistical models of society and statistical models of gases.

1 Cf. Bikas K. Chakrabarti, Anirban Chakraborti, Arnab Chatterjee, *Econophysics and Sociophysics. Trends and Perspectives*, Weinheim: Wiley-VCH 2006; *Science and Culture. Special issue on: Fifteen Years of Econophysics Research*, 76, 9–10, 2010.

D. Dieks et al. (eds.), *Probabilities, Laws, and Structures*, The Philosophy of Science in a European Perspective 3, DOI 10.1007/978-94-007-3030-4_32,

32.2 Adolphe Quetelet: Statistics and the Unity of Sciences

The Belgian astronomer and social statistician Adolphe Quetelet is a figure who has awoken especially in the 1980s, the interest of many historians of probability and statistics.[2]

One of Quetelet's most significant features is certainly his interdisciplinary outlook. Being the founder and director of the Royal Astronomical Observatory in Brussels and pursuing at the same time a brilliant career as a social statistician, he found himself at the intersection of different research areas which were developing at the same time, and to whose convergence Quetelet himself very much contributed: the classical probability calculus, and especially the newly developed "law of large numbers", the theory of observational errors in astronomy and social statistics.

A unitary conception of natural and social phenomena characterizes Quetelet's perspective and his "transversal" application of statistics. Significantly enough, his most famous work was entitled *Essai de physique sociale*.[3] In the "Preface" to the first English edition Quetelet explains: "In giving to my work the title of Social Physics, I have had no other aim than to collect, in a uniform order, the phenomena affecting man, nearly as physical science brings together the phenomena appertaining to the material world."[4]

Each of the following constitutive elements of Quetelet's *physique sociale* fills the gap between social and natural sciences and is, at the same time, intrinsically related to the application of probability and statistics.

32.2.1 Observation and Quantification of Facts

Quetelet's Social Physics starts with the observation of facts, the facts – Quetelet writes – that "society presents to our view".[5] This is the first, essential step towards talking about human beings scientifically, avoiding any speculative "Theory of Man"[6]. Not only "physical qualities" (births, deaths, stature, weight, strength, etc.), but also

---

2 For instance, Quetelet's work is a major topic of many contributions to the volume: Lorenz Krüger, Lorraine J. Daston, Michael Heidelberger (Eds.), *The Probabilistic Revolution, Vol. I: Ideas in History*. Cambridge, Mass.: MIT Press 1987. Cf. also Theodore M. Porter, *The Rise of Statistical Thinking 1820–1900*, Princeton: Princeton University Press 1986 (Part II); Stephen M. Stigler, *The History of Statistics*, Cambridge, Mass./London: Belknap Press of Harvard University Press 1986 (Part II, Ch. 5); Gerd Gigerenzer et al. (Eds.), *The Empire of Chance*, Cambridge: Cambridge University Press 1989 (Ch. 2); Ian Hacking, *The Taming of Chance*, Cambridge: Cambridge University Press 1990 (Chs. 13–15 and 20–21).

3 The first version of Quetelet's Social Physics was published in 1835 with the title *Sur l'homme et le développement de ses facultés ou Essai de physique sociale*. In this paper I refer to the first English translation of this edition: Adolphe Quetelet, *A Treatise on Man and the Development of his Faculties*, Edinburgh: William and Robert Chambers 1842. In 1869, Quetelet would publish a revised and enlarged edition of this work under the title *Physique sociale, ou Essai sur le développement des facultés de l'homme*.

4 Quetelet, *A Treatise on Man and the Development of his Faculties, op. cit.*, p. vii.

5 *Ibid.*, p. vii.

6 *Ibid.*, p. 8.

"moral" (dispositions to good or evil) and "intellectual" (intellectual power) qualities are conceived of as facts to be observed – if not "directly", then through their effects:

> The analysis of the moral man through his actions and of the intellectual man through his production [...] form[s] one of the most interesting parts of the sciences of observation, applied to anthropology. *It may be seen, in my work, that the course which I have adopted is that followed by the natural philosopher.*[7]

The "qualities of man" are expressed by facts, and these facts are illustrated by statistics: births, deaths, diseases, suicides, crimes, prostitution, production of works of literature, philosophy, science, etc. Statistics allows us to *measure, to quantify* the qualities of men and society exactly as we would measure the properties of a physical object.

32.2.2 The Law of Large Numbers and Other "Laws"

Once we have collected enough data, "a miracle occurs": out of the large numbers regularities emerge. According to Quetelet, for example, the number of murders committed in France every year – but also the percentage of these murders committed, say, by strangulation – converges toward a mean; furthermore, this mean remains stable in the course of the years, provided that the "organization of the social state"[8] remains the same. Also the physical traits of man, if they are measured sufficiently many times for a particular population, would converge toward a mean (so that we can speak, for example, of a French *homme moyen*): "It would appear, then, that moral phenomena, when observed on a great scale, are found to resemble physical phenomena".[9]

Relying on the regularities emerging out of the large numbers Quetelet can further look for statistical *correlations*, for example, between the "residence in town or country" and the "ratio of births of the two sexes", or between "the period of the maximum of conceptions" and "that of the greatest numbers of rapes".[10] Quetelet's aim is the same as the natural scientist's, to wit, "to discover the laws forming the connecting links of phenomena"[11]:

> Having...observed the progress made by astronomical science in regards to worlds, why should not we endeavour to follow the same course in respect to man? Would it not be an absurdity to suppose, that, whilst all is regulated by such admirable laws, man's existence alone should be capricious [...]?[12]

---

7 *Ibid.*, p. viii; my emphasis.

8 *Ibid.*, p. 6.

9 *Ibid.*, p. 6.

10 Cf. *Ibid.*, p. 12 and p. 22 respectively. The revised and enlarged edition of Quetelet's social physics (see above, n. 3.) entails a much greater variety of such correlations.

11 *Ibid.*, p. 8.

12 *Ibid.*, p. 9. Quetelet's "normal" distributions and "stable" means, as well as Quetelet's "laws", appear quite problematic to a modern eye: with respect to the former, cf. for instance Ian Hacking, *The Taming of Chance, op. cit.*, S. 113; with respect to the latter, cf. Bernard-Pierre Lécuyer, "Probability in Vital and Social Statistics: Quetelet, Farr, and the Bertillons", in: Krüger, Daston, Heidelberger (Eds.), *The Probabilistic Revolution, Vol. I: Ideas in History, op. cit.*, pp. 317–335 (see p. 321).

32.2.3 Causes

Quetelet also infers the existence of causes from statistical regularities. The model of causation found in Quetelet's work follows exactly what Lorenz Krüger – referring in general to classical probability in the age of determinism – has called "the deterministic account of statistical regularities". This account "was built on two complementary ideas: (i) the causal efficacy of structural conditions [...] and (ii) the mutual compensation of accidental causes".[13] Correspondingly, we find in Quetelet, on the one hand, the idea of a constant causal influence, for example, by "a given state of society"[14] or by a Nature's tendency to realize the "typical (e.g. French) man". On the other hand, Quetelet talks of "accidental causes" – for example, the free decisions or the accidental properties of single individuals – that compensate each other and happen to be normally distributed exactly like errors in a repeated measurement. This mutual compensation is the effect of what Quetelet calls the "law of accidental causes": "Variations, which arise from accidental causes, are regulated with such harmony and precision that we can classify them *in advance* numerically and by order of magnitude, within their limits".[15]

32.2.4 Predicting

Social Physics is not only a description of facts. It can tell us "in advance" something about future facts: like the natural sciences, it allows prediction; like the probability calculus, it suggests a rational degree of expectation. According to Quetelet, "we might even predict annually how many individuals will stain their hands with the blood of their fellow-men, how many will be forgers, how many will deal in poison, pretty nearly in the same way as we may foretell the annual births and deaths".[16]

It should be clear by now that in Quetelet's thought the application of statistical models within his social physics a unitary conception of science was intimately interwoven with: in particular, the application of statistics to the social domain let its kinship with the natural one emerge. But what kind of unity of science was advocated by Quetelet? Quetelet refused Comte's idea of a hierarchy of sciences and, of course, did not share Comte's dislike for the use of mathematics in social sciences. In fact, Quetelet was committed to the *methodological* unity of science.[17]

---

13 Lorenz Krüger, "The Slow Rise of Probablism: Philosophical Arguments in the Nineteenth Century", in: Krüger, Daston, Heidelberger (Eds.), *The Probabilistic Revolution, Vol. I: Ideas in History, op. cit.*, pp. 59–85 (see p. 71).

14 Quetelet, *A Treatise on Man and the Development of his Faculties, op. cit.*, p. vii.

15 Quetelet, cit. in Lécuyer, "Probability in Vital and Social Statistics: Quetelet, Farr, and the Bertillons", *op. cit.*, p. 320; my emphasis.

16 Quetelet, *A Treatise on Man and the Development of his Faculties, op. cit.*, p. 6.

17 Cf. Porter, *The Rise of Statistical Thinking* 1820–1900, *op. cit.*, p. 41–42: "Quetelet maintained that a single method was appropriate for every science". Porter deals here also with the relationship between Quetelet and Comte.

Furthermore, his talking of a *physique sociale* and even of a *mecanique sociale*, as well as his insistence on the law-like character of social phenomena (which thus resemble natural ones), both suggest that Quetelet also tended to support a *nomological* and an *ontological* unity of science.[18] Such a unitary conception of science and Quetelet's "transversal" use of statistics went hand in hand and supported each other.

32.3 Reactions to Quetelet's work in the Nineteenth Century

The philosophical issue about the unity/disunity of science played a big role also in the context of the reception of Quetelet's work.

Notwithstanding Quetelet's international reputation, his ideas about statistical laws and the unity of the sciences were not welcomed with enthusiasm in the German-speaking world. There the will to divorce *Naturwissenschaften* from *Sozial-* and *Geisteswissenschaften* went hand in hand with a different conception of statistics, one that rejected the notion of statistical law and any causal talk about society.[19] In the first place, German academic statisticians and social scientists resisted the identification of statistics with numbers until the 1860s. Later, after the 1860s, statistics was conceived of by most Germans as a method for mass observation and for description, but most German statisticians, like Engel, Fallati, Casper and Rümelin, questioned Quetelet's idea of statistical regularities being laws or symptoms of true causal relations. The German tendency to emphasize the role of history and culture in defining the identity of peoples and nations jarred with any attempt to apply to a society fixed and unhistorical laws; furthermore, the will to promote state-directed reforms clashed with the idea of a society intrinsically ruled by "spontaneous" laws.

---

18 Quetelet always defended himself from the charge of denying human free will by underlying that the statistical laws of his Social Physics do not apply to single individuals (cf. for instance Quetelet, *A Treatise on Man and the Development of his Faculties*, *op. cit.*, p. 7). Nevertheless, he does not seem to have considered the hypothesis that this limitation on the validity of statistical laws could imply an in-principle difference between his Social Physics and, say, Newtonian physics.

19 The first German translation of Quetelet's *Physique Sociale* appeared already in 1838. On the reception of Quetelet's work in the German speaking world, cf. Wilhelm Winkler, "Das Problem der Willensfreiheit in der Statistik", in: *Revue de l'Institut International de Statistique/Review of the International Statistical Institute*, Vol. 5, No. 2, 1937, p. 115–131 (see esp. p. 128–130); Paul F. Lazarsfeld, "Notes on the History of Quantification in Sociology – Trends, Sources and Problems", in: *Isis*, Vol. 52, No. 2, 1961, p. 277–333 (see p. 283–294 and pp. 309–310); Theodore M. Porter, "Lawless Society: Social Science and the Reinterpretation of Statistics in Germany, 1850–1880", in Krüger, Daston, Heidelberger (Eds.), *op. cit.*, pp. 351–375, and Hacking, *The Taming of Chance, op. cit.*, Ch. 5 and 15.

Most importantly, a statistical approach to society would neglect – according to the Germans – the single individual and his or her (free) will and motives. The German economist and statistician Georg Friedrich Knapp, for example, criticized in 1871–72 any approach which, like Quetelet's one, "explains from the outside to the inside; [...] sees the constancy of the whole and limits therefore the individual. The German school [...] explains from the inside to the outside; it takes the individual as he is and looks for reasons of the constancy of the whole."[20]

Diametrically opposite ideas *both* about statistics and about the relationship between natural and social sciences are at the bottom of the enthusiastic reaction to Quetelet's work by Thomas Buckle, the author of the gigantic, unfinished work *History of Civilization in England*.[21] Buckle appeals to statistics and to Quetelet's work in order to argue for the scientific nature of history. According to him, statisticians have been the first to deliver the "proofs of the regularity of human actions".[22] Consequently, he feels licensed to pursue his "study of the movements of Man" just like natural scientists study the "movements of nature"[23]: seeking laws[24] and causes and trying to predict. In fact Buckle makes an explicit plea for the unity of science: referring to the moral and the natural domains, he expresses the hope that his work "will at least have the merit of contributing something towards filling up that wide and dreary chasm, which, to the hindrance of our knowledge, separates subjects that are intimately related, and should never be disunited".[25]

For what concerns natural scientists, the British mathematician and astronomer John Herschel wrote a long, favorable commentary on the statistical work of his colleague Quetelet in 1850.[26] Herschel strongly supports the application of statistics and probability calculus to the inquiries in the social and in the political domains, and expresses this position in the same breath as his unitary conception of science:

> [Statistics] is the basis of social and political dynamics, and affords the only secure ground on which the truth or falsehood of the theories and hypotheses of this complicated science can be brought to the test. It is not unadvisedly that we use the term Dynamics as applied to the mechanism and movements of the social body; *nor it is by any loose metaphor or*

20 Cit. in Michael Heidelberger, "From Mill via von Kries to Max Weber: Causality, Explanation, and Understanding", in: Ulijana Fest (Ed), *Historical Perspectives on Erklären and Verstehen*, Dordrecht/Heidelberg/London/New York: Springer 2010, pp. 241–265.

21 Thomas Buckle, *History of Civilization in England*, vols. I-V, Leipzig: Brockhaus 1865.

22 *Ibid.*, vol. I, pp. 19-20.

23 *Ibid.*, vol. I, p. 7.

24 Cf. *Ibid.*, vol. I, p. 26.

25 *Ibid.*, vol. I, p. 33.

26 John Herschel, "Quetelet on Probabilities", in: *The Edinburgh Review*, July 1850. Quetelet would later use this comment as Introduction to his enlarged version of the *Physique Sociale*.

strained analogy that much of the language of mechanical philosophy finds a parallel meaning in the discussion of such subjects.[27]

Herschel takes here the applicability of statistics within social and political inquiries as indicating a kind of homogeneity between social and natural "Dynamics" which is more than a mere analogy.

On the contrary, according to Glenn Shafer,[28] Maxwell and Boltzmann were only using "analogies" or "didactic devices" as they – in turn – referred to social statistics in their foundational writings on statistical mechanics, as in the following passages:

The modern atomists have [...] adopted a method which is, I believe, new in the department of mathematical physics, though it has long been in use in the section of Statistics. When the working members of Section F get hold of a report of the Census, or any other document containing the numerical data of Economic and Social Science, they begin by distributing the whole population into groups, according to age, income-tax, education, religious belief, or criminal convictions. The number of individuals is far too great to allow of their tracing the history of each separately, so that, in order to reduce their labour within human limits, they concentrate their attention on a small number of artificial groups. The varying number of individuals in each group, and not the varying state of each individual, is the primary datum from which they work. [...] The smallest portion of matter which we can subject to experiment consists of millions of molecules, no one of which ever becomes individually sensible to us. We cannot, therefore, ascertain the actual motion of any one of these molecules; so that we are obliged [...] to adopt the statistical method of dealing with large groups of molecules.[29]

As is well known, Buckle has shown by statistics that if only we take a large enough number of people, then so long as external circumstances do not change significantly, there is complete constancy not only in the processes determined by nature, such as number of deaths, diseases and so on, but also of the relative number of so-called voluntary actions, such as marriage at a certain age, crime, suicide and the like. Likewise with molecules [...][30]

Shafer's idea that such references are only "analogies" and "didactic devices" is meant to undermine Porter's thesis according to which Quetelet's social statistics had inspired the probabilistic thinking and models of natural scientists like

27 *Ibid.*, pp. 434–435; my emphasis. See also p. 373 and 437.

28 Glenn R. Shafer, "Review of: T. M. Porter, *The Rise of Statistical Thinking 1820–1900*". In: *Annals of Science* 47, March 1990, pp. 207–209.

29 James Clerk Maxwell, "Molecules. A Lecture" [1873], in: W. D. Niven (Ed.), *The Scientific Papers of James Clerk Maxwell*, Vol. II, Cambridge: Cambridge University Press 1890, pp. 361–377 (see pp. 373–374).

30 Ludwig Boltzmann, "The Second Law of Thermodynamics" (Engl. transl. of: "Der zweite Hauptsatz der mechanischen Wärmetheorie", 1886), in: Boltzmann, *Theoretical Physics and Philosophical Problems: Selected Writings*, Dordrecht: Reidel 1974, pp. 13–32 (see p. 20).

Maxwell and Boltzmann, thus playing a significant role in the origins of statistical mechanics. Again, a certain wish to emphasize the gap between the social and natural sciences seems to be responsible for Shafer's aversion even to the purely historical arguments supporting the idea of a transfer of statistical methods from the social to the natural sciences.[31] Still, his suggestion should be taken seriously. An inquiry into Maxwell's and Boltzmann's respective conceptions of the relationship between the social and natural sciences would be necessary before one could take a stand on this issue, though. If Shafer were right, one could furthermore ask why, while importing statistical models from the natural into the social sciences had implied a unitary conception of the sciences, importing statistical models from social statistics to statistical mechanics would have amounted only to an analogy with didactical purposes. These issues cannot be solved within the limits of this paper. What I would like to suggest, instead, in the next section, is rather that Maxwell's and Boltzmann's "analogies" have had a greater impact and importance than Shafer is disposed to recognize.

32.4 Statistics and the Unity of Science in the Vienna Circle

The analogy between social statistics and statistical mechanics has had a significant resonance within the Vienna Circle, and in particular in some writings by Neurath, Frank and Zilsel. Considering the significance of Boltzmann for the Vienna Circle, it is possible that its members became acquainted with this analogy through him.

Philipp Frank, in his book on *The Law of Causality and its Limits*,[32] goes as far as to refer to a gas model in order to explain the "materialist conception of history" and to argue for its scientific nature. Single individuals – writes Frank – are like gas molecules, and in principle we could even assume that they behave according to deterministic, psychophysical micro laws. But, explains Frank,

> Historical and sociological sciences [...] do not deal with the psychological states of individuals; they speak of social conditions like density of population, diseases, political parties, constitutions of states, etc. We then often ask whether we can predict the state variables of the future if the present are known. [...] In principle we can always assume in the sense of classical physics that there are laws if we enter into ever finer structures. We have however to assume that all observable state variables define only a macrostate for which there can be no strict laws at all, but [...] only predictions about average conduct.[33]

---

31 Cf. Shafer's very polemical arguments at p. 208 of his "Review of: T.M. Porter, *The Rise of Statistical Thinking 1820–1900*", *op. cit.*

32 Philipp Frank, *The Law of Causality and Its Limits* (Engl. transl. of: *Das Kausalgesetz und seine Grenzen*, 1932), Robert S. Cohen (Ed.), Dordrecht: Reidel 1998 (see in particular Ch. 8).

33 *Ibid.*, p. 198.

Frank appeals here – like Quetelet had done in his *Physique Sociale* – to the applicability of statistical models to society as to something that would testify to the possibility of pursuing social sciences "scientifically", and thus speaks in favor of the continuity of these latter with the natural sciences. Indeed, the outright rejection of any in-principle distinction between social and natural sciences was a most important matter especially within the so-called "left-wing" of the Vienna Circle, which pursued the project of *Einheitswissenschaft*, or "Unity of Science". This commitment supports Neurath's contention that the Viennese Logical Empiricism was more kindred in spirit to the British and to the French philosophical traditions than to the German one.[34] The Vienna Circle's "left-wing" was closer to Quetelet and Buckle than to the nineteenth century German statisticians.

Still, from the last quotation from Frank, a much more "modest" attitude than Quetelet's becomes apparent: Frank places a new emphasis on the *limits* of predictions. By the time Frank had written his book, the development of statistical mechanics and quantum mechanics had yielded a most interesting and significant consequence for the Vienna Circle's unitary conception of science. While Quetelet and his followers pointed to statistics to argue that the social sciences resemble the natural sciences with respect to causality, lawfulness, prediction and – in sum – *determinacy*, the Vienna Circle members pointed to statistics to show that the natural sciences are not essentially different from the social sciences, since both are characterized by a certain degree of *in*determinacy, which however does not prevent the formulation of laws and predictions.

This new perspective repeatedly comes to the fore in Zilsel's writings, from the very beginning to the end of his life.[35] Zilsel appeals to the degree of indeterminacy in physics in order to contest the presumptive non-causal character of life sciences,[36] sociology and history.[37] If physics – he argues – delivers causal laws

34 Cf. Otto Neurath, "Die Entwicklung des Wiener Kreises und die Zukunft des Logischen Empirismus", in: Neurath, *Gesammelte philosophische und methodologischen Schriften*, Rudolf Haller and Heiner Rutte (Eds.), Vienna: Hölder-Pichler-Tempsky, Vol. 2, pp. 673–702 (see p. 676).

35 In his first book, *Das Anwendungsproblem*, Zilsel gave an indeterministic foundation to *all* scientific laws, which are conceived of as mere statistical regularities emerging from indeterminacy (Edgar Zilsel, *Das Anwendungsproblem*, Leipzig: Barth 1916). Towards the end of his life, in 1941, Zilsel would write: "historical phenomena are scarcely more difficult to predict than the weather and certainly not more difficult than volcanic eruptions and earthquakes. What would scientists think of a geophysicist who gives up the search for geophysical laws because of their inexactness?" (Edgar Zilsel, "Physics and the Problem of Historico-Sociological Laws", in: *Philosophy of Science*, Vol. 8, No. 4, 1941, pp. 567–579; see p. 570).

36 Cf. Edgar Zilsel, "Naturphilosophie" in: Franz Schnaß (Ed.), *Einführung in die Philosophie*, Osterwieck-Harz: Zickfeldt 1928, pp. 107–143 (see p. 138).

37 Cf. e.g. Zilsel, "Physics and the Problem of Historico-Sociological Laws", *op. cit.*, and Zilsel, "Problems of Empiricism", in: Neurath (Ed.), *Foundations of the Unity of Science: Towards an International Encyclopedia of Unified Science*, Vol. II, 8, Chicago: University of Chicago Press 1947 (first edition 1942), pp. 171–208 (see in particular

but nonetheless admits indeterminacy, a degree of indeterminacy in sociology and history cannot be taken as proof of their non-causal or non-explicative character.

Along the same lines, Neurath writes:

When [sociologists] plead their case for the inclusion of sociological predictions, like those of all the other sciences, into the unified science of Physicalism, they will be less inclined to claim that sociology achieves as much as the most successful sciences. Rather, they will point out that certain limitations, to which sociology most obviously is subject, also hold for all the other sciences to some degree and that sociological predictions are scientific predictions like all the others.[38]

Neurath's idea of a "Sociology in the Framework of Physicalism"[39] shows a significant resemblance to Quetelet's program. Neurath himself recognized it:

All empirical sciences are, in the end, physics in the widest sense. Quetelet speaks of 'social physics', when he derives his average man and then tries to ascertain how certain changes of social quantities are linked, for instance changes of criminality with changes in food prices. One might speak of the physics of society in the same way as of the physics of a machine.[40]

Still, a brief comparison between the already mentioned cornerstones of Quetelet's social physics and Neurath's meta-reflection on sociology brings to light Neurath's realization of the limitations to which *both* natural and social sciences appear to be subjected.

32.4.1 Observation and Quantification of Facts

To Quetelet's reliance on "social facts" corresponds Neurath's wish to trace back the statements of social science to observable "states of affairs"[41] or to spatio-temporal descriptions.[42] Neurath's "social behaviourism", and his dislike of any reference to "intentions", "introspection", "empathy", "comprehension" or other mental states in social science,[43] shows an interesting resemblance to Quetelet's idea of investigating moral and intellectual properties "through their products".

p. 195, where Zilsel also refers – like Frank in 1932 – to a gas model of society).

38 Neurath, "Sociological Predictions" (Engl. transl. of "Soziologische Prognosen", 1936), in: Neurath, *Economic Writings. Selections 1904–1945*, Dordrecht/Boston/London: Dordrecht 2004, pp. 506–512 (see pp. 511–512).

39 Neurath, "Sociology in the Framework of Physicalism", (Engl. transl. of "Soziologie im Physikalismus", 1931), in: Neurath, *Philosophical Papers 1913–1946*, Dordrecht/Boston/Lancaster: D. Reidel 1983, pp. 58–90.

40 Neurath, "Empirical Sociology" (Engl. transl. of *Empirische Soziologie*, 1931), in: Neurath, *Empiricism and Sociology*, Dordrecht/Boston: D. Reidel 1973, pp. 319–421 (see p. 390).

41 This expression recurs in Neurath, *Empiricism and Sociology, op. cit.*

42 Neurath "Sociology in the Framework of Physicalism", *op. cit.*, p. 61.

43 Cf. e.g. Neurath, "Empirical Sociology", *op. cit.*, p. 325 and "Sociology in the Framework of Physicalism", *op. cit.*, pp. 68ff.

However, Neurath does not share Quetelet's blind faith in "facts". With respect to statistics in particular – Neurath warns us – the precision and clarity of the mathematical form in which statistical "facts" are expressed should not distract from the *conventional* nature of the numerical indexes and of the reference classes we choose.[44]

32.4.2 The "Law" of Large Numbers

The belief in the emergence of stability out of the large numbers is still present in Neurath, and it is acknowledged as a heritage from Quetelet (note how Neurath formulates here exactly Porter's above mentioned thesis!):

> The scientific approach is most difficult to introduce wherever there is interest in the future fate of single individuals [...] Where the subject is masses and groupings of men, stability is larger, and the instability of the individual is less conspicuous. Therefore such questions are more amenable to scientific treatment, and the interest in such questions furthers the scientific attitude. The modern statistical approach, which has become so significant in physics, has its origins in sociological methods that were advocated about the middle of the nineteenth century and even earlier by Quetelet and others.[45]

32.4.3 Correlations Instead of Laws and Causes

Still, Neurath does not share Quetelet's belief in "statistical laws" and he does not like "the cause-effect phraseology".[46] All sciences – Neurath argues – just look for *correlations*.[47] The elimination of the reference to laws and causes, and the reliance on the "weaker" concept of "correlation" place Neurath in a better position than Quetelet's to argue in favor of the unity of science, since Neurath does not have to provide any deterministic account of statistical regularities in order to point out what sociology and physics have in common.

32.4.4 Prediction

In fact, Neurath shifts the main focus of attention from the concepts of laws and causes to the concept of prediction.[48] He warns against the many limits of sociological predictions,[49] but – as already mentioned – he also argues that these limits hold for every science: it is just a matter of degree.

---

44 Cf. Neurath, *Foundations of the Social Sciences*, Chicago: University of Chicago Press 1944, pp. 24–25 and 33.

45 Neurath, "Ways of the Scientific World-Conception" (Engl. transl. of: "Wege der wissenschaftlichen Weltauffassung", 1930), in: Neurath, *Philosophical Papers*, *op. cit.*, pp. 32–47 (see pp. 44–45).

46 Cf. Neurath, *Foundations of the Social Sciences*, *op. cit.*, pp. 20–21.

47 Cf. Neurath, "Sociology in the Framework of Physicalism", *op. cit.*, p. 68.

48 Cf. *Ibid.*, p. 61 and Neurath, "Sociological Predictions", *op. cit.*

49 Cf. Neurath, "Empirical Sociology", *op. cit.*, §10; "Sociological Predictions", *op. cit.*; *Foundations of the Social Sciences*, *op. cit.*, §12.

To sum up, a significant echo of Quetelet's unitary conception of the sciences and of his "transversal" use of statistical models can be found in Neurath, Frank and Zilsel's writings. Still, the important developments undergone in the meantime by science (e.g. the indeterministic turn in Physics) and by its philosophy (e.g. the impact of conventionalism and pragmatism) are reflected in an emergent awareness of the limitations to which any science is subjected and in a new deflationist attitude with respect to facts, laws and causes: these latter appear to have been de-ontologized and to some extent relativized,[50] so that any further account about their "mirroring" a deterministic world becomes meaningless and pointless.

32.5 Conclusion

Let me conclude by highlighting the main findings of my selective historical *tour de force* from Quetelet to Neurath with respect to the two main issues mentioned in the Introduction.

1. My reconstruction has shown how in many cases the readiness to "transfer" statistical thinking from natural to social sciences and vice versa has been (and still is[51]) related to the corresponding standpoint on the issue of the unity or disunity of science.

In the nineteenth century Adolphe Quetelet, perhaps the most important pioneer of the quantitative methods in social science, applied to society the same statistical methods he used to apply as astronomer, and expressed his unitary conception of the sciences by dubbing his inquiries into society "social physics". While authors like Thomas Buckle and John Herschel appreciated Quetelet's statistical work and explicitly shared his unitary conception of the natural and the social sciences, in Germany a conception of statistics different than Quetelet's typically went hand in hand with the conviction that there is an in-principle gap between the natural and the social sciences.

Interestingly enough, from the late 1920s until the 1940s some Vienna Circle members still invoked statistics to argue for of the unity of the social and the natural sciences – like Quetelet and his followers had done. Nevertheless, one can identify an interesting twist in Frank, Zilsel and Neurath's arguments. While

50 Concerning Neurath, cf. the incisive résumé of his epistemology in: Nancy Cartwright, Jordi Cat, Lola Fleck, Thomas Uebel, *Otto Neurath: Philosophy between Science and Politics*, Cambridge: Cambridge University Press 1996, p. 3: "Knowledge has no foundations. The things we believe can only be checked against other beliefs; nothing is certain; and all is historically conditioned".

51 Cf. the case of Shafer, *supra*, p. 8. Another, more recent example is Donald Gillies, who has argued for interpreting statistics and probability in the natural sciences differently than in the social sciences as natural and social sciences are in principle different (cf. Donald Gillies, *Philosophical Theories of Probabilities*, New York: Routledge 2000, pp. 187–200).

Quetelet and his followers pointed to statistics to argue that the social sciences resemble the natural sciences with respect to causality, lawfulness, prediction and – in sum – *determinacy*, the Vienna Circle members pointed to statistics to show that the natural sciences are not essentially different from the social sciences, since both are characterized by a certain degree of *in*determinacy, which however does not prevent the identification of significant correlations and the formulation of predictions.

2. The literature by Frank, Zilsel and Neurath which I have considered also provides new evidence for Theodore Porter's thesis according to which "a close and significant relationship between social statistics and the origins of probabilism in physics is apparent".[52] The analogy between statistical models of society and statistical models of gases – whose historical impact has been minimized by Shafer in the context of his criticism of Porter – seems in fact to have been well-known in the Vienna Circle. Furthermore, Neurath formulated already in 1930 exactly Porter's thesis.

Department of Philosophy
University of Vienna
Universitätsstraße 7
A-1010 Wien
Austria
donata.romizi@univie.ac.at

52 Porter, *The Rise of Statistical Thinking*, *op. cit.*, p. 192.

Chapter 33

Artur Koterski

The Backbone of the Straw Man
Popper's Critique of the Vienna Circle's Inductivism

> So, anti-positivism is in fashion. I do not think that it influenced you but rather that you yourself help to shape it even there where in principle you sympathize with logical empiricism.[1]

33.1 Introduction

In his monograph on the Vienna Circle Kraft writes that "one of the earliest and most fundamental insights of the Vienna Circle" was "that no deductive or logical justification of induction is at all possible."[2] In *Logik der Forschung* (hereafter: *LdF*), Popper developed his philosophical conception starting from a very emphatic critique of logical positivism and its alleged essential feature-inductivism. Although Kraft's assessment is essentially correct – as the present paper intends to show – Popper's opinion prevailed and came to dominate philosophical handbooks for decades. However, it must be admitted that the Vienna Circle's attitude towards induction might have been misleading, and in a sense invited misunderstandings. Whilst the members of the *Schlick-Kreis* clearly recognized the impossibility of any logical justification of induction, some of them believed that induction was a part and parcel of scientific conduct and instead of denying its existence they tried to change its epistemological status.

The aim of this paper is to display this evasive policy – how to keep induction rationally, nevertheless without justification – and demonstrate that Popper's criticism of the 1930s was already by then an anachronism.

---

1 Neurath to Popper, 1936-02-04 (Wiener Kreis Stichting [WKS]. Archive materials quoted by permission of the Wiener Kreis Stichting, Haarlem, Netherlands. All rights reserved.).

2 Victor Kraft, *The Vienna Circle*, New York: Philosophical Library 1953, p. 130.

D. Dieks et al. (eds.), *Probabilities, Laws, and Structures*, The Philosophy of Science in a European Perspective 3, DOI 10.1007/978-94-007-3030-4_33,

33.2 The Target of Popper's Criticism

LdF gives the impression that the object of Popper's critique is characterized in a clear and unambiguous manner. He criticizes inductivism and "especially those empiricists who follow the flag of positivism."[3] When the Viennese context of that time is taken into account, there are strong indications that the Vienna Circle must have been his target. However, when we try to justify this impression, we encounter a series of difficulties.

33.2.1 *Inductivism* selon *Popper*

According to Popper, inductivists claim that the "empirical sciences can be characterized by the fact that they use *inductive methods*."[4] Therefore, the problem of induction concerns the validity of inductive methods – "whether inductive methods are justified, or under what conditions."[5]

Induction may be understood as a method of generalization and/or a method of corroboration. Although Popper wants to maintain a sharp division between the context of discovery and the context of justification, he identifies both kinds of induction: "[…] to ask whether there are natural laws known to be true appears to be only another way of asking whether inductive inferences are logically justified."[6] The problem of induction thus identified receives a negative solution: we are not only unable to have any logic of discovery, we also have no method of verification/confirmation.

According to Popper, inductivism consists in the following claims:

I_1. There are inductive methods of discovery *and* of justification of universal laws;
I_2. These are inductive inferences;
I_3. They can be justified (and the justification is searched for) or they already are;
I_4. Philosophy should be or it already is turned into a logic of induction.

He opposed and rejected all of them.

33.2.2 *Who Are Popperian Inductivists?*

Only a few alleged inductivists are mentioned by name, among them Reichenbach and Richard von Mises. There should be, however, more of them, otherwise Popper would simply argue against the Berlin Group and Reichenbach's most outspoken local critic. But he never says anything like that.

Instead of listing his enemies by name, Popper invokes the already mentioned 'empiricists who follow the flag of positivism.' They can be old, he says, or modern. Undoubtedly, the modern ones are what we are looking for. So let us see their characteristics[7]:

3 Karl Popper, *The Logic of Scientific Discovery*, London: Hutchinson 1959, p. 34.
4 *Ibid.*, p. 27.
5 *Ibid.*, p. 28.
6 *Ibid.*, p. 28.
7 *Ibid.*, p. 35.

P_1. A modern positivist claims that science is a system of statements;
P_2. In his analysis of science, a positivist operates with expressions like 'protocol sentences' or 'reduction'; he thinks that 'scientific or legitimate' statements are reducible to atomic, elementary, or protocol sentences;
P_3. All those 'basic' statements are about private experiences, so he is a phenomenalist[8];
P_4. He claims that if an expression is not reducible to them, then it is meaningless;
P_5. He tries to prove that metaphysics is non-reducible, thus meaningless;
P_6. He claims that only expressions of science are meaningful;
P_7. Thus, philosophy is impossible and any reflection on science must be naturalized.

P_1-P_7 *jointly* define 'modern positivists.' Because of P_2 and P_3, the list also points to inductivists. A phenomenalist who accepted the existence of science must also accept a thesis of inductive discovery; the reductionist's claim that all proper propositions are reducible to 'basic' statements, must be – since such reduction is not a deductive procedure – about their inductive justification. Thus, inductivism is a part of 'modern positivism,' actually the most important one as it offers a solution to the problem of demarcation, albeit a wrongful one. And there comes the aforementioned impression: as the Vienna Circle was formed by modern positivists (without quotation marks), it was a center of inductivism.

Popper had many opportunities to learn about the views of his future opponents before his book was sent to the publisher, but – as the above specifications (i.e., P_1-P_7) show – he did not seize any of them: *even the leading figures of the Vienna Circle, like Schlick or Carnap, do not satisfy this description.*[9]

33.3 The Vienna Circle on Induction

Reconstructing the Viennese problem-situation for Popper's criticism of induction, we encounter two fundamental problems. The first one is that it is impossible to identify unambiguously against whom in the Vienna Circle Popper's criticism was directed unless we say it was a straw man.

As regards the Vienna Circle members themselves, they took *LdF* to be meant as a criticism of their views.[10] They stressed, however, that the criticism was

8 It seems Popper took Frank and Hahn to simply continue (what he though to be) Mach's phenomenalism (cf. *ibid.*, p. 94, fn. 3). He was, however, wrong (cf. Rudolf Haller, "Was Wittgenstein a Neopositivist?", in: Rudolf Haller, *Questions on Wittgenstein*, London: Routledge 1988, p. 39).

9 Cf. Artur Koterski, "Popper i Koło Wiedeńskie. Historyczna analiza sporu", in: *Przegląd Filozoficzny* 1, 1998, pp. 47–72.

10 Cf. Rudolf Carnap, "Intellectual Autobiography", in: Paul Schilpp (Ed.), *The Philosophy of Rudolf Carnap*, La Salle: Open Court 1963, p. 30. Neurath, for one, explicitly

hyperbolized and the only open question was to what a degree Popper overstated the differences between them and his own position.[11] If we adopt such a policy and examine Popper's critique with respect to induction, we encounter the second problem: overemphasizing the dissimilarities or attacking already abandoned ideas, Popper only hits a straw man again.

As rightly noticed by Malachi Hacohen, Popper had no understanding of the dynamics of the Vienna Circle and was not able to learn from them.[12] The image of the Vienna Circle he had, was not only highly fragmentary but practically also a still one.[13] The following survey outlines the views on induction held by the Vienna Circle members up to 1934 and illustrates Popper's misapprehension about their alleged inductivism.

33.3.1 Moritz Schlick

During his carrier Schlick proposed two approaches to the problem of induction. The first one was expounded in *Allgemeine Erkenntnislehre*. It was basically Humean. According to Schlick, the ampliative judgments are obtained because of habituation and rest on associations. Thus, the inductive reasoning is a subject of psychology and biology, not logic. The principle of induction, being itself a synthetic statement, should be investigated in empirical research.[14] The judgments derived by inductive reasoning are hypothetical and have only probabilistic validity. According to Schlick, the explanation of induction appeals to the same processes as the explanation of causality. It shows that: "This general connection of

---

complained: "You anyway do not treat very carefully [...] those you admittedly call positivists. [...] without indicating, which doctrines and persons come under it. Since they are pseudo-problems strugglers, therefore perhaps [it is] Carnap, Frank, Neurath, Schlick [...] – it is not quite clear how much you count as positivists Poincaré, Russell etc. [...]" (Neurath to Popper, 1935-08-24 [WKS]); cf. Otto Neurath, "Pseudorationalism of Falsification", in: Otto Neurath, *Philosophical Papers 1913–1946*, Dordrecht: D. Reidel 1983, p. 131.

11 Cf. Otto Neurath, "Pseudorationalism of Falsification", *op. cit.*, pp. 121–131; Rudolf Carnap, "Karl Popper: Logik der Forschung", in: *Erkenntnis* 5, 1935, p. 293; Kurt Grelling, "Karl Popper: Logik der Forschung", in: *Theoria* 3, 1937, p. 135; Carl Hempel, "Karl Popper: Logik der Forschung", *Deutsche Literaturzeitung* 8, 1937, p. 314; see also Victor Kraft, "Popper and the Vienna Circle", in: Paul Schilpp (Ed.), *The Philosophy of Karl Popper*, La Salle: Open Court 1974, pp. 187–188.

12 Cf. Malachi Hacohen, *Karl Popper: The Formative Years 1902–1945*, Cambridge: CUP 2001, pp. 209–210.

13 It seems Popper took the Vienna Circle to be a philosophical school. He wrote in one of his letters: "[...] I have a deep dislike of overly close scientific fraternité: scientific friends can and should argue objectively! I do not like 'schools'!" (Popper to Neurath, 1935-07-10 [WKS]).

14 Cf. Moritz Schlick, *General Theory of Knowledge*, Wien: Springer-Verlag 1974, p. 115.

habituation […] is nothing other than the causal connection, or rather its subjective mirror image."[15]

Induction is rooted in causality: it is a necessary condition for making inductive inferences. If the causes and effects are discoverable and differentiable from each other, then we are able to apply tools like Mill's methods.[16] However, causality cannot serve as the ultimate justification for induction: "That causality and hence inductive inference cannot be established by a rational proof was perceived quite early with the aid of an empiricist line of argument."[17]

The only acceptable way is to postulate the principle of causality and *a fortiori* the principle of induction: "In the case of these and similar attempts at a foundation […] the strict validity of the causal principle and of […] inductively obtained truths figures as a *postulate*."[18] Schlick's position, accordingly, allows only for a practical justification of induction – it cannot be a replacement for a theoretical one but it is enough in life and for science.

In the period under heavy influence of Wittgenstein and his verificationism, Schlick takes a more radical approach according to which the question of induction is a pseudo-problem. Laws, having strictly universal character, are not conclusively verifiable, therefore, they are not proper sentences. The question of justification of the inference from the particular to the unrestricted general does not occur any longer simply because there is no such inference: "[…] the so-called problem of "induction" is […] rendered vacuous."[19] Laws are just prescriptions how to obtain singular statements from other singular statements. If we ask, however, how we get those prescriptions, the answer is that we get them… inductively – we guess. The success of this 'method' is possible because scientific guessing is 'methodologically guided'.[20] Because of this claim Schlick may be charged with I_1-inductivism. He explicitly denied all other kinds by 1931.

15 *Ibid.*, p. 388.

16 Cf. Moritz Schlick, "Philosophical Reflections on the Causal Principle", in: Moritz Schlick, *Philosophical Papers*, Vol. I, Dordrecht: D. Reidel 1979, p. 298; cf. also Herbert Feigl, "Zufall und Gesetz", in: Rudolf Haller/Th. Binder (Eds.), *Zufall und Gesetz*, Atlanta: Rodopi, p. 179.

17 Moritz Schlick, *General Theory of Knowledge*, *op. cit.*, p. 394. This said, Schlick proceeds to criticism and rejects some common attempts at justifying induction – those arguments were repeated later by Popper.

18 Moritz Schlick, *General Theory of Knowledge*, *op. cit.*, p. 395.

19 Moritz Schlick, "Causality in Contemporary Physics", in: Moritz Schlick, *Philosophical Papers*, Vol. II, *op. cit.*, p. 197.

20 Moritz Schlick, "On the Foundation of Knowledge", in: Moritz Schlick, *Philosophical Papers*, Vol. II, *op. cit.*, p. 380–381.

33.3.2 Herbert Feigl

In his dissertation,[21] Feigl objects that we must be content with psychological and biological account of induction. Even if it is good enough for scientists or everyday agents, it is not satisfactory for a philosopher of science. He attempts to justify induction by taking determinism as a necessary presupposition in the theory of science.

The analysis of inductive reasoning in the context of natural science requires two questions. (1) *Does nature posses the strict lawfulness*? If the answer is affirmative, there is another one: (2) *Is it possible at the present stage of scientific research to establish that a lawful relation valid within a limited scope is also universally valid*? Feigl labels the first question 'the problem of the general induction' and the other 'the problem of the special induction'. The affirmative answer to the problem of general induction expresses only a postulate of searching for the universal lawfulness. General induction is, therefore, a heuristic principle.

It makes sense to pose the first question only once the second has been answered affirmatively. However, the search for laws within the limited scope seems to make sense only if we assume the strict lawfulness of nature. Even talking about probabilities of physical statements is possible only when some kind of regularity is assumed (thus, special induction assumes validity of general induction). This assumption is determinism. The possibility of induction requires the assumption of causality; and if the principle of causality holds, then we must live in a world that is determined to *some degree*.

In 1929, Feigl drops the hypothesis of determinism and instead focuses on another question from *Zufall und Gesetz*: what methods lead to discoveries of natural laws. These are inductive methods – i.e. extrapolation and interpolation in the sense of the most simplifying generalization established with the use of tools like Mill's method, the method of least squares etc.

Although he points to some methods of discovery, Feigl denies the possibility of logic of induction – unless some *Obersätze*, i.e. superordinated statements are assumed. However, they all are doubtful and not needed. Theories, in their origins, depend on the creativity and inventive power of scientists. This cannot be reconstructed and captured by any scheme ready for future application.

Feigl separates the contexts of discovery and of justification. The origin of a theory is rather not an inductive process, and though inductive methods may be applied during the creation of a theory, there is no logic of discovery. However, when the theory is rationally reconstructed we can induce it from the set of its confirmed predictions by the use of the most simplifying generalization: "[…] the *validity* of theories can only be founded inductively."[22]

21 Cf. Herbert Feigl, "Zufall und Gesetz", pp. 169ff.

22 Herbert Feigl, "Meaning and Validity of Physical Theories", in: Herbert Feigl, *Inquiries and Provocations*, Dordrecht: D. Reidel Publishing Company 1981, p. 129.

Feigl's views on induction were developed further in 1930–1931 under Wittgenstein's influence. As did Schlick, he also denied that the principle of induction was a (declarative) sentence: "[…] *it is not a proposition at all. It is, rather, the principle of a procedure, a regulative maxim, an operational rule.*"[23]

At that time Feigl rejected once again the possibility of and the very need for replacing or supplementing deductive logic with a 'logic of probability'.[24] He also denies the possibility of application of what Carnap later called probability$_1$ to empirical questions.[25]

Feigl's views on induction kept changing. In the years 1927–1930, they could be related with I_1–I_3 claims. However, in 1931 his position was in principle indistinguishable from that of Schlick, so he could be possibly accused of being an I_1-inductivist.

33.3.3 Marcel Natkin

One year after Feigl's dissertation was defended, another student of Schlick's presented his thesis. In it, Natkin disagreed with Schlick and Feigl as regards induction and its role in science. While Schlick tried to root induction in causality, Natkin considered such a solution unsatisfactory. If we assume that the principle of causality is valid, we still do not know all initial conditions, so our inductions may turn out to be false:

> It is not only that we cannot infer the validity of the laws we found from the validity of the law of causality; we cannot even infer the probability of correct predictions from the assumption of validity of our laws. We learn from this that the principle of causality by no means coincides with the principle of induction […].[26]

Thus, causality is not a foundation for the principle of induction.

While Feigl believed that inductive reasoning is the most important feature of empirical science, Natkin removed it from science. The principle of induction is not a part of science and it is not necessary to know it in order to grasp the essence of science. It is just an instruction how to make use of scientific cognition.[27]

---

23 Herbert Feigl, "The Logical Character of the Principle of Induction", in: *Philosophy of Science*, 1 (1934), p. 27 (written in 1931); cf. Victor Kraft, "The Problem of Induction", in: Paul Feyerabend, Grover Maxwell (Eds.), *Mind, Matter, and Method*, Minneapolis: University of Minnesota Press 1966, pp. 310–311.

24 Herbert Feigl, "Probability and Experience", in: Feigl, *Inquiries and Provocations*, *op. cit.*, p. 107.

25 Cf. *ibid.*, p. 108; Herbert Feigl, "The Logical Character of the Principle of Induction", *op. cit.*, p. 23.

26 Marcel Natkin, "Einfachheit, Kausalität und Induktion", in: Haller, Binder (Eds.), *Zufall und Gesetz*, *op. cit.*, p. 293.

27 Cf. *ibid.*, pp. 294–295.

33.3.4 Otto Neurath

Those passages of the *Manifesto* where induction is discussed, seem to be written or outlined by Neurath. There, the validity of induction is relative to regularity in nature. But instead of looking for the foundation of this regularity, as it happened in the cases of Schlick and Feigl, the *Manifesto* encourages to use induction, as well as *any* other method, if it is fruitful – even if it is not theoretically justified: "The scientific world-conception will not condemn the success of a piece of research because it has been gathered by means that are inadequate, logically unclear or empirically unfounded."[28] It is a question of our decision whether we use such a method.[29]

In *Empirical Sociology* Neurath repeats this thesis in the context of Duhemian underdetermination:

> More than one system of theorems satisfies the conditions of consistency and of compatibility with the observation statements. Moreover, we know which description we lack, quite apart from the uncertainty which attaches to any induction from the outset. *Induction itself is based on a decision* [...].[30]

On account of underdetermination of hypotheses and theories, and fallibility of induction, the latter resembles guessing. Neurath's remark about the decision to use induction must be linked with the specific character of that guessing: it is not an arbitrary procedure but, as Schlick put it, a methodologically guided one. Of course, Neurath agrees with Mach and Einstein that the possibility of making a successful conjecture depends on inventive power of the respective researcher. There cannot be any automated discovery.[31]

Until 1935, Neurath had no significant reservations about induction. If it is fruitful and as far as it does not bring in metaphysics, simply make the decision to use it. And this is what actually happens: "Within the physicalist sphere, induction always leads to meaningful statements."[32] In 1935, however, he left himself the

28 Hans Hahn, Otto Neurath, Rudolf Carnap, "The Scientific Conception of the World: The Vienna Circle", in: Otto Neurath, *Empiricism and Sociology*, Dordrecht: D. Reidel Publishing Company 1973, p. 313.

29 Cf. Otto Neurath, "Diskussion über Wahrscheinlichkeit", in: *Erkenntnis* 1, 1930/1931, p. 277; Otto Neurath, "Physicalism", in: Neurath, *Philosophical Papers 1913–1946*, *op. cit.*, p. 53; see also Otto Neurath, "Universal Jargon and Terminology", in: Neurath, *Philosophical Papers 1913–1946*, *op. cit.*, p. 222.

30 Otto Neurath, "Empirical Sociology. The Scientific Content of History and Political Economy", in: Neurath, *Empiricism and Sociology*, *op. cit.*, p. 407.

31 Cf. Otto Neurath, "The Unity of Science as a Task", in: Neurath, *Philosophical Papers 1913–1946*, *op. cit.*, p. 116; Otto Neurath, "Prognosen und Terminologie in Physik, Biologie, Soziologie", in: Otto Neurath, *Gesammelte philosophische und methodologische Schriften*, Wien: Hölder-Pichler-Tempsky 1981, p. 789.

32 Otto Neurath, "Sociology in the Framework of Physicalism", in: Neurath, *Philosophical Papers 1913–1946*, *op. cit.*, p. 74.

possibility of removing 'induction', along with words like 'true' and 'false' from the scientific language.[33] From 1935 on, and in a direct reference to Popper's book, he points out the limits of empirical methods, including inductive ones.[34] Doing so, he rejects Popper's thesis that ascribes inductivism to logical empiricism; the demand of justification for inductive reasoning is, as he explicitly says, pseudo-rationalistic.

In later years Neurath unwillingly notices that some of his philosophical allies try to establish rules for induction.[35] In a paper published only posthumously, Neurath specifies whom his worries concern, and, of course, it is Carnap who just started his work on probability$_1$.[36]

Popper believed that Neurath was a phenomenalist (cf. P_3 in section 2.2), thus an I_1-inductivist; but in actual fact he was not a phenomenalist.[37] Nonetheless, according to him, the use of inductive methods (educated guessing) was legitimate in particular cases, so the anti-I_1-inductivism charge might be upheld. Neurath overtly rejected I_2 and I_3 and replaced them with his decisionism. Finally, he was a strong adversary of turning methodology into logic of induction.

33.3.5 Rudolf Carnap

Before 1934 Carnap, who was later to become the main target of Popperian anti-inductivism crusade, was quite laconic about induction. He does not discuss induction but rather takes it as a matter of course. Therefore, there are only some short slip-in passages where induction is mentioned *en passant*.

Induction – a process that consists in assembling and processing facts – is a method of discovery: a method for obtaining universal statements, including natu-

33 Cf. Otto Neurath, "Zur Induktionsfrage", in: Neurath, *Gesammelte philosophische und methodologische Schriften*, *op. cit.*, p. 631. However, at the end of his encyclopedic contribution on foundations of social sciences he attached a list of words used and avoided there – and induction belongs to the former (cf. Otto Neurath, "Foundations of the Social Sciences", in: Otto Neurath, Rudolf Carnap, Charles Morris (Eds.), *Foundations of the Unity of Science. Toward an International Encyclopedia of Unified Science*, Vol. II, Chicago: The University of Chicago Press 1970).

34 Cf. Otto Neurath, "Pseudorationalism of Falsification", *op. cit.*, p. 123; Otto Neurath, "Die Entwicklung des Wiener Kreis und die Zukunft des Logischen Empirismus", in: Neurath, *Gesammelte philosophische und methodologische Schriften*, *op. cit.*, pp. 700-701; Otto Neurath, "Individual Sciences, Unified Science, Pseudorationalism", in: Neurath, *Philosophical Papers 1913–1946*, *op. cit.*, p. 136.

35 Otto Neurath, "Prediction and Induction", *op. cit.*, p. 244.

36 Cf. Otto Neurath, "After Six Years", in: Neurath, *Economic Writings*, Dordrecht: Kluwer 2004, p. 553.

37 Popper upheld his charge in the English version of his book (cf. §26) disregarding Neurath's protests: "I think you misunderstood the thesis of physicalism. The point is, however, that only physicalistic terms enter protocol sentences – contrary to the earlier [thesis] when there was separate experience language etc." (Neurath to Popper, 1935-01-22 [WKS]).

ral laws.[38] It does not provide unassailable conclusions, but it is used in science anyway, and it is even a guarantee of empiricism.[39] There is no reason for despair because of lack of logical foundation of induction. It simply works: "[...] *induction has no strict logical justification*. However, it can adduce as credentials its experimental confirmation."[40]

Induction may seem to be a neglected topic. However, in the discussion during the Prague conference (1929), Carnap replied to Kurt Grelling that he was wrong in thinking that in Vienna the problem of induction had been pushed aside. Quite to the contrary, as Carnap continues, "the problem is extraordinarily important."[41] But a longer and more illuminating passage on induction was published only in 1934 (still before Popper's book was printed):

> [...] it is not possible to lay down any set rules as to how new primitive laws are to be established on the basis of actually stated protocol-sentences. One sometimes speaks in this connection of the method of so-called *induction*. Now this designation may be retained so long as it is clearly seen that it is not a matter of a regular method but only one of a practical procedure which can be investigated solely in relation to expedience and fruitfulness. That there can be no rules of induction is shown by the fact that the L-content of a law, by reason of its unrestricted universality, always goes beyond the L-content of every finite class of protocol-sentences.[42]

This passage, where Carnap reaffirms the views of Schlick and Feigl, on one hand, and Neurath's, on the other, is also an expression of anti-inductivism in Popper's sense (I_1–I_4).[43] There is no method of discovery, and induction is not a proper

38 Cf. Rudolf Carnap, "Psychology in Physical Language", in: Alfred Ayer (Ed.), *Logical Positivism*, New York: The Free Press 1959, p. 169; Rudolf Carnap, "The Physical Language as the Universal Language of Science", in: William Alston, George Nakhnikian (Eds.), *Readings in Twentieth-Century Philosophy*, New York: The Free Press 1963, p. 398.

39 Cf. Rudolf Carnap, *Der Raum*, Berlin: Reuther & Reichard 1922, p. 63; Rudolf Carnap, "The Elimination of Metaphysics through Logical Analysis of Language", in: Ayer (Ed.), *Logical Positivism*, *op. cit.*, p. 77.

40 Rudolf Carnap, *Physikalische Begriffsbildung*, Karlsruhe: G. Braun 1926, p. 8.

41 Rudolf Carnap, "Diskussion über Wahrscheinlichkeit", *Erkenntnis* 1, 1930/1931, pp. 282–283.

42 Rudolf Carnap, *The Logical Syntax of Language*, London: Kegan Paul, Trench, Trubner & Co. 1937, pp. 317–318.

43 Popper's anti-inductivist critique from the 1950s and 60s refers mainly to the logic of induction in the sense of probability$_1$. Recollecting the Vienna Circle period with respect to interpretations of probability, Carnap notices that at that time 'we took for granted the frequency conception' (Carnap, "Intellectual Autobiography", *op. cit.*, p. 70). However, not all of them did. Waismann, supported by Schlick, were the exceptions; he tried to develop Wittgensteinian conception of probability where probability statements are analytical. He presented his paper at the 1929 conference in Prague and the very idea of analytic probability was received *quite critically*. However, with an exception again. During the discussion Carnap says: 'in the outlines the argument of

method. Like Schlick, Feigl, Natkin and Neurath, Carnap treats it as a practical procedure that we use because it is fruitful, however without having any logical justification.

Carnap's view on induction in the 1920s reflects that of Poincaré: induction is used in science as a method of discovery, and perhaps it would be mad to deny it. Nonetheless, it hangs in the air above the theoretical ground and there is not much to say to support it. At the beginning of the 1930s Carnap rejects induction in the context of discovery. He remains convinced that successful predictions inductively support a scientific system of statements.[44]

In 1932, being on holidays in Tyrolean Alps, Carnap was joined by Feigl and Popper. At that time, Popper was in the middle of writing *Die beiden Grundprobleme* and he was very eager to discuss it – and the anti-inductivism advocated in it. So we may ask whether Carnap changed his mind, when he learnt more about Popper's approach. If so, Popper should have known that such a change had taken place and when he was dictating the first chapter of *Logik,* he should not have had Carnap in mind as his anti-inductivist target[45]. Or, perhaps – *à rebours* Hacohen – it was Popper who learnt that his own views were not so different from Carnap's and Feigl's?

In both cases, Carnap would not be among the indictees of Popper's book. And he does not seem to be. Firstly, Carnap is listed there as a kind of proponent of hypothetico-deductive conception of science,[46] alongside with Kraft[47] – another

---

Mr. Waismann is right' (Carnap, "Diskussion über Wahrscheinlichkeit", *op. cit.*, pp. 268–269; cf. Friedrich Waismann, "A Logical Analysis of the Concept of Probability", in: Friedrich Waismann, *Philosophical Papers*, Dordrecht: D. Reidel 1977, pp. 4–21; Schlick, "Causality in Contemporary Physics", p. 201).

44 See. fn. 48. We may note here that in the context of justification at that time 'inductive support' was a pleonasm. Thus, to deny a possibility of 'inductive support' was to deny that successful predictions might support a hypothesis under a test.

45 Cf. Karl Popper, "The Demarcation between Science and Metaphysics", in: Schilpp (Ed.), *The Philosophy of Rudolf Carnap*, *op. cit.*, p. 184.

46 At the end of 1931, i.e. several months before the "Tyrolese summit", Carnap already held that scientific theories cannot be inferred from the protocols, that universal statements always remain hypotheses with respect to protocol sentences. From those hypotheses *plus* singular sentences expressing appropriate initial conditions we may deduce a prediction which in turn is to be tested; if the outcome is positive it supports given system of statements. It is not, however, conclusive verification (cf. Carnap, "The Physical Language as the Universal Language of Science", *op. cit.*, p. 403).

47 In his habilitation thesis, ten years before *LdF* – i.e., when Popper preached verificationism and inductivism – Kraft wrote: "There is no generalization by 'inductive inference' from the singular to the general. Further discoveries and assumptions are necessary for generalization. Thus, induction cannot constitute any specific method of generalization. [...] There is, therefore, only practical conduct for *trial and error that is modified by successful and unsuccessful attempts by which one learns how to adopt himself.* [...] Perhaps it sufficiently explains that knowledge of natural laws does not come from any specific procedure of induction, by "inductive inference" [...], but [it comes] solely on

member of the Circle.[48] Secondly, in another footnote Popper admits that his criticism of positivism and naturalism no longer applies to Carnap and his *Syntax*.[49]

33.4 Conclusions

Popper's description of inductivism contains several points (cf. 2.1 above). If we understand I_1–I_4 as a conjunction then Popper's description is *obviously* inadequate as a characterization of the Vienna Circle's views as of 1934. Therefore, I_1–I_4 must be understood disjunctively. If so, claims I_2–I_4 have to be dropped as they are openly denied by the Vienna Circle members. Nevertheless, they still would be inductivists for Popper, if they support I_1 at least. If Popper's criticism is historically adequate at all, it is adequate to I_1-degree. However, if guessing counts as a method, then even Popper, for whom guessing is the only 'tool' for finding new ideas in science, is an inductivist too.[50]

Popper never corrected his criticism of the Vienna Circle (quite the contrary), though he was informed more than once about its misleading character. The clearing of mistakes and misrepresentations would expose the architectural weaknesses of Popper's book where criticism of 'inductivists' served as a spring-board for his falsificationism.

Faculty of Philosophy
Maria Curie-Sklodowska University
Pl. Marii Curie-Sklodowskiej 4/207
20-031, Lublin
Poland
artur.koterski@poczta.umcs.lublin.pl

the deductive way" (Victor Kraft, *Die Grundformen der wissenschaftlichen Methoden*, Wien: Verlag der Österreichischen Akademie der Wissenschaften 1973, p. 53, italics added; cf. Kraft, "The Problem of Induction", *op. cit.*, p. 317; Karl Popper, "Intellectual Autobiography", in: Schilpp (Ed.), *The Philosophy of Karl Popper*, *op. cit.*, pp. 64–65).

48 Karl Popper, *The Logic of Scientific Discovery*, *op. cit.*, p. 30, fn. 5

49 *Ibid.*, p. 53, fn. 6.

50 Cf. *ibid.*, p. 278; Hans Reichenbach, "Induction and Probability. Remarks on Popper's 'The Logic of Scientific Discovery'", in: Hans Reichenbach, *Selected Writings 1909-1953*, Vol. II, Dordrecht: Reidel 1978, p. 385.

CHAPTER 34

THOMAS UEBEL

CARNAP'S LOGIC OF SCIENCE AND PERSONAL PROBABILITY

The aim of the present paper is to consider how Rudolf Carnap's later preoccupation with inductive logic fits into the framework of philosophy of science as a bipartite metatheory, a framework for which the members of the so-called left wing of the Vienna Circle can be seen to have provided a blueprint.[1] The bipartite metatheory comprises both an a priori logic of science, analysing the structure of scientific theories and the entailment relations between its propositions and exploring the expressive powers of logically possible languages, and an empiricist pragmatics of science, comprising the psychology, sociology and history of science and investigating the practical utility of possible language forms. The issue is this: while the theory of logical probability provided by Carnap in *Logical Foundations of Probability* fits rather neatly into the bipartite schema, the more personalist form of inductive logic developed in "A Basic System of Inductive Logic" appears to raise difficulties for such an integration.[2] By pointing to concurrent developments in Carnap's understanding of what's involved in pragmatics, I hope to show that such an integration can after all be effected. My aim is not, however, to save or revitalise inductive logic but to confront a difficulty in the interpretation of Carnap's work.[3]

---

1 On the bipartite metatheory conception, see T. Uebel, "Some Remarks on Current History of Analytical Philosophy of Science." In F. Stadler et al., *The Present Situation in Philosophy of Science*, Dordrecht: Springer 2010, 13–28. The present paper investigates a possible objection not considered there.

2 R. Carnap, *Logical Foundations of Probability*. Chicago: University of Chicago Press, 1950 (hereafter "LFP"); "A Basic System of Inductive Logic, Part 1." In R. Carnap and R. Jeffrey (Eds.), *Studies in Inductive Logic and Probability Vol. I*, Berkeley: University of California Press 1971, 33–166 (hereafter "BS1"); "A Basic System of Inductive Logic, Part 2." In R. Jeffreys (Ed.), *Studies in Inductive Logic and Probability Vol. II*, Berkeley: University of California Press 1980, 7–155 (hereafter "BS2").

3 See also P. Wagner, "Carnap's Theories of Confirmation", in: D. Dieks et al. (Eds.), *Explanation, Prediction, and Confirmation*, Dordrecht: Springer, 2011, 477-486. On the possible further development of Carnap's inductive logic, see R. Jeffrey, "Carnap's Inductive Logic." In J. Hintikka (Ed.), *Rudolf Carnap, Logical Empiricist*. Dordrecht: Reidel 1975, 325–332; on its legacy see S. Zabell, "Carnap on Probability and Induction." In M. Friedman, R. Creath (Eds.), *The Cambridge Companion to Carnap*. Cambridge: Cambridge University Press 2007, 273–294 (2007). For criticisms of it see, e.g., C. Howson and P. Urbach, *Scientific Reasoning. The Bayesian Approach*, 2nd ed., Chicago: Open Court 1993, 66–72.

D. Dieks et al. (eds.), *Probabilities, Laws, and Structures*, The Philosophy of Science in a European Perspective 3, DOI 10.1007/978-94-007-3030-4_34,

34.1 The Logic of Science and Pragmatics

Abstracting here from its origins, we may note that the perspective of a division of labour between the logic of science and the empirical sciences of science found clear expression in Carnap's introductory essay for the *International Encyclopedia of Unified Science*.

> The task of analyzing science may be approached from various angles. ... We may, for instance, think of an investigation of scientific *activity*. ... These investigations of scientific activity may be called history, psychology, sociology, methodology of science. The subject matter of such studies is science as a body of actions carried out by certain persons under certain circumstances. Theory of science in this sense ... is certainly an essential part of the foundation of science.[4]

By contrast, the logic of science does not focus on the activity but its theoretical results, for "it is possible to abstract in an analysis of the statements of science from the persons asserting the statements and from the psychological and sociological conditions of such assertions. The analysis of the linguistic expressions of science under such an abstraction is *logic of science*."[5]

Note that the logic of science abstracts from all psychological issues, consistently with Carnap's declaration to have abandoned not only metaphysical philosophy but also what he deemed unduly psychologistic epistemology. Carnap moved from rational reconstructions of subject-based beliefs to logical explications of knowledge claims independent of particular epistemic subjects. The logic of science was concerned no longer with doxastic but with a kind of propositional justification, justification not of individual believings but of propositions in light of available evidence – where evidence is conceived of independently of its appreciation by a subject. The question arises whether Carnap's later understanding of inductive logic as "a theory of logical probability providing rules for inductive thinking" and of "personal probability" as "the probability assigned to a proposition or event *H* by a person *X*" remained consistent with the subjectless approach to epistemological issues that was characteristic of all inquiries in the logic of science.[6] Before we can begin to answer this question we must consider Carnap's views of pragmatics.

---

4 R. Carnap, "Logical Foundations of the Unity of Science." In O. Neurath et al., *Encyclopedia and Unified Science*. Chicago: University of Chicago Press 1938, 42–62, at 42, orig. emphasis.

5 *Ibid.*, 43, orig. emphasis.

6 R. Carnap, "Inductive Logic and Rational Decisions." In R. Carnap and R. Jeffrey (Eds.), *Studies in Inductive Logic and Probability Vol. 1*, Berkeley: University of California Press 1971, 2–32, at 7–8. (Hereafter "BS Intro". Prev. publ. as "The Aim of Inductive Logic." In E. Nagel, P. Suppes, A. Tarski (Eds.), *Logic, Methodology and Philosophy of Science. Proceedings of the 1960 International Congress*, Stanford: Stanford University Press 1962, 303–318.)

The same subjectless approach as to the logic of science is evident in Carnap's adaptation of Charles Morris's proposal to conceive of all of philosophy of science as inquiries into different aspects of the semiotics of the language of science.[7] In his Encyclopedia monograph Carnap adopted the division of pragmatics, semantics and syntax, according to which the first concerned "the action, state, and environment of a man who speaks or hears" a certain linguistic expression.[8] Carnap recognised the fact that linguistic signs are produced in "order to be perceived by other members of the group and to influence their behavior", but warned that since his interest "concerns the language of science" he restricted the investigation to "the theoretical side of language, i.e., to the use of language for making assertions".[9] The logic of science has no interest in pursuing psychological or sociological aspects of the use of the language of science but considers that language only as far as syntactic constraints and designation relations are concerned: pragmatics was excluded and banished to the empirical sciences of science.

Carnap also drew a distinction between "pure" and "descriptive" inquiries which was superimposed on the categories of (logical) syntax and semantics.[10] Early on, however, he did not entertain the possibility of dividing pragmatics in a similar way. Whereas logical syntax and semantics had a pure, analytical core pursued by a priori reasoning, which underwrote their descriptive, empirical variants, pragmatics was assigned no such analytical "pure" core. Unlike logical syntax and semantics, pragmatics was an essentially empirical discipline for Carnap. Whereas descriptive semantics and descriptive syntax not only abstracted from pragmatics, but were "strictly speaking part of pragmatics", pure semantics and syntax were "independent of pragmatics".[11] The very "abstraction" from language users and historically given languages that syntax and semantics were capable of (in their pure variants) reflected, it seems, the independent a priori status of their conceptual framework. Pragmatics, by contrast, was descriptive and linked exclusively to historically given languages. This was, to be sure, not Morris's view of the matter who long urged that pure pragmatics be recognised.[12]

By the mid-1950s, however, when Carnap was engaged in defending intensionalist semantics against Quine, he set out to "clarify the nature of the pragmatical concept of intension in natural languages" in order to give a "practical

7 See C. Morris, "Scientific Empiricism." In O. Neurath et al., *Encyclopedia and Unified Science*, *op. cit.*, 63–75.

8 R. Carnap, *Foundations of Logic and Mathematics*, Chicago: University of Chicago Press 1939, 4.

9 *Ibid.*, 3.

10 R. Carnap, *The Logical Syntax of Language*. Chicago: Open Court 2002, R. Carnap §2; *Introduction to Semantics*, Cambridge, MA: Harvard University Press 1942, §5.

11 R. Carnap, *Introduction to Semantics*, *op. cit.*, 13.

12 See C. Morris, *Foundations of the Theory of Signs*. Chicago: University of Chicago Press 1938, 9.

vindication for the semantical intension concepts".[13] To do so, he elaborated the "conceptual framework of theoretical pragmatics" by giving formal explications of an expanded notion of the pragmatical concept of intension, making use of the also formally explicated concepts of belief and holding true and adding formal explications of the concepts of assertion and utterance.[14] Carnap now agreed with Morris who argued that "if we are to develop a language to talk about the users of signs, then we need a body of terms to do so, and the introduction of these terms and the study of their relations seems as 'pure' as is the development of languages to talk about the structures and significations of signs."[15] Carnap conceded that there could be such a thing as pure pragmatics separate from descriptive pragmatics (he called it "theoretical" pragmatics) that abstracted from individual users and occasions of use and instead gave the logic of the concepts involved.[16] Ever so carefully delimiting the starting point of theoretical or pure pragmatics in this way "to small groups of concepts", Carnap expressed hope for "tentative outlines of pragmatical systems" which, more fully developed, would "include all those concepts needed for discussions in the theory of knowledge and the methodology of science".[17] With the recognition of theoretical pragmatics and its abstention from empirical theses about uses that were actually made, pragmatics in its "pure" form now found its place in the logic of science.

34.2 Carnap's Normative Decision Theory in Bipartite Metatheory

We are now ready to consider whether Carnap's personalist version of the theory of logical probability coheres with the subjectless approach of the logic of science. Note, to begin with, that Carnap's explications of knowledge claims independent of particular epistemic subjects found a ready exemplification in his *Logical Foundations of Probability*. There Carnap argued for a logical understanding of the notion of probability that differed from the objective statistical frequency conception but also avoided the subjectivism associated with traditional inductivism – and so found its place in the logic of science. Before he settled on this logical conception, however, the metatheoretical landscape had looked different. Still in his *Introduction to Semantics,* Carnap had distinguished the semantical concept of truth from "fundamentally different … concepts like 'believed', 'verified', 'highly con-

13 R. Carnap, "Meaning and Synonymy in Natural Language." In R. Carnap, *Meaning and Necessity*. 2nd ed. with supplementary essays. Chicago: University of Chicago Press, 1956, 233–247, at 235.

14 R. Carnap, "On Some Concepts of Pragmatics." In R. Carnap, *Meaning and Necessity, op. cit.*, 248–250, at 248–249.

15 C. Morris, "Pragmatism and Logical Empiricism." In P. A. Schilpp (Ed.), *The Philosophy of Rudolf Carnap*. La Salle, Ill.: Open Court 1963, 87–98, at 88–89.

16 R. Carnap, "Replies." In Schilpp. *op. cit.*, 859–1015, at 861–862.

17 R. Carnap, "On Some Concepts of Pragmatics", *op. cit.*, 250.

firmed', etc." which "belong to pragmatics and require a reference to a person".[18] There he distinguished the concepts of statistical probability from that of degrees of confirmation with the latter, unlike the former, being designated pragmatical, like the concept of degree of belief.[19] Likewise, still earlier he had stated that "a statement of a degree of confirmation does not characterise an objective situation but rather the state of knowledge of a certain person with respect to a certain situation, while a statement of probability in the statistical sense characterises an objective situation. The first belongs to pragmatics, the second to science itself, if expressed with respect to events, or to semantics if expressed with respect to the sentences describing the events."[20] According to this characterisation, confirmation theory and inductive logic must be excluded from the logic of science on account of their use of pragmatical concepts. Yet as Carnap began to explore the concept of logical probability, he found that he could characterise probability in terms of the evidential support given to propositions by other propositions – independently of anybody's belief in these propositions. That of course meant that logical probability was not a pragmatical concept, unlike the concept of degree of confirmation he had entertained in 1939–1942.

As Carnap put it when he introduced the concept, he was concerned "with what may be called the logical side of confirmation, namely, with certain logical relations between sentences", adding that "both parts of logic" – deductive and inductive – "belong to semantics".[21] This logical conception of probability had to be distinguished sharply from the frequency conception of probability.[22] The former designates "the degree of confirmation of a hypothesis *h* with respect to an evidence statement *e*", the latter "the relative frequency (in the long run) of one property of events or things with respect to another".[23] Another explication of logical probability that Carnap endorsed regards them as rational degrees of belief or fair betting quotients such that "a bet on *h* with a betting quotient *q* for the two bettors whose knowledge is *e* is a fair bet".[24] Yet Carnap's explication of logical probability in terms of the semantic concepts of confirmation and rational degrees of belief still had to be distinguished from what, earlier, he had rejected from the logic of science as merely pragmatic. Carnap opposed Ramsey's char-

18 R. Carnap, *Introduction to Semantics*, *op. cit.*, 28.

19 *Ibid.*, 244–245.

20 R. Carnap, "Science and Analysis of Language." in: *Journal of Unified Science (Erkenntnis)* 9 (1939), 221–226, at 225.

21 R. Carnap, "Two Concepts of Probability." In H. Feigl and W. Sellars (Eds.), *Readings in Philosophical Analysis*, New York: Appleton-Century-Crofts, 1949, 330–348, at 330–331.

22 *Ibid.*, 333–334.

23 R. Carnap, *LFP*, *op. cit.*, 19.

24 *Ibid.*, 166.

acterisation of the theory of probability as "the logic of partial belief" as unduly "psychological and subjectivistic"[25] and added:

It cannot, of course, be denied that there is also a subjective, psychological concept for which the term "probability" sometimes is used. This is the concept of the degree of actual, as distinguished from rational belief: "the person *X* at time *t* believes in *h* to the degree *r*". This concept is of importance for the theory of human behavior, hence for psychology, sociology, economics, etc. But it cannot serve as a basis for inductive logic or a calculus of probability applicable as a general tool of science.[26]

So delimited, inductive logic as a theory of logical probability was restored to the logic of science (while the frequency conception was assigned to use by first-order sciences).

Carnap did not stay wholly untouched by psychological questions raised by his theory of probability – and it is this fact that threatens to confuse our characterisation of metatheoretical inquiries: first, by challenging the neat taxonomy developed so far and, second, by casting doubt on its antipsychologistic credentials. Having settled in *Logical Foundations of Probability* on a particular numerical function (a *c*-function) as explicating logical probability (namely *c**) – albeit warning against thinking it "perfectly adequate …, let alone … the only adequate one"[27] – Carnap soon came to recognise a plurality of such *c*-functions each of which represented "the optimum method" for different circumstances.[28] However, in still later work Carnap rejected the view that the choice of an inductive method or *c*-function was objectively determined and declared it instead to be dependent on personality traits of different inquirers – this is the "personalist point of view" that occasions our worry. In his final "Basic System of Inductive Logic", Carnap sought to delimit this freedom somewhat by means of further constraints, but he had to admit that all his inductive logic was able to furnish were "some general rules, each of which warns … against certain unreasonable steps", "certain features of a general policy", all of which left "some freedom for choice within certain limits" in picking a particular *c*-function.[29] Let's specify our problem further.

In his "Basic System" Carnap noted that he understood by "'inductive logic' … a theory of logical probability providing rules for inductive thinking".[30] Accordingly he took an interest in "normative decision theory" as a theory that "states conditions for the rationality of decisions".[31] Normative decision theory served as the "connecting link between descriptive decision theory and inductive

25 *Ibid.*, 45–46.

26 *Ibid.*, 51.

27 *Ibid.*, 563.

28 R. Carnap, *The Continuum of Inductive Methods*. Chicago: University of Chicago Press, 1952, 56.

29 R. Carnap, BS2, *op. cit.*, 106.

30 R. Carnap, BS Intro, *op. cit.*, 7.

31 *Ibid.*, 8.

logic" because it was concerned not with a person's actual degrees of belief in a proposition but with what their degree of belief should be, in other words, "not with actual credence but with rational credence".[32] Normative decision theory applies inductive logic to decision making. (It is distinct from descriptive decision theory which deals in actual degrees of belief, not rational ones.) The concepts of normative decision theory are "quasi-psychological" because they are idealised counterparts to the concepts of descriptive decision theory: they are "assigned to an imaginary subject *X* supposed to be equipped with perfect rationality and an unfailing memory".[33] Due to that idealising supposition the concepts of a rational initial credence function (Cr_o) and a rational stable credibility function (*Cred*) can be regarded as instantiating the values of their corresponding concepts of the inductive measure function (*M*) and the inductive confirmation function (*C*).[34] The latter, of course, are purely logical concepts having "nothing to do with observers and agents, whether natural or constructed, real or imaginary"[35]. Given that "inductive logic studies those *M*-functions that correspond to rational Cr_o-functions, and those *C*-functions that correspond to rational *Cred*-functions",[36] Carnap's claim that inductive logic provides "rules for inductive thinking" is quite unobjectionable since the inductive logic itself dealing with *M*- and *C*-functions remains purely logical.[37]

Yet Carnap now also noted that the "methodological status" of normative decision theory "is in fact somewhat problematic" but did not elaborate.[38] It may

---

32 *Ibid.*, 13.

33 *Ibid.*, 25.

34 Carnap defined these functions as follows: "I call a system of degrees of belief for a given field of propositions a *credence function*. We wish to distinguish between reasonable and non-reasonable credence functions. … [T]he *credibility function* … is defined as follows. The credibility of a proposition *H*, with respect to another proposition *A*, for a person *X* means the degree of belief that *X* would have in *H* if and when his total observational knowledge of the world was *A*. … While the credence functions merely reflect his momentary beliefs at various times, his credibility function expresses his underlying permanent disposition for forming and changing beliefs under the influence of his observations." ("Inductive Logic and Inductive Intuition." In I. Lakatos (Ed.), *The Problem of Inductive Logic*. Amsterdam: North Holland 1968, 258–267, at 260 and 262).

35 *Ibid.*

36 R. Carnap, BS Intro, *op. cit.*, 25.

37 "Inductive logic … may be regarded as as a part of logic in view of the fact that the concepts occurring are logical concepts." (*Ibid.*, 26) Carnap went on parenthetically in the revised version: "Exactly speaking, this holds only for *pure* inductive logic, not for *applied* inductive logic." (*Ibid.*)

38 *Ibid.*, 13. Already in *LFP* Carnap had noted that "a rule which tells man *X*, with the help of inductive logic, which decisions it would be reasonable for him to make in view of his past experiences … does not belong to inductive logic itself but involves the methodology of induction and of psychology" (1950a, 252–253) and he stressed that "the problems and difficulties here involved belong to the methodology of a special

appear that this unclarity could be overcome by declaring that normative decision theory is an application of inductive logic. With the theory of logical probability part of the logic of science, one might thus place normative decision theory in the pragmatics of science.

Yet putting matters this way overlooks several things. First, that Carnap's "Basic System of Inductive Logic" employed additional psychological-empirical parameters, for instance, so-called Lambda-families of functions intended to reflect the rate of "learn[ing] … from experience".[39] Carnap's later view of inductive logic itself appears to be not a purely logical theory of probability but one employing quasi-psychological notions. Should Carnap's worry about the methodological status of normative decision theory perhaps extend to his personalist inductive logic itself? Its status as part of the logic of science appears to be called into doubt. Second, there is the issue as to which type of pragmatics the normative decision theory should be assigned to. Applications of logical theories are often, or even typically, assigned to pragmatics as a descriptive empirical inquiry, but normative decision theory patently is not empirical. So it cannot be an application in the sense in which the descriptive syntax of L_{II} of *Logical Syntax* represents an application of its theory of pure syntax. In addition a problem arises that goes beyond issues of classification, to wit, the question of what this personalist turn of his inductive logic means for Carnap's anti-psychologism. Has Carnap not abandoned the very principles on which his logic of science was built – and so undermined the bipartite metatheory conception?

The second of these three questions is perhaps the easiest to answer. We can take a leaf out of Carnap's book and make use of his category of theoretical or pure pragmatics. As we saw, Carnap expressed the hope that theoretical pragmatics would give logical explications of families of pragmatical concepts still beyond those related to the concept of intension in natural languages. The concepts involved in normative decision theory are natural candidates for this extension. As Carnap explained, normative decision theory is closely connected with but clearly separable from logic:

> It is an interesting result that this part of normative decision theory, namely, the logical theory of the *M*- and *C*-functions, can thus be separated from the rest. We should note, however, that this logical theory deals only with the abstract, formal aspects of probability, and that the full meaning of (personal) probability can be understood only in the wider context of decision theory through the connections between probability and the concepts of utility and rational action.[40]

---

branch of empirical science, the psychology of valuations as a part of the theory of human behavior, and that therefore they should not be regarded as difficulties of inductive logic" (*ibid.*, 254).

39 R. Carnap, BS2, *op. cit.*, 95.

40 R. Carnap, BS Intro, *op. cit.*, 26.

Normative decision theory – precisely due to the connection of its logical concepts with those of utility and rational action – can be counted into pure pragmatics. What renders it pure is that no particular individuals are mentioned, but only variables in their place. This coheres with Carnap's categorisation of "applied inductive logic":

The relation between pure and applied IL [inductive logic] is somewhat similar to that between pure (mathematical) and empirical (physical) geometry. ... The situation in IL is analogous. In *applied* IL, we give an interpretation of the language. ... In contrast, in *pure* IL we describe a language system in an abstract way, without giving an interpretation of the nonlogical constants (individual and predicate constants). Strictly speaking, we merely deal with unspecified individuals a_1, a_2, and so on, a family of, say, six unspecified attributes $P_1, P_2, \ldots, P_6$, with corresponding regions $X_1, X_2, \ldots, X_6$ in an abstract space U, and with functions d and w.[41]

Applied inductive logic comes under the heading of descriptive pragmatics since it mentions particular individuals. Normative decision theory belongs to pure pragmatics since no particular individuals are mentioned.

Yet what are we to make of an inductive logic that includes parameters for the rate of learning from experience? Note that the task of inductive logic was that "of telling us how to arrive at values for our degree of belief which we can defend as rational".[42] Note also that the choice of a particular lambda-function, of a particular rate of learning from experience, was left open by the inductive logic itself. Thus one might hold that "although certain boundaries for lambda can be determined objectively by the consideration of rationality requirements, within these limits, everyone is free to make this choice as he pleases".[43] Carnap himself once preferred what he called the "personalist point of view": "we might regard *X*'s choice of a lambda value (and likewise his decisions in other respects in the process of constructing a *C*-function for some form of language) as determined by, and therefore symptomatic of, certain features of *X*'s personality".[44] Accordingly, "the difference may be attributed to their different inductive inertia."[45] Yet Carnap also considered the choice of the lambda-value "from an objectivist point of view", namely as determined by the curve of the eta-function. He concluded that

there need not be a controversy between the objectivist point of view and the personalist or subjectivist point of view. Both have a legitimate place in the context of our work, that is, the construction of a system of rules for determining probability values with respect to possible evidence. At each step in the construction, a choice is to be made; the choice is

41 R. Carnap, BS1, *op. cit.*, 69–70.

42 R. Carnap, "Inductive Logic and Inductive Intuition", *op. cit.*, 259–260.

43 R. Carnap , BS2, *op. cit.*, 111–112.

44 *Ibid.*, 112.

45 *Ibid.*, 114.

not completely free but is restricted by certain boundaries. Basically, there is a mere difference in attitude or emphasis between the subjectivist tendency to emphasize the existing freedom of choice, and the objectivist tendency to stress the existence of limitations. I give more attention to the latter because in my world I am mainly interested in discovering new rationality requirements which lead to narrower boundaries.[46]

Carnap was resigned to finding that not inconsiderable leeway remained in the construction of a system of rules for determining probability values with respect to possible evidence.

So inductive logic did not determine the rate of learning from experience: it held a place for it, but allowed for different determinations of it. Yet given that such psychological parameters were an inherent part of it, what becomes of the place of inductive logic in the taxonomy of the pure and descriptive subdisciplines of the semiotics of the language of science? As we learnt, Carnap assigned logical probability to semantics. That over the years Carnap's favoured explication of logical probability shifted from that of indicating the degree of confirmation of *h* given *e* to that of indicating the degree of rational belief in *h* given *e* did not change its basic semantic nature. If anything, that shift in interpretation gives grounds to assimilate something like the rate of learning from experience to the concepts closely related to the semantic notion of probability so understood. (Carnap was, after all, interested in a rational rate of learning from experience even though no unique value can be determined for it.) Since, moreover, the rate was not only not determined by the inductive logic nor was any specific learner indicated as such, it is clear that this inductive logic was not to be counted into descriptive semantics. Within the logic of science it remained a branch of pure semantics, albeit one that was informed – unlike deductive logic – by pure pragmatics.

It is important to notice clearly the following distinction. While the *axioms* of inductive logic themselves are formulated in purely logical terms and do not refer to any contingent matters of fact, the *reasons* for our choice of the axioms are not purely logical. … In order to give my reasons for the axiom [of symmetry: *M* is invariant with respect to any finite permutation of individuals], I move from pure logic to the context of decision theory and speak about beliefs, actions, possible losses and the like. However, these considerations are not in the field of descriptive, but of normative decision theory. Therefore, in giving my reasons, I do not refer to particular empirical results concerning particular agents or particular states of nature and the like. Rather, I refer to a *conceivable* series of observations by *X*, to conceivable sets of possible acts, to possible states of nature, to possible outcomes of the acts, and the like. These features are characteristic for an analysis of the *reasonableness* of a given function Cr_0, in contrast with an investigation of the *successfulness* of the (initial or later) credence function of a given person in the real world. Success depends upon the particular contingent circumstances, rationality does not.[47]

46 *Ibid.*, 119.

47 R. Carnap, BS Intro, *op. cit.*, 26.

This allows us, finally, to give fairly quick answer to the third question. Did Carnap's personalist turn in inductive logic betray the anti-psychologism on which his logic of science was built – and so undermined the bipartite metatheory conception? All the reasons we gave that allowed both inductive logic itself and normative decision theory to remain within the logic of science also dispell this threat.

34.3 Conclusion

With normative decision theory pursued in pure pragmatics and inductive logic remaining part of pure semantics, the architecture of Carnap's, Neurath's and Frank's replacement of traditional philosophy by a bipartite metatheory of science stays intact. It is not threatened by Carnap's personalist turn in inductive logic. Importantly, however, the bipartite metatheory conception is not committed to conceiving of the theory of probability in just the way Carnap did. If indeed "the probability calculus corresponds to some quite objective feature of subjective uncertainty"[48], then Bayesianism too could find its place therein.

School of Social Science
University of Manchester
Oxford Road
M13 9PL, Manchester
United Kingdom
thomas.uebel@manchester.ac.uk

48 C. Howson and P. Urbach, *Scientific Reasoning*, *op. cit.*, 95.

CHAPTER 35

MICHAEL STÖLTZNER

ERWIN SCHRÖDINGER, VIENNA INDETERMINIST

Whenever Erwin Schrödinger wrote about causality and determinism, he acknowledged Franz Serafin Exner as the first to have advocated a genuinely indeterminist conception of physics. Today best known is his 1922 Zurich inaugural address "What is a Law of Nature?", that was published only in 1929 with an introductory note stating that the "development of quantum mechanics has brought Exner's sphere of ideas into the focus of scientific interest." (1929, p. 9/133)

Scholars substantially disagree about the content and import of Schrödinger's philosophy of physics. To some, he repeatedly changed his mind about fundamental issues, among them causality and realism; to others, he tenaciously pursued a complex philosophical program on various levels that was notoriously misunderstood by his Copenhagen opponents. In a classic paper, Paul Forman (1971) has attributed Schrödinger's endorsement of indeterminism in the Zurich speech to the influence of the anti-causal and anti-scientific milieu of the early Weimar Republic, while his discovery of a seemingly causal quantum mechanics in 1926, the Schrödinger wave equation, prompted him to abandon this position. Forman and, more explicitly, Paul Hanle have criticized Schrödinger's – and Exner's – "failure to distinguish between indeterminacy in principle and the practical inability to analyze the determinate causes in an aggregation of micro-physical events." (Hanle 1979, p. 227) Yemina Ben-Menahem instead holds "that Schrödinger did not change his views in any substantial way with regard to causality. To the end of his life he was ready to entertain the idea that some of the fundamental laws of nature are merely statistical laws." (1989, p. 309) In the same vein, Michel Bitbol considers Schrödinger's writings of the 1920s as "an early and simplistic way of coming close to the interpretation of the 1950s" and concludes from a retrospective analysis that Schrödinger developed "by fits and starts … a coherent methodological program" (1996, p. vii).

Interestingly, interpreters' stand on continuity versus vacillation strongly depends upon how much importance they assign to philosophy within Schrödinger's overall work. For Forman, his philosophical convictions were determined by the milieu. To Mara Beller, "Schrödinger was no less a philosophical 'opportunist' than his Göttingen-Copenhagen opponents" (1999, p. 284), while Ben-Menahem, Bitbol, and Henk de Regt take his philosophy very seriously and associate its core traits not only with the local Mach-Boltzmann tradition and Exner's indeterminism, but also

D. Dieks et al. (eds.), *Probabilities, Laws, and Structures*, The Philosophy of Science in a European Perspective 3, DOI 10.1007/978-94-007-3030-4_35,

with a long list of classical and contemporary views on causality and realism, among them Schopenhauer (de Regt 1997; Bitbol 1996), neo-Kantianism (Beller 1999), Husserlian and post-modern phenomenology (Bitbol 1996), van Fraassen's constructive empiricism (de Regt 1997), and Putnam's internal realism (Bitbol 1996).

The aim of this paper is to investigate Schrödinger's views about causality and indeterminism by embedding them into the thought-style characteristic of the Vienna physicists around Boltzmann and Exner. At least until the famous cat paper (Schrödinger 1935) changed the focus of the debates about quantum mechanics to issues of realism *tout court*, Schrödinger, I shall argue, remained committed to a specific Viennese brand of indeterminism that remained agnostic about the alternative 'in principle' vs. 'in practice' which Hanle, and many other contemporary readers, considered impeccable. The tradition that I have called Vienna Indeterminism (Stöltzner 1999) objected to the metaphysical alternative between determinism and indeterminism, but favored indeterminism on epistemological and methodological grounds. Having thus shown that Schrödinger accepted the indeterministic nature of the basic laws of nature already in 1914, what were then his problems with the Copenhagen Interpretation? Ironically, one might say, it was a kind of static positivism that imposed a priori limits of meaning to basic physical concepts, against which Schrödinger insisted on the openness of the scientific enterprise expressed in Boltzmann's conception of theories as universal pictures. There is of course a development in Schrödinger's attempts to elaborate a more precise formulation of Vienna Indeterminism in the light of changing physical theories. While he initially entertained the prospect of a definitive empirical resolution in favor of indeterminism, he later came to consider it as a matter of convention and ontological parsimony.

35.1 A Brief of Vienna Indeterminism

The tradition of Vienna Indeterminism began with Exner's 1908 inaugural address as Rector of the University of Vienna titled "On Laws in the Sciences and Humanities". Exner combined core traits of Mach's empiricism and Boltzmann's philosophical justification of statistical mechanics to argue that chance was the basis of all natural laws and that accordingly the apparent determinism in the macroscopic domain emerged only as the thermodynamic limit of many random microscopic events. To those who took Mach and Boltzmann as the principal foes in the struggle about atomism – as most German physicists did –, such a synthesis seemed surprising. But it was characteristic for the thinking of the Viennese physicists, as Schrödinger explained in a letter to Arthur S. Eddington in 1940.

> [W]e did not consider them irreconcilable. Boltzmann's ideal consisted in forming absolutely clear, almost naively clear and detailed 'pictures' – mainly in order to be quite sure of avoiding contradictory assumptions. Mach's ideal was the cautious synthesis of observational facts that can, if desired, be traced back till the plain, crude sensual perception. …

However, ... one was quite permitted to follow one or the other [method of attack] provided one did not lose sight of the important principles ... of the other one. (from Moore 1989, p. 41)

Exner's speech prompted a staunch criticism by Max Planck (1914), and thus became the starting point of a debate about the relationship between causality and physical ontology between Vienna and Berlin that, with a series of new thematic twists, lasted until the 1930s. (See Stöltzner 2003) The basic alternative was this: Either one followed Kant, as did Planck, by holding that to stand in a causal relationship was a condition for the possibility of the reality of a physical object (Kant's 'empirical realism'), or one agreed with Mach that causality consisted in functional dependencies between the determining elements and that physical ontology was about 'facts' (*Tatsachen*), i.e., in stable complexes of such dependencies. To those standing in the Kantian tradition, the Machian stance fell short of the aims of scientific inquiry. Those standing in the Machian tradition, however, had more leeway in searching for an ontology suitable for a new scientific theory.

Based upon this basic distinction, Vienna Indeterminism – as pronounced by Exner – can be characterized by the following three commitments: (i) The highly *improbable events* admitted by Boltzmann's statistical derivation of the second law of thermodynamics exist. (ii) In a consistently empiricist perspective, the burden of proof rests with the determinist who must provide a sufficiently specific theory of microphenomena before claiming victory over a merely statistical theory. (iii) The only way to arrive at an empirical notion of objective probability is by way of the limit of relative frequencies. While many physical systems can practically be treated as close to this limit, it is illusory for processes in the descriptive sciences and the humanities. Exner's endorsement of the relative frequency interpretation implied that there existed a region of transition between the microscopic and the macroscopic; in the years to come it would be filled by an increasing number of fluctuation phenomena.

35.2 Searching for Indeterministic Phenomena: Schrödinger's Vienna Years

Schrödinger's "first outstanding paper" (Moore 1989, p. 75) harked back to Boltzmann's atomism. To find an explicit example where atomism and continuum physics yielded diverging scenarios, one encountered a two-fold task: "First, all those differential equations first derived by consideration of a continuous medium as differential equations in the strict sense, now must instead be derived in the above sense as difference equations on the basis of a model constructed of molecules." (Schrödinger 1914, p. 916f.) Second and more importantly, one had to search and predict "*such* conditions under which the differential equation based on a continuum actually leads to an incorrect result because of the truly atomistic structure of matter." (*Ibid.*, p. 917) Schrödinger thus compared a one-dimensional atomistic model of a string in which one elongates a finite number of isolated atoms with the familiar d'Alembert differential equation of the vibrating string. While for functions which correspond to averages over a sufficiently large number of atoms, both approaches were equivalent, the phenomenon of thermal disturbance (*Wärmestörung*), which corresponded to internal differences of elongation, could be described by the atomistic model only.

Thermal disturbance represented a fluctuation phenomenon. At that time, the Viennese already considered them as physical phenomena in their own right that had been discovered at the beginning of the century simultaneously in the fields of radioactivity research, Brownian motion (the zig-zag motion of suspended particles in a liquid visible under the microscope), and atmospheric electricity. In 1905, Egon von Schweidler, then Exner's assistant, had shown that the phenomenological law of radioactive decay, the Rutherford-Soddy law $N(t) = N_{t=0}e^{-t}$, was only valid for a large number of decaying atoms, while for a small number of atoms the decay constant exhibited fluctuations. In 1906 and independently of Albert Einstein, Exner's former assistant Marian von Smoluchowski derived the formula for the position fluctuations of a Brownian particle. (See Stöltzner 2011)

While most physicists quickly accepted Brownian motion as an experimental proof of atomism, Schweidler fluctuations, until the 1920s, were typically considered as a convenient phenomenological regularity still to be explained by the hitherto unknown laws governing the decay of the single atom. The Viennese thought differently (Cf. Coen 2002), as can also be seen from a long paper in which Schrödinger undertook a detailed statistical analysis of the measurement of radioactive fluctuations. He emphasized that Schweidler's seminal discovery "does *not* consist, as some believe, in any $\sqrt{z}$- or $\sqrt{i}$- dependence, but in the *fundamental recognition of the probabilistic character of the decay constant.*" (1919, p. 179) He developed a theory of the preferred measuring device, the electroscope, treating the motion of its pointer as a Brownian process, or as he put it, a Smoluchowski motion. (Cf. *ibid.*, p. 184) Thus he used a phenomenon with a deterministic foundation as the theory of measurement of a phenomenon without such a foundation. The important point is that Schrödinger did not distinguish between indeterminacy in principle and indeterminacy in practice at all, but took both kinds of fluctuations as physical phenomena. For the Vienna Indeterminist did not entertain any a priori preference for an indeterminist theory with deterministic foundations at the micro-level over one without such foundations. In any case, the burden of proof was with the determinist.

35.3 Zurich 1922: "What is a Law of Nature?"

In his 1922 Zurich inaugural address, Schrödinger specified the historical scale on which this view had emerged. It harked back, beyond Exner, to the days when

Boltzmann had just begun to develop his statistical theory of the second law of thermodynamics and Ernst Mach had given a new version of Hume's empiricist notion of causality. "Within the past four or five decades physical research has clearly and definitely shown that *chance* is the common root of all the strict regularity that has been observed, at least in the overwhelming majority of natural processes, the regularity and invariability of which have led to the establishment of the postulate of universal causality." (1929, p. 9/136)

There was something ironic to the history of this postulate. For, the deeply engrained 'habit of thought' of presupposing causality had emerged "from observing … precisely *those* regularities [*Gesetzmäßigkeiten*] in nature which, in the light of our present knowledge, are most certainly not *causal*, or at least not directly causal, but *directly statistical* regularities." (*Ibid*., p. 11/144) What we observe on the macroscopic scale already involves such a huge number of individual events that the statistical laws appear as strict regularities. Although this guarantees a practical value for the principle of causality, the inference to a causal behavior on the molecular scale is unwarranted. In gas theory,

> we generally assume the validity of the mechanical laws for the single event, the collision. But this is *not* at all necessary. It would be quite sufficient to assume that at each individual collision an increase in mechanical energy and mechanical momentum is just *equally probable* as a decrease, so that taking the *average of a great many* collisions, these quantities remain constant in much the same way as two dice cubes, if thrown a million times, will yield the average 7 whereas the result of each single throw is purely a matter of chance. (*Ibid*., p. 10/138f.)

Brownian motion and radioactive fluctuations were the crucial experiments to establish the statistical character of natural laws in the sense of the program he had depicted in 1914. "More than by however many examples, our conviction of the statistical character of physical laws is strengthened by the fact that the second law of thermodynamics, or law of entropy, *which plays a role in positively every real physical process,* has clearly proved to be the *prototype* of statistical law." (*Ibid*., p. 10/140f.) The reason for the universality of the second law is its intimate connection with the direction of time and the tendency towards more probable states. This had also been Exner's (1909) point. Thus "the assertion of determinism was certainly *possible*, yet by no means *necessary*, and when more closely examined *not at all very probable*." (Schrödinger 1929, p. 10/142f.) Schrödinger was aware that this conclusion was controversial and dependent on the respective state of physics. Even though the problem of causality remained empirically open, in virtue of Occam's razor, "*[t]he burden of proof falls on those who champion absolute causality, and not on those who question it*." (*Ibid*., p. 11/147). Sticking with causality come what may, would yield to a "duplication of natural law [that] so closely resembles the animistic duplication of natural *objects,* that I cannot regard it as at all tenable." (*Ibid*., p. 11/145)

Schrödinger followed Exner and Boltzmann in extending the statistical viewpoint to the most basic principles of physics. "Naturally we *can* explain the theorem of energy conservation on the large scale by its already holding true in the small [that is, for the single events]. But I do not see that we *must* do so." (*Ibid.*, p. 10/143) Even the law of gravitation could be of statistical origin, although Schrödinger admitted that "Einstein's theory strongly suggests the *absolute validity of the theories of energy and momentum conservation.*" (*Ibid.*, p. 11/146) To avoid a conflict, Schrödinger made a surprising move and considered relativity theory as virtually irrelevant for the issue of causality. For, "the whole theory of gravitation can be considered as the reduction of *gravitation* to the *law of inertia.* That *under certain conditions nothing changes* is surely the simplest law that can be conceived, and hardly falls within the concept of causal determination. It may after all be equally reconcilable with a strictly acausal view of nature." (*Ibid.*, p. 11/146)

35.4 The BKS-Theory of 1924: Was the Decisive Example Found?

In 1924, Niels Bohr, Hendrik A. Kramers, and John C. Slater proposed a new quantum theory in which energy conservation held only on average, but not for the individual atomic processes. Schrödinger was enthusiastic about the BKS-theory and wrote to Bohr on 24 May, 1924: "As a pupil of the old Franz Exner, I have long ago become accustomed to the idea that the basis of our statistics is probably not microscopic 'regularity', but perhaps 'pure chance' and that perhaps even the laws of energy and momentum have only statistical validity." (in Bohr 1984, p. 490) Schrödinger's reading of the new theory corresponded to the second of the two tasks he had laid out in 1914. In a survey article in *Die Naturwissenschaften* he emphasized that in the new theory, the conception "that the individual molecular process is not causally determined by 'laws' in a unique fashion, for the first time attains a tangible form." (1924, p. 720) If this theory were true, it would confirm the "Exner-Bohr conception" (*Ibid.*, p. 724) according to which energy conservation is only of statistical validity, and accordingly put it alongside Schweidler's fluctuations and Brownian motion as a factual demonstration of indeterminism. Schrödinger provided some rough estimates to show that the new theory did not contradict present experiences. However only a year later, Geiger and Bothe showed that the energy was conserved in each individual process.

At the end of the paper, Schrödinger argued that a merely statistical validity of the theorem of energy conservation would have "much deeper theoretical consequences than in the case of the entropy theorem." (*Ibid.*, p. 724) While in the latter case a closed system approaches the exact thermodynamic laws in the limit of infinite observation time, in the BKS theory it exhibits an average behavior only for relatively short times.

In the limit $t \rightarrow \infty$ its behavior becomes *completely undetermined.* ... We can reduce the deviation only by increasing the *size* of the system, or by considering it as a subsystem of a more extended system ("heat bath"). The *exact* validity of thermodynamics now could perhaps be maintained at most ... for the double limit $t \rightarrow \infty$ *and* heat bath$\rightarrow \infty$. But this double limit poses much bigger conceptual difficulties than the single one. ... The separated individual systems would be, from the standpoint of unity, a chaos. It requires the connection [with the rest of the world] as a permanent *regulator*, without which, energetically considered, it would wander about at random. (*Ibid.*, p. 724)

Whereas Schrödinger's article was unambiguously positive about the BKS theory, his above-quoted letter continued with some criticism against Bohr's reality criterion.

Your new account to a large extent signifies a return to the classical theory, as far as radiation is concerned. I cannot completely go along with you when you keep calling this radiation 'virtual' ... For what is the 'real' radiation if it is not that which 'causes' transitions, i.e., which creates the transition probabilities? Moreover, another sort of radiation is surely not assumed. Indeed, if one adopts a purely philosophical standpoint, one might even dare to doubt which electron system has a greater reality – the 'real one' which describes the stationary trajectories or the 'virtual one' that emits virtual radiation and scatters impinging virtual radiation. (from Bohr 1984, p. 490)

Interpreting this passage, scholar's have largely followed Linda Wessels' view that "Schrödinger was enthusiastic about the assumption of irreducibly statistical processes, but objected to the authors' reluctance to give a coherent physical picture for the theory." (1977, p. 313) De Regt even concludes that "[p]recisely because his epistemological position amounts to Machian anti-realism, Schrödinger is in a position to object to calling some terms in the theory 'virtual'. If he had adhered to a hard-headed correspondence realism, he would have dismissed the BKS-theory out of hand." (1997, p. 473) De Regt's assessment is essentially correct, with the qualification that the stability of functional dependencies served as Mach's empiricist reality criterion for the basic facts. Accordingly, Mach's whole conception involved a holistic stance, such that no entities were designated in advance as 'real' without their standing in causal relations. Darrigol thus rightly surmises that "Schrödinger would not have dared such a loose speculation in a scientific journal ... had not he been very eager to connect two of his main favorite themes, holism and acausality." (1992, p. 268) In this reading of the BKS-theory, Boltzmann's program had become even more Machian.

35.5 Alleged Counterevidence: The 1926 Letters to Wien

The *locus classicus* for claims that Schrödinger at least temporarily changed his mind in favor of deterministic causality, is a letter he sent to Wilhelm Wien on 25

August 1926. He apparently abrogated the main thrust of his 1922 inaugural address.

> [T]oday I no longer like to assume with Born that an individual event of this kind is "absolutely random", i.e., completely undetermined. I no longer believe today that there is much to be gained from this conception (which I championed so enthusiastically four years ago). ... the *waves* must be strictly causally determined through field laws, the *wavefunctions* on the other hand have only the meaning of probabilities for the *actual* motion of light- or material-particles. I believe that Born thereby overlooks that ... it would still depend on the taste of the observer *which* he now wishes to regard as *real*, the particle or the guiding field. There exists really no philosophical criterion for reality [*Realität*] if one does not want to say: the *real* is only the complex of sense impressions, all the rest are only pictures.

At face value, Schrödinger rejected indeterminism and Born's positivism, and advocated a spatio-temporal description instead. But what to make of the claim that there simply is no other philosophical criterion of reality than Mach's amended by Boltzmann's pictures? Bohr and Born's returning to a pure Machian ontology happened in such a way that, to Schrödinger's mind, the theoretical pictures became entirely detached from any possible realities in space and time. Or in more historical terms, Born's positivism on the basis of a still classical particle ontology endangered the subtle equilibrium between the teachings of Mach and Boltzmann which Schrödinger had imbibed at the Vienna Institute of Physics and, since 1914, elaborated into a joint advocacy of continuous pictures and indeterminism.

Ben-Menahem additionally points to the fact that for Schrödinger, "[c]ausality and continuity were independent." (Ben-Menahem 1989, p. 321) Her main evidence is an earlier letter to Wien written on 18 June 1926. "It appears, to be sure, that at present not all parties are convinced that the renunciation of the basic discontinuities, *if* possible, is to be absolutely welcomed. But I have always wholeheartedly wished that it would be possible, and would have seized the opportunity with both hands – as I did with Bohr-Kramers-Slater." Here, "Schrödinger himself regarded his earlier response to the Bohr-Kramers-Slater paper as fully consistent with the views he held in 1926 when working on wave mechanics." (*Ibid.*, p. 322) Because of the interdependence of causality and continuity, the apparently odd "claim that there is 'not much to be gained' by a probabilistic interpretation ... makes perfect sense. In the BKS paper causality was renounced but continuity rescued. In Born's case, however, there was no such pay-off." (*Ibid.*, p. 326) Darrigol's interpretation is similar: "one theory [wave mechanics] offered a fairly detailed space-time *picture* of radiation processes, despite the quantum jumping, while the other [matrix mechanics] explicitly denied the possibility of representing quantum processes in space and time. What Schrödinger could not accept was the mutual destruction of the claims of causality and visualizability." (1992, p. 268) In one of Schrödinger's early papers on wave mechanics, Bitbol (1996, p. 17) rightly finds the same motive at work against Born's probabilistic interpretation of wave mechanics.

I am flinching from this conception [*Begriffsbildung*], not so much on account of its complexity as on account of the fact that a theory which postulates an absolute primary probability as a law of nature should at least repay us by freeing us from the old 'ergodic difficulties' and establishing us to understand the unidirectionality of natural processes without further supplementary assumptions. (Schrödinger 1927, p. 968)

If one adopts a fully probabilistic approach it should at least eliminate the, back then notoriously unclear, ergodic hypothesis which arises from the combination of a deterministic theory of the microphenomena and a statistical theory at the macroscopic level. So in the end, Copenhagen was simply too classical in ontological matters. Or as Bitbol puts it, "Schrödinger did not consider it satisfactory to add an empirically void 'clothing' to the structure of quantum mechanics just for the sake of recovering the classical ontology or for the sake of satisfying the desire for pictures. What he wished to demonstrate was rather that there exists an adequate picture and a (non-classical) ontology which arises quite naturally from unmodified quantum mechanics itself." (1996, p. 68) Indeterminism by itself, it becomes clear, was not Schrödinger's problem.

35.6 Continuing the Debate with Planck: Berlin 1929

In 1927, Schrödinger had assumed Planck's former chair at the University of Berlin. Planck still hoped that Schrödinger's formulation of quantum mechanics, albeit meanwhile proven equivalent to matrix mechanics, promised a return to deterministic physics and a realist physical world view. Yet his 1929 inauguration speech as a member of the Prussian Academy of Sciences continued the debate between Exner and Planck on his teacher's side. He contemplated whether the development of quantum mechanics forced us to abandon

the maxim that fixed laws together with random initial conditions uniquely determine the happenings in each individual case. It is the question about the purposivity [*Zweckmäßigkeit*] of the unswerving postulate of causality. It is true, in practice we had had to forgo causality already within the classical mechanical explanation of nature. (Schrödinger 1929, p. 732)

[T]he probabilistic conception of the laws of nature … by itself does not really contradict the causal postulate. Uncertainty in this case arises only from the practical impossibility of determining the initial state of a body composed of billions of atoms. Today, however, the doubt as to whether the processes of nature are uniquely determined is of quite a different character. The difficulty of ascertaining the initial state is supposed to be not one of practice but of principle. (*Ibid.*, p. 732/xvi)

Had Schrödinger now, as a consequence of the quantum revolution, ultimately accepted the distinction between "in practice" and "in principle", rather than insisting that Brownian motion and radioactivity stood on a par as empirical demonstrations of the indeterministic character of a certain domain of phenomena? Not quite, but he provided a new philosophical justification as to why the alternative was not as fundamental as most physicists thought.

> Franz Exner … was the first to mention the possibility and the advantages of an acausal conception of nature. … But I do not believe that in this form [this fundamental question] will ever be answered. In my opinion this question does not involve a decision as to what the real constitution of nature is, but rather as to whether the one or the other predisposition of mind be the more purposive and convenient one with which to approach nature. Henri Poincaré has illustrated that we are free to apply Euclidean or any kind of non-Euclidean geometry we like to real space, without having to fear the contradiction of facts. But the physical laws we discover are a function of the geometry which we apply, and it may be that the one geometry entails complicated laws, the other much simpler ones. In that case the former geometry is inconvenient, the latter is convenient, but the words "right" or "wrong" are unsuitable. The same probably applies to the postulate of rigid causality. One can hardly imagine empirical facts which ultimately decide on whether the natural phenomena are in reality absolutely determined or partially indetermined, but at best on whether the one or the other conception permits a simpler survey of what is observed. Even this question will probably take a long time to decide. (*Ibid.*, p. 732/xvii f.)

The argument from simplicity was not new; recall Schrödinger's criticism of the duplication of natural law. Moreover, the reference to Poincaré made clear that the choice of the proper conception was not empirically void, but guided by simplicity.[1] Yet Schrödinger's (1924) optimism that a decision in favor of indeterminism was close has faded away. Admittedly, also Exner had remained open with respect to the alternative between determinism and indeterminism. But he had preferred the latter in virtue of manifold supportive evidence and because of its more unified character. Schrödinger's own works, particularly his proof of the equivalence between wave mechanics and matrix mechanics, substantially changed the nature of the alternative. There was, on the one hand, a beautifully deterministic differential equation the application or interpretation of which permitted only statistical predictions. There was, on the other hand, an abstract and openly indeterministic theory which nonetheless integrated the whole conceptual apparatus of classical mechanics in a quantized form. What Schrödinger established with his equivalence proof corresponded to the systematic classification of all possible geometries achieved at the end of the nineteenth century which had constituted the basis of Poincaré's conventionalism. In contrast to a Machian view which took all theoretical descriptions just as mere economizations, conventionalist choice required a precise formal characterization of the alternatives. But it did not reintroduce a principal divide between fluctuation phenomena and atomic physics. For

---

1 In notes written in 1918 and titled "Kausalität", Schrödinger had already "quoted Poincaré's statement about principles: 'They are neither true nor wrong, they are expedient [commodes]' and he commented: 'This is certainly entirely true of the causality principle." (Darrigol 1992, p. 264)

ultimately, indeterminism and ontology were intermingled to an even larger extent than in statistical mechanics.

35.7 Schrödinger's Indeterminism after 1930

In 1932, Schrödinger assembled two lectures into a small booklet dedicated to the memory of Franz Exner. In the first, titled "On Indeterminism in Physics", Schrödinger argued that while one and a half decades ago nobody doubted the dogma of determinism, now many physicists believed that the repeated failures to understand the experimental results of the preceding three decades by means of deterministic pictures had lead to a dismissal of determinism in the sense of classical mechanics. But repeated failure by itself could not be decisive.

It will be difficult to ever *prove* that no determined [*bestimmtes*] picture can be found which equally does justice to the facts. But what makes these modern attempts to abandon determinism nonetheless very interesting is that their declarations of a lack of determination are not at all *vague and undetermined*, but entirely precise, quantitative, expressible in cm, g, s (Schrödinger 1932, p. 3/55).

Since a "comprehensive and definitive judgment about these matters *does not at all exist at the present moment*" (*Ibid*., p. 7/59), Schrödinger added three "in part loosely connected" (*Ibid*., p. 6/59) remarks in which he nonetheless defended quite specific theses. The conventionalist considerations of the "Antrittsrede" were, however, not taken up again.

Interestingly, the first rehearsed Schrödinger's 1914 concern (and Boltzmann's teaching) that the issue between determinism and indeterminism had to be decided by the more adequate mathematical description. As mechanical motions are determined by the accelerations, we got used to count the initial velocity among the initial conditions. But this is, strictly speaking, not correct because the definition of velocity by means of a differential quotient involves two moments in time of which one imagines that they can be made to coincide in the limit. "[P]erhaps this mathematical limit ... is inadmissible. Perhaps the thought machinery [*Denkapparat*] invented by Newton is *not sufficiently adapted to nature*. The modern claim, that for sharply defined position in space the concept of velocity becomes meaningless points strongly in that direction." (*Ibid*., p. 9/62)

Schrödinger declared, in the second remark, that "the overwhelming majority [of natural laws are statistical] because the course of nature is essentially irreversible, *one-sided*," (*Ibid*., p. 11/64) perhaps except for gravitation. While before the advent of quantum mechanics, "the abandonment of determinacy was merely of a *practical* kind, *today* one assumes that it is theoretical." (*Ibid*., p. 12f./66–68) But this did not warrant a radical shift. "It was said, and sometimes it is said still today, that ... without a strictly deterministic background our *picture* of nature would degenerate into a complete chaos and thus would not fit to our *given* nature because nature is in fact not completely chaotic. This is certainly *not* correct." (*Ibid.*, p. 14/68)

In his third remark, Schrödinger criticized the ontology of the Göttingen-Copenhagen picture by claiming "that the concepts 'position', 'trajectory' [*Bahn, Bahnkurve*] are exaggerated when applied to such small [atomic] spatial and temporal dimensions" (*Ibid.*, p. 22/77). The same held true for a material point whose motion constituted a trajectory. "To speak of electrons and protons as material points but to deny nevertheless that they have definite trajectories seems to be contradictory and rather crazy. … [F]rom atomistics one can quite well understand, or at least conjecture, that the concept of *trajectory* is lost at very small dimensions." (*Ibid.*, p 17/72)

At this point Schrödinger turned Boltzmann's atomism against the Göttingen-Copenhagen picture according to which material particles are the basis of quantum mechanical ontology without having well-defined trajectories. If we depart from how we actually observe natural phenomena, it seems to be clear that "[e]very quantitative, measuring observation is discontinuous by its very nature" (*Ibid.*, p. 17/72) because it ultimately represents nature's answer to a finite number of yes-no question. We complete this finite raw material by interpolation and in this way arrive at a continuous trajectory, which in itself is not directly observable. This procedure, however, is admissible only if all such measurements could in principle be performed by really existing apparatus. To be sure, "we continuously have to complete what is directly observed; otherwise there would be no picture of nature but only an inextricable patchwork of individual findings [*Einzelfeststellungen*]." (*Ibid.*, p. 21/76f.) By inferring from a finite set of observations to a continuum in this way, we run the risk to erroneously complete our factual observations and "mess up our picture of nature" (*Ibid.*, p. 21/77) by employing a concept, such as 'trajectory', outside its domain of validity.

Two years later, Schrödinger intensified his criticism that the concepts of classical point mechanics were still applied albeit with absolute limits of precision. "The *concepts* must be abandoned, not their sharp definitiveness. One tries to get around the monstrosity of unsharply defined concepts by hundred thought experiments," (1934, p. 519) among them the Heisenberg microscope. "Among the concepts to be abandoned is also position. But this means: *geometry*." (*Ibid.*, p. 519) The reason was that geometry was based on congruence the empirical realization of which presupposed the existence of rigid bodies. According to Schrödinger, the application of geometry to real objects represented a gedanken experiment which had to be consistent with the laws of nature. The classical solution to approximate rigid connections by potentials was impossible due to the finite distance between energy levels. Thus there could be only approximately rigid bodies. Schrödinger concluded that "the spatial structure derived from the group of translations fit to nature only approximately – and not merely that there do not exist sufficiently precise material measuring rods to measure it. The true geometry of physics is … the four-dimensional one of relativity theory. ... The difficulty to adapt to the requirement of relativity is a well-known crux of quantum mechanics." (*Ibid.*, p. 520) In short, geometry was inapplicable to small distances. And consequently, quantum mechanics was at odds with relativity theory. The speculations that even the theory of gravity could not escape indeterminism apparently had not gone away. But a suitable ontology for such a future theory was not in sight.

Searching for such ontology would become Schrödinger's main concern in the two decades to come from the cat paper (Schrödinger 1935) onward. Carrying out what Bitbol (1996) has convincingly reconstructed as a complex, yet largely coherent interpretational program involved additional philosophical ideas that transcend the local Viennese tradition analyzed in the present paper, among them the measurement problem and the subject-object divide inherent by the Copenhagen interpretation.

References

Beller, Mara (1999), *Quantum Dialogue. The Making of a Revolution,* Chicago: The University of Chicago Press.

Ben-Menahem, Yemina (1989), "Struggling with Causality: Schrödinger's Case", *Studies in History and Philosophy of Modern Science* 20, 307–334.

Bitbol, Michel (1996), *Schrödinger's Philosophy of Quantum Mechanics*, Dordrecht: Kluwer.

Bohr, Niels (1984), *Collected Works. Volume 5: The Emergence of Quantum Mechanics (Mainly 1924–1926),* edited by Klaus Stolzenburg, North-Holland, Amsterdam.

Coen, Deborah R. (2002), "Scientists' Errors, Nature's Fluctuations, and the Law of Radioactive Decay, 1899–1926", *Historical Studies in the Physical Sciences* 32, 179–205.

Darrigol, Olivier (1992), "Schrödinger's Statistical Physics and Some Related Themes", in Michel Bitbol, and Olivier Darrigol, (Eds.), *Erwin Schrödinger. Philosophy and the Birth of Quantum Mechanics*, Gif-sur Yvette: Editions Frontières, pp. 237–276.

De Regt, Henk W. (1997), "Erwin Schrödinger, *Anschaulichkeit*, and Quantum Theory", *Studies in History and Philosophy of Modern Physics* 28, 461–481.

Exner, Franz S. (1909), *Über Gesetze in Naturwissenschaft und Humanistik*, Wien-Leipzig: Alfred Hölder.

Forman, Paul (1971), "Weimar Culture, Causality, and Quantum Theory, 1918-1927: Adaption by German Physicists and Mathematicians to a Hostile Intellectual Environment", *Historical Studies in the Physical Sciences* 3, 1–114.

Hanle, Paul. A. (1979), "Indeterminacy before Heisenberg: The Case of Franz Exner and Erwin Schrödinger", *Historical Studies in the Physical Sciences* 10, 225–269.

Moore, Walter (1989), *Schrödinger – Life and Thought,* Cambridge: Cambridge University Press.

Planck, Max (1914), *Dynamische und statistische Gesetzmäßigkeit,* Leipzig: Barth.

Schrödinger, Erwin (1914), "Zur Dynamik elastisch gekoppelter Punktsysteme", *Annalen der Physik* 4, 916–934.

Schrödinger, Erwin (1919), "Wahrscheinlichkeitstheoretische Studien betreffend Schweidler'sche Schwankungen, besonders die Theorie der Meßanordnung", *Sitzungsberichte der Österreichischen Akademie der Wissenschaften, Mathematisch-naturwissenschaftliche Klasse*, Abt. IIa, 128, 177–237.

Schrödinger, Erwin (1929), "Was ist ein Naturgesetz?", *Die Naturwissenschaften* 17, 9-11; English translation by J. Murphy and W. H. Johnston in *Science and the Human Temperament,* New York: W. W. Norton & Co., 1939, 133–147.

Schrödinger, Erwin (1924), "Bohrs neue Strahlungshypothese und der Energiesatz", *Die Naturwissenschaften* 12, 720–724.

Schrödinger, Erwin (1927), "Energieaustausch nach der Wellenmechanik", *Annalen der Physik* IV.83, 956–968.

Schrödinger, Erwin (1929), "Aus der Antrittsrede des neu in die Akademie eintretenden Herrn Schrödinger", *Die Naturwissenschaften* 17, 732; partial English translation in the Introduction to *Science and the Human Temperament, op. cit.,* xiii–xviii.

Schrödinger, Erwin (1932), *Über Indeterminismus in der Physik. Ist die Naturwissenschaft milieubedingt? Zwei Vorträge zur Kritik der naturwissenschaftlichen Erkenntnis*, Leipzig: J.A. Barth. English translation in *Science and the Human Temperament, op. cit.*, 52–80.

Schrödinger, Erwin (1934), "Über die Unanwendbarkeit der Geometrie im Kleinen", *Die Naturwissenschaften* 22, 518–520.

Schrödinger, Erwin (1935), "Die gegenwärtige Situation in der Quantenmechanik", *Die Naturwissenschaften* 23, 807–812 & 823–828 & 844–849.

Stöltzner, Michael (1999), "Vienna Indeterminism: Mach, Boltzmann, Exner", *Synthese* 119, 85–111.

Stöltzner, Michael (2003), *Vienna Indeterminism. Causality, Realism and the Two Strands of Boltzmann's Legacy,* Ph.D.- dissertation, University of Bielefeld, March 2003; accessible under http://bieson.ub.uni-bielefeld.de/volltexte/2005/694/

Stöltzner, Michael (2011), "Zur Genese der Schweidlerschen Schwankungen und der Brownschen Molekularbewegung", in Silke Fengler and Carola Sachse (Eds.), *Kernforschung in Österreich*, Wien: Böhlau, forthcoming.

Wessels, Linda (1977), "Schrödinger's Route to Wave Mechanics", *Studies in History and Philosophy of Modern Science,* 311–340.

Department of Philosophy
University of South Carolina
SC 29208, Columbia
USA
stoeltzn@mailbox.sc.edu

CHAPTER 36

MIKLÓS RÉDEI[1]

SOME HISTORICAL AND PHILOSOPHICAL ASPECTS OF QUANTUM PROBABILITY THEORY AND ITS INTERPRETATION

36.1 THE MAIN CLAIMS

This paper argues that von Neumann's work on the theory of 'rings of operators' has the same role and significance for quantum probability theory that Kolmogorov and his work represents for classical probability theory: Kolmogorov established classical probability theory as part of classical measure theory (Kolmogorov 1933); von Neumann established quantum probability theory as part of non-classical (non-commutative) measure theory based on von Neumann algebras (1935–1940). Since the quantum probability theory based on general von Neumann algebras contains as a special case the classical probability theory (Sect. 36.2), there is a very tight conceptual-structural *similarity* between classical and quantum probability theory. But there is a major interpretational *dissimilarity* between classical and quantum probability: a straightforward frequency interpretation of non-classical probability is not possible (Sect. 36.3). A possible way of making room for a frequency interpretation of quantum probability theory is to accept the so-called Kolmogorovian Censorship Hypothesis, which can be shown to hold for quantum probability theories based on the theory of von Neumann algebras (Sect. 36.4), which however has both technical weaknesses and philosophical ramifications that are unattractive, as will be seen in Sect. 36.4.

36.2 QUANTUM PROBABILITY THEORY

A general quantum probability space is the triplet

$$(\mathcal{N}, \mathcal{P}(\mathcal{N}), \phi) \tag{36.1}$$

where $\mathcal{N}$ is a von Neumann algebra, $\mathcal{P}(\mathcal{N})$ is the lattice of projection of $\mathcal{N}$ and ϕ is a normal state on $\mathcal{N}$. (See Kadison and Ringrose 1986 or Takesaki 1979 for the operator algebraic notions, or Rédei 1998 for a brief review of the basic concepts.)

1 Work supported in part by the Hungarian Scientific Research Found (OTKA), contract number: K68043.

D. Dieks et al. (eds.), *Probabilities, Laws, and Structures*, The Philosophy of Science in a European Perspective 3, DOI 10.1007/978-94-007-3030-4_36,

Standard Hilbert space quantum probability theory is a particular case of (36.1): taking as von Neumann algebra the set $\mathcal{B}(\mathcal{H})$ of *all* bounded operators on a Hilbert space $\mathcal{H}$ the projection lattice $\mathcal{P}(\mathcal{B}(\mathcal{H}))$ of $\mathcal{B}(\mathcal{H})$ is the set $\mathcal{P}(\mathcal{H})$ of *all* projections on $\mathcal{H}$ (Hilbert lattice) and ϕ is a normal state given by some density matrix. $(\mathcal{H}, \mathcal{P}(\mathcal{H}), \phi)$ refers to this situation.

Classical probability theory also can be regarded as a particular case of (36.1) by taking the von Neumann algebra $\mathcal{N}$ to be commutative: a commutative von Neumann algebra is isomorphic to the set $L^\infty(X, \mathcal{S}, \mu)$ of essentially bounded measurable functions on a set X with a bounded measure μ on some Boolean algebra $\mathcal{S}$ of subsets of X. Elements of X are the elementary random events, the projections $\mathcal{P}(L^\infty(X, \mathcal{S}, \mu))$ in $L^\infty(X, \mathcal{S}, \mu)$ can be identified with the characteristic functions of the subsets of X belonging to $\mathcal{S}$ (and thereby they can be identified with general random events), and the functions in $L^\infty(X, \mathcal{S}, \mu)$ are the classical random variables. The normal state ϕ is a σ additive measure on $\mathcal{P}(L^\infty(X, \mathcal{S}, \mu))$. Thus a classical, Kolmogorovian probability measure space $(X, \mathcal{S}, p)$ can be recovered as a particular case of quantum probability theory, showing that the conceptual structure of classical and quantum probability theory is the same (see Rédei and Summers 2007 for some more details about how classical probability theory is contained in quantum probability theory). This is not to say that quantum probability theory does not have features that are not present in classical probability theory (for instance entanglement) but the basic structure is the same. The crucial difference between classical and quantum probability theory, which is the source of all interpretational difficulties, is that the set of quantum random variables, the von Neumann algebra $\mathcal{P}(\mathcal{N})$, is non-commutative (equivalently: the von Neumann lattice $\mathcal{P}(\mathcal{N})$ is an orthomodular but not distributive lattice), whereas the algebra of classical random variables $L^\infty(X, \mathcal{S}, \mu)$ is commutative (equivalently: $\mathcal{S}$ is an orthocomplemented distributive lattice, Boolean algebra).

The theory of von Neumann algebras was established by von Neumann (partly in collaboration with J. Murray) during the 1930s in a series of ground-breaking papers (Murray and von Neumann 1936, 1937; von Neumann 1940; Murray and von Neumann 1943). Originally, von Neumann algebras were called 'rings of operators', it was Dieudonné who suggested in 1954 to call 'rings of operators' von Neumann algebras to acknowledge that it was von Neumann who established the field (see von Neumann's letter to Dixmier, June 18, 1954, Rédei (2005)). The motivation to develop the theory of von Neumann algebras (and in particular classifying them) was not probability theory as such but the decomposition of quantum systems into independent subsystems and the intention to show that there is essentially only one way to do the decomposition (see Sect. 3. of the Introduction in Rédei (2005) for historical comments on the development of von Neumann algebras).

A major result of Murray and von Neumann (1936) was the classification of von Neumann algebras. They proved that the factor von Neumann algebras (the ones in which there is no non-trivial element commuting with every element in the algebra and from which non-factor von Neumann algebras can be put together) can be classified on the basis of the type of range of a so-called dimension function d defined on the lattice of projections of the algebra. The types are summarized in the table below. In the table the following notations are used. $\dim(\mathcal{H})$ is the dimension of Hilbert space $\mathcal{H}$. If $\mathcal{H}$ is N dimensional with some natural number N, then this is indicated by writing $\mathcal{H} = \mathcal{H}_N$. Tr denotes the trace functional on the set of bounded operators $\mathcal{B}(\mathcal{H})$. When $\mathcal{N}$ is a type $\mathbf{II}_1$ von Neumann algebra then τ denotes the normalized tracial dimension function on the projection lattice $\mathcal{P}(\mathcal{N})$. In the left column of the table typical classical measure spaces and measures μ are listed that correspond to the von Neumann algebra types.

From the point of view of probability theory the significance of the classification (and of the fact there exist examples for each type), is that quantum probability theory based on the theory of von Neumann algebras can model probabilistically all types of quantum physical systems, even ones that cannot be described by standard Hilbert space probability theory, such as infinite lattice gases and relativistic quantum fields (see Rédei and Summers 2007 for a concise review of these cases).

Table 36.1: Types of von Neumann algebras

Classical measure spaces	von Neumann algebra types	Name of type
$X = \{x_1, x_2, \ldots x_N\}$ $\mathcal{S} = P(X)$ $\mu(x_i) = 1 \quad (i = 1, \ldots N)$ Range of $\mu = \{1, 2, \ldots N\}$ Finite, discrete measure	$\mathcal{H}_N$, $\dim\mathcal{H}_N = N$, $\mathcal{N} = \mathcal{B}(\mathcal{H}_N)$ $\mathcal{P}(\mathcal{N}) = \mathcal{P}(\mathcal{H}_N)$ $d = Tr$ Range of $d = \{1, 2, \ldots N\}$ Finite, discrete measure (type)	type I_N
$X = \{x_1, x_2, \ldots x_N, \ldots\}$ $\mathcal{S} = P(X)$ $\mu(x_i) = 1 \quad (i = 1, \ldots)$ Range of $\mu = \{1, 2, \ldots\}$ Non-finite, discrete measure	$\mathcal{H}$, $\dim\mathcal{H} = \infty$, $\mathcal{N} = \mathcal{B}(\mathcal{H})$ $\mathcal{P}(\mathcal{N}) = \mathcal{P}(\mathcal{H})$ $d = Tr$ Range of $d = \{1, 2, \ldots\}$ Non-finite, discrete measure (type)	type I_∞
$X = [0, 1]$ $\mathcal{S}$ = Borel σ-algebra of $[0, 1]$ μ = Lebesgue measure on $[0, 1]$ Range of $\mu = [0, 1]$ Finite, continuous measure	$\mathcal{N}$ $\mathcal{P}(\mathcal{N})$ $d = \tau$ dimension function Range of $\tau = [0, 1]$ Finite, continuous measure (type)	type II_1
$X = \mathbb{R}$ $\mathcal{S}$ = Borel σ-algebra of $\mathbb{R}$ μ = Lebesgue measure on $\mathbb{R}$ Range of $\mu = \mathbb{R}$ Non-finite, continuous	$\mathcal{N}$ $\mathcal{P}(\mathcal{N})$ d dimension function Range of $d = \mathbb{R}$ Non-finite continuous (type)	type II_∞
	$\mathcal{N}$, $\mathcal{P}(\mathcal{N})$ d dimension function Range of $d = \{0, \infty\}$ Very non-finite	type III

36.3 Interpretation of Probability

The general problem of interpretation of any probability theory is the problem of specifying the relation between the mathematical structure $(\mathcal{N}, \mathcal{P}(\mathcal{N}), \phi)$ and elements of reality. Specifically, one has to answer these two questions:

1. What elements of reality correspond to elements of $\mathcal{P}(\mathcal{N})$ and their relations as expressed by the lattice operations?
2. What is the meaning of $\phi(A) = r$?

There are several possible answers to these questions but in application of probability theory in physics probabilities are typically tested by counting (relative) frequencies, and the Boolean operations are interpreted according to a natural intuition about events:

1. Every A either happens or does not happen.
2. If A happens then $A^{\perp}$ does not happen.
3. If A happens and B happens then $A \wedge B$ happens.
4. If $A \vee B$ happens then either A or B happens.

(1)–(4) mean that, under a natural interpretation of what events are, if $\mathcal{P}(\mathcal{N})$ represents the event structure, then there exists a Boolean algebra homomorphism h from $\mathcal{P}(\mathcal{N})$ into the two element Boolean algebra $\{0, 1\}$.

According to the relative frequency interpretation (worked out first systematically by Richard von Mises[2] 1919, 1928), the probability space $(X, \mathcal{S}, p)$ has a relative frequency interpretation if there exists a *fixed* statistical ensemble $\mathcal{E} = \{e_1, e_2, \ldots\}$ of (countably infinite) elementary events such that the following hold:

- For every event A, it can be decided unambiguously and *without* changing the ensemble whether e_i is in A or not (whether e_i realizes A or not).
- For every $A \in \mathcal{S}$ the number $p(A)$ is equal to the limit of relative frequency of event A in finite initial segments of $\{e_1, e_2, \ldots\}$.

Such a frequency interpretation of a general, genuinely quantum (i.e. non-commutative) probability space $(\mathcal{N}, \mathcal{P}(\mathcal{N}), \phi)$ does not seem feasible however. We have seen that under a natural interpretation of what events are, the assumption that the von Neumann lattice $\mathcal{P}(\mathcal{N})$ is an event structure entails the existence of a Boolean algebra homomorphism from $\mathcal{P}(\mathcal{N})$ into the two element Boolean algebra. But no such homomorphism exists in the case of a genuinely non-classical $\mathcal{P}(\mathcal{N})$, as was shown by Döring (2005):

2 Apart from the two requirements detailed here, von Mises requires the ensemble to be *random*, which is a problematic requirement but does not play a role from the perspective of this paper, thus it is neglected here.

Proposition 1 *Under weak assumptions on the von Neumann algebra $\mathcal{N}$ ($\mathcal{N}$ must not contain direct summand of type* $\mathbf{I}_1$ *and* $\mathbf{I}_2$*, see Table 36.1. For the types* $\mathbf{I}_1$ *and* $\mathbf{I}_2$*), there exists no partial Boolean algebra homomorphism from the von Neumann lattice $\mathcal{P}(\mathcal{N})$ into the two element Boolean algebra; hence there exists no Boolean algebra homomorphism from $\mathcal{P}(\mathcal{N})$ into the two element Boolean algebra either.*

A map h from $\mathcal{P}(\mathcal{N})$ is a partial Boolean algebra homomorphism if it is a Boolean algebra homomorphism on every distributive sublattice of $\mathcal{P}(\mathcal{N})$.

Another obstacle standing in the way of the frequency interpretation of quantum probabilities along the lines of the frequency interpretation of classical probabilities is the following. The so-called "general additivity rule":

$$p(A) + p(B) = p(A \vee B) + p(A \wedge B), \tag{36.2}$$

which holds for a classical probability measure, is a *necessary* condition for a relative frequency interpretation in a fixed ensemble (with the understanding that $A \vee B$ and $A \wedge B$ denote the events "either A or B happens" and "both A and B happen"). This is because the function $\#$ defined by

$$\{e_1, e_2, \dots e_N\} \supseteq A \mapsto \#(A) \;=\; \text{number of elements } e_i \ (i \leq N) \text{ in } A, \tag{36.3}$$

in terms of which the relative frequencies are computed, behaves like an ordinary measure, for which (36.2) holds. In general, a quantum probability measure is *not* additive in the sense of Eq. 36.2 however, as the following Proposition proved by Petz and Zemanek (1988) shows.

Proposition 2 *A normal state τ on a von Neumann algebra $\mathcal{N}$ satisfies the general additivity rule (36.2) if and only if it is a trace.*

The state τ is a trace by definition if

$$\tau(XY) = \tau(YX) \qquad \text{for all } X, Y \in \mathcal{N} \tag{36.4}$$

This means that only those quantum probabilities can be interpreted as relative frequencies that are given by a tracial state. Since, however, a tracial state is precisely the state that disregards the non-commutativity of the algebra (in the sense of Eq. 36.4), this indicates that genuinely quantum states and quantum probabilities cannot be interpreted as relative frequencies.

Note that there exists no tracial state whatsoever in the Hilbert space formalism: the only tracial functional on $\mathcal{B}(\mathcal{H})$, the standard Tr, is not bounded if $\mathcal{H}$ is infinite dimensional. There exist tracial states on type $\mathbf{II}_1$ von Neumann algebras however (the dimension function d in terms of which the von Neumann factors are classified, is the restriction of a tracial state on $\mathcal{N}$ (see the Table 36.1.)). This was the reason why von Neumann preferred the type $\mathbf{II}_1$ von Neumann algebras to the Hilbert space formalism (see the papers (Rédei 1996, 1999, 2001, 2007) for a detailed analysis of von Neumann's preference of the type $\mathbf{II}_1$ algebras.)

36.4 Kolmogorovian Censorship

It is a fact however that many probabilistic statements of quantum theory are tested experimentally by counting frequencies. How is this compatible with the difficulties outlined in the previous section? One answer to this question is the so-called Kolmogorovian Censorship hypothesis.

The idea of Kolmogorovian Censorship hypothesis is that there are in fact no genuinely non-classical probabilities: quantum probabilities are always classical *conditional* probabilities of outcomes of measurements of quantum observables, where the conditioning events are the events of choosing to set up a measuring device to measure a certain observable.

To maintain such an interpretation, one has to prove formally that quantum probabilities can in fact be viewed as classical conditional probabilities. A Kolmogorovian censorship proposition was first proved rigorously by Bana and Durt (1997) for non-classical probability theories based on finite dimensional Hilbert spaces. Subsequently, Szabó (2001) proved a Kolmogorovian Censorship proposition for quantum probability spaces $(\mathcal{H}, \mathcal{P}(\mathcal{H}), \phi)$ with an infinite dimensional Hilbert space $\mathcal{H}$. As it turns out, Szabó's proof does not depend in any way on any particular features of the non-classical probability theory $(\mathcal{H}, \mathcal{P}(\mathcal{H}), \phi)$ (with an infinite dimensional Hilbert space $\mathcal{H}$) and could be carried over to von Neumann algebras without any modification. This was shown in Rédei (2010). The next Proposition formulates the Kolmogorovian Censorship for general non-classical probability spaces given by arbitrary von Neumann algebras.

Proposition 3 (Kolmogorovian Censorship for von Neumann algebras)

1. *Let $(\mathcal{N}, \mathcal{P}(\mathcal{N}), \phi)$ be a non-commutative probability space with ϕ being a normal state on the von Neumann algebra $\mathcal{N}$.*

2. *Let Γ be a countable set of selfadjoint operators in $\mathcal{N}$ such that*

$$[Q,R] \neq 0 \quad \text{if} \quad Q \neq R, \quad 0 \neq Q, R \in \Gamma. \tag{36.5}$$

3. *For every $Q \in \Gamma$, let $\mathcal{P}(Q)$ be a maximal Abelian sublattice of $\mathcal{P}(\mathcal{N})$ containing all the spectral projections of Q.*

4. *Let a map $p_0 \colon \Gamma \to [0,1]$ be such that*

$$\sum_{Q \in \Gamma} p_0(Q) = 1 \qquad p_0(Q) > 0 \quad \text{if } Q \neq 0. \tag{36.6}$$

Then there exists a classical probability space $(X, \mathcal{S}, p)$ with the following properties:
For every projection A^Q in any $\mathcal{P}(Q)$ there exist events A^Q_{cl} and a^Q_{cl} in $\mathcal{S}$ such that

$$A^Q_{cl} \subset a^Q_{cl} \tag{36.7}$$

$$a^Q_{cl} \cap a^R_{cl} = 0 \quad \text{if} \quad Q \neq R \tag{36.8}$$

$$\phi(A^Q) = p(A^Q_{cl} | a^Q_{cl}) \tag{36.9}$$

The interpretation of the assumptions and of the conclusion of Proposition 3 is the following.

1. Intuitively, the set Γ of non-commuting selfadjoint operators is the set of observables that are selected for measurement. The measurement device to measure $Q \in \Gamma$ is set up with probability $p_0(Q)$ specified in Eq. 36.6.

2. Events A^Q_{cl} and a^Q_{cl} are classical random events; intuitively A^Q_{cl} is the event representing a certain *outcome* of a measurement of Q; event a^Q_{cl} represents the ordinary, classical event of setting up the measurement device that measures the value of Q.

3. Condition (36.7) expresses that no outcome is possible without the event of setting up a measuring device to measure observable Q.

4. Condition (36.8) expresses that incompatible observables Q and R cannot be simultaneously measured; hence the events a^Q_{cl} and a^R_{cl} representing the setting up the measuring devices measuring Q and R, respectively, cannot happen jointly, they are disjoint.

5. Condition (1) states that quantum probabilities can be written as classical conditional probabilities: conditional probabilities of outcomes of measurements on condition that the appropriate measuring device has been set up to measure observables.

The Kolmogorovian Censorship Hypothesis makes it possible in principle to interpret quantum probabilities in terms of relative frequencies *in a given, fixed experimental situation* because it reduces the quantum probabilities to classical probabilities; however, the Kolmogorovian censorship hypotheses is not without problems. The problems are both technical and conceptual-philosophical.

The technical problem is that while Proposition 3 shows that quantum probabilities can indeed be regarded as conditional probabilities in a classical probability space, the assumption of countability of the set Γ of observables to be measured is a serious restriction. In principle, *any* selfadjoint operator in $\mathcal{N}$ can be selected for measurement and there are an uncountable number of incompatible observables in $\mathcal{N}$, even if $\mathcal{N}$ is the set of all bounded observables on a *finite* dimensional Hilbert space; so a countably infinite Γ is a very "small" set. In other words, one the basis of Proposition 3, one cannot claim that the whole non-commutative probability theory $(\mathcal{N}, \mathcal{P}(\mathcal{N}), \phi)$ can be interpreted according to the idea of Kolmogorovian Censorship – not even in the case of finite dimensional Hilbert space probability theory.

One could try to generalize Proposition 3 by allowing Γ to be the whole set of observables; the problem with such an attempt is that even in this case at most a countably infinite number of incompatible observables can be chosen to be measured with non-zero probability because by the definition of p and α_{a^Q} one has of course

$$p_0(Q) = p(\alpha_{a^Q}) = p(a^Q) \tag{36.10}$$

and it follows that if there are an uncountable number of observables in Γ then there are an uncountable number of mutually disjoint measuring-device-set-up events a^Q and the σ-additivity of p excludes all having non-zero probability – at most a countably infinite number of mutually disjoint events all having non-zero probabilities can exist in a classical probability space. Thus there is no hope of a generalization of Proposition 3 by allowing Γ to be the whole set of observables, and this fact limits severely the significance of the Kolmogorovian Censorship Hypothesis as stated in Proposition 3.

Another, more philosophical problem with Kolmogorovian Censorship is its instrumental character: accepting this "deconstruction" of quantum probabilities, we are forced to acknowledge that it is meaningless to talk about quantum probabilities without actually measuring them. Probabilities are thus not features of quantum systems in and of themselves, they are features that only manifest themselves upon measurement. Philosophers (or physicists) with a robust realist conviction may find unattractive this strongly instrumentalist flavor of interpretation of quantum probability forced upon us by the Kolmogorovian Censorship Hypothesis.

To sum up. The theory of von Neumann algebras (1935–1940) provides a general non-classical (non-commutative) measure theoretical framework that can accommodate both classical probability theory as this was formulated by Kolmogorov in terms of classical measure theory (1933) and standard Hilbert space quantum probability theory summarized by von Neumann (1932).While there is thus a very strong structural similarity between classical and quantum probability theory, a straightforward frequency interpretation of quantum probability theory is not possible. The Kolmogorovian Censorship Hypothesis re-interprets a general quantum probability theory in terms of classical conditional probabilities and this makes it possible to view quantum probabilities as relative frequencies in principle; however, this comes at the price of having to accept an instrumentalist view of quantum probabilities. Thus the problem of how to interpret quantum probability theory remains a conceptually intriguing issue.

References

G. Bana and T. Durt. Proof of Kolmogorovian censorship. *Foundations of Physics*, 27:1355–1373, 1997.

A. Döring. Kochen-Specker theorem for general von Neumann algeberas. *International Journal of Theoretical Physics*, 44:139–160, 2005.

R. V. Kadison and J. R. Ringrose. *Fundamentals of the Theory of Operator Algebras*, volume I. and II. Academic Press, Orlando, 1986.

A. N. Kolmogorov. *Grundbegriffe der Wahrscheinlichkeitsrechnung*. Springer, Berlin, 1933. English translation: *Foundations of the Theory of Probability*, (Chelsea, New York, 1956).

F. J. Murray and J. von Neumann. On rings of operators. *Annals of Mathematics*, 37:116–229, 1936. Reprinted in Taub (1961) No. 2.

F. J. Murray and J. von Neumann. On rings of operators, II. *American Mathematical Society Transactions*, 41:208–248, 1937. Reprinted in Taub (1961) No. 3.

F. J. Murray and J. von Neumann. On rings of operators, IV. *Annals of Mathematics*, 44:716–808, 1943. Reprinted in Taub (1961) No. 5.

D. Petz and J. Zemanek. Characterizations of the trace. *Linear Algebra and its Applications*, 111:43–52, 1988.

M. Rédei. Why John von Neumann did not like the Hilbert space formalism of quantum mechanics (and what he liked instead). *Studies in the History and Philosophy of Modern Physics*, 27:1309–1321, 1996.

M. Rédei. *Quantum Logic in Algebraic Approach*, volume 91 of *Fundamental Theories of Physics*. Kluwer Academic Publisher, 1998.

M. Rédei. 'Unsolved problems in mathematics' J. von Neumann's address to the International Congress of Mathematicians Amsterdam, September 2-9, 1954. *The Mathematical Intelligencer*, 21:7–12, 1999.

M. Rédei. Von Neumann's concept of quantum logic and quantum probability. In M. Rédei and M. Stöltzner, editors, *John von Neumann and the Foundations of Quantum Physics*, Institute Vienna Circle Yearbook, pages 153–172. Kluwer Academic Publishers, Dordrecht, 2001.

M. Rédei, editor. *John von Neumann: Selected Letters*, volume 27 of *History of Mathematics*, Rhode Island, 2005. American Mathematical Society and London Mathematical Society.

M. Rédei. The birth of quantum logic. *History and Philosophy of Logic*, 28:107–122, May 2007.

M. Rédei. Kolmogorovian Censorship Hypothesis for general quantum probability theories. *Manuscrito - Revista Internacional de Filosofia*, 33:365–380, 2010.

M. Rédei and S. J. Summers. Quantum probability theory. *Studies in the History and Philosophy of Modern Physics*, 38:390–417, 2007.

L. E. Szabó. Critical reflections on quantum probability theory. In M. Rédei and M. Stöltzner, editors, *John von Neumann and the Foundations of Quantum Physics*, Institute Vienna Circle Yearbook, pages 201–219. Kluwer Academic Publishers, Dordrecht, 2001.

M. Takesaki. *Theory of Operator Algebras*, volume I. Springer Verlag, New York, 1979.

A. H. Taub, editor. *John von Neumann: Collected Works*, volume III. Rings of Operators, New York and Oxford, 1961. Pergamon Press.

R. von Mises. Grundlagen der Wahrscheinlichkeitsrechnung. *Mathematische Zeitschrift*, 5:52–99, 1919.

R. von Mises. *Probability, Statistics and Truth*. Dover Publications, New York, 2nd edition, 1981. Originally published as 'Wahrscheinlichkeit, Statistik und Wahrheit' (Springer, 1928).

J. von Neumann. On rings of operators, III. *Annals of Mathematics*, 41:94–161, 1940. Reprinted in Taub (1961) No. 4.

Department of Philosophy, Logic and Scientific Method
London School of Economics and Political Science
Houghton Street
WC2A 2AE, London
UK
M.Redei@lse.ac.uk

Index of Names

Not included are footnotes, figures, tables, notes and references.

D. Dieks et al. (eds.), *Probabilities, Laws, and Structures*, The Philosophy of Science in a European Perspective 3, DOI 10.1007/978-94-007-3030-4,